Complete Version

ANATOMY &
PHYSIOLOGY
Laboratory Textbook

Seventh Edition

Complete Version

ANATOMY & PHYSIOLOGY
Laboratory Textbook

Seventh Edition

Harold J. Benson
Pasadena City College

Stanley E. Gunstream
Pasadena City College

Arthur Talaro
Pasadena City College

Kathleen P. Talaro
Pasadena City College

Boston   Burr Ridge, IL   Dubuque, IA   Madison, WI   New York   San Francisco   St. Louis
Bangkok   Bogotá   Caracas   Lisbon   London   Madrid   Mexico City   Milan
New Delhi   Seoul   Singapore   Sydney   Taipei   Toronto

WCB/McGraw-Hill

A Division of The McGraw·Hill Companies

ANATOMY AND PHYSIOLOGY Laboratory Textbook, Complete Version,
SEVENTH EDITION

 This book is printed on recycled, acid-free paper containing 10% postconsumer waste.

5 6 7 8 9 0 QPD/QPD 04

ISBN 0-697-28255-4

Vice President and editorial director: *Kevin T. Kane*
Executive editor: *Colin Wheatley*
Sponsoring editor: *Kristine Tibbetts*
Developmental editor: *Brittany J. Rossman*
Marketing manager: *Heather Wagner*
Senior project manager: *Kay J. Brimeyer*
Production supervisor: *Sandra Hahn*
Coordinator of freelance design: *Michelle D. Whitaker*
Art editor: *Joyce Watters*
Compositor: *Carlisle Communications Ltd.*
Typeface: *11/12 Times Roman*
Printer: *Quebecor Printing Book Group/Dubuque, IA*

Freelance cover designer: *Heidi J. Baughman*
Covor photo: © *Dr. Dennis Kunkel/Phototake NYC*

www.mhhe.com

Contents

In this seventh edition of the *Anatomy and Physiology Laboratory Textbook* we have done the following: (1) expanded the Histology Atlas to incorporate new material, (2) provided four histology self-quizzes, (3) expanded the instructions for using Intelitool equipment, (4) upgraded many of the anatomical illustrations, (5) updated certain concepts that shed new light on previously little understood theories, (6) shortened some exercises and transferred some material to the Instructor's Handbook to make room for new material, and (7) changed many illustrations to improve visualization.

The Histology Atlas has been expanded to forty pages from thirty-six. Although much of the Atlas remains essentially the same, additions and alterations were made that are related to the skin, tongue, respiratory passages, lungs, female reproductive organs, and spermatogenesis.

One of the shortcomings of previous editions has been the inability to measure the extent of a student's comprehension of histology. Although some questions in the Laboratory Reports do address this problem, we have felt for some time that more testing was needed.

At many institutions students are subjected to laboratory practical examinations in which microscopes are set up with various types of tissues displayed for identification by the student. This type of exam can be quite traumatic to the student the first time it is experienced. To give the student an opportunity to see how he or she might do on such an exam, we have developed four "Self-Quizzes." In laboratories where practical exams are used, these self-quizzes can do much to show the student ahead of time what type of questions might be asked. Even if no laboratory practical is used, the self-quizzes are helpful review tests.

The four Self-Quizzes are located at the end of Exercise 89. They encompass eighty microscope setups with 261 questions. Answers to the quizzes are located on pages 414 and 456. At appropriate places in the manual students are prompted to take each quiz.

Another shortcoming of the previous edition was that we didn't have complete instructions for performing Intelitool experiments on all computer platforms. At the time of publication of the sixth edition, only instructions for the Apple II were available. Since then software and instructions for IBM (or compatible PC), and Macintosh have been developed. On one system (Spirocomp™) instructions for Windows 95 are also available. In this edition we have included instructions for running all the Intelitool experiments on all platforms.

One battle we always seem to lose in coming out with a new edition is the "battle of the bulge." Keeping the book from getting too big is a constant struggle. In this edition we have added 32 pages of new material, yet we have increased the size of the book by only six pages. This achievement was made possible by (1) eliminating some less important material, (2) compacting some exercises, and (3) transferring some material to the Instructor's Handbook. The bar codes for the *Slice of Life* disks fall in this last category. Instructors who use these bar codes may wish to provide students with photocopies of the codes.

One other item that was transferred from the manual to the Instructor's Handbook was the former Appendix B, which contained recipes for the various solutions and reagents that are used in the experiments. Since students don't generally need this information, it was felt that placement in the Instructor's Handbook was appropriate.

In previous editions, a set of $2'' \times 2''$ Kodachrome slides pertaining to the Histology Atlas illustrations was made available to users of the manual. They are still available to those who do not have a set. A legend explaining the slides, which is provided in the back of the Instructor's Handbook, is helpful when projecting the slides to groups of students. The slides can be had at no cost by simply contacting the Educational Services Department at McGraw-Hill in Dubuque, Iowa.

The changes in this edition are partially the result of suggestions made by the following individuals who currently use the complete and short versions of this manual: James Ezell and Barbara L. Stewart of J. Sargeant Reynolds Community College, Richmond, VA; Ralph E. Reiner of College of the Redwoods, Eureka, CA; A. Scott Helgeson of Des Moines Area Community College, Des Moines, IA; Juville Dario-Becker of Central Virginia Community College, Lynchburg, VA; Donna Sasnow of South Suburban College, South Holland, IL; and Marian G. Langer of St. Francis College, Loretto, PA. Our sincerest gratitude is extended here for their valuable assistance. Although we haven't been able to incorporate all the changes requested, most ideas have been included in this edition.

Introduction

These laboratory exercises have been developed to provide you with a basic understanding of anatomical and physiological principles that underlie medicine, nursing, dentistry, and other related health professions. Laboratory procedures that reflect actual clinical practices are included wherever feasible. In each exercise you will find essential terminology that will become part of your working vocabulary. Mastery of all concepts, vocabulary, and techniques will provide you with a core of knowledge crucial to success in your chosen profession.

During the first week of this course your instructor will provide you with a schedule of laboratory exercises in the order of their performance. There is an implied expectation that you will have familiarized yourself with the content of each experiment before that week's session, thus ensuring that you will be properly prepared so as to minimize disorganization and mistakes.

The *Laboratory Reports* coinciding with each exercise are located at the back of the book. They are perforated for easy removal; be sure to remove each sheet as necessary. Removing them will facilitate data collection, completion of answers, and grading. Your instructor may give further procedural details on the handling of these reports.

The exercises in this laboratory guide consist essentially of four kinds of activities: (1) illustration labeling, (2) anatomical dissections, (3) physiological experiments, and (4) microscopic studies. The following suggestions should be helpful in performing these assignments.

Labeling   The activity of labeling illustrations is essentially a determination of your understanding of the written text. Since all labeled structures are explicitly described in the manual, all that is necessary is to read the manual very carefully. Incorrectly labeled illustrations usually indicate a lack of comprehension.

Once the illustrations are labeled, they can be useful to you in two ways. First, the illustrations may be used for reference purposes in dissections or examinations of anatomical specimens, particularly in the skeletal and nervous systems. Second, the illustrations can be used for review purposes. If the number legend of the labels is covered over as you mentally attempt to name the structures on the illustration, you can easily determine your level of understanding. Periodic reviews of this type during the semester will be very helpful.

In general, the labeling of illustrations will be performed *before* you come to the laboratory. In this way the laboratory time will be used primarily for dissections, experimentation, or microscopic examinations.

Dissections   Although cadavers provide the ideal dissection specimens in human anatomy, they are expensive and thus may be impractical for large classes in introductory anatomy and physiology. Since much of cat anatomy is similar to human anatomy, the cat has been selected as the primary dissection specimen in this manual. The rat will also be used. Occasionally, sheep and beef organs will be studied. Frogs will frequently be used in physiological experiments.

When using live animals in experimental procedures it is imperative that they be handled with great care. Consideration must be exercised to minimize pain in all experiments on vertebrate animals. Inconsiderate or haphazard treatment of any animal will not be tolerated.

Physiological Experiments   Before performing any physiological experiments be sure that you understand the overall procedure. Reading the experiment before entering the laboratory will help a great deal.

Handle all instruments carefully. Most pieces of equipment are expensive, may be easily damaged, and are often irreplaceable. The best insurance against breakage or damage is to thoroughly understand how the equipment is expected to function.

Maintain astuteness in observations and record keeping. Record data immediately; postponement detracts from precision. Insightful data interpretation will also be expected.

Microscopic Studies   Cytological and histological studies will be made to lend meaning to text descriptions. Familiarize yourself with the contents of the Histology Atlas, which includes photomicrographs of most of the tissues you will study in this course. Note that the Atlas is in the middle of the book and is readily located by the color band on the edges of its pages. If drawings are required execute them with care, and label those structures that are significant.

Histology Self-Quizzes Periodically you will be prompted to test your comprehension of histological studies by taking Histology Self-Quizzes, which are located on pages 483 to 498. Note that there are four of them. The answers to each self-quiz are included on designated pages so that you can determine for yourself your degree of understanding. You should find them quite helpful, particularly if lab practical exams are a part of this course.

Laboratory Efficiency Success in any science laboratory requires a few additional disciplines:

1. Always follow the instructor's verbal comments at the beginning of each laboratory session. It is at this time that difficulties will be pointed out, group assignments will be made, and procedural changes will be announced. Take careful notes on substitutions or changes in methods or materials.
2. In view of the above statement, it is obvious that the beginning of each laboratory period is a critical time. It is for this reason that tardiness is intolerable. If you are late to class don't expect your instructor to be very helpful.
3. Keep your work area tidy at all times. Books, bags, purses, and extraneous supplies should be located away from the work area. Tidiness should also extend to assembly of all apparatus.
4. Abstain from eating, drinking, or smoking within the confines of the laboratory.
5. Report immediately to the instructor any injuries that may occur.
6. Be serious-minded and methodical. Horseplay, silliness, and flippancy will not be tolerated during experimental procedures.
7. Work independently, but cooperatively, when performing team experiments. Attend to your assigned responsibility, but be willing to lend a hand to others when necessary. Participation and development of laboratory techniques are integral parts of the course.

Laboratory Reports When seeking answers to the questions and problems on the Laboratory Reports, work independently. The effort you expend to complete these reports is as essential as doing the experiment. The easier route of letting someone else solve the problems for you will handicap you at examination time. You are taking this course to learn anatomy and physiology. No one else can learn it for you.

Some of the laboratory experiments included in this text may be hazardous if materials are handled improperly or if procedures are conducted incorrectly. Safety precautions are necessary when you are working with chemicals, glass test tubes, hot-water baths, sharp instruments, and the like, or for any procedures that generally require caution. Your school may have set regulations regarding safety procedures that your instructor will explain to you. Should you have any problems with materials or procedures, please ask your instructor for help.

PART 1

Some Fundamentals

This unit, which consists of only three exercises, attempts to accomplish two things: (1) to equip you with some of the basic anatomical terminology that you will need to get started and (2) to present an overview of the systems and organs of the body that are going to be studied in greater detail as we progress through the book.

When you start Exercise 1 scan the entire exercise before attempting to label any of the illustrations. Note that the first two pages consist of descriptive terminology, followed by an Assignment, which describes what you are to label. This, in turn, is followed by additional descriptive text, and another Assignment on page 5. After all the illustrations in the exercise have been labeled, you will transfer the label numbers from the legends of each illustration to the Laboratory Report sheet, which is located in the back of the manual beginning on page 499. Once the Laboratory Report is completed, it can be turned in to your instructor for evaluation or self-graded if a key is made available.

1

Anatomical Terminology

Anatomical description would be extremely difficult without specific terminology. A consensus prevails among many students that anatomists synthesize multisyllabic words in a determined conspiracy to harass the beginner's already overburdened mind. Naturally, nothing could be further from the truth.

Scientific terminology is created out of necessity. It functions as a precise tool that allows us to say a great deal with a minimum of words. Conciseness in scientific discussion not only saves time, but it promotes clarity of understanding as well.

Most of the exercises in this laboratory manual employ the terms defined in this exercise. They are used liberally to help you locate structures that are to be identified on the illustrations. If you do not know the exact meanings of these words, obviously you will be unable to complete the required assignments. Before you attempt to label any of the illustrations in this exercise, read the text material first.

Relative Positions

Descriptive positioning of one structure with respect to another is accomplished with the following pairs of words. Their Latin or Greek derivations are provided to help you understand their meanings.

Superior and Inferior   These two words are used to denote vertical levels of position. The Latin word *super* means *above;* thus, a structure that is located above another one is said to be superior. Example: The nose is *superior* to the mouth.

The Latin word *inferus* means *below* or *low;* thus, an inferior structure is one that is below or under some other structure. Example: The mouth is *inferior* to the nose.

Anterior and Posterior   Fore and aft positioning of structures are described with these two terms. The word *anterior* is derived from the Latin, *ante,* meaning *before.* A structure that is anterior to another one is in front of it. Example: Bicuspids are *anterior* to molars.

Anterior surfaces are the most forward surfaces of the body. The front portions of the face, chest, and abdomen are anterior surfaces.

Posterior is derived from the Latin *posterus,* which means *following.* The term is the opposite of anterior. Example: The molars are *posterior* to the bicuspids.

When these two terms are applied to the surfaces of the hand and arm, it is assumed that the body is in the *anatomical position,* which is as shown in figures 1.1 and 1.2. In the anatomical position the palms of the hands face forward.

Cranial and Caudal   When describing the location of structures of four-legged animals, these terms are often used in place of anterior and posterior. Since the word *cranial* pertains to the skull (Greek: *kranion,* skull), it may be used in place of anterior. The word *caudal* (Latin: *cauda,* tail) may be used in place of posterior.

Dorsal and Ventral   These terms, as used in comparative anatomy of animals, assume all animals, including humans, to be walking on all fours. The dorsal surfaces are thought of as *upper* surfaces, and the ventral surfaces as *underneath* surfaces.

The word *dorsal* (Latin: *dorsum,* back) not only applies to the back of the trunk of the body but may also be used in describing the back of the head and the back of the hand.

Standing in a normal posture, a human's dorsal surfaces become posterior. A four-legged animal's back, on the other hand, occupies a superior position.

The word *ventral* (Latin: *venter,* belly) generally pertains to the abdominal and chest surfaces. However, the underneath surfaces of the head and feet of four-legged animals are also often referred to as ventral surfaces. Likewise, the palm of the hand may also be referred to as being ventral.

Proximal and Distal   These terms are used to describe parts of a structure with respect to its point of attachment to some other structure. In the case of the arm or leg, the point of reference is where the limb is attached to the trunk of the body. In the case of a finger, the point of reference is where it is attached to the palm of the hand.

Proximal (Latin: *proximus,* nearest) refers to that part of the limb nearest to the point of attachment. Example: The upper arm is the *proximal* portion of the arm.

Distal (Latin: *distare,* to stand apart) means just the opposite of proximal. Anatomically, the distal portion of a limb or other part of the body is that portion that is most remote from the point of reference (attachment). Example: The hand is *distal* to the arm.

Medial and Lateral   These two terms are used to describe surface relationships with respect to the median line of the body. The *median line* is an imaginary line on a plane that divides the body into right and left halves.

The term *medial* (Latin: *medius,* middle) is applied to surfaces of structures that are closest to the median line. The medial surface of the arm, for example, is the surface next to the body because it is closest to the median line.

As applied to the appendages, the term *lateral* is the opposite of medial. The Latin derivation of this word is *lateralis,* which pertains to *side.* The lateral

surface of the arm is the outer surface, or that surface farthest away from the median line. The sides of the head are said to be lateral surfaces.

Body Sections

To observe the structure and relative positions of internal organs it is necessary to view them in sections that have been cut through the body. Considering the body as a whole, there are only three planes to identify. Figure 1.1 shows these sections.

Sagittal Sections   A section parallel to the long axis of the body (longitudinal section) that divides the body into right and left sides is a *sagittal section.* If such a section divides the body into equal halves, as in figure 1.1, it is said to be a *midsagittal section.*

Frontal Section   A longitudinal section that divides the body into front and back portions is a *frontal* or *coronal* section. The other longitudinal section seen in figure 1.1 is of this type.

Transverse Sections   Any section that cuts through the body in a direction that is perpendicular to the long axis is a *transverse* or *cross section.*

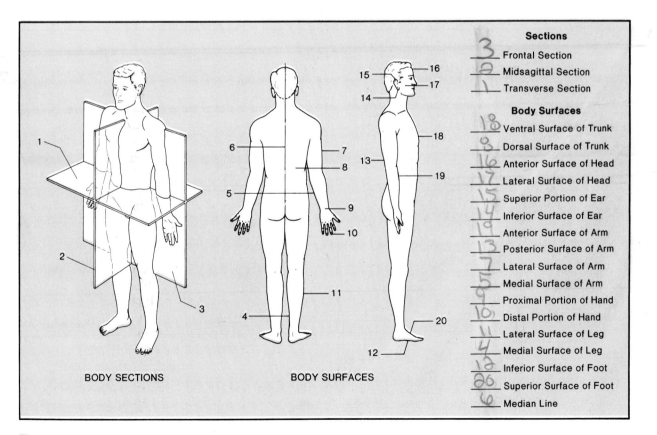

	Sections
3	Frontal Section
2	Midsagittal Section
1	Transverse Section
	Body Surfaces
18	Ventral Surface of Trunk
3	Dorsal Surface of Trunk
16	Anterior Surface of Head
17	Lateral Surface of Head
15	Superior Portion of Ear
14	Inferior Surface of Ear
19	Anterior Surface of Arm
13	Posterior Surface of Arm
7	Lateral Surface of Arm
5	Medial Surface of Arm
9	Proximal Portion of Hand
10	Distal Portion of Hand
11	Lateral Surface of Leg
4	Medial Surface of Leg
12	Inferior Surface of Foot
20	Superior Surface of Foot
6	Median Line

BODY SECTIONS   BODY SURFACES

Figure 1.1   Body sections and surfaces.

This is the third section shown in figure 1.1. In this case it is parallel to the ground.

Although these sections have been described here only in relationship to the body as a whole, they can be used on individual organs such as the arm, finger, or tooth.

Assignment:

To test your understanding of the above descriptive terminology, identify the labels in figure 1.1 by placing the correct numbers in front of the terms to the right of the illustrations. Also, record these numbers on the Laboratory Report.

Regional Terminology

Various terms such as *flank, groin, brachium,* and *hypochondriac* have been applied to specific regions of the body to facilitate localization. Figures 1.2 and 1.3 pertain to some of the more predominantly used terminology.

Trunk

The anterior surface of the trunk may be subdivided into two pectoral, two groin, and the abdominal regions. The upper chest region may be designated as **pectoral** or **mammary** regions. The anterior trunk region not covered by the ribs is the **abdominal** region. The depressed area where the thigh of the leg meets the abdomen is the **groin.**

The posterior surface, or dorsum, of the trunk can be differentiated into the costal, lumbar, and buttocks regions. The **costal** (Latin: *costa,* rib) portion is the part of the dorsum that lies over the rib cage. The lower back region between the ribs and hips is the **lumbar** or **loin** region. The **buttocks** are the rounded eminences of the rump formed by the gluteal muscles; this is also called the **gluteal** region.

The side of the trunk that adjoins the lumbar region is called the **flank.** The armpit region that is between the trunk and the arm is the **axilla.**

Upper Extremities

To differentiate the parts of the upper extremities, the term **brachium** is used for the upper arm and **antebrachium** for the forearm (between the elbow and wrist). The elbow area on the posterior surface of the arm is the **cubital** area. That area on the opposite side of the elbow is the **antecubital area.** It is also correct to refer to the entire anterior surface of the antebrachium as being antecubital.

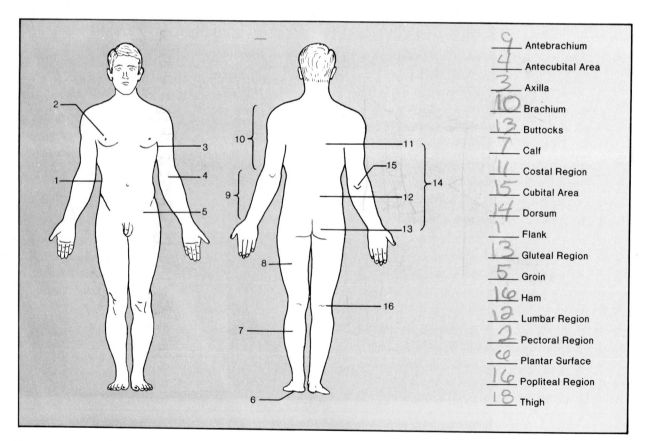

Figure 1.2 Regional terminology.

Lower Extremities

The upper portion of the leg is designated as the **thigh,** and the lower fleshy posterior portion is called the **calf.** Between the thigh and calf on the posterior surface, opposite to the knee, is a depression called the **ham** or **popliteal** region. The sole of the foot is the **plantar** surface.

Abdominal Divisions

The abdominal surface may be divided into quadrants or into nine distinct areas. To divide the abdomen into nine regions one must establish four imaginary planes: two that are horizontal and two that are vertical. These planes and areas are shown in figure 1.3. The **transpyloric plane** is the upper horizontal plane, which would pass through the lower portion of the stomach (pyloric portion). The **transtubercular plane** is the other horizontal plane, which touches the top surfaces of the hipbones (iliac crests). The two vertical planes, or **right** and **left lateral planes,** are approximately halfway between the midsagittal plane and the crests of the hips.

The above planes describe the umbilical, epigastric, hypogastric, hypochondriac, and lumbar regions. The **umbilical** area lies in the center, includes the naval, and is bordered by the two horizontal and two vertical planes. Immediately above the umbilical area is the **epigastric,** which covers much of the stomach. Below the umbilical zone is the **hypogastric,** or *pubic area.* On each side of the epigastric are a right and left **hypochondriac** area, and beneath the hypochondriac areas are the right and left **lumbar** areas. (Note that although we tend to think of only the lower back as being the lumbar region, we see here that it extends around to the anterior surface as well.)

Assignment:
Label figures 1.2 and 1.3 and transfer these numbers to the Laboratory Report.

Laboratory Report

After transferring all the labels from figures 1.1 through 1.3 to the proper columns on Laboratory Report 1,2, answer the questions that pertain to this exercise.

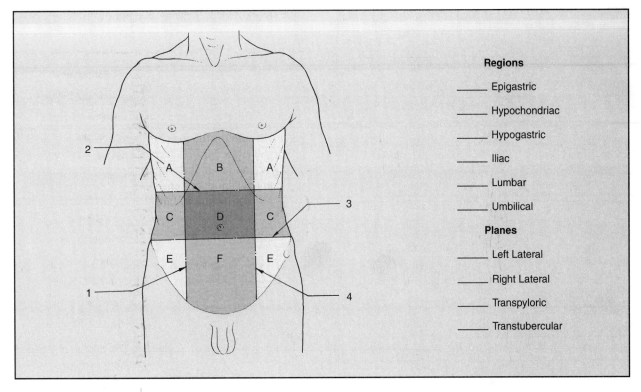

Regions

_____ Epigastric

_____ Hypochondriac

_____ Hypogastric

_____ Iliac

_____ Lumbar

_____ Umbilical

Planes

_____ Left Lateral

_____ Right Lateral

_____ Transpyloric

_____ Transtubercular

Figure 1.3   Abdominal regions.

2   Body Cavities and Membranes

All the internal organs *(viscera)* are contained in body cavities, which are completely or partially lined with smooth membranes. The relationships of these cavities to each other, the organs they contain, and the membranes that line them will be studied in this exercise.

Body Cavities

Figure 2.1 illustrates the seven principal cavities of the body. The two major cavities are the dorsal and ventral cavities. The **dorsal cavity,** which is nearest to the dorsal surface, includes the cranial and spinal cavities. The **cranial cavity** is the hollow portion of the skull that contains the brain. The **spinal cavity** is a long tubular canal within the vertebrae that contains the spinal cord. The **ventral cavity** is the largest cavity and encompasses the chest and abdominal regions.

The superior and inferior portions of the ventral cavity are separated by a dome-shaped thin muscle, the **diaphragm.** The **thoracic cavity,** which is that part of the ventral cavity superior to the diaphragm, is separated into right and left compartments by a membranous partition or septum called the **mediastinum.** The lungs are contained in these right and left compartments. The heart, trachea, esophagus, and thymus gland are enclosed within the mediastinum.

Figure 2.2 reveals the relationship of the lungs to the structures within the mediastinum. Note that within the thoracic cavity there exists a pair of right and left **pleural cavities** that contain the lungs and a **pericardial cavity** that contains the heart.

The **abdominopelvic cavity** is the portion of the ventral cavity that is inferior to the diaphragm. It consists of two portions: the abdominal and pelvic

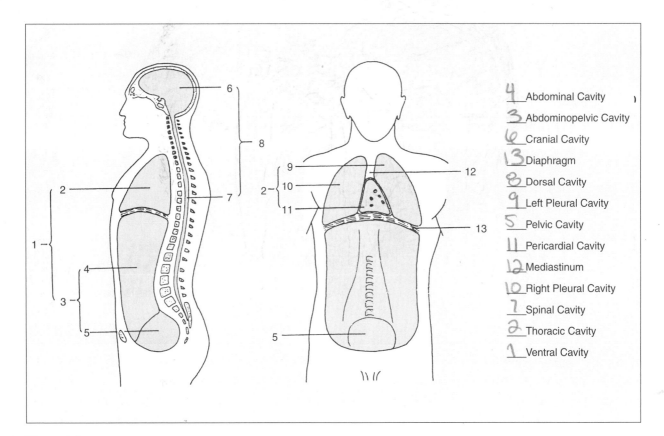

4	Abdominal Cavity
3	Abdominopelvic Cavity
6	Cranial Cavity
13	Diaphragm
8	Dorsal Cavity
9	Left Pleural Cavity
5	Pelvic Cavity
11	Pericardial Cavity
12	Mediastinum
10	Right Pleural Cavity
7	Spinal Cavity
2	Thoracic Cavity
1	Ventral Cavity

Figure 2.1   Body cavities.

cavities. The **abdominal cavity** contains the stomach, liver, gallbladder, pancreas, spleen, kidneys, and intestines. The **pelvic cavity** is the most inferior portion of the abdominopelvic cavity and contains the urinary bladder, sigmoid colon, rectum, uterus, and ovaries.

Body Cavity Membranes

The body cavities are lined with serous membranes that provide a smooth surface for the enclosed internal organs. Although these membranes are quite thin, they are strong and elastic. Their surfaces are moistened by a self-secreted *serous fluid* that facilitates ease of movement of the viscera against the cavity walls.

Thoracic Cavity Membranes

The membranes that line the walls of the right and left thoracic compartments are called **parietal pleurae** (*pleura,* singular). The lungs, in turn, are covered with **visceral (pulmonary) pleurae.** Note in Figure 2.2 that these pleurae are continuous with

each other. The potential cavity between the parietal and visceral pleurae is the **pleural cavity.** Inflammation of the pleural membranes results in a condition called *pleurisy.*

Within the broadest portion of the mediastinum lies the heart. It, like the lungs, is covered by a thin serous membrane, the **visceral pericardium,** or **epicardium.** Surrounding the heart is a double-layered fibroserous sac, the **parietal pericardium.** The inner layer of this sac is a serous membrane that is continuous with the epicardium of the heart. Its outer layer is fibrous, which lends considerable strength to the structure. A small amount of serous fluid produced by the two serous membranes lubricates the surface of the heart to minimize friction as it pulsates within the parietal pericardium. The potential space between the visceral and parietal pericardia is called the **pericardial cavity.**

Abdominal Cavity Membranes

The serous membrane of the abdominal cavity is the peritoneum. It does not extend deep down into the pelvic cavity, however; instead, its most inferior

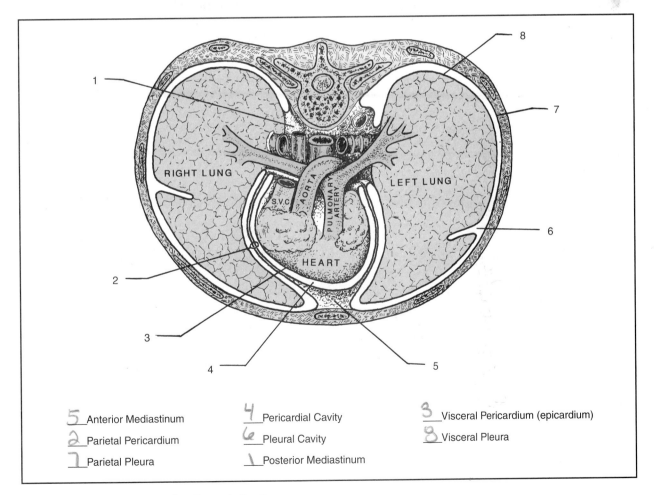

5 Anterior Mediastinum

2 Parietal Pericardium

7 Parietal Pleura

4 Pericardial Cavity

6 Pleural Cavity

1 Posterior Mediastinum

3 Visceral Pericardium (epicardium)

8 Visceral Pleura

Figure 2.2 Transverse section through the thorax.

boundary extends across the abdominal cavity at a level that is just superior to the pelvic cavity. The top portion of the urinary bladder is covered with the peritoneum.

In addition to lining the abdominal cavity, the peritoneum has double-layered folds called **mesenteries,** which extend from the dorsal body wall to the viscera, holding these organs in place. These mesenteries contain blood vessels and nerves that supply the viscera enclosed by the peritoneum.

That part of the peritoneum attached to the body wall is the **parietal peritoneum.** The peritoneum that covers the visceral surfaces is the **visceral peritoneum.** The potential cavity between the parietal and visceral peritoneums is called the **peritoneal cavity.**

Extending downward from the inferior surface of the stomach is a large mesenteric fold called the **greater omentum.** This double-membrane structure passes downward from the stomach in front of the intestines, sometimes to the pelvis, and back up to the transverse colon, where it is attached. Because it is folded upon itself it is essentially a double mesentery consisting of four layers. Protuberance of the abdomen in obese individuals is due to fat accumulation in the greater omentum.

A smaller mesenteric fold, the **lesser omentum,** extends between the liver and the superior surface of the stomach and a short portion of the duodenum. Illustration B of figure 2.3 shows the relationship of these two omenta to the abdominal organs.

Assignment:
Label figures 2.1, 2.2, and 2.3

Laboratory Report

Complete Laboratory Report 1,2 by answering the questions that pertain to this exercise.

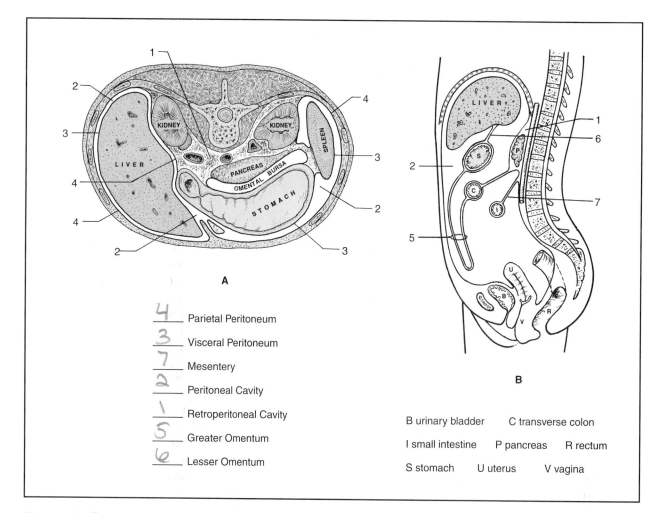

<u>4</u> Parietal Peritoneum

<u>3</u> Visceral Peritoneum

<u>7</u> Mesentery

<u>2</u> Peritoneal Cavity

<u>1</u> Retroperitoneal Cavity

<u>5</u> Greater Omentum

<u>6</u> Lesser Omentum

B urinary bladder   C transverse colon

I small intestine   P pancreas   R rectum

S stomach   U uterus   V vagina

Figure 2.3   Transverse and longitudinal sections of the abdominal cavity.

3

Organ Systems:
Rat Dissection

During this laboratory period we will dissect a freshly killed rat to perform a cursory study of the majority of the organ systems. Since rats and humans have considerable anatomical and physiological similarities, much will be learned here about human anatomy.

Before beginning the dissection, however, it will be necessary to review the eleven systems of the body. A brief description of each system follows. Keep in mind that an **organ** is defined as a structure composed of two or more tissues that performs one or more physiological functions. A **system,** on the other hand, is a group of organs that directly relate to each other functionally. To make the dissection more meaningful, answer the questions on the Laboratory Report *before* doing the dissection.

The Integumentary System

The Latin word *integumentum* means covering. The body surface covering that makes up this system includes the skin, hair, and nails.

The skin's principal function is to prevent bodily invasion by harmful microorganisms. In addition to being a mechanical barrier, the skin produces sweat and sebum (oil), which contain antimicrobial substances for further protection.

The skin also aids in temperature regulation and excretion. The evaporation of perspiration cools the body. The fact that perspiration contains many of the same excretory products found in urine indicates that the kidneys are aided by the skin in the elimination of water, salts, and some nitrogenous wastes from the blood.

To some extent, the skin also plays a role in nutrition. Evidence of this is seen in the way vitamin D forms in the skin's deeper layers when a vitamin D precursor in those cells is exposed to ultraviolet rays of sunlight.

As long as the skin remains intact, the internal environment is protected. However, serious skin injury, such as deep burns, may result in excessive serous fluid loss and electrolytic imbalances.

The Skeletal System

The skeletal system forms a solid framework around which the body is constructed. It consists of bones, cartilage, and ligaments. This system provides support and protection for the softer parts of the body. Delicate organs such as the lungs, heart, brain, and spinal cord are protected by the bony enclosure of the skeletal system.

In addition to protection, the bones provide points of attachment for muscles, which act as levers when the muscles contract. This arrangement makes movement possible.

Two other important functions of the skeletal system are mineral storage and blood cell production. The mineral component of bones provides a pool of calcium, phosphorous, and other ions that may be utilized to stabilize the mineral content of the blood. With respect to blood cell genesis, both red and white blood cells are formed in the red marrow of certain bones of the body.

The Muscular System

Attached to the skeletal framework of the body are muscles that make up nearly half the weight of the body. The skeletal muscles consist primarily of long multinucleated cells. The ability of these cells to shorten when stimulated by nerve impulses enables the muscles to move parts of the body in walking, eating, breathing, and other activities.

Two other kinds of muscle tissue exist in the body: smooth and cardiac. Smooth muscle tissue is the type found in the walls of the stomach, intestines, arteries, veins, urinary bladder, and other organs. Cardiac muscle tissue is found in the walls of the heart. Both types function like skeletal muscle fibers in that they perform work by shortening (contraction). However, whereas skeletal muscles are voluntarily controlled, smooth and cardiac muscle tissues are involuntarily regulated. Although skeletal muscles are primarily involved in limb movement, some skeletal muscles function in the reproductive, respiratory, and digestive systems.

The Nervous System

The nervous system consists of the brain, spinal cord, nerves, and receptors. Although the nervous system has many functions, such as muscular control and regulation of circulation and breathing, its basic function is to facilitate adaptability; i.e., the ability to adjust to external and internal environmental changes.

To be able to adapt to environmental changes there are, first of all, a multitude of different kinds of **receptors** throughout the body that are activated by various kinds of stimuli. A receptor may be stimulated by changes in such things as temperature, pressure, chemicals, sound waves, or light. Once activated, nerve impulses pass along **conduction pathways** (nerves and spinal cord) to **interpretation centers** in the brain, where recognition and evaluation of stimuli occur. Correct responses, which may be muscular, are then achieved by the nervous system via outgoing conduction pathways.

The Cardiovascular System

The cardiovascular system consists of the heart, arteries, veins, capillaries, blood, and spleen. The **heart** is a muscular pump that moves blood throughout the body. **Arteries** are thick-walled vessels that carry blood from the heart to the microscopic **capillaries** that permeate all the tissues. **Tissue fluid,** containing nutrients and oxygen, leaves the blood through the capillary walls and passes into the spaces between the cells. **Veins** are large blood vessels that convey blood from the capillaries back to the heart.

Thus, we see that this system provides transportation of various materials from one part of the body to another. In addition to carrying nutrients, oxygen, and carbon dioxide, the cardiovascular system transports hormones from glands, metabolic wastes from cells, and excess heat from muscles to the skin.

Another very important function of the blood is protection against microbial invasion. The presence of phagocytic (cell-eating) white blood cells, antibodies, and special enzymes in the blood prevents invading microorganisms from destroying the body.

The **spleen** is an oval structure on the left side of the abdominal cavity that acts to some extent as a blood reservoir. It also plays an important role in the removal of fragile red blood cells.

Even more significantly, this organ is probably the source of B-lymphocytes. Evidence of this is the presence of a large number of these antibody-producing cells in the spleen. This organ is part of the reticuloendothelial system, which is described below under the lymphatic system.

The Lymphatic System

The lymphatic system is a network of lymphatic vessels that returns tissue fluid from the intercellular spaces of tissues to the blood. This system is also responsible for the absorption of fats from the intestines. Although carbohydrates and proteins are absorbed directly into the blood through the intestinal wall, fats must pass first into the lymphatic system and then into the blood.

Once tissue fluid enters the lymphatic vessels it is called **lymph.** As this fluid moves through the lymphatic vessels it passes through nodules of lymphoid tissue called **lymph nodes.** Stationary phagocytic cells in these nodes remove bacteria and other foreign material, purifying the lymph before it is returned to the blood. Lymphocytes are also produced here.

Lymphoid tissue, as seen in the nodes, is also seen in the thymus gland, liver, spleen, tonsils, adenoids, appendix, Peyer's patches (in the digestive tract), and bone marrow. This diverse collection of lymphoidal tissue is collectively referred to as the **reticuloendothelial system,** an important component of the immune system.

The Respiratory System

The respiratory system consists of two portions: the air passageways and the respiratory portion. The actual exchange of gases between the blood and the air occurs in the respiratory portion. The lungs contain many tiny sacs called **alveoli,** which greatly increase the surface area for the transfer of oxygen and carbon dioxide in breathing. The passageways consist of the **nasal cavity, nasopharynx, larynx, trachea,** and **bronchi.**

The Digestive System

The digestive system includes the **mouth, salivary glands, esophagus, stomach, small intestine, large intestine, rectum, pancreas, liver,** and **gallbladder.** Its function is to convert ingested food to molecules that are small enough to pass through the intestinal lining into the capillaries and lymphatic vessels of the intestinal wall. This digestive process is achieved by hydrolysis of food molecules by **enzymes** produced in the digestive tract.

In addition to ingestion, digestion, and absorption of food, this system functions in the elimination of undigested materials and protection against infection. Since the contents of the digestive tract contain many foreign, potentially harmful microorganisms, the digestive tract lining, like the integument, acts as a barrier to invasion of microorganisms.

The Urinary System

Cellular metabolism produces waste materials such as carbon dioxide, excess water, nitrogenous products, and excess metabolites. Although the skin, lungs, and large intestine assist in the removal of some of these wastes from the body, the majority of these products are removed from the blood by the **kidneys.** During this process of waste removal, the kidneys function to regulate the chemical composition of all body fluids to maintain homeostasis.

To assist the kidneys in excretion are the **ureters,** which drain the kidneys, the **urinary bladder,** which stores urine, and the **urethra,** which drains the bladder.

The Endocrine System

This system consists of a number of widely dispersed **endocrine glands** that dispense their secretions directly into the blood. These secretions, which are absorbed directly into capillaries within the glands, are called **hormones.**

Hormones perform such functions as integrating various physiological activities (metabolism and growth), directing the differentiation and maturation of the ovaries and testes, and regulating specific enzymatic reactions.

The endocrine system includes the following glands: **pituitary, thyroid, parathyroid, thymus,** **adrenal, pancreas,** and **pineal.** In addition, the **ovaries, testes, stomach lining, placenta,** and **hypothalamus** also produce important hormones.

The Reproductive System

Continuity of the species is the function of the reproductive system. Spermatozoa are produced in the testes of the male, and ova are produced by the ovaries of the female. Male reproductive organs include the **testes, penis, scrotum, accessory glands,** and various ducts. The female reproductive organs include the **ovaries, vagina, uterus, uterine** (Fallopian) **tubes,** and **accessory glands.**

Laboratory Report

Complete the Laboratory Report for this exercise. It is due at the beginning of the laboratory period in which the rat dissection is scheduled.

Rat Dissection

You will work with a laboratory partner to perform this part of the exercise. Your principal objective in the dissection is *to expose the organs for study, not to simply cut up the animal.* Most cutting will be performed with scissors. Whereas the scalpel blade will be used only occasionally, the flat blunt end of the handle will be used frequently for separating tissues.

Materials:
> freshly killed rat, dissecting pan with wax bottom, dissecting kit, dissecting pins

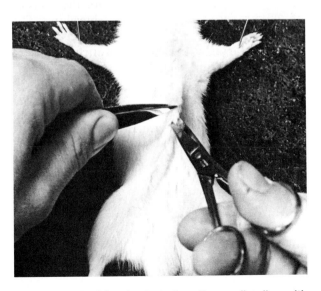

Figure 3.1 Incision is started on the median line with a pair of scissors.

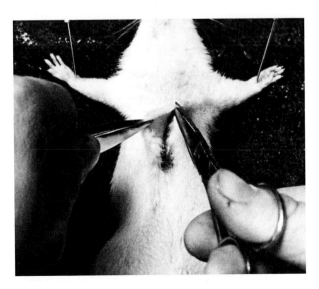

Figure 3.2 The first cut is extended up to the lower jaw.

Skinning the Ventral Surface

1. Pin the four feet to the bottom of the dissecting pan as illustrated in figure 3.1. Before making any incision examine the oral cavity. Note the large **incisors** in the front of the mouth that are used for biting off food particles. Force the mouth open sufficiently to examine the flattened **molars** at the back of the mouth. These teeth are used for grinding food into small particles.

   Note that the **tongue** is attached at its posterior end. Lightly scrape the surface of the tongue with a scalpel to determine its texture. The roof of the mouth consists of an anterior **hard palate** and a posterior **soft palate.** The throat is the **pharynx,** which is a component of both the digestive and respiratory systems.

2. Lift the skin along the midventral line with your forceps and make a small incision with scissors as shown in figure 3.2. Cut the skin upward to the lower jaw, turn the pan around, and complete the incision to the anus, cutting around both sides of the genital openings. The completed incision should appear as in figure 3.3.

3. With the handle of the scalpel, separate the skin from the musculature as shown in figure 3.4. The fibrous connective tissue that lies between the skin and musculature is the **superficial fascia** (Latin: *fascia,* band).

4. Skin the legs down to the "knees" or "elbows" and pin the stretched-out skin to the wax. Examine the surfaces of the **muscles** and note that **tendons,** which consist of tough fibrous connective tissue, attach the muscles to the skeleton.

   Covering the surface of each muscle is another thin gray feltlike layer, the **deep fascia.** Fibers of the deep fascia are continuous with fibers of the superficial fascia, so considerable force with the scalpel handle is necessary to separate the two membranes.

5. At this stage your specimen should appear as in figure 3.5. If your specimen is a female, the mammary glands will probably remain attached to the skin.

Opening the Abdominal Wall

1. As shown in figure 3.5, make an incision through the abdominal wall with a pair of scissors. To make the cut it is necessary to hold the muscle tissue with a pair of forceps.

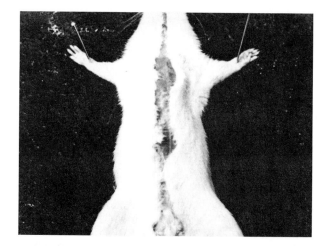

Figure 3.3 Completed incision from the lower jaw to the anus.

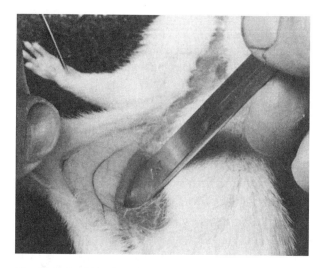

Figure 3.4 Skin is separated from musculature with scalpel handle.

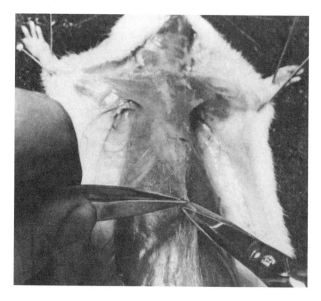

Figure 3.5 Incision of musculature is begun on the median line.

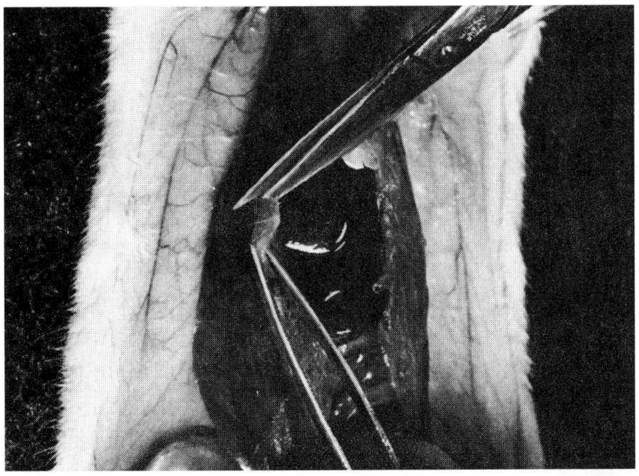

Figure 3.6 Lateral cuts at base of rib cage are made in both directions.

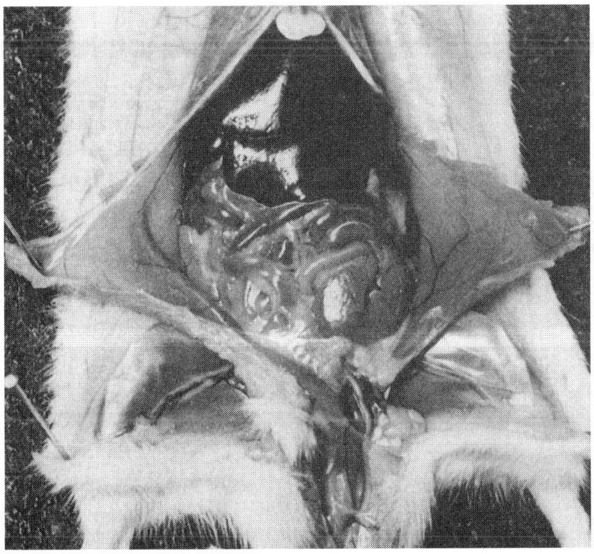

Figure 3.7 Flaps of abdominal wall are pinned back to expose the viscera.

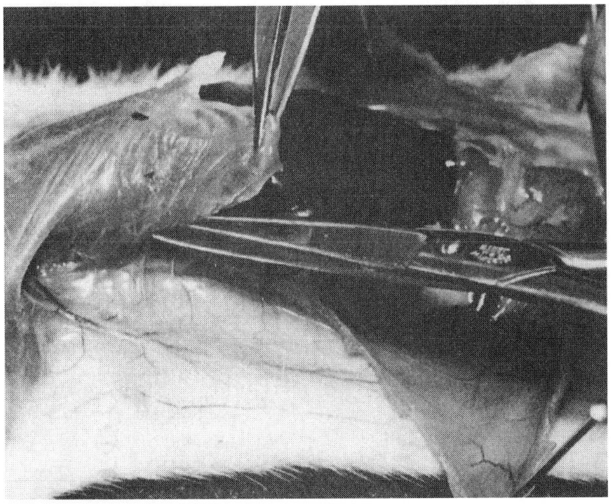

Figure 3.8 Rib cage is severed on each side with scissors.

Caution: Avoid damaging the underlying viscera as you cut.

2. Cut upward along the midline to the rib cage and downward along the midline to the genitalia.
3. To completely expose the abdominal organs make two lateral cuts near the base of the rib cage—one to the left and the other to the right. See figure 3.6. The cuts should extend all the way to the pinned-back skin.
4. Fold out the flaps of the body wall and pin them to the wax as shown in figure 3.7. The abdominal organs are now well exposed.
5. Using figure 3.9 as a reference, identify all the labeled viscera without moving the organs out of place. Note in particular the position and structure of the **diaphragm.**

Examination of Thoracic Cavity

1. Using your scissors, cut along the left side of the rib cage as shown in figure 3.8. Cut through all of the ribs and connective tissue. Then, cut along the right side of the rib cage in a similar manner.
2. Grasp the xiphoid cartilage of the sternum with forceps as shown in figure 3.10 and cut the diaphragm away from the rib cage with your scissors. Now you can lift up the rib cage and look into the thoracic cavity. See figure 3.11.
3. With your scissors, complete the removal of the rib cage by cutting off any remaining attachment tissue.
4. Now examine the structures that are exposed in the thoracic cavity. Refer to figure 3.9 and identify all the structures that are labeled.
   Note the pale-colored **thymus gland,** which is located just above the heart. Remove this gland.
6. Carefully remove the thin **pericardial membrane** that encloses the **heart.**
7. Remove the heart by cutting through the major blood vessels attached to it. Gently sponge away pools of blood with Kimwipes or other soft tissues.
8. Locate the **trachea** in the throat region. Can you see the **larynx** (voice box), which is located at the anterior end of the trachea? Trace the trachea posteriorly to where it divides into two **bronchi** that enter the **lungs.** Squeeze the lungs with your fingers, noting how elastic they are. Remove the lungs. See figure 3.12.
9. Probe under the trachea to locate the soft tubular **esophagus** that runs from the oral

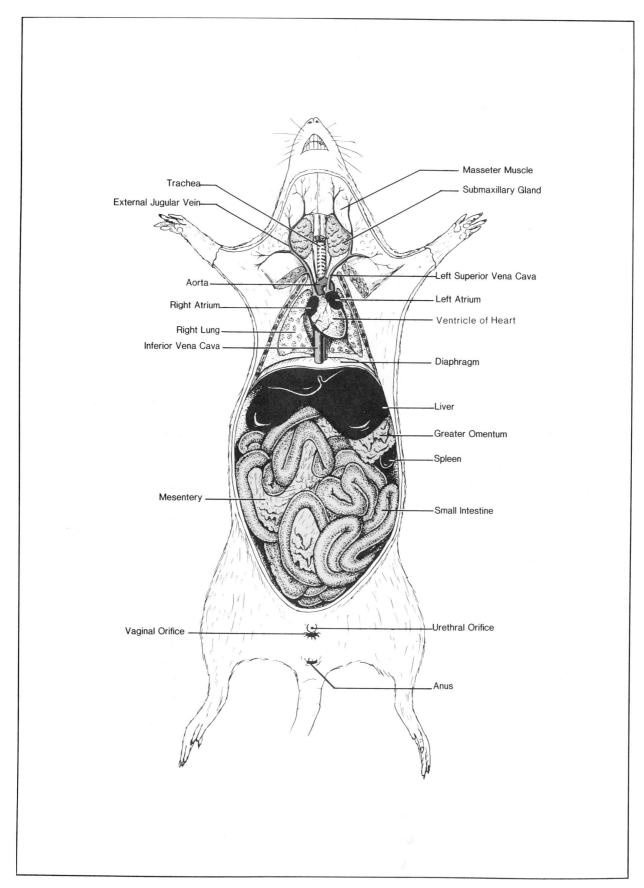

Figure 3.9 Viscera of a female rat.

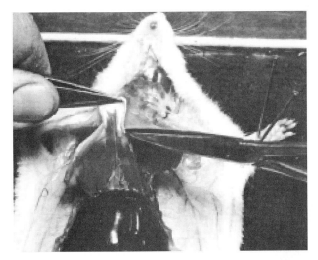

Figure 3.10 Diaphragm is cut free from edge of the rib cage as it is held up by xiphoid process of sternum.

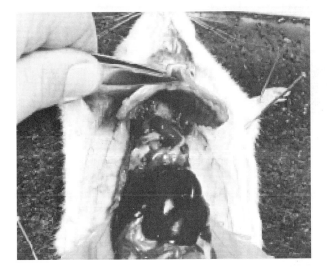

Figure 3.11 Thoracic organs are exposed once the rib cage has been cut loose.

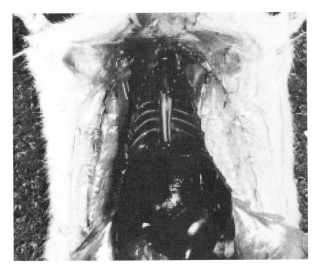

Figure 3.12 Specimen with heart, lungs, and thymus gland removed.

cavity to the stomach. Excise a section of the trachea to reveal the esophagus as illustrated in figure 3.13.

Deeper Examination of Abdominal Organs

1. Lift up the lobes of the reddish brown liver and examine them. Note that rats lack a **gallbladder.** Carefully excise the liver and wash out the abdominal cavity. The stomach and intestines are now clearly visible.

2. Lift out a portion of the intestines and identify the membranous **mesentery,** which holds the intestines in place. It contains blood vessels and nerves that supply the digestive tract. If your specimen is a mature healthy animal, the mesenteries will contain considerable fat.

3. Now lift the intestines out of the abdominal cavity, cutting the mesenteries, as necessary, for a better view of the organs. Note the great length of the **small intestine.** Its name refers to the diameter, not its length.

   The first portion of the small intestine, which is connected to the stomach, is called the **duodenum.** At its distal end the small intestine is connected to a large saclike structure, the **cecum.** The *appendix* in humans is a vestigial portion of the cecum.

   The cecum communicates with the **large intestine.** The latter structure consists of the **ascending, transverse, descending,** and **sigmoid** divisions. The sigmoid colon, in turn, empties into the **rectum.**

4. Try to locate the **pancreas,** which is embedded in the mesentery alongside the duodenum. It is often difficult to see. Pancreatic enzymes enter the duodenum via the **pancreatic duct.** See if you can locate this minute tube.

5. Locate the **spleen,** which is situated on the left side of the abdomen near the stomach. It is reddish brown and held in place with mesentery. Do you recall the functions of this organ?

6. Remove the stomach, small intestine, and large intestine after severing the distal end of the esophagus and the distal end of the sigmoid colon.

   Removal of these organs enables you to see the **descending aorta** and the **inferior vena cava.** The aorta carries blood posteriorly to the body tissues. The inferior vena cava is a vein that returns blood from the posterior regions to the heart.

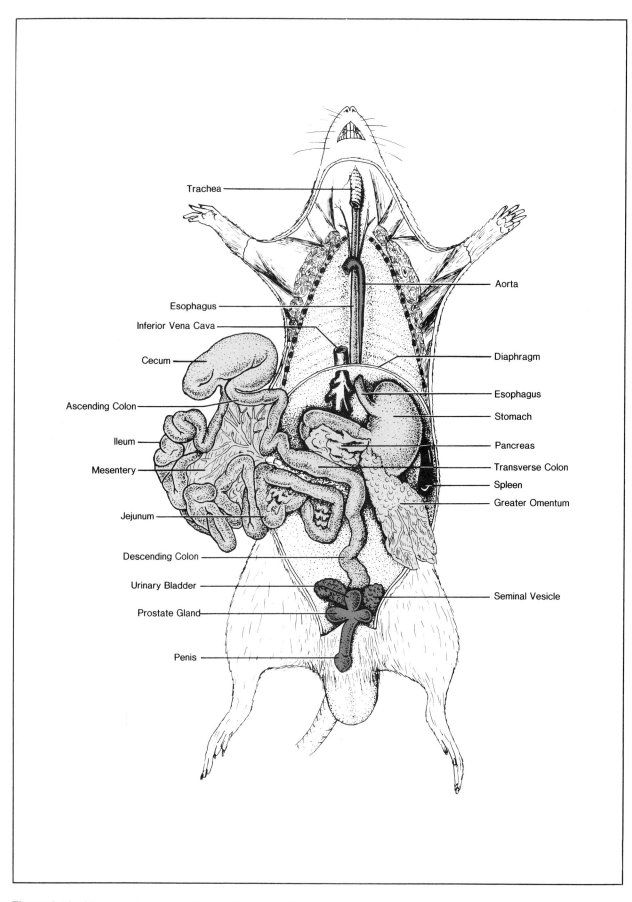

Figure 3.13 Viscera of a male rat (heart, lungs, and thymus removed).

7. Peel away the peritoneum and fat from the posterior wall of the abdominal cavity. *Removal of this fat will require special care to avoid damaging important structures.* Removing fat will make the kidneys, blood vessels, and reproductive structures more visible.

Locate the two **kidneys** and **urinary bladder.** Trace the two **ureters,** which extend from the kidneys to the bladder. Examine the anterior surfaces of the kidneys and locate the **adrenal glands,** which are important components of the endocrine gland system.

8. **If your specimen is a female:** Compare it with figure 3.14. Locate the two **ovaries,** which lie lateral to the kidneys. From each ovary a **uterine tube** leads posteriorly to join the **uterus.**

Note that the uterus is a Y-shaped structure joined to the vagina.

If your specimen appears to be pregnant, open up the uterus and examine the developing embryos. Note how they are attached to the uterine wall.

9. **If your specimen is a male:** Compare your specimen with figure 3.15. The **urethra** is located in the **penis.** Apply pressure to one of the testes through the wall of the **scrotum** to see if it can be forced up into the **inguinal canal.**

Open up the scrotum on one side to expose a **testis,** the **epididymis,** and the **vas deferens.** Trace the vas deferens over the urinary bladder to where it penetrates the **prostate gland** to join the urethra.

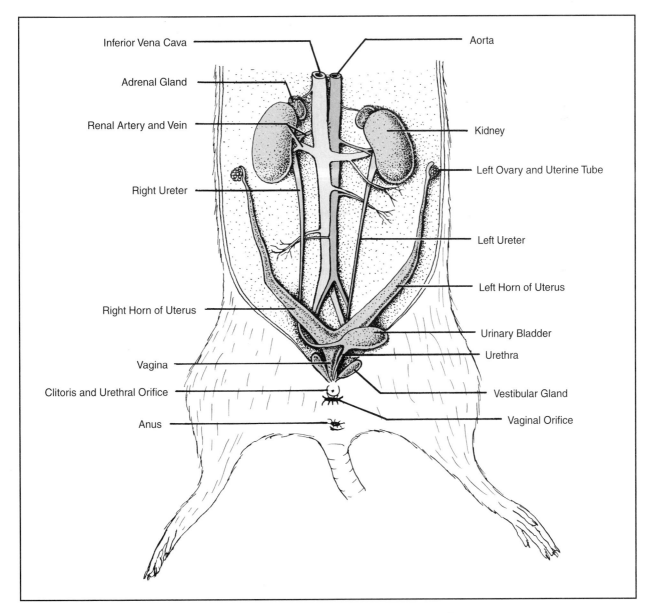

Figure 3.14 Abdominal cavity of a female rat (digestive tract and liver removed).

Summarization In this cursory dissection you have become acquainted with the respiratory, cardiovascular, digestive, urinary, and reproductive systems. Portions of the endocrine system have also been observed. Five systems (integumentary, skeletal, muscular, lymphatic, and nervous) have been omitted at this time. These will be studied later.

If you have done a careful and thoughtful rat dissection, you should have a good general understanding of the basic structural organization of the human body. Much that we have seen here has its human counterpart.

Cleanup Dispose of the specimen as directed by your instructor. Scrub your instruments with soap and water, rinse, and dry them.

Wash your hands with soap and water, rinse, and dry thoroughly

Laboratory Report

Answer the questions on the Laboratory Report to test your understanding of the organ systems.

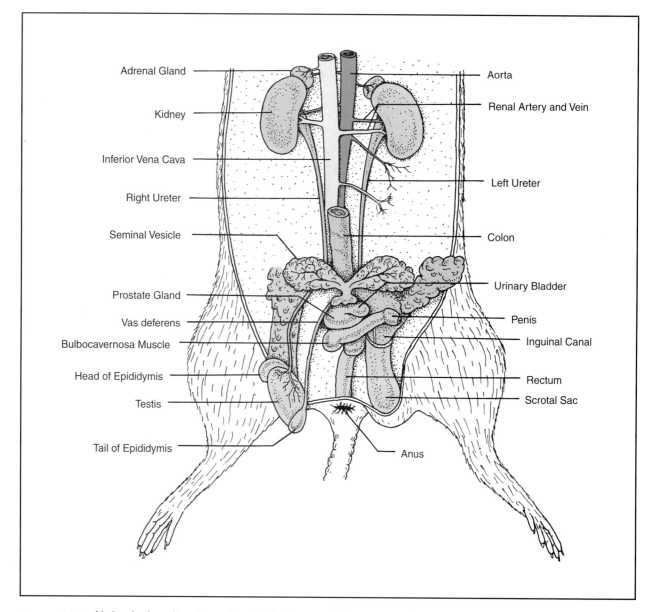

Figure 3.15 Abdominal cavity of a male rat (intestines and liver removed).

PART 2

Cells and Tissues

This unit consists of nine exercises that pertain, primarily, to the study of cells and tissues. Since considerable emphasis will be placed on histological studies throughout the course, this unit, which is very basic to anatomy and physiology, is extremely important.

The first exercise, which pertains to microscopy, should be studied before going to the laboratory. It is of a general nature and attempts to provide some of the physical and mathematical reasons why certain techniques in microscopy work and others do not.

The next five exercises pertain to cytological studies. Cellular structure, physiology, and reproduction are studied in Exercises 5 and 6. In Exercises 7, 8, and 9 our concerns are related to the immediate environment of cells, as we study the role of osmosis, pH, and buffers on cellular physiology.

A survey of epithelial and connective tissues is covered in Exercises 10 and 11. Microscope slides will be examined in the laboratory to note the distinct differences of the various types of tissues. Exercise 12, which pertains to the skin, is included in this unit because the skin contains both epithelial and connective tissues.

4

Microscopy

Many types of microscopes are available today to one who wishes to study cells and tissues: the electron microscope, phase-contrast, interference, and fluorescence microscopes are just a few examples. In this laboratory we will use the conventional brightfield microscope for our studies. This instrument, despite its limitations, is still the most widely used type of microscope in hospitals and clinical laboratories. In this exercise our concern will be with highlighting accepted procedures for using this instrument.

Microscopy can be fascinating or frustrating. Success in this endeavor depends to a large extent on how well you understand the mechanics and limitations of your microscope. It is the purpose of this exercise to outline procedures that should minimize difficulties and enable you to get the most out of your subsequent attempts at cytological microscopy.

Before entering the laboratory to do any of the studies on cells and tissues, it would be desirable if you answered all the questions on the Laboratory Report after reading over this entire exercise. Your instructor may require that the Laboratory Report be handed in before doing any laboratory work.

Care of the Instrument

Microscopes represent considerable investment and can be damaged rather easily if certain precautions are not observed. The following suggestions cover most hazards.

Transport When carrying your microscope from one part of the room to another, use both hands when holding the instrument (see figure 4.1). If it is carried with only one hand and allowed to dangle at your side, there is always the possibility of collision with furniture or some other object. And, incidentally, under no circumstances should one attempt to carry two microscopes at one time.

Clutter It is very important that you keep your workstation uncluttered at all times while doing microscopy. Nonessential books, lunches, and other unneeded supplies should be stashed away in a drawer or some other out-of-the-way place. A clear work area promotes efficiency and results in fewer accidents.

Electric Cord Unbelievably, dangling microscope cords can result in catastrophic accidents. In crowded quarters a dangling cord has been known to become entangled in a student's foot. Result: a damaged microscope on a concrete floor. The best precaution is to make sure your microscope cord is kept out of harm's way.

Lens Care At the beginning of each laboratory period check the lenses to make sure they are clean. At the end of each lab session be sure to wipe any immersion oil off the immersion lens if it has been used. More specifics about lens care are provided on pages 22 and 23.

Dust Protection In most laboratories dustcovers are used to protect the instruments during storage. If one is available, place it over the microscope at the end of the period.

Components

Before we discuss the procedures for using a microscope, let's identify the principal parts of the instrument as illustrated in figure 4.2.

Framework All microscopes have a basic frame structure that includes the **arm** and **base**. To this framework all other parts are attached. On many of the older microscopes the base is not rigidly

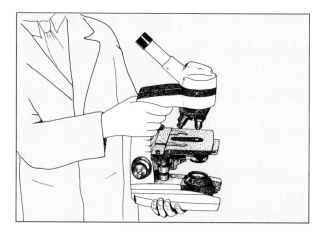

Figure 4.1 The microscope should be held firmly with both hands when it is carried.

attached to the arm as is the case in figure 4.2; instead, a pivot point is present that enables one to tilt the arm backward to adjust the eyepoint height.

Stage   The horizontal platform that supports the microscope slide is called the *stage*. Note that it has a clamping device, the **mechanical stage,** which is used for holding and moving the slide around on the stage. Note, also, the location of the **mechanical stage control** in figure 4.2.

Light Source   In the base of most microscopes is positioned some kind of light source. Ideally, the lamp should have a **voltage control** to vary the intensity of light. The microscope in figure 4.2 has a knurled wheel on the right side of its base to regulate the voltage supplied to the lightbulb The microscope base in figure 4.4 has a knob (the left one) that controls voltage.

Most microscopes have some provision for reducing light intensity with a **neutral density filter.**

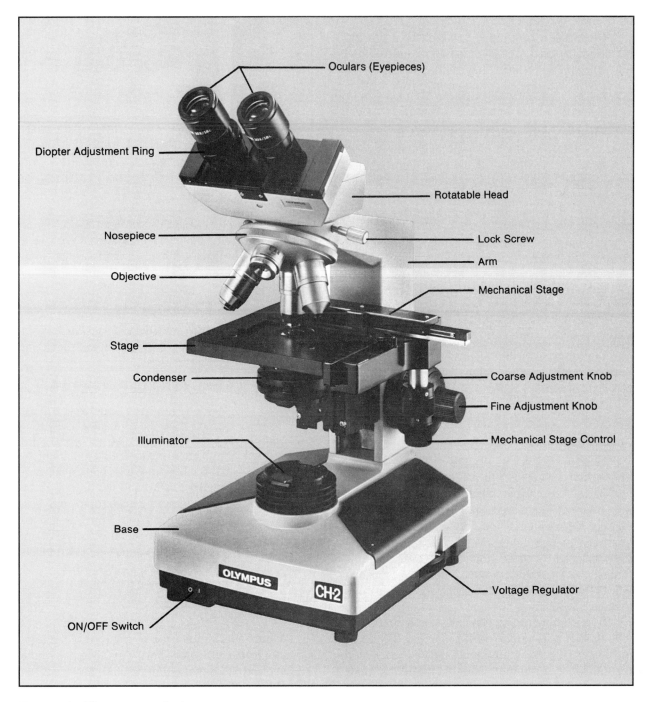

Figure 4.2   The compound microscope.
Courtesy of the Olympus Corporation, Lake Success, N.Y.

Such a filter is often needed to reduce the intensity of light below the lower limit allowed by the voltage control. On microscopes such as the Olympus CH-2, one can simply place a neutral density filter over the light source in the base. On some microscopes a filter is built into the base (see figure 4.4).

Lens Systems All microscopes have three lens systems: the oculars, the objectives, and the condenser. Figure 4.3 illustrates the light path through these three systems.

The **ocular,** or eyepiece, which is at the top of the instrument, consists of two or more internal lenses and usually has a magnification of 10×. Although the microscope in figure 4.2 has two oculars (binocular), a microscope often has only one.

Three or more **objectives** are usually present. Note that they are attached to a rotatable **nosepiece,** that makes it possible to move them into position over a slide. Objectives on most laboratory microscopes have magnifications of 10×, 45×, and 100×, designated as **low power, high-dry,** and **oil immersion,** respectively.

The third lens system is the **condenser,** which is located under the stage. It collects and directs the light from the lamp to the slide being studied. The condenser can be moved up and down by a knob under the stage. A **diaphragm** within the condenser regulates the amount of light that reaches the slide. On the Olympus microscope in figure 4.2 the diaphragm is controlled by turning a knurled ring. On some microscopes a diaphragm lever is present. Figure 4.3 illustrates best the position of the condenser and diaphragm.

Focusing Knobs The concentrically arranged **coarse adjustment** and **fine adjustment knobs** on the side of the microscope are used for bringing objects into focus when studying an object on a slide.

Ocular Adjustments On binocular microscopes one must be able to change the distance between the oculars and to make diopter changes for eye differences. On most microscopes the interocular distance is changed by simply pulling apart or pushing together the oculars.

To make diopter adjustments, one focuses first with the right eye only. Without touching the focusing knobs, diopter adjustments are then made on the left eye by turning the knurled **diopter adjustment ring** (figure 4.2) on the left ocular until a sharp image is seen. One should now be able to see sharp images with both eyes.

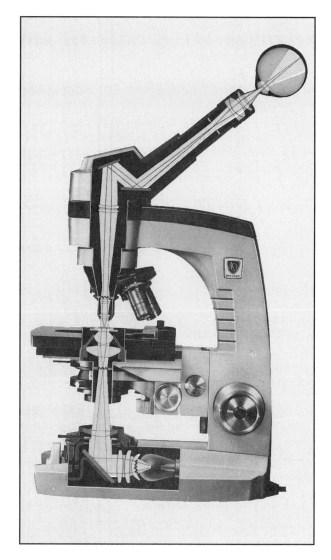

Figure 4.3 The light pathways of a microscope.

Resolution

The resolution limit, or **resolving power,** of a microscope lens system is a function of its numerical aperture, the wavelength of light, and the design of the condenser. The maximum resolution of the best microscopes with oil immersion lenses is around 0.2 μm. This means that two small objects that are 0.2 μm apart will be seen as separate entities; objects closer than that will be seen as a single object.

To get the maximum amount of resolution from a lens system, the following factors must be taken into consideration:

- A **blue filter** should be in place over the light source because the short wavelength of blue light provides maximum resolution.
- The **condenser** should be kept at its highest position where it allows a maximum amount of light to enter the objective.

- The **diaphragm** should not be stopped down too much. Although stopping down improves contrast, it reduces the numerical aperture.
- **Immersion oil** should be used between the slide and the 100× objective.

Lens Care

Keeping the lenses of your microscope clean is a constant concern. Unless all lenses are kept free of dust, oil, and other contaminants, they are unable to achieve the degree of resolution that is intended. Consider the following suggestions for cleaning the various lens components:

Cleaning Tissues Only lint-free, optically safe tissues should be used to clean lenses. Tissues free of abrasive grit fall in this category. Booklets of lens tissue are most widely used for this purpose. Although several types of boxed tissues are also safe, *use only the type of tissue that is recommended by your instructor.*

Solvents Various liquids can be used for cleaning microscopic lenses. Green soap with warm water works very well. Xylene is universally acceptable. Alcohol and acetone are also recommended, but often with some reservations. Acetone is a powerful solvent that could possibly dissolve the lens mounting cement in some objective lenses if it were used too liberally. When it is used it should be used sparingly. Your instructor will inform you as to what solvents are best for the lenses of your microscope.

Oculars The best way to determine if your eyepiece is clean is to rotate it between the thumb and forefinger as you look through the microscope. A rotating pattern will be evidence of dirt.

If cleaning the top lens of the ocular with lens tissue fails to remove the debris, one should try cleaning the lower lens with lens tissue and blowing off any excess lint with an air syringe or gas cannister. *Whenever the ocular is removed from the microscope, it is imperative that a piece of lens tissue be placed over the open end of the microscope as illustrated in figure 4.5.*

Objectives Objective lenses often become soiled by materials from slides or fingers. A piece of lens tissue moistened with green soap and water, or one of the acceptable solvents mentioned above, will usually remove whatever is on the lens. Sometimes a cotton swab with a solvent will work better than lens tissue. At any time that the image on the slide is unclear or cloudy, assume at once that the objective you are using is soiled.

Condenser Dust often accumulates on the top surface of the condenser; thus, wiping it off occasionally with lens tissue is desirable.

Procedures

If your microscope has three objectives you have three magnification options: (1) low-power, or 100× magnification, (2) high-dry magnification, which is 450× with a 45× objective, and (3) 1000× magnification with a 100× oil immersion objective. Note that the total magnification seen through an objective is calculated by simply multiplying the power of the ocular by the power of the objective.

Whether you use the low-power objective or the oil immersion objective will depend on how much magnification is necessary. Generally speaking, however, it is best to start with the low-power

Figure 4.4 On this older type of microscope the left knob controls voltage, the other knob controls the neutral density filter.

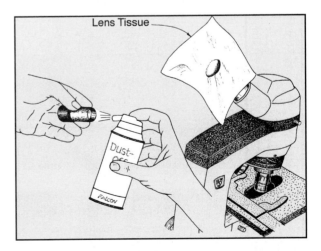

Figure 4.5 When oculars are removed for cleaning, cover the ocular opening with lens tissue. A blast from an air syringe or gas cannister removes dust and lint.

objective and progress to the higher magnifications as your study progresses. Consider the following suggestions for setting up your microscope and making microscopic observations:

Viewing Setup   If your microscope has a rotatable head, such as the ones being used by the two students in figure 4.6, there are two ways that you can use the instrument. Note that the student on the left has the arm of the microscope *near* him, and the other student has the arm *away from* her. With this type of microscope, the student on the right has the advantage in that the stage is easier to observe. Note also that when focusing the instrument she is able to rest her arm on the table. The manufacturer of this type of microscope intended that the instrument be used in the way demonstrated by the young lady. If the microscope head is not rotatable, it will be necessary to use the other position.

Low-Power Examination   The main reason for starting with the low-power objective is to enable you to explore the slide to look for the object you are planning to study. Once you have found what you are looking for, you can proceed to higher magnifications. Use the following steps when exploring a slide with the low-power objective:

1. Position the slide on the stage with the material to be studied on the *upper* surface of the slide. Figure 4.7 illustrates how the slide must be held in place by the mechanical stage retainer lever.
2. Turn on the light source, using a *minimum* amount of voltage. If necessary, reposition the slide so that the stained material on the slide is in the *exact center* of the light source.
3. Check the condenser to see that it has been raised to its highest point.

4. If the low-power objective is not directly over the center of the stage, rotate it into position. Be sure that as you rotate the objective into position it clicks into its locked position.
5. Turn the coarse adjustment knob to lower the objective *until it stops*. A built-in stop will prevent the objective from touching the slide.
6. While looking down through the ocular (or oculars), bring the object into focus by turning the fine adjustment focusing knob. Don't readjust the coarse adjustment knob. If you are using a binocular microscope it will also be necessary to adjust the interocular distance and diopter adjustment to match your eyes.
7. Manipulate the diaphragm control to reduce or increase the light intensity to produce the clearest, sharpest image. Note that as you close down the diaphragm to reduce the light intensity, the contrast improves and the depth of field increases. Stopping down the diaphragm when using the low-power objective does not decrease resolution.
8. Once an image is visible, move the slide about to search out what you are looking for. The slide is moved by turning the knobs that move the mechanical stage.
9. Check the cleanliness of the ocular, using the procedure outlined above.
10. Once you have identified the structures to be studied and wish to increase the magnification, you may proceed to either high-dry or oil immersion magnification. However, before changing objectives, *be sure to center the object you wish to observe.*

High-Dry Examination   To proceed from low-power to high-dry magnification, all that is necessary is to rotate the high-dry objective into position and open up the diaphragm somewhat. It may be

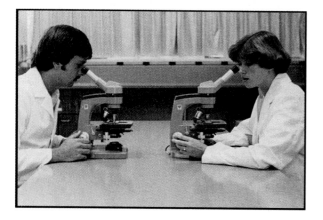

Figure 4.6   The microscope position on the right has the advantage of stage accessibility.

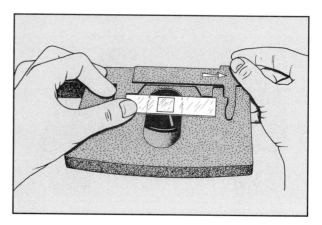

Figure 4.7   The slide must be properly positioned as the retainer lever is moved to the right.

necessary to make a minor adjustment with the fine adjustment knob to sharpen up the image, but *the coarse adjustment knob should not be touched.*

If a microscope is of good quality, only minor focusing adjustments are needed when changing from low-power to high-dry because all the objectives will be **parfocalized.** Nonparfocalized microscopes do require considerable refocusing when changing objectives.

High-dry objectives should only be used on slides that have cover glasses; without them, images are usually unclear. When increasing the lighting, be sure to open up the diaphragm first instead of increasing the voltage on your lamp; reason: *lamp life is greatly extended when used at low voltage.* If the field is not bright enough after opening the diaphragm, feel free to increase the voltage. A final point: keep the condenser at its highest point.

Oil Immersion Techniques The oil immersion lens derives its name from the fact that a special mineral oil is interposed between the lens and the microscope slide. The oil is used because it has the same refractive index as glass, which prevents the loss of light due to the bending of light rays as they pass through air. The use of oil in this way enhances the resolving power of the microscope. Figure 4.8 illustrates this phenomenon.

With parfocalized objectives one can go to oil immersion from either low-power or high-dry. Once the microscope has been brought into focus at one magnification, the oil immersion lens can be rotated into position without fear of striking the slide.

Before rotating the oil immersion lens into position, however, a drop of immersion oil is placed on the slide. If the oil appears cloudy it should be discarded.

When using the oil immersion lens it is best to open the diaphragm as much as possible. Stopping down the diaphragm tends to limit the resolving

power of the optics. In addition, the condenser must be kept at its highest point. If different-colored filters are available for the lamp housing, it is best to use blue or greenish filters to enhance the resolving power.

Using this lens takes a little practice. The manipulation of lighting is critical. Before returning the microscope to the cabinet, all oil must be removed from the objective and stage. Slides with oil should be wiped clean with lens tissue, also.

Putting It Away

When you take a microscope from the cabinet at the beginning of the period, you expect it to be clean and in proper working condition. The next person to use the instrument after you have used it will expect the same consideration. A few moments of care at the end of the period will ensure these conditions. Check over this list of items at the end of each period before you return the microscope to the cabinet.

1. Remove the slide from the stage.
2. If immersion oil has been used, wipe it off the lens and stage with lens tissue. If the slide has a cover glass on it remove the oil with tissue. Slides that lack a cover glass (such as blood smears) should not be wiped dry. Simply place these slides in a slide box and let the oil drain off.
3. Rotate the low-power objective into position.
4. If the microscope has been inclined, return it to an erect position.
5. If the microscope has a built-in movable lamp, raise the lamp to its highest position.
6. If the microscope has a long attached electric cord, wrap it around the base.
7. Adjust the mechanical stage so that it does not project too far on either side.
8. Replace the dustcover.
9. If the microscope has a separate transformer, return it to its designated place.
10. Return the microscope to its correct place in the cabinet.

Laboratory Report

Before the microscope is to be used in the laboratory, answer all the questions on the Laboratory Report. Preparation on your part before going to the laboratory will greatly facilitate your understanding. Your instructor may wish to collect this report at the *beginning of the period* on the first day that the microscope is to be used in class.

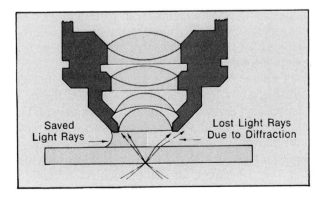

Figure 4.8 Immersion oil, having the same refractive index as glass, prevents light loss due to diffraction.

Saved Light Rays

Lost Light Rays Due to Diffraction

Cell Study Proceed to the cellular studies of Exercise 5.

5

Basic Cell Structure

As a prelude to the study of cellular physiology, it is essential to review cellular anatomy. Structural visualization of the plasma membrane, nucleus, and organelles such as the endoplasmic reticulum, centrosome, and lysosomes goes hand in hand with understanding the physiological activities of the cell as a whole.

In the first portion of this exercise, a summary of the present knowledge of cellular structure and function is presented. This is followed by laboratory studies of epithelial cells and living microorganisms.

Since much of the discussion in this exercise pertains to cellular structure as seen with an electron microscope, and since our laboratory observations will be made with a light microscope, there will be some cellular structures that cannot be seen in great detail in the laboratory.

To get the most out of this exercise *it is recommended that the questions on the Laboratory Report be answered before entering the laboratory for microscopic studies.*

The Basic Design

The study of cellular structure (**cytology**) has advanced at a rapid rate over the past decades with the use of electron microscopy and special techniques in the study of cellular physiology. The old notion of the cell as a "sac of protoplasm" has been supplanted by a much more dynamic model.

A cell is now viewed as a highly complex miniature computer with an integrated, compartmentalized ultrastructure capable of communicating with its environment, altering its shape, processing information, and synthesizing a great variety of substances. It is largely through our advancing knowledge of the cellular organelles that this new viewpoint has emerged.

Figure 5.1 is a diagrammatic version of a pancreatic cell, incorporating the various organelles that can be seen with an electron microscope. A pancreatic cell has been chosen for our study here because it lacks the degree of specialization that is seen in many other cells. In addition, it contains a majority of the organelles that are present in most body cells.

Although it is not possible to see the fine structure of many organelles with our ordinary light microscope, we are including descriptions of their appearance and current theories of function. As the various cell structures are discussed in the following text, identify them in figure 5.1.

The Plasma Membrane

The outer surface of every cell consists of an extremely thin, delicate cell membrane that is often referred to as the *plasma membrane*. Chemical analyses of plasma membranes indicate that phospholipids, glycolipids, and proteins are the principal constituents. Lipids account for one-half the mass of plasma membranes; proteins make up the other half.

Examination with an electron microscope reveals that all cell membranes have a similar basic trilaminar structure, the so-called *unit membrane*. The current interpretation of the molecular organization of this membrane proposes that **lipid molecules** form a double fluid layer in which **protein** and **glycoprotein molecules** are embedded. On the left side of figure 5.1 is an enlarged portion of this membrane illustrating the relationship of these various components to each other. Note that some of the protein molecules (color code: brown) protrude externally, others protrude internally, and some extend through both sides of the membrane. This asymmetrical architecture appears to account for the external and internal differences in receptor sites that affect the permeability characteristics of the membrane. The tubular nature of protein molecules that extend completely through the membrane provides a convenient passageway for certain types of ions and molecules.

Cell membranes are *selectively* permeable in that they allow certain molecules to pass through easily and prevent other molecules from gaining entrance to the cell. Although cell membranes play both active and passive roles in molecule movements, experimental evidence seems to indicate that the movement of all molecules, including water, is assisted by the membrane. In Exercise 7 we will study some of the forces that are involved in permeability.

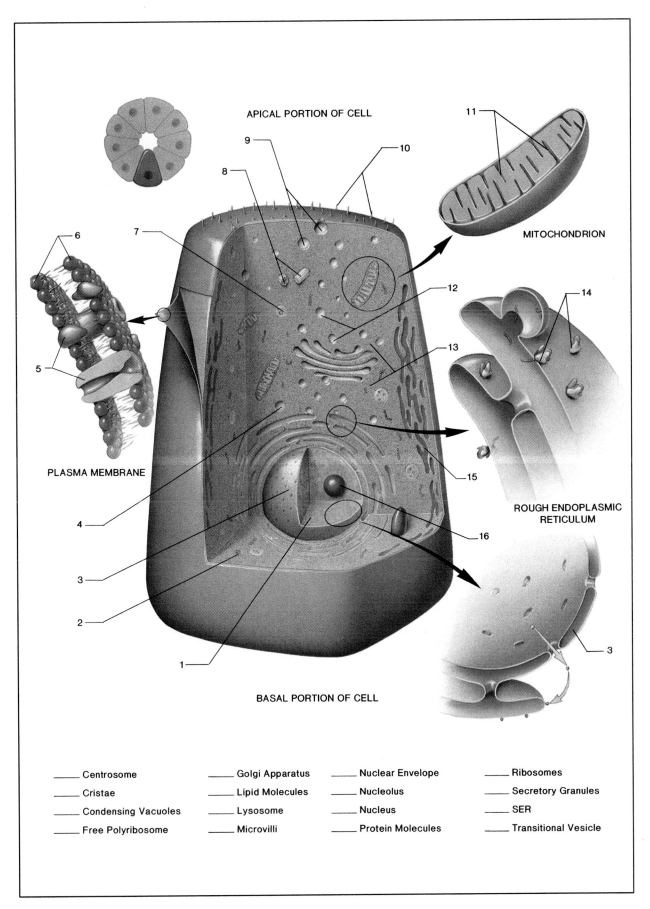

APICAL PORTION OF CELL

MITOCHONDRION

PLASMA MEMBRANE

ROUGH ENDOPLASMIC
RETICULUM

BASAL PORTION OF CELL

_____ Centrosome   _____ Golgi Apparatus   _____ Nuclear Envelope   _____ Ribosomes

_____ Cristae   _____ Lipid Molecules   _____ Nucleolus   _____ Secretory Granules

_____ Condensing Vacuoles _____ Lysosome   _____ Nucleus   _____ SER

_____ Free Polyribosome   _____ Microvilli   _____ Protein Molecules   _____ Transitional Vesicle

Figure 5.1   The microstructure of a cell.

The Nucleus

Near the basal portion of the cell is a large spherical body, the *nucleus*. It is shown in figure 5.1 with a portion cut out of it. The inner substance *(nucleoplasm)* of this body is surrounded by a double-layered **nuclear envelope** that, unlike other membranes, is perforated by pores of significant size. These openings provide a viable passageway between the nucleoplasm and cytoplasm.

A cell that has been stained with certain dyes will exhibit darkly stained regions in the nucleus called *chromatin granules* (not shown in figure 5.1). These granules are visible even with a light microscope. They represent highly condensed DNA molecules that constitute part of the chromosomes. Current theories suggest that increased condensed chromosomal material indicates a metabolically less active cell.

The most conspicuous spherical structure within the nucleus is the **nucleolus.** Although nucleoli of different cells vary considerably in size, number, and structure, they all consist primarily of RNA and function chiefly in the production of ribonucleoprotein for the ribosomes. The arrow shown in the magnified view in the lower right-hand corner of figure 5.1 illustrates the passage of RNA from the nucleoplasm through a pore of the nuclear envelope.

The Cytoplasmic Matrix

Between the plasma membrane and nucleus is an area called the *cytoplasmic matrix,* or *cytoplasm.* This region is a heterogeneous aggregation of many components involved in cellular metabolism. The cytoplasm is structurally and functionally compartmentalized by the numerous organelles, many of which are hollow and enclosed by unit membranes. A description of each of these cytoplasmic organelles follows.

Endoplasmic Reticulum The most extensive structure within the cytoplasm is a complex system of tubules, vesicles, and sacs called the *endoplasmic reticulum* (ER). It is a double-layered unit membrane that is somewhat similar to the nuclear and plasma membranes; in some instances it is continuous with those membranes (see lower right-hand enlarged view of figure 5.1). It appears to be a manufacturing organelle where many of the essential proteins of the cell are produced. It also acts as a microcirculatory system in that it transports newly formed molecules to the Golgi apparatus.

Detailed studies of cells have shown that ER may have surfaces that are smooth or rough. ER that is designated as **rough endoplasmic reticu-**

lum **(RER)** is a fluid-filled canalicular system studded with ribosomes. It is the type illustrated in the magnified view on the right side of figure 5.1. RER is believed to be an important site of protein synthesis and storage. It is more developed in cells that are primarily secretory in function.

ER that lacks ribosomes is designated as **smooth endoplasmic reticulum (SER).** Some layers of SER are shown in the cytoplasm near the plasma membrane. It is thought that SER is active in various types of biosynthesis.

Ribosomes As stated above, ribosomes are small bodies attached to the surface of the RER. They are also scattered throughout the cytoplasm in rosettes and chains called **free polyribosomes** (label 2). Since they are only 170 Ångstrom units (17 nanometers) in diameter they can be resolved only by electron microscopy.

The components for each ribosome originate in the nucleolus of the nucleus. They pass from the nucleolus through the pores of the nuclear membrane into the cytoplasm, where they unite to form the completed ribosomal particle. Each ribosome consists of about 60% RNA and 40% protein. Their proposed structure is shown in the enlarged view of the RER in figure 5.1.

Ribosomes are sites of protein synthesis and are particularly numerous in cells that are actively synthesizing proteins. Free polyribosomes appear to be involved in synthesis of proteins for endogenous use; attached ribosomes on RER, on the other hand, are implicated in the synthesis of protein to be transported extracelluarly (i.e., secretion).

Mitochondria These large (1×2–3 μm) bean-shaped organelles have double-layered membranous walls with inward-protruding partial partitions called **cristae** that serve to increase the total surface area.

Mitochondria contain DNA and are able to replicate themselves. They also contain their own ribosomes. They are sometimes referred to as the "power plants" of the cell, since it is here that the energy-yielding reactions of oxidative respiration occur. Simply stated, these reactions oxidize glucose ($C_6H_{12}O_6$) to yield energy in the form of ATP:

$$C_6H_{12}O_6 + 6O_2 \rightarrow 6CO_2 + 6H_2O + ATP + Heat$$

Golgi Apparatus This unit membrane organelle is quite similar in basic structure to SER. In figure 5.1 it appears as a layered stack of vesicles between the nucleus and the apical end of the cell.

Its primary role is related to the packaging, movement, and completed synthesis of products to be released by secretory cells. The protein molecules produced by the RER are transported to the Golgi apparatus via **transitional vesicles** (color code: blue). These vesicles form on the surface of the RER, move toward the Golgi apparatus, and then coalesce with the surface membrane of the Golgi apparatus to deliver the protein molecules. Inside the Golgi apparatus additional biosynthesis occurs, often with the addition of sugars.

The completely processed materials accumulate at the apical end of the organelle and form **condensing vacuoles.** These membrane-bound vesicles move to the apex of the cell, where, as **secretory granules,** they are ultimately exocytosed. Figure 5.1 illustrates how the granules fuse with the plasma membrane. The fate of the exocytosed product depends upon the cell type.

Lysosomes   Saclike structures in the cytoplasm that act like recycling centers are called *lysosomes.* These sacs (color code: turquoise) have membranes that consist of a single membrane and are believed to form by pinching off of sacs from the Golgi apparatus. The contents of lysosomes vary from cell to cell, but typically they include cellular debris and enzymes that hydrolyze proteins and nucleic acids.

Portions of mitochondria, dead microorganisms, worn-out red blood cells, and other debris have been identified in these sacs. It appears that the expired proteins of these waste products are refashioned here into proteins needed by the cell. Lysosomes that cease to be of value to a cell are exocytosed from the cell through the cell membrane.

When cells die, the membranes surrounding the lysosomes disintegrate, releasing enzymes into the cytoplasm. The hydrolytic action of the enzymes on the cell hastens the death of the cell. It is for this reason that these organelles have been called "suicide bags."

The Centrosome   Near the apex of the cell is seen a pair of microtubule bundles that are arranged at right angles to each other. Each bundle, called a *centriole,* consists of nine triplets of microtubules arranged in a circle. The two centrioles are collectively referred to as the *centrosome.* Because of their small size, centrosomes appear as very tiny dots when observed with a light microscope. In most cells the centrosome is located closer to the nucleus than shown in figure 5.1.

During mitosis and meiosis the centrioles play a role in the formation of spindle fibers that aid in the

Figure 5.2 A computer-generated visualization of microtubules, vesicles, and mitochondria in the axon of a nerve cell. Vesicles and mitochondria are being propelled by the microtubules.

separation of chromatids. They also function in the formation of cilia and flagella in certain types of cells.

Cytoskeletal Elements

It now appears that a living cell has an extensive permeating network of fine fibers that contribute to cellular properties such as contractility, support, shape, and translocation of materials. Microtubules and microfilaments fall into this category.

Microtubules   Molecules of protein *(tubulin),* arranged in submicroscopic cylindrical hollow bundles are called *microtubules.* They are found dispersed throughout the cytoplasm of most cells and converge especially around the centrioles to form *aster fibers* and *spindles* that are seen during cell division (Exercise 6). They also play a significant role in the transport of vesicles from the Golgi apparatus to other parts of the cell.

Nerve cells that have long processes, called axons, utilize microtubules to transport vesicles and mitochondria. Movement of these organelles is from the cell body to the axon terminal, as well as from the terminal back to the cell body.

Figure 5.2 is a computer-generated image of how these microtubules might look where they enter an axon of a nerve cell. Note that various-sized vesicles are propelled along the microtubules and that these vesicles move along the microtubules in both directions *simultaneously.* The energy for vesicle transport is supplied by mitochondria: a portion of one is seen clearly in the lower right quadrant; another is barely discernible in the upper left quadrant of figure 5.2.

Microfilaments All cells show some degree of contractility. Cytoskeletal elements responsible for this phenomenon are the *microfilaments*. These filaments are composed of protein molecules arranged in solid parallel bundles. They are most highly developed in muscle cells that are adapted specifically for contraction. In other types of cells, they are present as interwoven networks of the cytoplasm and are relatively inconspicuous even in electron photomicrographs. Besides aiding in contractility, they also appear to be involved in cell motility and support.

Cilia and Flagella

Hairlike appendages of cells that function in providing some type of movement are either *cilia* or *flagella*. Cilia are usually less than 20 μm long, but flagella may be thousands of micrometers in length. Cells that line the respiratory tract and uterine tubes have cilia (see illustration D in figure HA-2 and illustration C in figure HA-35 of the Histology Atlas). The only human cells with flagella are the spermatozoa. Both cilia and flagella originate from centrioles, which explains why they both have microtubular internal structure similar to the centrioles.

The Microvilli

Certain cells, such as the one in figure 5.1, have tiny protuberances known as *microvilli* on their apical or free surfaces. They may appear as a fine **brush border** (illustration C, figure HA-2) or as **stereocilia** (illustration B, figure HA-38). Both of these may be mistaken for cilia with a light microscope.

Each microvillus is an extension of cytoplasm enclosed by the plasma membrane. Microvilli are very common on cells lining the intestine and kidney tubules, where they function to increase the surface area of the cells for absorption of water and nutrients. Microvilli are not motile.

Laboratory Assignment

After labeling figure 5.1 and answering the questions on the Laboratory Report pertaining to cell structure and function, do the following two cellular studies in the laboratory. The epithelial cell will be removed from the inner surface of the cheek and examined with high-dry optics. The study of microorganisms (primarily protozoa) is performed here to observe the activity of living cells. Ciliary and flagellar action, amoeboid movement, secretion and excretion of cellular products, and cell division can be observed in viable cultures containing a mixture of protozoans.

Materials:
    microscope slides
    cover glasses
    toothpicks
    IKI solution
    mixed cultures of protozoa
    medicine droppers
    microscope

Epithelial Cell Prepare a stained wet mount slide of some cells from the inside surface of your cheek as follows:

1. Wash a microscope slide and cover glass with soap and water.
2. Gently scrape some cells loose from the inside surface of your cheek with a clean toothpick. It is not necessary to draw blood!
3. Mix the cells in a drop of IKI solution on the slide.
4. Cover with a cover glass.
5. After locating a cell under low power of your microscope, make a careful examination of the cell under high-dry magnification. Identify the *nucleus, cytoplasm,* and *cell membrane.*
6. Draw a few cells on the Laboratory Report, labeling the above structures.

Microorganisms Prepare a wet mount of some protozoans by placing a drop of the culture on a slide and covering it with a cover glass. Be sure to insert the medicine dropper all the way to the bottom of the jar for your study sample since most protozoans will be found there.

Examine the slide first under low power, then under high-dry. Study individual cells carefully, looking for nuclei, cilia, flagella, vacuoles of ingested food, excretory vacuoles, and so on.

Laboratory Report

Answer the questions on Laboratory Report 5,6 that pertain to this exercise.

6

Mitosis

Growth of body structures and the repair of specific tissues occur by cell division. Although many cells in the body are able to divide, many do not beyond a certain age. Cell division is characterized by the equal division of all cellular components to form two daughter cells that are identical in genetic composition to the original cell. Cleavage of the cytoplasm is referred to as *cytokinesis;* nuclear division is *karyokinesis,* or *mitosis.* Cells that are not undergoing mitosis are said to be in a "resting stage," or *interphase.*

During this laboratory period we will study the various phases of mitosis on a microscope slide of hematoxylin-stained embryonic cells of *Ascaris,* a roundworm. The advantage of using *Ascaris* over other organisms is that it has so few chromosomes.

To facilitate clarity in understanding this process it has been necessary to combine photomicrographs and diagrams in figures 6.1 and 6.2; however, before we get into the mitotic stages, let's see what takes place in the interphase.

The Interphase

Although a cell in the interphase doesn't appear to be active, it is carrying on the physiological activities that are characteristic of its particular specialization. The cell may be in one of two different phases: the G_1 phase or the S phase (G_1 for growth, S for synthesis).

After a previous division, cells in the *S phase* begin to synthesize DNA, with each double-stranded DNA molecule replicating itself to produce another identical molecule. Most cells in the S phase will accomplish this replication process within twenty hours and then divide.

If division does not occur within the twenty-hour time frame, the cell is said to be in the *G phase* and will not divide again. Note in illustration 1, figure 6.1, that cells in the interphase stage exhibit a more or less translucent nucleus with an intact nuclear membrane.

Stages of Mitosis

The process of mitotic cell division is a continuous one once the cell has started to divide; however, for discussion purposes, the process has been divided

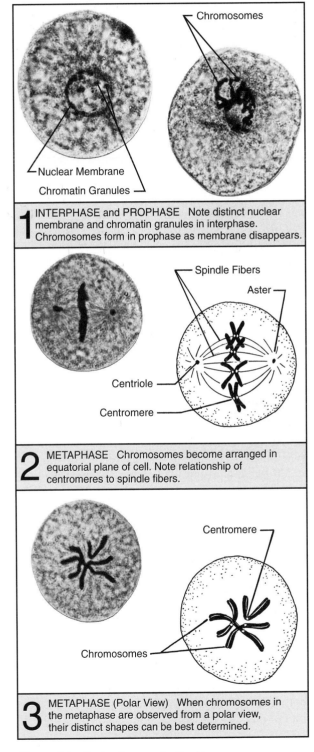

1 INTERPHASE and PROPHASE   Note distinct nuclear membrane and chromatin granules in interphase. Chromosomes form in prophase as membrane disappears.

2 METAPHASE   Chromosomes become arranged in equatorial plane of cell. Note relationship of centromeres to spindle fibers.

3 METAPHASE (Polar View)   When chromosomes in the metaphase are observed from a polar view, their distinct shapes can be best determined.

Figure 6.1   Early stages of mitosis.

31

into a series of four recognizable stages: prophase, metaphase, anaphase, and telophase. A description of each phase follows.

Prophase   In this first stage of mitosis the nuclear membrane disappears, **chromosomes** become visible, and each **centriole** of the centrosome moves to opposite poles of the cell to become an **aster.** The dark-staining chromosomes are made up of DNA and protein. Each aster consists of a centriole and astral rays (microtubules). By the end of the prophase some of the astral rays from each aster extend across the entire cell to become the **spindle fibers.**

From the early prophase to the late prophase the chromosomes undergo thickening and shortening. During this time each chromosome has a double nature, consisting of a pair of **chromatids.** This dual structure is due to the replication of the DNA molecules during the interphase.

Metaphase   During this stage the chromosomes migrate to the equatorial plane of the cell. Note in illustration 2, figure 6.1, that each chromosome has a **centromere,** which is the point of contact between each pair of chromatids and a spindle fiber.

Anaphase   As indicated in illustration 1 of figure 6.2, the individual chromatids of each chromosome are separated from each other and move apart along a spindle fiber to opposite poles of the cell.

Telophase   The telophase begins with the appearance of a **cleavage furrow** and the formation of a new **plasma membrane** between the two portions of the dividing cell. This stage is much like the prophase in reverse in that the chromosomes become less distinct and a new nuclear membrane begins to form around each set of chromosomes.

In the late telophase the spindle fibers disappear and the centrioles duplicate themselves. When the nuclear membranes and newly formed cell membrane are complete the two new structures are referred to as **daughter cells.**

Laboratory Assignment

Examine a prepared slide of *Ascaris* mitosis (Turtox slide E6.24). It will be necessary to use both the high-dry and oil immersion objectives. Identify all of the stages and make drawings, if required. Identifiable structures should be labeled. Answer the questions on Laboratory Report 5, 6.

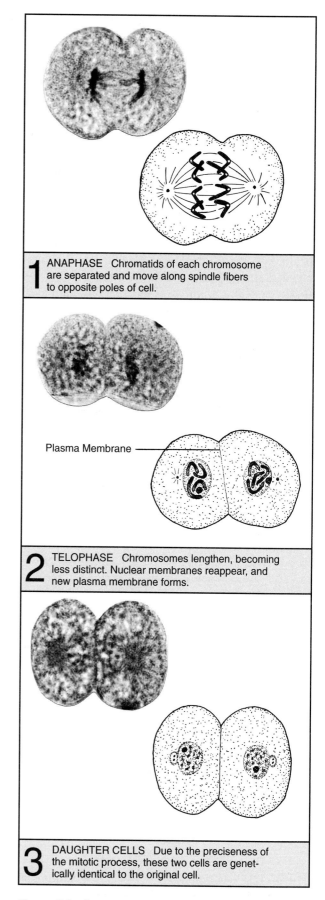

1 ANAPHASE   Chromatids of each chromosome are separated and move along spindle fibers to opposite poles of cell.

Plasma Membrane ———

2 TELOPHASE   Chromosomes lengthen, becoming less distinct. Nuclear membranes reappear, and new plasma membrane forms.

3 DAUGHTER CELLS   Due to the preciseness of the mitotic process, these two cells are genetically identical to the original cell.

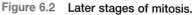
Figure 6.2   Later stages of mitosis.

7 Osmosis and Cell Membrane Integrity

Every cell of the body is bathed in a watery fluid that contains a mixture of molecules that are essential to its survival. This fluid may be the plasma of blood or the tissue fluid in the interstitial spaces. In either case, these molecules, whether water, nutrients, gases, or ions, pass into and out of the cell through the plasma membrane.

Some molecules, usually of small size, are able to diffuse *passively* through the cell membrane from areas of high concentration to low concentration. Larger organic molecules, such as glucose and amino acids and certain ions, move through the plasma membrane either with or against a concentration gradient by *active transport,* which requires energy expenditure by the cell. Movement of all molecules occurs only if the membrane is viable and undisturbed.

The integrity of plasma membranes of cells throughout the body is due largely to the fact that body fluids are isotonic. To see what happens to cell membranes when isotonic conditions do not exist, we will study here the mechanics of **molecular diffusion** and **osmosis.**

Demonstration Setups:
  Brownian movement (india ink preparation)
  Osmosis (thistle tube setup)

Molecular Movement

All molecules, whether in a gas or liquid state, are in constant motion due to the release of kinetic energy. As the molecules bump into each other, directions are changed, causing random dispersal. Although direct observation of molecular movement is impossible due to molecules' small size, one can observe this activity indirectly with both microscopic and macroscopic techniques. Two such observations are related to Brownian movement and molecular diffusion.

Brownian Movement

A Scottish botanist, Robert Brown, in 1827 observed that extremely small particles in suspension in the protoplasm of plant cells are in constant vibration. He erroneously assumed, at first, that this movement was a characteristic peculiar to living cells. Subsequent studies revealed, however, that this vibratory movement was caused by invisible water molecules bombarding the small visible particles. This type of movement became known as *Brownian movement.*

Demonstration:
The simplest way to observe this phenomenon is to examine a wet mount slide of india ink under the oil immersion objective of a microscope. India ink is a colloidal suspension of carbon particles in water, alcohol, and acetone.

Examine the demonstration setup that has been provided in the laboratory. This type of movement is of particular interest to microbiologists, since it must be differentiated from true motility in bacteria. Nonmotile bacteria are small enough to be displaced in this manner by moving water molecules.

Diffusion

If a crystal of some soluble compound is placed on the bottom of a container of water, molecules will disperse from the crystal to all parts of the container. The molecules are said to have spread throughout the water by *diffusion.* Movement of the molecules away from the original site of high concentration to low concentration is due to the fact that the reduced number of obstructing molecules in the lower concentration areas favors movement of molecules into those areas.

The rate of diffusion is variable and depends on ambient temperature and molecular size. Given sufficient time, diffusion eventually achieves even dispersal of all molecules throughout the container. This slow dispersal of molecules can be considerably augmented by convection currents caused by temperature differences.

To observe the phenomenon of diffusion, one can use crystals of colored compounds such as potassium permanganate (purple) and methylene blue. Figure 7.1 illustrates how to set up such a demonstration.

Methylene blue has a molecular weight of 320. Potassium permanganate has a molecular weight of 158. A petri plate of 1.5% agar will be used. This agar medium of 98.5% water allows freedom of movement of molecules. Prepare such a plate as follows and record your observations on the Laboratory Report.

Materials:

crystals of potassium permanganate and methylene blue
petri plate with about 12 ml of 1.5% agar
plastic ruler with metric scale

1. Select one crystal of each of the two different chemicals and place them on the surface of the agar medium about 5 cm apart. Try to select crystals of similar size.
2. Every 15 minutes, for one hour, examine the plate and measure the distance of diffusion (in millimeters) from the center of each crystal. Use a small plastic metric ruler.
3. Record the measurements on the Laboratory Report.

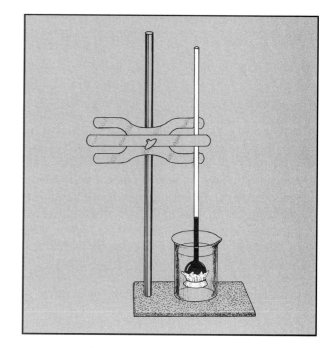

Figure 7.2   Osmosis setup.

Osmotic Effects

Water molecules, because of their small size, move freely through cell membranes into and out of the cell. Water acts as a vehicle that enables ions and other molecules to move through the membrane. This movement of water molecules through a semipermeable membrane is called *osmosis.*

Although water molecules are always moving both ways through the membrane, the predominant direction of flow is determined by the concentrations of solutes on each side of the membrane. If a funnel containing a sugar solution is immersed in a beaker of water and separated from the water with a semipermeable membrane, as in Figure 7.2, more water molecules will enter the funnel through the membrane than will leave the sugar solution to enter the beaker of water. If a demonstration of this phenomenon is set up by your instructor in the laboratory, observe the upward movement that occurs during the period.

As was observed in the diffusion experiment, molecules tend to move from an area of high concentration to areas of low concentration. In this case, the water molecules are obviously of greater concentration in the beaker than in the sugar solution. The inward flow of water molecules produces an upward movement of the sugar solution in the tube. The force that would be required to restrain the upward flow is called the *effective osmotic pressure.*

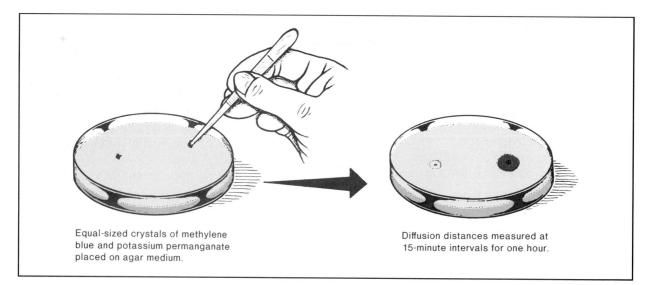

Equal-sized crystals of methylene blue and potassium permanganate placed on agar medium.

Diffusion distances measured at 15-minute intervals for one hour.

Figure 7.1   Comparing areas of diffusion.

Solutions that contain the same concentration of solutes as cells are said to be **isotonic solutions.** Blood cells immersed in such solutions gain and lose water molecules at the same rate, establishing an *osmotic equilibrium.* See Figure 7.3.

If a solution contains a higher concentration of solutes than is present in cells, water leaves the cells faster than it enters, causing the cells to shrink and develop cupped (crenated) edges. This shrinkage is called *plasmolysis,* or *crenation.* Such solutions of high solute concentration are called **hypertonic solutions** and are said to have a higher osmotic potential than isotonic solutions.

A solution that has a lower solute concentration than is present in cells is said to be a **hypotonic solution.** In such solutions water flows rapidly into the cells, causing them to swell *(plasmoptysis)* and burst *(lyse)* as the plasma membrane disintegrates. Lysis of red blood cells is called *hemolysis.*

To observe the effects of the various types of solutions on red blood cells, we will follow the procedures outlined in Figure 7.4. Blood cells will be added to various concentrations of solutions. The effects of the solutions on the cells will be determined macroscopically and microscopically. Proceed as follows:

Materials:

5 serological test tubes and tube rack
small beaker of distilled water (50 ml size)
2 depression microscope slides, cover glasses
wax pencil, Vaseline, and toothpicks
1 serological pipette (5 ml size)
cannister for used pipettes

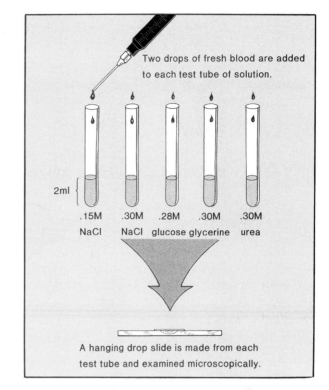

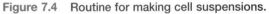

Figure 7.4 Routine for making cell suspensions.

syringes, needles, Pasteur pipettes
mechanical pipetting device
fresh blood
solutions of: 0.15M NaCl, 0.30M NaCl, 0.28M
    glucose, 0.30M glycerine, 0.30M urea

1. Label five clean serological tubes **1** to **5** and arrange them sequentially in a test-tube rack.
2. With a 5 ml pipette, deliver 2 ml of each solution to the appropriate tube. Refer to Figure 7.4

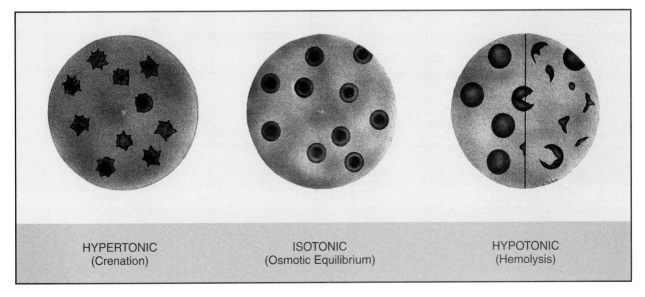

Figure 7.3 Effects of solutions on red blood cells.

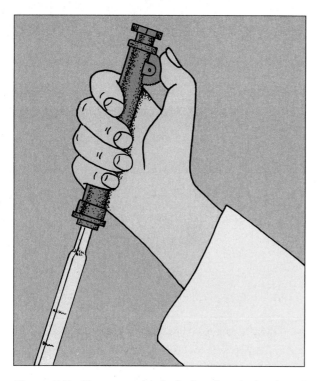

Figure 7.5 **Use a mechanical pipetting device for all pipetting.**

to determine which solution goes into which tube and refer to Figure 7.5 for delivery method. Be sure to rinse out the pipette with distilled water after each delivery.

3. Dispense 2 drops of blood to each tube, using a syringe. Shake each tube from side to side to mix, then let stand for **5 minutes.**

4. Hold the rack of tubes up to the light and compare them. If the solution is transparent, **hemolysis** has occurred. If you are unable to see through the tube, **no hemolysis** has occurred and the cells should be intact.

   If crenation has occured, the appearance will be somewhat between the clarity of hemolysis and the opacity of osmotic equilibrium.

   Record your results on the Laboratory Report.

5. Make a hanging drop slide from each tube as follows:

   a. Place a tiny speck of Vaseline near each corner of a cover glass. See Illustration 1, Figure 7.6.

   b. With a Pasteur pipette transfer a drop of the cell suspension to the center of the cover glass.

   c. Place a depression slide on the cover glass with the depression facing downward.

   d. With the cover glass held to the slide by the Vaseline, quickly invert.

6. Examine each slide under high-dry and record your results on the Laboratory Report.

Laboratory Report

After recording your results on the Laboratory Report, complete the report by answering all the questions.

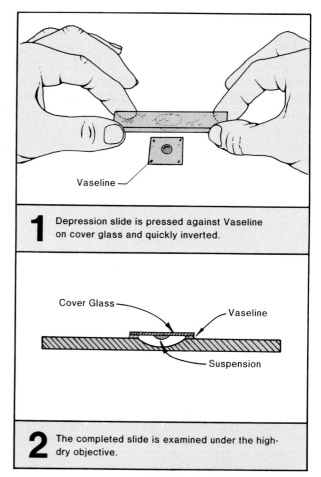

1 Depression slide is pressed against Vaseline on cover glass and quickly inverted.

2 The completed slide is examined under the high-dry objective.

Figure 7.6 **Procedure for making a hanging drop slide.**

8 Hydrogen Ion Concentration

In addition to being isotonic, extracellular fluid must have a hydrogen ion concentration that is compatible with the physiology of the cell. Hydrogen ions (protons) readily diffuse from the extracellular fluid into the cell through the plasma membrane and have a profound effect on the enzymatic reactions occurring within the cell. If the concentration of these ions is too great or too low, the physiology of the cell is adversely affected.

The hydrogen ion concentration of extracellular fluid throughout the body is very close to pH 7.36. Arterial blood has a pH of 7.41. Venous blood is close to the pH of extracellular fluid. The pH of body fluids rarely falls below 7.0 or rises above 7.8, even under extremely abnormal conditions. As the hydrogen ion concentration of extracellular fluid deviates from the norm of pH 7.36, certain metabolic reactions in cells are accelerated or depressed.

Due to the significance of hydrogen ion concentration in physiological processes, our concern here will pertain to two aspects of pH: (1) its mathematical derivation and (2) the relative merits of colorimetric versus electronic methods of measurement. Unless both aspects are clearly understood, subsequent experiments may be difficult to comprehend.

A Review of pH

The pH scale was developed by the Danish chemist Sørensen to provide a simplified numerical system for expressing the concentration of hydrogen (H^+) and hydroxyl (OH^-) ions. The concentration of these ions is a function of the degree of ionization that occurs. For example, pure distilled water dissociates (comes apart) to produce these two ions, as follows:

$$HOH \rightleftharpoons H^+ + OH^-$$

The frequency of dissociation of water occurs at a very low rate: 1×10^{-7} mols per liter. This means that 1 out of 10,000,000 molecules will be in the ionized state at a given time and that there will be an equal amount of hydrogen and hydroxyl ions present at the same time.

Strength of Acids and Bases

When acids are added to water, they undergo a variable amount of dissociation. *Strong acids,* such as hydrochloric acid (HC1), dissociate almost completely to produce a maximum of free hydrogen ions. *Weak acids,* on the other hand, may dissociate only 1% or so to release very few ions by comparison.

The same thing is true about bases. Most of the molecules of a *strong base,* such as sodium hydroxide (NaOH), will ionize to produce an alkaline solution of high OH^- content; *weak bases,* such as ammonium hydroxide (NH_4OH), produce fewer OH^- ions.

Definition of pH

To do away with the cumbersome way of expressing these ion concentrations in terms of 1×10^{-7}, or 1 part in 10,000,000. Sørensen devised the pH or *potential hydrogen* scale. He defined pH as *the logarithm of the reciprocal of the hydrogen ion concentration expressed as grams per liter, or,*

$$pH = log \frac{1}{(H^+)}$$

Knowing that the H^+ concentration of distilled water is 1×10^{-7}, we arrive at pH 7 as follows:

$$pH = log \frac{1}{1 \times 10^{-7}}$$

$$= log \frac{10^7}{1.0}$$

$$log\ 10^7 = 7.0000\ and$$

$$log\ 1.0 = 0.0000$$

$$\therefore pH = 7.0000 - 0.0000$$

$$= 7$$

To apply this equation to another problem, consider this example: If the hydrogen ion concentration of a sample of arterial blood were

0.0000000389 gm H^+ per liter, would this concentration be near the expected value of pH 7.41?

$$pH = \log \frac{1}{3.89 \times 10^{-8}}$$

$$= \log \frac{10^8}{3.89}$$

$$\log 10^8 = 8.0000 \ and$$

$$\log 3.89 = 0.5899$$

$$\therefore pH = 8.0000 - 0.5899$$

$$= 7.41$$

The answer to the above question is, obviously, in the affirmative.

Reference to the logarithm tables in Appendix A will confirm the log values for the above two examples. The log tables should be consulted if the above sequence is unfamiliar.

Normality and pH

Table 8.1 illustrates that the pH scale extends from 0 to 14. Note that a solution of pH 0 contains 1 gram of hydrogen ions per liter. Such a solution is also designated as being a 1 N solution (N = normal), since by definition a normal solution contains 1 gram equivalent of hydrogen ions per liter of water.

The upper limit of the pH scale, which is 14, represents a hydrogen ion concentration of 10^{-14}, or 0.000,000,000,000,01 N. Since the hydroxyl ions vary inversely with the number of hydrogen ions, there is 1 gram of hydroxyl ions in a solution that is pH 14.

pH Measurements

Two methods are available for determination of pH: one is colorimetric and the other is electronic. In this portion of the exercise, a comparison will be made of the two methods. The pH of certain body fluids will be measured.

The **colorimetric method** utilizes color indicators, either in solution or in papers. To cover a broad pH range, you must use several indicators, as shown in table 8.2. The range here is from 1.2 through 8.8. Only if color standards are used, however, can you determine pH with any degree of accuracy by this method. Instead of using indicator solutions in this experiment, we will use papers, which are more convenient.

The **electronic method** utilizes a pH meter that is activated by a glass electrode placed in the solution. When using one of these instruments you must standardize (calibrate) the unit for the pH range in which it is to be used. An adjustment must also be made for the temperature of the solution. When determining the pH of several different solutions, it is imperative that the electrode be cleansed with a spray of distilled water between measurements. These instruments are extremely sensitive to pH changes.

The hydrogen ion concentrations of twelve different phosphate solutions will be determined here.

Table 8.1 Ion concentrations as related to pH.

HYDROGEN ION CONCENTRATION (Grams H^+ per Liter)		pH		HYDROXYL ION CONCENTRATION (Grams OH^- per Liter)
1.0	$= 10^0$	0	Acid	0.00000000000001
0.1	$= 10^{-1}$	1		0.0000000000001
0.01	$= 10^{-2}$	2	↑	0.000000000001
0.001	$= 10^{-3}$	3		0.00000000001
0.0001	$= 10^{-4}$	4		0.0000000001
0.00001	$= 10^{-5}$	5		0.000000001
0.000001	$= 10^{-6}$	6		0.00000001
0.0000001	$= 10^{-7}$	7	Neutral	0.0000001
0.00000001	$= 10^{-8}$	8		0.000001
0.000000001	$= 10^{-9}$	9		0.00001
0.0000000001	$= 10^{-10}$	10		0.0001
0.00000000001	$= 10^{-11}$	11		0.001
0.000000000001	$= 10^{-12}$	12		0.01
0.0000000000001	$= 10^{-13}$	13	↓	0.1
0.00000000000001	$= 10^{-14}$	14	Alkaline	1.0

Table 8.2 pH indicators (range of 1.2–8.8).

INDICATOR	COLOR		pH RANGE
	ACID	ALKALINE	
Thymol blue	Red	Yellow	1.2–2.8
Bromphenol blue	Yellow	Blue	3.0–4.6
Methyl red	Red	Yellow	4.4–6.4
Bromcresol purple	Yellow	Purple	5.2–6.8
Bromthymol blue	Yellow	Blue	6.0–7.6
Phenol red	Yellow	Red	6.8–8.4
Cresol red	Yellow	Red	7.2–8.8

pH measurements of each solution will be made, first with pH papers and then with a pH meter. If only a small number of pH meters are available, it may be necessary to divide the class into teams that will be assigned a certain number of solutions. If the team approach is used, results will be posted on the chalkboard so that all members of the class will have a complete set of results. If teams are unnecessary, students will work in pairs to perform the entire experiment.

Materials:

per pair of students:
 4 small beakers (30–50 ml size)
 pH indicator paper
 pH meter
 onion
 china marking pencil
 wash bottle of distilled water
 large beaker (250 ml size)
 Kimwipes

on demonstration table:
 12 phosphate mixture solutions in squeeze
  bottles (labeled A through L)

1. Label four small beakers A through D and dispense solutions A through D into the appropriate beakers. Each beaker should receive about ½″ solution (approximately 15–20 ml).
2. Determine the pH of each solution with pH paper, recording the pH in the table on the Laboratory Report.

3. Measure the pH of each solution with the pH meter. Be sure to rinse off the electrode with wash bottle and dry with Kimwipes between tests. See figure 8.1. Record your results in the table on the Laboratory Report.
4. Discard solutions A through D. Rinse out the beakers and relabel the four beakers E through H.
5. Dispense solutions E, F, G, and H to their respective containers.
6. Determine the pH of each solution by both methods and record the results on the Laboratory Report.
7. Clean out the four beakers again and repeat the above steps for solutions I, J, K, and L.
8. Determine the pH of saliva, urine, blood, tears, and perspiration. When testing tears, do not place pH indicator paper into the eye. Instead, with the onion, induce tear flow onto the cheek or slide for testing. Record your results.

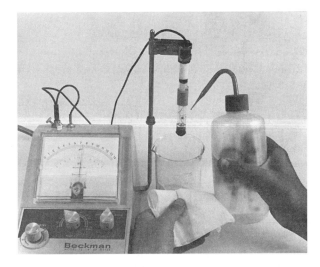

Figure 8.1 Electrode tip should be rinsed with distilled water and blotted with tissue between solutions.

Laboratory Report

After recording your results on Laboratory Report 8,9, do the pH conversions in Part C on the Laboratory Report.

9 Buffering Systems

As stated previously, body cells function best in a very narrow pH range. When the arterial blood becomes less than 7.40, **acidosis** occurs. In severe acidosis (below pH 7.0), one passes into a coma and death occurs within minutes.

When the pH of arterial blood exceeds pH 7.41, **alkalosis** occurs. Severe alkalosis (above pH 7.8) will result in convulsions and death in an equally short period of time.

The energy-yielding processes that occur in all cells of the body result in the production of large quantities of carbon dioxide and organic acids. The carbon dioxide combines with water to form carbonic acid, which in turn, dissociates to produce hydrogen and bicarbonate ions. Although carbonic acid is a weak acid, so much of it is produced from carbon dioxide that it forms a major source of H^+ for the body fluids. These reactions are reversible and appear as follows:

$$CO_2 + H_2O \rightarrow H_2CO_3 \rightleftharpoons H^+ + HCO_3^-$$

The removal of these excess hydrogen ions is regulated by three mechanisms: the kidneys, the respiratory system, and buffering systems. Of the three, the kidneys are most effective in maintaining acid-base balance; however, they work very slowly, requiring *up to twenty-four hours* to make adjustments.

The respiratory system regulates H^+ concentration by controlling carbon dioxide levels in the blood. Approximately *one to three minutes* are required for this adjustment.

The most rapid adjustments to pH change are made by the buffer systems. These systems make adjustments within *fractions of a second*. It is these systems that concern us here.

All **buffering systems** consist of a weak acid and its conjugate base. Each of these compounds has a common anion but different cations. The anion of the base will react with excess H^+ to form the weak acid.

The most important buffering system in the blood is the bicarbonate system. In addition to this system, the blood contains proteins and inorganic phosphates that act as buffers. Even the erythrocytes

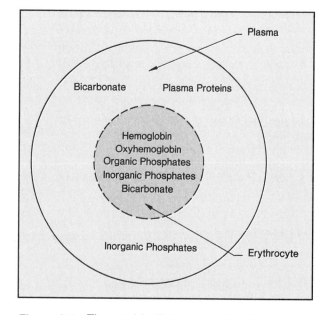

Figure 9.1 The total buffering capacity of the blood includes the buffers within erythrocytes as well as those in the plasma.

act as buffers due to the fact that they contain bicarbonate, phosphates, and hemoglobin. Figure 9.1 illustrates, diagrammatically, the number of buffering systems that exist in the blood.

The **bicarbonate system** utilizes carbonic acid as the weak acid and sodium bicarbonate as the conjugate base. Dissociation of these two compounds illustrates how they both produce the same anion (HCO_3^-):

weak acids
$$H_2CO_3 \rightleftharpoons H^+ + HCO_3^-$$

conjugate base
$$NaHCO_3 \rightarrow Na^+ + HCO_3^-$$

When hydrogen ions are added to this system, the anion combines with hydrogen ions to form the weak acid, thus preventing the pH from falling precipitously:

$$NaOH + H_2CO_3 \rightarrow NaHCO_3 + H_2O$$

When a strong base, such as NaOH is added to this system, sodium bicarbonate and water form to prevent the pH from rising too much.

$$NaOH + H_2CO_3 \rightarrow NaHCO_3 + H_2O$$

The ratio of H_2CO_3 to $NaHCO_3$ in extracellular fluid is approximately 1:20. The extra HCO_3^- provides an alkali reserve, or base excess, to react with H^+ produced under most normal activity.

In this exercise, a comparison will be made of the relative effectiveness of five different buffering systems. Hydrochloric acid will be added drop by drop to each solution to determine how much H^+ each buffer can take up in lowering the pH from 7.4 to 7.0. The solutions to be tested are (1) blood, (2) plasma, (3) albumin suspension, (4) a phosphate system, and (5) a bicarbonate system. The blood and plasma will be diluted 1:10, so any value arrived at for these two solutions will be multiplied by ten to get a correct comparison. A sample of distilled water will also be included for comparison. Proceed as follows:

Materials:
per pair of students:
  pH meter or pH paper
  6 graduated beakers (30–50 ml size)
  wax pencil
  0.0001N HCl in plastic squeeze bottle

Test solutions (all at pH 7.4):
  A—distilled water
  B—bicarbonate buffer
  C—phosphate buffer
  D—albumin suspension
  E—blood (1:10 saline dilution)
  F—plasma (1:10 dilution)

1. Label six clean beakers A, B, C, D, E, and F and dispense 15 ml of the appropriate solution to each beaker. If beakers are graduated, fill to the 15 ml mark. If beakers are not calibrated, use a graduate.
2. With the pH electrode (or pH paper) in the beaker of test solution A (distilled water), add 0.0001N HCl slowly, drop by drop, counting until the pH reaches 7.0. Shake the beaker frequently for dispersal. Record the number of drops on the Laboratory Report.
3. Repeat the above procedure for each beaker of solution. Be sure to rinse off the electrode with distilled water after acidifying each solution.

Laboratory Report

Complete Laboratory Report 8,9 by answering the questions that pertain to this exercise.

10

Epithelial Tissues

Although all cells of the body share common structures such as nuclei, centrosomes, Golgi, etc., they differ considerably in size, shape, and structure according to their specialized functions. An aggregate of cells that are similar in structure and function is called a **tissue.** The science that is related to the study of tissues is called **histology.**

In this exercise we will study the different types of epithelial tissues. Prepared microscope slides from portions of different organs will be available for study. Learning how to identify specific epithelial tissues on slides that have several types of tissue will be part of the challenge of this exercise.

Epithelial tissues are aggregations of cells that perform specific protective, absorptive, secretory, transport, and excretory functions. They often serve as coverings for internal and external surfaces and they rest upon a bed of connective tissue. Characteristics common to all epithelial tissues are as follows:

• The individual cells are closely attached to each other at their margins to form tight sheets of cells lacking in extracellular matrix and vascularization.

• The cell groupings are oriented in such a way that they have an apical (free) surface and a basal (bound) region.

The basal portion is closely anchored to underlying connective tissue. This thin adhesive margin between the epithelial cells and connective tissue is called the **basement lamina.** Although this structure was formerly referred to as the "basement membrane," it is not a true membrane. Unlike true membranes, the basement lamina is acellular. In reality, it is a colloidal complex of protein, polysaccharide, and reticular fibers.

Differentiation of the various epithelia is illustrated in the separation outline, figure 10.1. The basic criteria for assigning categories are cell shape, surface specializations, and layer complexity. The three divisions (*simple, pseudostratified,* and *stratified*) are based on layer complexity. A discussion of each follows.

Simple Epithelia

Epithelial tissues that fall in this category are composed of a single cell layer that extends from the basement lamina to the free surface. Figure 10.2 illustrates three basic kinds of simple epithelia.

Simple Squamous Epithelia   These cells are very thin, flat, and irregular in outline. An example of this type is seen in illustration 1, figure 10.2. They form a pavementlike sheet in various organs that

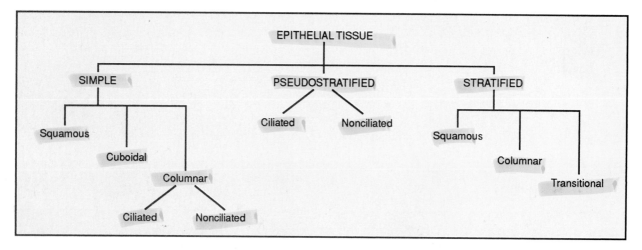

Figure 10.1   A morphological classification of epithelial types.

Epit
Exercise 10 • Epithe
Stratified Col
shown in illu
although th
nar, the
No

perform filtering or
walls, alveolar walls
the pleurae, and blood
ple squamous epithelia

Cuboidal Epithelia C
tion 2, figure 10.2) are s
section and hexagonal fr

This type of tissue is
as the thyroid, salivary, an
in the ovaries and the caps
of the eye.

Simple Columnar Epithelia   Illustration 3, figure 10.2, reveals that cells of this type are elongated between their apical and basal surfaces. When the free surfaces are observed, the cells have a polygonal configuration.

Functionally, this type of tissue may be specialized for protection, secretion, or absorption. Note in illustration 3 that certain secretory cells, called **goblet cells,** may be present that produce **mucus,** a protective glycoprotein. Also, note that the cells may be *ciliated* or not ciliated *(plain).*

Plain columnar epithelial tissue lines the stomach, intestines, and kidney collecting tubules. Ciliated columnar epithelia can be found in the lining of the respiratory tract, uterine tubes, and portions of the uterus. Goblet cells can be present in both the plain and ciliated types.

Another modification of the free surfaces of columnar cells is the presence of **microvilli.** These structures are seen as a **brush border** when observed with a light microscope under oil immersion. Columnar cells that line the small intestine and make up the walls of the kidney collecting tubules exhibit distinct brush borders.

Stratified Epithelia

Squamous, columnar, and cuboidal epithelia that exist in layers are referred to as *stratified.* Figure 10.3 illustrates four layered epithelia that have distinct differences.

Stratified Squamous   Illustration 1, figure 10.3 is of this type. Note that, although the superficial cells are distinctly squamous, the deepest layer is columnar; in some cases this layer is cuboidal. In between the basal cell layer and the squamous cells are successive layers of irregular and polyhedral cells.

Protection is the chief function of this type of tissue. Exposed inner and outer surfaces of the body, such as the skin, oral cavity, esophagus, vagina, and cornea consist of stratified squamous epithelia.

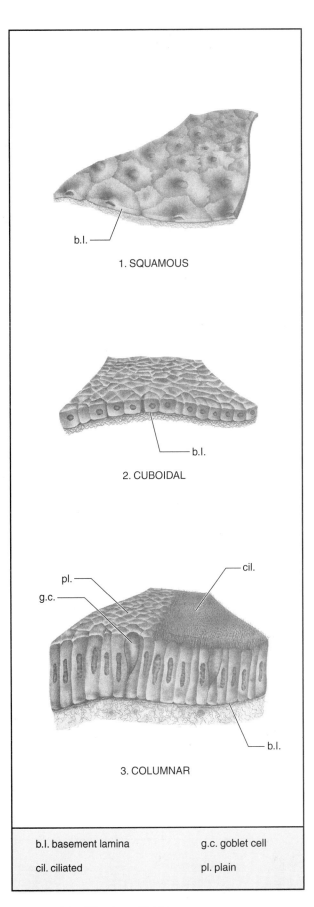

1. SQUAMOUS

2. CUBOIDAL

3. COLUMNAR

| b.l. basement lamina | g.c. goblet cell |
| cil. ciliated | pl. plain |

Figure 10.2   Simple epithelia.

...nar This type of epithelium is ...stration 2, figure 10.3. Observe that, ...e superficial cells are distinctly colum-... deeper cells are irregular or polyhedral. ...e also that the superficial columnar cells are ...variable in height.

Protection and secretion are the chief functions of this type of tissue. Distribution of stratified columnar is limited to some glands, the conjunctiva, the pharynx, a portion of the urethra, and lining of the anus.

Stratified Cuboidal Although no stratified cuboidal tissue is listed in figure 10.1, tissue of this configuration is seen in the ducts of sweat glands of the skin. The ducts in illustration C,

figure HA-9, shows what stratified cuboidal epithelium looks like.

Transitional A unique characteristic of this type of stratified epithelium is that the surface layer consists of large, round, dome-shaped cells that may be binucleate. The deeper strata are cuboidal, columnar, and polyhedral.

Note in illustration 3, figure 10.3, that the deeper cells are not as closely packed as in other stratified epithelia. Distinct spaces can be seen between the cells. This looseness of cells imparts a certain degree of elasticity to the issue. Organs such as the urinary bladder, ureters, and kidneys (the calyces), contain transitional epithelium that enables distension due to urine accumulation.

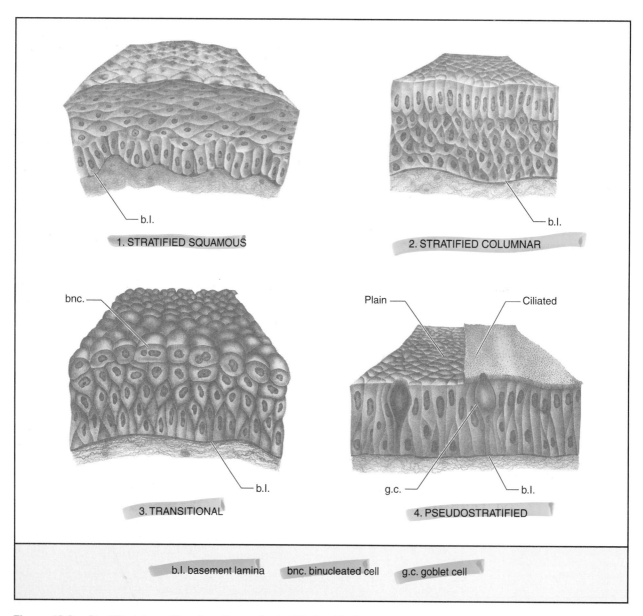

1. STRATIFIED SQUAMOUS

2. STRATIFIED COLUMNAR

3. TRANSITIONAL

4. PSEUDOSTRATIFIED

b.l. basement lamina   bnc. binucleated cell   g.c. goblet cell

Figure 10.3 Stratified, transitional, and pseudostratified epithelia.

Pseudostratified Epithelia

This epithelium presents a superficial stratified appearance because of the staggered nuclei as well as two different cell orientations. It is only one layer thick, however. Close examination will reveal that every cell is in contact with the basal lamina, but only the columnar types extend to the free surface. The smaller cells wedged between them have no free surface.

Two types are shown in Illustration 4 of figure 10.3: plain and ciliated. The *plain,* or nonciliated, is found in the male urethra and parotid gland. The *ciliated* type lines the trachea, bronchi, auditory tube, and part of the middle ear. Since both types frequently produce mucus, they possess goblet cells.

Laboratory Assignment

Do a systematic study of each kind of epithelial tissue by examining prepared slides that should show the kinds of tissue you are looking for. Remembering that epithelial cells have a free surface and a basement lamina, always look for the tissue near the edge of a structure.

Note in the materials list below that preferred and optional lists of slides are given. The *preferred* list is of slides that are limited to the type of tissue being studied. If these are available, use them. If they are not available, use the *optional* slides, which are very good but more difficult to use because they usually have several kinds of tissue on each slide.

Note, also, that after each slide is a number in parenthesis. These numbers, which are Turtox identification numbers, will not apply if slides are available that were made by some other biological supply company. Since Turtox slides were used almost exclusively when making the photomicrographs for the Histology Atlas, the numbers are included here to assist those laboratories where Turtox slides are available.

Materials:
preferred slides:
    squamous (H1.1)
    stratified squamous (H1.14)
    cuboidal (H1.21)
    pseudostratified ciliated (H1.32)
    transitional (H1.41)

optional slides:
    skin (H11.12 or H11.14)
    trachea (H6.41)
    stomach (H5.415 or H5.425)
    ileum (H5.531)
    kidney (H9.11 or H9.15)
    thyroid gland (H14.11, H14.12, or H14.13)

Squamous Epithelium Look for the flattened cells that are representative of this kind of tissue. Use Figure HA-1 in the Histology Atlas for reference. The exfoliated cells are the same ones you studied in Exercise 5.

Explore each slide first with the low-power objective before using the high-dry or oil immersion objectives. Make drawings if required.

Columnar Epithelium Consult figure HA-2 in the Histology Atlas for representatives of this group of epithelial cells. To see the **brush border** or **cilia** on these cells it will be necessary to study the cells with the high-dry or oil immersion objectives. Can you identify the **basement lamina** on each tissue?

The **lamina propria** consists of connective tissue, vascular and lymphatic channels, lymphocytes, plasma cells, eosinophils, and mast cells. All these structures will be studied later.

Cuboidal Epithelium Illustrations A and B, figure HA-3, reveal the appearance of cuboidal tissue. The preferred slide (H1.21) is usually made from the uterine lining of a pregnant guinea pig. If the preferred slide is unavailable, use a slide of the thyroid gland for this type of tissue.

Transitional Tissue Illustrations C and D of figure HA-3 and Illustration C, figure HA-33, provide good examples of this tissue. Note that the cells seem to be loosely arranged. Look for **binucleate cells.**

Ciliated Pseudostratified Columnar Epithelium
The best place to look for this type of tissue is in a cross section of the trachea. Illustration D, figure HA-2 is the epithelium of the trachea.

Laboratory Report

Answer the questions on Laboratory Report 10,11 that pertain to the epithelial tissues.

11

Connective Tissues

Connective tissues include those tissues which perform binding, support, transport, and nutritive functions for organs and organ systems. Characteristically, all connective tissues have considerable amounts of nonliving extracellular substance that holds and surrounds various specialized cells.

The extracellular material, or **matrix,** is composed of fibers, fluid, organic ground substance, and/or inorganic components intimately associated with the cells. This matrix is a product of the cells in the tissue. The relative proportion of cells to matrix will vary from one tissue to another.

Several systems of classification of connective tissues have been proposed. Figure 11.1 reveals the system that we will use here. It is based on the nature of the extracellular material; in some cases there is a certain amount of overlapping of categories.

Connective Tissue Proper

Histologically, *connective tissue proper* is composed, primarily, of protein fibers, special cells, and a ground substance that varies among the several types. The fibers may differ in protein composition and density. From the standpoint of composition, fibers consist of either *collagenous* or *elastin* protein.

There are three basic types of fibers: collagenous, reticular, and elastic. **Collagenous** (white) **fibers,** are relatively long, thick bundles seen in most ordinary connective tissues in varying amounts. **Reticular fibers** constitute minute networks of very fine threads. Although both collagenous and reticular fibers are composed of collagen, the reticular fibers stain more readily with silver dyes (i.e, they are *argyrophilic*).

Elastic (yellow) **fibers,** are often found in connective tissue stroma of organs that must yield to changes in shape. These are the only fibers to contain elastin protein. All three types of fibers are seen in loose (areolar) tissue (figure 11.2).

The ground substance usually consists of complex peptidoglycans that form an amorphous solution or gel around the cells and fibers. Chondroitin sulfate and hyaluronic acid are frequent components.

Various types of cells are seen in the matrix of connective tissue proper. Examples include fibroblasts, adipose cells, mast cells, plasma cells, macrophages, and other types of blood cells.

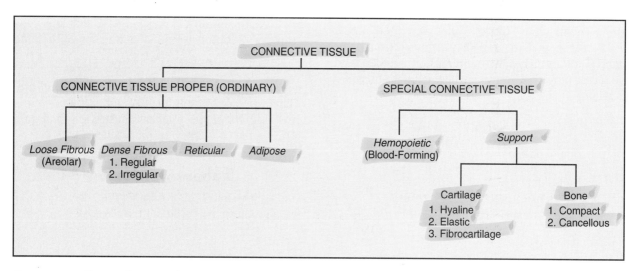

Figure 11.1 Types of connective tissue.

Loose Fibrous (Areolar) Connective Tissue

This tissue is found throughout the body, interwoven into the stroma of many organs. The cells and extracellular substances of this tissue are very loosely organized. Because of the preponderance of spaces in this tissue, it is often designated as *areolar* (a small space). It is highly flexible and capable of distension when excess extracellular fluid is present. The left-hand illustration in figure 11.2 depicts its generalized structure. Note the presence of **mast cells, macrophages,** and **fibroblasts.**

Three important functions are served by loose connective tissue: (1) it provides flexible support and a continuous network within organs, (2) it furnishes nutrition to cells in adjacent areas due to its capillary network, and (3) it provides an arena for activities of the immune system.

This type of tissue is found beneath epithelia, around and within muscles and nerves, and as part of the serous membranes. The deep and superficial *fasciae* encountered in the rat dissection are of this type.

Adipose Tissue

Fat or adipose cells can be found in small groupings throughout the body. However, as they constitute a storage depot for fat, they frequently accumulate in

large areas to mak[...] latter areas are m[...] *sue.* There are va[...] *ordinary adipose* [...] humans. Refer to [...]

Fat cells are [...] a spherical or po[...] their mass may [...] deposition ens[...] reduced and thi[...] appears as a lar[...] of cytoplasm. T[...] producing a sig[...] of reticular fibers exists between the cells.

Fat tissue serves as a protective cushion and insulation for the body, in addition to being a potential source of energy and heat generator. Its rich vascular supply points to its relatively high rate of metabolism and turnover.

Reticular Tissue

This tissue (figure 11.3) is generally regarded as a network of reticular fiber elements within certain organs. Since the fibers show a unique pattern and staining reaction, they can be identified as the supporting framework of many vascular organs, such as the liver, lymphatic structures, hemopoietic tissue, and basement lamina. The left-hand illustration

Con[...]

Connec[...]

Exercise 11   •

in figure 11.3 reveal[...] lum of a lymph no[...]

Dense F[...] Conne[...]

Tiss[...] va[...]

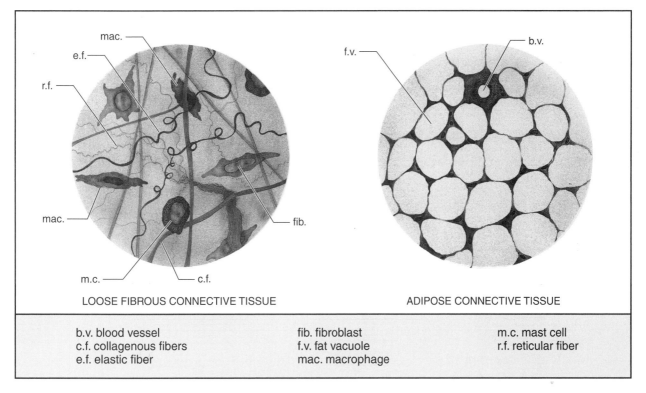

LOOSE FIBROUS CONNECTIVE TISSUE   ADIPOSE CONNECTIVE TISSUE

b.v. blood vessel	fib. fibroblast	m.c. mast cell
c.f. collagenous fibers	f.v. fat vacuole	r.f. reticular fiber
e.f. elastic fiber	mac. macrophage	

Figure 11.2 Loose fibrous and adipose tissues.

...s the appearance of the reticu-
...de.

...brous (White)
...ctive Tissue

...e in this category differs from the loose
...iety in having a predominance of fibrous ele-
...ments and a sparseness of cells, ground substance, and capillaries. According to the arrangement of fibers this tissue can be separated into
two groups: *regular dense* and *irregular dense*
tissues. The right-hand illustration in figure 11.3
reveals both types.

Dense regular connective tissue is the type
found in tendons, ligaments, fascia, and aponeuroses. It consists of thick groupings of longitudinally
organized collagenous fibers and some elastic fibers.
These tough strands have enormous tensile strength
and are capable of withstanding strong pulling forces
without stretching. Fibroblasts are the principal cells,
although they are quite scarce.

Dense irregular connective tissue has elastic
fibers that are woven into flat sheets to form capsules around certain organs. The sheaths that encircle nerves and tendons are of this type. A large
portion of the dermis also consists of the dense
irregular type.

Special Connective Tissue

As indicated in the categorization chart in figure
11.1, this type of tissue includes two divisions:
hemopoietic and support tissues. *Hemopoietic tissue*
consists of red bone marrow connective tissue that
produces the formed elements of blood (red blood
cells, white blood cells, and platelets). Although this
tissue will not be studied here, we will study the
products of this tissue in Exercises 52 and 53.

Our principal concern here will pertain to the
support tissues, cartilage and bone. These two tissues are adapted for the bearing of weight.
Structural characteristics shared by these tissues
include: (1) a solid, flexible, yet *strong extracellular matrix*, (2) cells contained in matrix cavities
called *lacunae*, and (3) an external covering
(periosteum or *perichondrium)* that is capable of
generating new tissue.

Cartilage

In general, cartilage consists of a stiff, plastic matrix
that has lubricating as well as weight-bearing capability. As a result, it is found in areas that require
support and movement (skeleton and joints).

Basically, all three types of cartilage in humans
consist of cartilage cells (**chondrocytes**), embedded within a matrix that contains ground substance

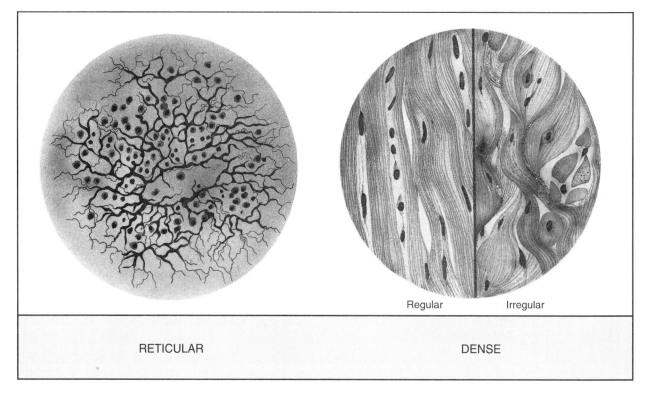

RETICULAR

Regular   Irregular

DENSE

Figure 11.3 Reticular and dense fibrous connective tissue.

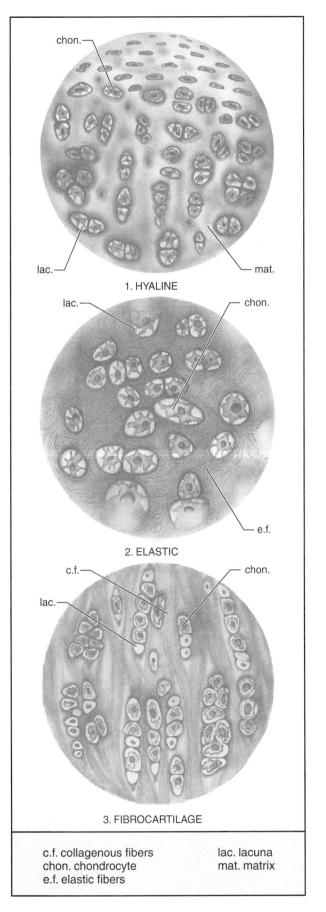

chon.

lac. —

— mat.

1. HYALINE

lac. —

— chon.

— e.f.

2. ELASTIC

c.f. —

— chon.

lac. —

3. FIBROCARTILAGE

c.f. collagenous fibers
chon. chondrocyte
e.f. elastic fibers

lac. lacuna
mat. matrix

Figure 11.4   Types of cartilage.

and fibers. The proportion devoted to matrix is much greater than that for chondrocytes. Unlike other connective tissues, cartilage is devoid of a vascular supply and receives all nutrients through diffusion. The three classes of cartilage are shown in figure 11.4.

Hyaline cartilage (illustration 1) is a pearly glasslike tissue that makes up a very large part of the fetal skeleton. During fetal development it is gradually replaced by bone, except for certain areas in the joints, ear, larynx, trachea, and ribs.

The matrix is a firm homogenous gel made up of chondroitin sulfate and collagen. Cells occur singly or in "nests" of several cells, the sides of which may be flattened. Often the areas directly around lacunae appear denser in sections.

Elastic cartilage (illustration 2) is similar in basic structure to hyaline cartilage, but it differs in that it contains significant amounts of elastic fibers in the matrix. The result is a tissue that has highly developed flexibility and elasticity. The pinna of the external ear, the epiglottis, and the auditory tube are reinforced with this type of cartilage.

Fibrocartilage (illustration 3) differs from the other two types in that its chondrocytes are arranged in groupings between bundles of collagenous fibers. It serves a useful cushioning function in strategic joint ligaments and tendons. It is the major component, for instance, of the intervertebal disks of the vertebral column and the pubic symphysis.

Bone Tissue

The tissue that makes up the bones of the skeleton meets all the criteria of connective tissue, yet it has many striking and unique features. It is hard, unyielding, very strong, and relatively lightweight. It is found in those parts of the anatomy that require maximum weight-bearing capacity, protection, and storage of certain materials.

In a sense, bone can be visualized as a living organic cement. It consists of specialized cells, blood vessels, and nerves that are reinforced within a hard ground substance made up of organic secretions impregnated with mineral salts. It arises during embryonic and fetal development from cartilage and/or fibrous connective tissue precursors. The initial organic matrix consists of collagenous fibers that serve as a framework for the gradual deposition of calcium and phosphate salts by special bone cells, the *osteoblasts*. Far from being an inactive tissue, it is continuously being modified and reconstructed by both metabolic and external influences.

In macroscopic sections of bones, two frameworks are apparent: the solid, dense **compact bone,** which makes up the outermost layer; and the more porous **spongy** *(cancellous)* **bone,** that is located internally. Each of these variations has a distinctive histological character.

Compact Bone   Upon close inspection of a cut section of the sternum viewed from a three-dimensional perspective (figure 11.5, in illustration A), it is apparent that compact bone (label 2) is permeated by a microscopic framework of tunnels, channels, and interconnecting networks that are surrounded by a hard matrix. Within this hollow network exist the living substances of bone that facilitate its nourishment and maintenance.

The functional and structural unit of compact bone is a cylindrical component called the **osteon,** or Haversian system (label 9, illustration B). In the center of each osteon is a hollow space, the **central (Haversian) canal,** which contains one or two blood capillaries. Surrounding each central canal are several concentric rings of matrix called the **lamellae. Osteocytes,** which are responsible for secreting the lamellae, can be seen within small hollow cavities, the **lacunae.** Note that these cavities, which are oriented between the lamellae, have many minute hollow tunnels, called **canaliculi,** that radiate outward, imparting a spiderlike appearance. The canaliculi contain protoplasmic processes of the osteocytes.

If one follows the various levels of structure shown in figure 11.5, it should become evident that the network comprises a continuous communication system from the central canal to the lacunae and between adjacent lacunae via the canaliculi. The entrapped osteocytes, thus, can receive nourishment and exchange materials within the hard space of the matrix. Intimate contact between the protoplasmic processes of adjacent osteocytes through the canaliculi makes all of this possible.

Groups of osteons lie in vertical array with adjacent lamellae separated at lines of demarcation called **cement lines.** Continuity between the central canals of adjacent osteons is achieved by **perforating** (Volkmann's) **canals,** that penetrate the bone obliquely or at right angles.

Spongy Bone   This type of bone lies adjacent to compact bone and is continuous with it; there is no distinct line of demarcation between the two regions. Histologically, it presents a lesser degree of organization than compact bone. Illustration A reveals that its outstanding feature is a series of branching, overlapping plates of matrix called **tra-** beculae. These plates are oriented so as to produce large, interconnecting cavelike spaces. These spaces function well in storage and as pockets to hold the bloodforming cells (hemopoietic tissue) of the bone marrow. They also function in weight reduction.

Note that the trabeculae are also randomly punctuated by the osteocyte-holding spaces, or lacunae, and that blood vessels meander through the large spaces between the trabeculae, bringing nourishment to nearby osteocytes.

The Periosteum   Contiguous with the outer layer of compact bone and tightly adherent to it, is a thick, tough membrane called the *periosteum.* It is composed of an outer layer of fibrous connective tissue and an inner *osteogenic layer,* which serves as a source of new bone-forming cells and provides an access for blood vessels. The periosteum is anchored tightly to compact bone by bundles of collagen fibers that perforate and become firmly embedded within the outer lamellae. These minute attachments are called **Sharpey's fibers.**

Assignment:
Label figure 11.5.

Histological Study

Do a systematic study of each type of connective tissue by examining prepared slides that are available. Note that Histology Atlas references are indicated for each type of tissue.

Materials:
prepared slides of:
    loose fibrous (H2.13)
    adipose (H2.51)
    dense fibrous (H2.115)
    yellow (elastic) fibrous (H2.125)
    reticular (H2.31)
    hyaline cartilage (H2.61)
    fibrocartilage (H2.63)
    elastic cartilage (H2.62)
    bone, cross section (H2.735)
    developing bone (H2.79)

Connective Tissue Proper   (figure HA-4) Examine slides of loose fibrous, adipose, dense fibrous, and reticular connective tissues, identifying all the structures shown in figure HA-4.

Note the comments that are made in the legend pertaining to the function of mast cells that are seen in loose fibrous connective tissue.

Cartilage (figure HA-5) Study the three different types of cartilage, noting their distinct differentiating characteristics.

Bone (figure HA-6) When studying a slide of developing membrane bone try to differentiate the osteoblasts from the osteoclasts. Note that an osteo-clast is a large multinucleate cell with a clear area between it and the bony matrix.

Laboratory Report

Answer all the questions on combined Laboratory Report 10,11.

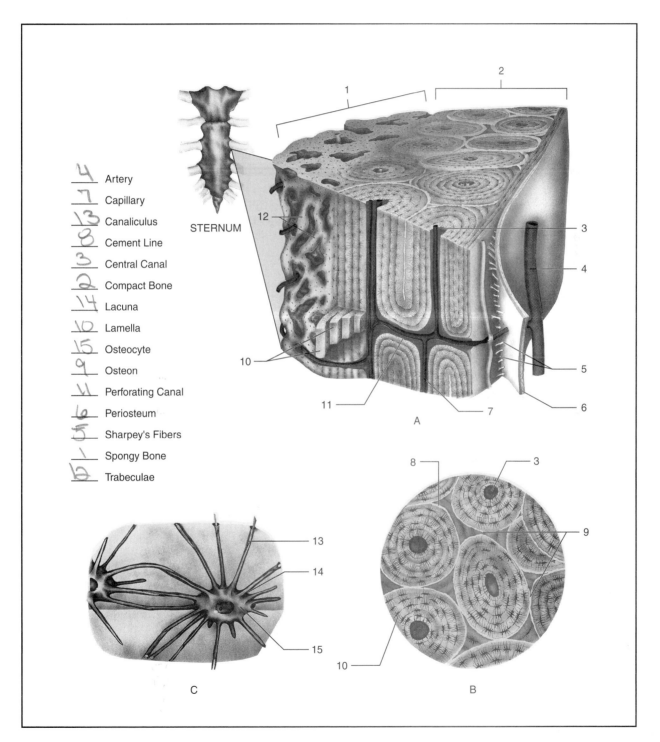

4   Artery
7   Capillary
13   Canaliculus
8   Cement Line
3   Central Canal
2   Compact Bone
14   Lacuna
10   Lamella
15   Osteocyte
9   Osteon
11   Perforating Canal
6   Periosteum
5   Sharpey's Fibers
1   Spongy Bone
12   Trabeculae

Figure 11.5   Bone tissue.

12

The Integument

Since it is constructed of epithelial and connective tissues, a study of the skin at this time provides one with an opportunity to review experiences of the last two laboratory periods. During this laboratory period prepared slides of the skin will be available for study. Before examining the slides, however, label figures 12.1 and 12.2. Note that the skin consists of two layers: an outer multilayered **epidermis** and a deeper **dermis.**

The Epidermis

The enlarged section of skin on the right side of figure 12.1 is the *epidermis.* Note that it consists of four distinct layers: an outer **stratum corneum,** a thin translucent **stratum lucidum,** a darkly stained **stratum granulosum,** and a multilayered **stratum spinosum** (stratum mucosum).

All four layers of the epidermis originate from the deepest layer of cells of the stratum spinosum. It is called the **stratum basale** (stratum germinativum). The columnar cells of this deep layer are constantly dividing to produce new cells that move outward to undergo metamorphosis at different levels. The brown skin pigment *melanin,* which is produced by stellate *melanocytes* of the stratum basale, is responsible for skin color. Skin color differences are due to the amount of melanin present.

The stratum corneum of the epidermis consists of many layers of the scaly remains of dead epithe-

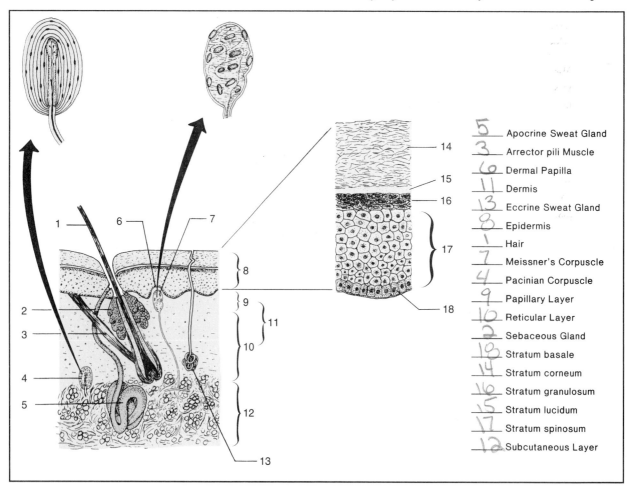

5 Apocrine Sweat Gland
3 Arrector pili Muscle
6 Dermal Papilla
11 Dermis
13 Eccrine Sweat Gland
8 Epidermis
1 Hair
7 Meissner's Corpuscle
4 Pacinian Corpuscle
9 Papillary Layer
10 Reticular Layer
2 Sebaceous Gland
18 Stratum basale
14 Stratum corneum
16 Stratum granulosum
15 Stratum lucidum
17 Stratum spinosum
12 Subcutaneous Layer

Figure 12.1　Skin structure.

lial cells. This protein residue of dead cells is primarily *keratin,* a water-repellent material. As the cells of the stratum spinosum are pushed outward, they move away from the nourishment of the capillaries, die, and undergo *keratinization.* The *eleidin* granules of the stratum basale are believed to be an intermediate product of keratinization. The translucent stratum lucidum consists of closely packed cells with traces of flattened nuclei.

The Dermis

This layer is often referred to as the "true skin." It varies in thickness of less than a millimeter to over 6 millimeters. It is highly vascular and provides most of the nourishment for the epidermis. It consists of two strata, the papillary and reticular layers.

The outer portion of the dermis, which lies next to the epidermis, is the **papillary layer.** It derives its name from numerous projections, or **dermal papillae,** which extend into the upper layers of the epidermis. In most regions of the body these papillae form no pattern; however, on the fingertips, palms, and soles of the feet they form regularly arranged patterns of parallel ridges that improve frictional characteristics in these areas.

The deeper portion, or **reticular layer,** contains more collagenous fibers than the papillary layer. These fibers greatly enhance the strength of the skin. The surface texture of suede leather is, essentially, the reticular layer of animal hides.

Subcutaneous Tissue

Beneath the dermis lies the **subcutaneous layer,** or *hypodermis;* it is also referred to as the *superficial fascia.* It consists of loose connective tissue, nerves, and blood vessels. One of its prime functions is to provide attachment for the skin to underlying structures.

Hair

Hair *(pili)* consists
pactly cemented to
1, figures 12.1 and
epithelial cells, the
of the hair shaft, or
onion-shaped region
12.2). Within the bulb
nective tissue called th
is through the latter
enters the shaft. The ro
internal root sheath a   . ..eath.

Extending diagonal., ..um the wall of the hair follicle to the epidermis is a band of smooth muscle fibers, the **arrector pili muscle.** Contraction of these muscle fibers causes the hair to move to a more perpendicular position, causing elevations on the skin surface, commonly referred to as "goose pimples."

Glands

Two kinds of glands are present in the skin: sebaceous and sweat.

Sebaceous Glands These glands are located within the epithelial tissues that surround each hair follicle. An oily secretion, called *sebum,* is secreted by these glands into the hair follicles and out onto the skin surface. Secretion is facilitated to some extent by the force of the arrector pili muscles during contraction. Sebum keeps hair pliable and helps to waterproof the skin.

Sweat Glands Sweat glands are of two types: eccrine and apocrine. The small sweat glands that empty directly out through the surface of the skin are **eccrine sweat glands.** These glands are simple tubular structures that have their coiled basal portions

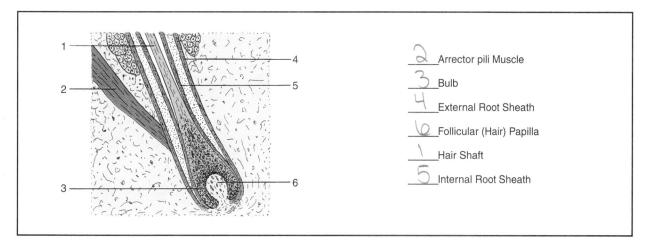

2	Arrector pili Muscle
3	Bulb
4	External Root Sheath
6	Follicular (Hair) Papilla
1	Hair Shaft
5	Internal Root Sheath

Figure 12.2 Structure of the hair root.

ermis. Except on the lips, glans
ris, they are widely distributed
e body.

gh all eccrine gland secretions are simi-
omposition, there are two different control-
stimuli. Almost everyone is aware that the
weat glands of some parts of the body, such as the
palms and axillae, are affected by emotional factors. Glands in other regions, however, such as the
forehead, neck, and back are regulated, primarily,
by thermal stimuli.

Apocrine sweat glands are much larger than
the eccrine type and have their secretory coiled portions located in the subcutaneous layer. Instead of
emptying out onto the surface of the epidermis, all
apocrine glands empty directly into a hair follicle
canal.

These glands are found in the axillae, scrotum
of the male, female perigenital region, external ear
canal, and nasal passages. Whereas eccrine sweat is
watery, the secretion of apocrine glands is thick
white, gray, or yellowish. Malodorous substances in
apocrine sweat are the principal contributors to
body odors. Psychic factors, rather than temperature
changes, contribute mostly to apocrine secretions.

Receptors

The receptors shown in figure 12.1 are Meissner's
and Pacinian corpuscles. **Meissner's corpuscles**
are located in the papillary layer of the dermis, projecting up into papillae of the epidermis. They function as receptors of touch.

Pacinian corpuscles are spherical receptors
with onionlike laminations. They lie deep in the
reticular layer of the dermis. Pacinian corpuscles
are sensitive to variations in sustained pressure.

Laboratory Assignment

After labeling figures 12.1 and 12.2 and answering
the questions on the Laboratory Report, proceed as
follows:

Materials:

> prepared slide of the skin (H11.11, H11.12,
> or H11.14)

Examine prepared slides of sections through the
skin and identify the structures seen in figures 12.1,
12.2, HA-9, HA-10, HA-11, and HA-12. Make
drawings if required.

PART 3

The Skeletal System

Many phases of medical science demand a thorough understanding of the skeletal system. Since most of the muscles of the body are anchored to specific loci on bones, it is imperative that the student of muscular anatomy know the names of many processes, ridges, and grooves on individual bones.

The study of the nervous and circulatory systems also depends to some extent on one's comprehension of the structure of the skeletal system. This is particularly true when studying the passage of nerves and blood vessels through openings in bones of the skull. Each passageway (foramen or meatus) has a name. A complete comprehension of the skeletal system includes a knowledge of the names of all these openings.

X-ray technology is an important branch of medical practice that relies heavily on osteology. In dentistry, the structure of the mandible and surrounding facial bones is of particular importance in taking X rays of the teeth. In medicine, all parts of the skeleton must be thoroughly understood.

13

The Skeletal Plan

In this exercise we will study the structure of the skeleton as a whole and the anatomy of a typical long bone. Detailed examination of individual parts will follow in subsequent exercises.

Materials:

fresh beef bones, sawed longitudinally
articulated human skeleton

The adult skeleton is made up of 206 named bones and many smaller unnamed ones. They are classified as being long, short, flat, irregular, or sesamoid.

The *long* bones include the bones of the arm, leg, metacarpals, metatarsals, and phalanges. The *short* bones are seen in the wrist and ankle. In addition to being shorter, the short bones differ from the long ones in another respect: they are filled with cancellous bone instead of having a medullary cavity.

The *flat* bones are the protective bones of the skull. Bones that are neither long, short, nor flat are classified as being *irregular.* The vertebrae and bones of the middle ear fall into this category.

Round bones embedded in tendons are called *sesamoid* bones (shape resembles sesame seeds). The kneecap is the most prominent bone of this type.

Bone Structure

A long bone, such as the femur shown in figure 13.1, is a complex structure made up of many different tissues. It performs many functions other than just supporting the body.

Before identifying the structures in figure 13.1, be aware of the following terms that apply to depressions and cavities on all bones of the skeleton:

Foramen: An opening in a bone that provides a passageway for nerves and blood vessels.
Fossa: A shallow depression in a bone. In some instances the fossa is a socket into which another bone fits.
Sulcus: A groove or furrow.
Meatus: A canal or long tubelike passageway.
Fissure: A narrow slit.
Sinus (antrum): A cavity in a bone.

The following terms apply to different types of *processes* on bones:

Condyle: A rounded knucklelike eminence on a bone that articulates with another bone.
Tuberosity: A large roughened process on a bone that serves as a point of anchorage for a muscle.
Tubercle: A small rounded process.
Trochanter: A very large process on a bone.
Head: A portion of a bone supported by a constricted part, or *neck.*
Crest: A narrow ridge of bone.
Spine: A sharp slender process.

Figure 13.1 We see in figure 13.1 a diagram of the femur cut open longitudinally to reveal its internal structure. Note that it has a long shaft called the **diaphysis** and two enlarged ends, the **epiphyses.** Where the epiphyses meet the diaphysis are growth zones called **metaphyses.**

During the growing years a plate of hyaline cartilage, the *epiphyseal disk,* exists in each of these growth areas. As new cartilage forms on the epiphyseal side, it is destroyed and replaced by bone on the diaphyseal side. The metaphysis during the growing years consists of the epiphyseal disk and calcified cartilage; at maturity the area becomes completely ossified, and linear growth ceases.

Note that the central portion of the diaphysis is a hollow chamber, the **medullary cavity.** The compact bone tissue of the shaft provides ample strength, obviating the need for central bone tissue. Observe that this cavity is lined with a membrane called the **endosteum.** This membrane is continuous with the linings of the central canals of the osteons.

The entire medullary cavity and much of the cancellous tissue of the bone extremities contain **yellow marrow,** a fatlike substance. The cancellous bone of the epiphyses of the femur (and humerus) contains **red marrow** in the adult. Other long bones of the skeleton contain only yellow marrow. Most of the red marrow in adults is contained in the ribs, sternum, and vertebrae.

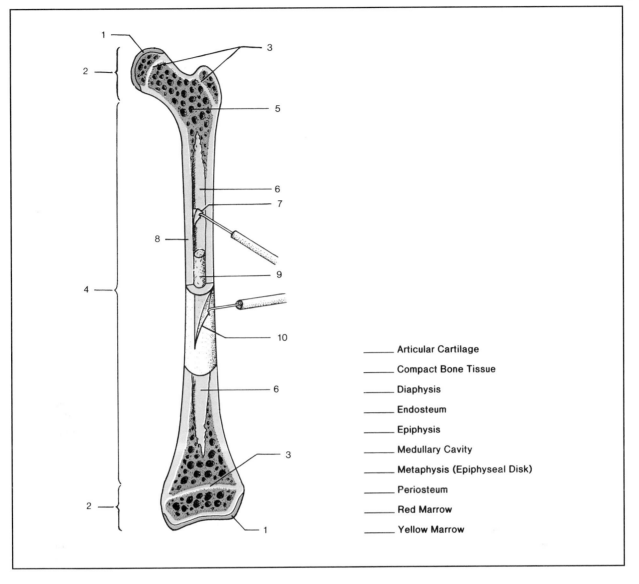

Figure 13.1 Long bone structure.

_____ Articular Cartilage
_____ Compact Bone Tissue
_____ Diaphysis
_____ Endosteum
_____ Epiphysis
_____ Medullary Cavity
_____ Metaphysis (Epiphyseal Disk)
_____ Periosteum
_____ Red Marrow
_____ Yellow Marrow

A tough covering, the **periosteum,** envelops the surfaces of the entire bone except for the areas of articulation. This covering consists of fibrous connective tissue that is quite vascular. The surfaces of each epiphysis that contact adjacent bones are covered with smooth **articular cartilage** that is of the hyaline type.

Assignment:
Identify the labels in figure 13.1.

Beef Bone Study Examine a freshly cut section of bone. Identify all structures shown in figure 13.1. Probe into the periosteum near a torn ligament or tendon; note the continuity of fibers between the periosteum and these structures. Probe into the marrow and note its texture.

Bone Names

The bones of the skeleton fall into two main groups: those that make up the axial skeleton and those forming the appendicular skeleton.

The Axial Skeleton The parts of the axial skeleton are the **skull, hyoid bone, vertebral column** (spine), and **rib cage.** The hyoid bone is a horseshoe-shaped bone that is situated in the neck under the lower jaw. The rib cage consists of twelve pairs of **ribs** and a **sternum** (breastbone).

The Appendicular Skeleton This portion of the skeleton includes the upper and lower extremities.
Each upper extremity consists of a pectoral girdle, arm, and hand. The **pectoral girdle** consists of

a **scapula** (shoulder blade) and **clavicle** (collar-bone). Each arm consists of an upper portion, the **humerus,** and two forearm bones, the **radius** and **ulna.** The radius is lateral to the ulna. The **hand** includes the bones of the fingers and wrist.

The lower extremities consist of the pelvic girdle and legs. The **pelvic girdle** is formed by two bones, the **ossa coxae,** which are attached posteriorly to the sacrum of the vertebral column and anteriorly to each other. The anterior joint where the ossa coxae are united on the median line is the **symphysis pubis.**

Each leg consists of four bones: the **femur** in the upper leg, a **tibia** (shinbone), a thin, long **fibula** parallel to the tibia, and the **patella,** or kneecap.

Assignment:
Label the parts of the skeleton in figure 13.2.

Bone Fractures

When bones of the body are subjected to excessive stress, various kinds of fractures occur. The type of fracture that results will depend on the nature and direction of forces that are applied. Some of the more common types are illustrated in figure 13.3.

If the fracture is contained in the soft tissues and does not communicate in any way with the skin or mucous membranes it is considered to be a **closed,** or **simple,** fracture. A majority of the fractures in figure 13.3 would probably fall in this category. If the fracture does communicate with the external surfaces, however, it is called an **open,** or **compound,** fracture. Fractures of this type may become considerably more difficult to treat because bone marrow infections (osteomyelitis) may result.

Fractures may be complete or incomplete, also. **Incomplete** fractures are the type in which the bone is split, splintered, or only partially broken. Illustrations A, B, and C in figure 13.3 are of this type. When a bone breaks through on only one side as a result of bending, it is often referred to as a **greenstick** fracture. These fractures are most common among the young. Linear splitting of a long bone may be referred to as a **fissured** fracture.

Complete fractures are those in which the bone is broken clear through. If the break is at right angles to the long axis, it is considered to be a **transverse** fracture. Breaks that are at an angle to the long axis are termed **oblique** fractures. If a fracture results from torsional forces, it may be referred to as a **spiral** fracture.

If a piece of bone is broken out of the shaft it is a **segmental** fracture. More extensive fractures, in

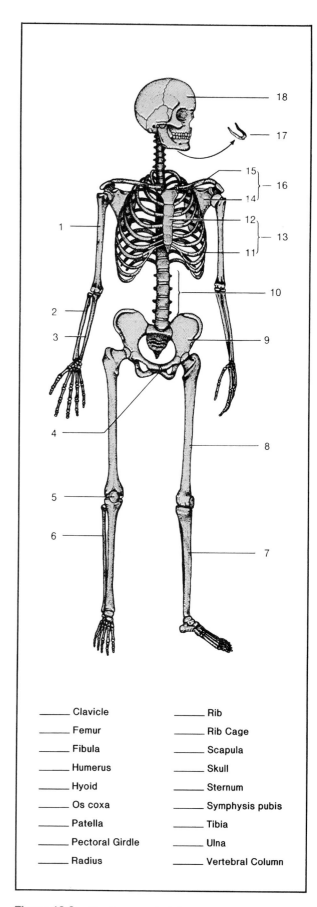

_____ Clavicle	_____ Rib
_____ Femur	_____ Rib Cage
_____ Fibula	_____ Scapula
_____ Humerus	_____ Skull
_____ Hyoid	_____ Sternum
_____ Os coxa	_____ Symphysis pubis
_____ Patella	_____ Tibia
_____ Pectoral Girdle	_____ Ulna
_____ Radius	_____ Vertebral Column

Figure 13.2   The human skeleton.

which two or more fragments are seen, are designated as **comminuted** fractures. When bone fragments have been moved out of alignment, as in illustration G, the fracture may also be referred to as being **displaced.**

Severe vertical forces can result in compacted or compression bone fractures. If a broken portion of bone is driven into another portion of the same bone, it is referred to as a **compacted** fracture. This is often seen in femur fractures, such as in illustration J. **Compression** fractures (not shown) often occur in the vertebral column when vertebrae are crushed due to falls from excessive heights.

Assignment:
Identify the types of fractures shown in figure 13.3.
　　　Complete the Laboratory Report for this exercise.

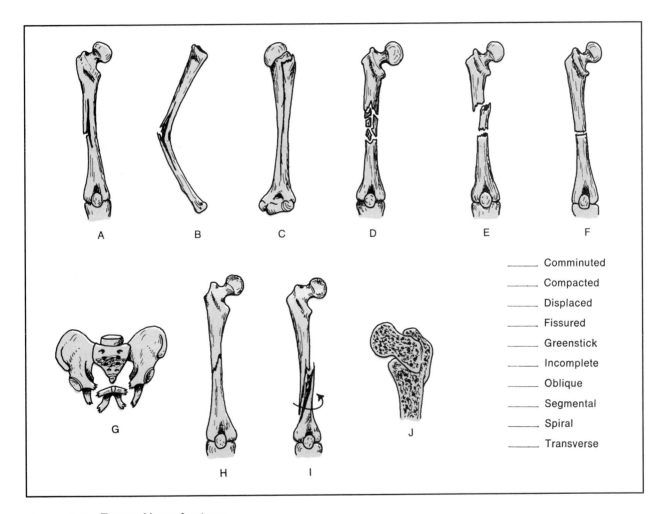

_____ Comminuted
_____ Compacted
_____ Displaced
_____ Fissured
_____ Greenstick
_____ Incomplete
_____ Oblique
_____ Segmental
_____ Spiral
_____ Transverse

Figure 13.3   Types of bone fractures.

14

The Skull

For this study of the skull, specimens will be available in the laboratory. As you read through the discussion of the various bones, identify them first in the illustrations and then on the specimens. Compare the specimens with the illustrations to note the degree of variance.

Care of Skulls

When handling laboratory skulls be very careful to avoid damaging them. **Never use a pencil as a pointer.** Pencil marks must not be made on the bones. A metal probe or a pipe cleaner should be used instead. If a metal probe is used, **touch the bones very gently to avoid bone perforation** where bone is thin.

Materials:

 whole and disarticulated skulls
 fetal skulls
 metal probe or pipe cleaner

The Cranium

That portion of the skull that encases the brain is called the *cranium.* It consists of the following bones: a single frontal, two parietals, one occipital, two temporals, one sphenoid, and an ethmoid. All these bones are joined together at their margins by irregular interlocking joints called *sutures.* The lateral and inferior aspects of the cranium are illustrated in figures 14.1 and 14.2. A sagittal section is seen in figure 14.6

Frontal The anterior superior portion of the skull consists of the *frontal* bone. It forms the eyebrow ridges and the ridge above the nose. The most inferior edge of this bone extends well into the orbit of the eye to form the **orbital plates** of the frontal bone. On the superior ridges of the eye orbits are a pair of foramina, the **supraorbital foramina.** See figure 14.3.

Parietals Directly posterior to the frontal bone on the sides of the skull are the *parietal* bones. The lateral view of the skull actually shows only the left parietal bone. The right parietal is on the other side of the skull.

The right and left parietals meet on the midline of the skull to form the **sagittal suture.** Between the frontal and each parietal bone is another suture, the **coronal suture.** Two semicircular bony ridges that extend from the forehead (frontal bone) and over the parietal bone are the **superior temporal line** and **inferior temporal line.** These ridges form the points of attachment for the longest muscle fibers of the *temporalis* muscle. Figure 36.1, page 161, shows the position of the muscle (label 2). It is the upper extremity of the muscle that falls on the superior temporal line.

Temporals On each side of the skull, inferior to the parietal bones, are the *temporals.* The left temporal is colored yellow in figure 14.1. Each temporal is joined to its adjacent parietal by the **squamosal suture.** A depression, the **mandibular** *(glenoid)* **fossa,** on this bone provides a recess into which the lower jaw articulates. Pull the jaw away from the skull to note the shape of this fossa. The rounded eminence of the mandible that fits into this depression is the **mandibular condyle.** Just posterior to the mandibular fossa is the ear canal, or **external acoustic meatus** *(acoustic:* hearing; *meatus:* canal or passage).

The temporal bone has three significant processes: the zygomatic, styloid, and mastoid. The **zygomatic process** is a long slender process that extends forward to articulate with the zygomatic bone. The zygomatic process and a portion of the zygomatic bone constitute the *zygomatic arch.* The **styloid process** is a slender spinelike process that extends downward from the bottom of the temporal bone to form a point of attachment for some muscles of the tongue and pharynx. This process is often broken off on laboratory specimens. The **mastoid process** is a rounded eminence on the inferior surface of the temporal just posterior to the styloid process. It provides anchorage for the *sternocleidomastoideus* muscle of the neck. Middle ear infections that spread into the cancellous bone of this process are referred to as *mastoiditis.*

Sphenoid The pink-colored bone seen in the lateral view of the skull, figure 14.1, is the *sphenoid* bone. Note in the inferior view, figure 14.2, that this bone extends from one side of the skull to the other.

Identify the two greater wings, the orbital surfaces, and the pterygoid processes of the sphenoid. The portions that are on the sides of the skull in the "temple" region are the **greater wings of the sphenoid.** On the ventral surface (figure 14.2) are the **pterygoid processes of the sphenoid** to which the pterygoid muscles of mastication are attached. The **orbital surfaces of the sphenoid** (figure 14.3) make up the posterior walls of each eye orbit.

Ethmoid On the medial surface of each orbit of the eye is seen the *ethmoid* bone (label 6, figure 14.1). The bone forms a part of the roof of the nasal cavity and closes the anterior portion of the cranium.

Examine the upper portion of the nasal cavity of your skull. Note that the inferior portion of the ethmoid has a downward extending **perpendicular plate** on the median line. Refer to figure 14.6 (label 4). This portion articulates anteriorly with the nasal and frontal bones. Posteriorly, it articulates with the sphenoid and vomer. On each side of the perpendicular plate are

irregular curved plates, the **superior** and **middle nasal conchae.** They provide bony reinforcement for the fleshy **upper nasal conchae** of the nasal cavity.

Occipital The posterior inferior portion of the skull consists primarily of the *occipital* bone. It is joined to the parietal bones by the **lambdoidal suture** (see figure 14.1).

Examine the inferior surface of your laboratory skull and compare it with figure 14.2. Note the large **foramen magnum,** which surrounds the brain stem in real life. On each side of this opening is seen a pair of **occipital condyles** (label 25). These two condyles rest on fossae of the *atlas,* the first cervical vertebra of the spinal column.

Near the base of each occipital condyle are two passageways, the condyloid and hypoglossal canals. The opening to the **condyloid canal** is posterior to the occipital condyle. The **hypoglossal canal** is seen with a piece of wire passing through it. The hypoglossal canal provides a passageway for the hypoglossal (twelfth) cranial nerve.

Three prominent ridges form a distinctive pattern posterior to the foramen magnum on the occipital bone. The **median nuchal line** is a ridge that

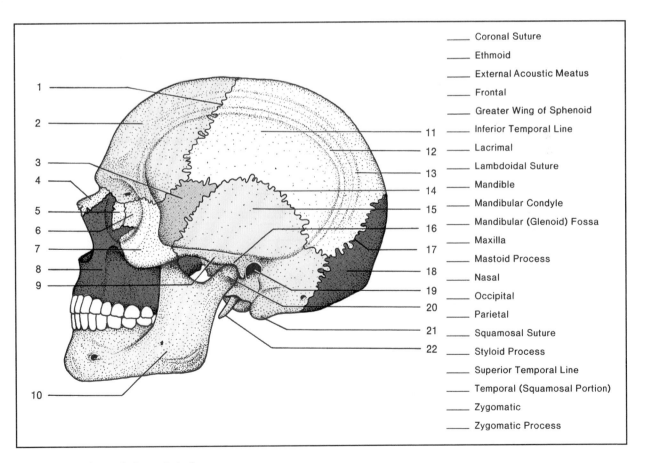

_____ Coronal Suture
_____ Ethmoid
_____ External Acoustic Meatus
_____ Frontal
_____ Greater Wing of Sphenoid
_____ Inferior Temporal Line
_____ Lacrimal
_____ Lambdoidal Suture
_____ Mandible
_____ Mandibular Condyle
_____ Mandibular (Glenoid) Fossa
_____ Maxilla
_____ Mastoid Process
_____ Nasal
_____ Occipital
_____ Parietal
_____ Squamosal Suture
_____ Styloid Process
_____ Superior Temporal Line
_____ Temporal (Squamosal Portion)
_____ Zygomatic
_____ Zygomatic Process

Figure 14.1 Lateral view of skull.

extends posteriorly from the foramen magnum. It provides a point of attachment for the *ligamentum nuchae*. The **inferior** and **superior nuchal lines** lie parallel to each other. The inferior one lies between the foramen magnum and the superior nuchal line. These ridges form points of attachment for various neck muscles. Note that where the superior nuchal and medial nuchal lines meet is a distinct prominence, the **external occipital protuberance.**

Foramina on Inferior Surface Examine the inferior surface of your laboratory skull and compare it with figure 14.2 to identify the following foramina. The significance of these openings will become more apparent when the circulatory and nervous systems are studied.

Two foramina, the foramen ovale and foramen spinosum, are seen on each half of the sphenoid bone. The **foramen ovale** (label 6) is a large elliptical foramen that provides a passageway for the mandibular branch of the trigeminal nerve. Slightly posterior and lateral to it is a smaller **foramen spinosum,** which allows a small branch of the mandibular nerve and middle meningeal blood vessels to pass through the skull.

The temporal bone has five openings on its inferior surface. The **foramen lacerum** (label 21) is characterized by having a jagged margin. If a piece of wire (bent paper clip) is carefully inserted into the foramen lacerum as shown in figure 14.2 at the left, a passageway, the **carotid canal,** can be observed. It is through this canal that the internal carotid artery supplies the brain with blood.

Just posterior to the carotid canal opening is an irregular slitlike opening, the **jugular foramen,** which allows drainage of blood from the cranial cavity via the inferior petrosal sinus. A depression, the **jugular fossa** (label 10), is adjacent to it. The **stylomastoid foramen** is a small opening at the base of the styloid process. The **mastoid foramen** is the most posterior foramen on the temporal bone. This foramen is sometimes absent.

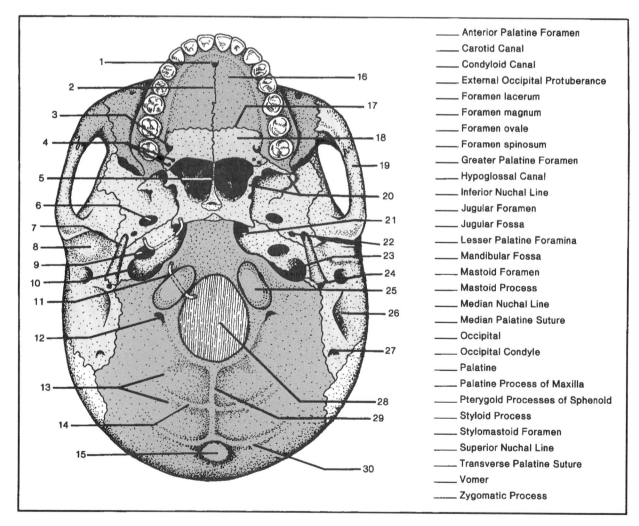

_____ Anterior Palatine Foramen
_____ Carotid Canal
_____ Condyloid Canal
_____ External Occipital Protuberance
_____ Foramen lacerum
_____ Foramen magnum
_____ Foramen ovale
_____ Foramen spinosum
_____ Greater Palatine Foramen
_____ Hypoglossal Canal
_____ Inferior Nuchal Line
_____ Jugular Foramen
_____ Jugular Fossa
_____ Lesser Palatine Foramina
_____ Mandibular Fossa
_____ Mastoid Foramen
_____ Mastoid Process
_____ Median Nuchal Line
_____ Median Palatine Suture
_____ Occipital
_____ Occipital Condyle
_____ Palatine
_____ Palatine Process of Maxilla
_____ Pterygoid Processes of Sphenoid
_____ Styloid Process
_____ Stylomastoid Foramen
_____ Superior Nuchal Line
_____ Transverse Palatine Suture
_____ Vomer
_____ Zygomatic Process

Figure 14.2 Inferior surface of skull.

Assignment:
Label all bones of the cranium in figure 14.1. Facial bones will be studied next.

Except for the hard palate and vomer, label all structures in figure 14.2. Labeling of other structures will be completed later.

The Face

The face of the skull consists of thirteen bones fused together, plus a movable mandible. Of the thirteen fused bones, only one bone, the vomer, is not paired. Figure 14.3 reveals a majority of the facial bones.

Maxillae The upper jaw consists of two maxillary bones (maxillae) that are joined by a suture on the median line. Remove the mandible from your laboratory skull and examine the hard palate. Compare it with figure 14.2. Note that the anterior portion of the hard palate consists of two **palatine processes of the maxillae.** A **median palatine suture** joins the two bones on the median line.

The maxillae of an adult support sixteen permanent teeth. Each tooth is contained in a socket, or *alveolus.* That portion of the maxillae that contains the teeth is called the *alveolar process.*

Three significant foramina are seen on the maxillae: two infraorbital and one anterior palatine. The **infraorbital foramina** are situated on the front of the face under each eye orbit. Nerves and blood vessels emerge from each of these foramina to supply the nose. The **anterior palatine** (incisive) **foramen** is seen in the anterior region of the hard palate just posterior to the central incisors.

Palatines In addition to the horizontal palatine processes of the maxillae, the hard palate also consists of two palatine bones. These bones form the posterior third of the palate.

Locate them on your laboratory specimen. Note that each palatine bone has a large **greater palatine foramen** and two smaller **lesser palatine foramina.**

Assignment:
Label the parts of the hard palate in figure 14.2.

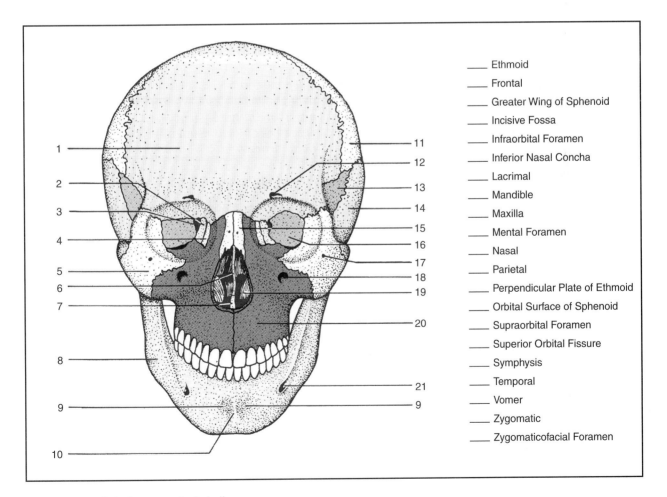

_____ Ethmoid
_____ Frontal
_____ Greater Wing of Sphenoid
_____ Incisive Fossa
_____ Infraorbital Foramen
_____ Inferior Nasal Concha
_____ Lacrimal
_____ Mandible
_____ Maxilla
_____ Mental Foramen
_____ Nasal
_____ Parietal
_____ Perpendicular Plate of Ethmoid
_____ Orbital Surface of Sphenoid
_____ Supraorbital Foramen
_____ Superior Orbital Fissure
_____ Symphysis
_____ Temporal
_____ Vomer
_____ Zygomatic
_____ Zygomaticofacial Foramen

Figure 14.3 Anterior aspect of skull.

Zygomatics On each side of the face are two *zygomatic* (malar) bones. They form the prominence of each cheek and the inferior, lateral surface of each eye orbit. Each zygomatic has a small foramen, the **zygomaticofacial foramen.**

Lacrimals Between the ethmoid and upper portion of the maxillary bones are a pair of lacrimal (Latin: *lacrima,* tear) bones—one in each orbit.

Locate a groove on the surface of each *lacrimal* that is continuous with a groove on the maxilla. This groove provides a recess for the lacrimal duct, through which tears flow from the eye into the nasal cavity.

Nasals The ridge of the nose is formed by a pair of thin, rectangular *nasal* bones.

Vomer This thin bone is located in the nasal cavity on the median line. Its posterior upper edge articulates with the back portion of the perpendicular plate of the ethmoid and the rostrum of the sphenoid. The lower border of the vomer is joined to the maxillae and palatines. The *septal cartilage* of the nose extends between the anterior margin of the vomer and the perpendicular plate of the ethmoid. Locate the vomer on figures 14.2, 14.3, and 14.6 as well as on your laboratory skull.

Inferior Nasal Conchae The inferior nasal conchae are curved bones attached to the walls of the nasal fossa. They are situated beneath the superior and middle nasal conchae, which are part of the ethmoid bone.

Assignment:
Label the vomer in figure 14.2.

Except for the parts of the mandible, label all structures in figure 14.3 and the remaining facial bones in figure 14.1.

Floor of the Cranium

Now that all bones of the skull have been identified let's examine the inside of the cranium to note the significant structural details. Remove the top of your laboratory skull and compare the floor of the cranium with figure 14.4 to identify the following structures.

Cranial Fossae As you look down on the entire floor of the cranium note that it is divided into three large depressions called *cranial fossae.* The one formed by the orbital plates of the frontal is called the **anterior cranial fossa.** The large depression formed within the occipital bone is the **posterior**

cranial fossa. It is the deepest fossa. In between these two fossae is the **middle cranial fossa,** which is at an intermediate level. The latter involves the sphenoid and temporal bones.

Ethmoid The *ethmoid* bone in the anterior cranial fossa is seen as a pale yellow structure between the orbital plates of the frontal bone. Note that it consists of a perforated horizontal portion, the **cribriform plate,** and an upward-projecting process, the **crista galli** (cock's comb). The holes in the cribriform plate allow branches of the olfactory nerve to pass from the brain to the nasal cavity. The crista galli serves as an attachment for the *falx cerebri* (label 1, figure 45.1).

Locate the small **foramen cecum,** which perforates the frontal bone just anterior to the ethmoid bone. This foramen provides a passageway for a small vein.

Sphenoid Note the batlike configuration of this bone, with the greater wings extending out on each side. The anterior leading edges of the wings are called the **small wings of the sphenoid.**

Observe that on the median line of the sphenoid there is a deep depression called the **hypophyseal fossa.** This depression contains the pituitary gland *(hypophysis)* in real life. Posterior to this fossa is an elevated ridge called the **dorsum sella.** The two spinelike processes anterior and lateral to the hypophyseal fossa that project backward are the **anterior clinoid processes.** The outer spiny processes of the dorsum sella are the **posterior clinoid processes.** The hypophyseal fossa, dorsum sella, and clinoid processes, collectively, make up the **sella turcica,** or *Turkish saddle.*

Just anterior to the sella turcica are a pair of openings called the **optic foramina.** These openings lead into a pair of short **optic canals.** Locate the latter on your laboratory skull. The optic nerves pass through these canals from the eyes to the brain.

Extending from one optic foramen to the other is a narrow shelf called the **chiasmatic groove.** Locate it on your specimen. It is on this bony ledge that fibers of the right and left optic nerves cross over in the optic chiasma (label 3, figure 45.9).

Lying under each anterior clinoid process of the sphenoid is a **superior orbital fissure.** Very little of it is shown in figure 14.4. Locate it on your laboratory specimen. Note that it is also seen in figure 14.3 (label 3). It enables the third, fourth, fifth, and sixth cranial nerves to enter the eye orbit from the cranial cavity. Certain blood vessels also pass through it.

Just posterior to the superior orbital fissure is the **foramen rotundum.** It provides a passageway for the maxillary branch of the fifth cranial (trigeminal) nerve. Just posterior and slightly lateral to the foramen rotundum is the large **foramen ovale.** The small foramen posterior and lateral to the foramen ovale is the **foramen spinosum.** The **foramen lacerum** (figure 14.4, label 4) is located between the margins of the sphenoid and temporal bones. Note the proximity of the **carotid canal** (label 7) to the foramen lacerum.

Temporals The significant parts of the temporal bone to identify in figure 14.4 are the petrous, squamous, and mastoid portions. The **squamous portion** of the temporal is that thin portion that forms a part of the side of the skull. The **petrous portion** (label 8) is probably the hardest portion of the skull. The medial sloping surface of the petrous portion has an opening to the **internal acoustic meatus.**

This canal contains the facial and vestibulocochlear cranial nerves. The latter nerve passes from the inner ear to the brain.

The most posterior part of the temporal bone is the **mastoid portion.** It has an enlargement, the **mastoid process,** and a **mastoid foramen,** which is posterior to the process.

Occipital Observe that this large bone has a semicircular groove called the **depression of the transverse sinus** (label 28). Blood of the brain collects in a large vessel in this groove. Locate the **jugular foramen,** where the internal jugular vein takes its origin. It appears as an irregular slit between the anterolateral margin of the occipital bone and the petrous portion of the temporal bone. Cranial nerves IX, X, and XI also pass through this foramen.

Identify the openings to the **condyloid** and **hypoglossal canals.** Reference to figure 14.2 establishes that the hypoglossal canal is the one that

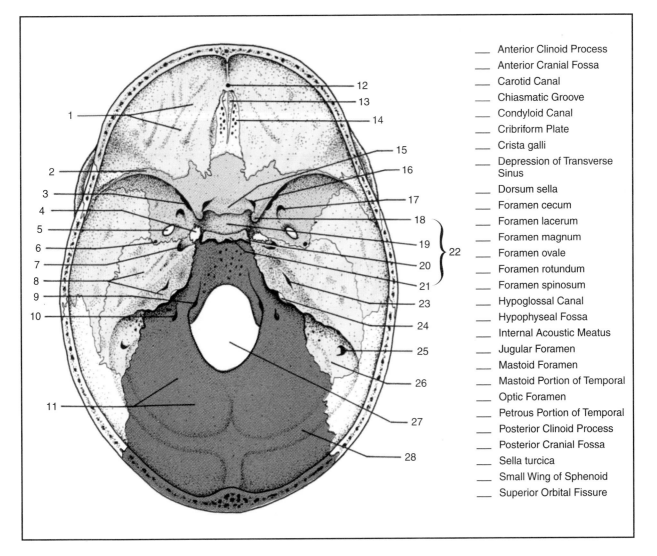

____ Anterior Clinoid Process
____ Anterior Cranial Fossa
____ Carotid Canal
____ Chiasmatic Groove
____ Condyloid Canal
____ Cribriform Plate
____ Crista galli
____ Depression of Transverse Sinus
____ Dorsum sella
____ Foramen cecum
____ Foramen lacerum
____ Foramen magnum
____ Foramen ovale
____ Foramen rotundum
____ Foramen spinosum
____ Hypoglossal Canal
____ Hypophyseal Fossa
____ Internal Acoustic Meatus
____ Jugular Foramen
____ Mastoid Foramen
____ Mastoid Portion of Temporal
____ Optic Foramen
____ Petrous Portion of Temporal
____ Posterior Clinoid Process
____ Posterior Cranial Fossa
____ Sella turcica
____ Small Wing of Sphenoid
____ Superior Orbital Fissure

Figure 14.4 Floor of cranium.

passes through the base of the occipital condyle. Explore these foramina with a slender probe or pipe cleaner to differentiate them.

Foramina Summary

Table 14.1 summarizes the more important foramina of the skull discussed in this exercise. Use this table for quick referencing.

Assignment:
Label figure 14.4.

The Paranasal Sinuses

Some of the bones of the skull contain cavities, the *paranasal sinuses,* which reduce the weight of the skull without appreciably weakening it. All of the sinuses have passageways leading into the nasal cavity and are lined with a mucous membrane similar to the type that lines the nasal cavities.

The paranasal sinuses are named after the bones in which they are situated. Figure 14.5 shows the location of these cavities. Above the eyes in the forehead are the **frontal sinuses.** The largest sinuses are the **maxillary sinuses,** which are in the maxillary bones. These sinuses are also referred to as the *antrums of Highmore.* The **sphenoidal sinus** is the most posterior sinus seen in figure 14.5. Between the frontal and sphenoidal sinuses are a group of small spaces called the **ethmoid air cells.**

Assignment:
Label figure 14.5.

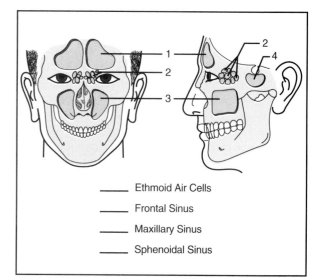

_____ Ethmoid Air Cells

_____ Frontal Sinus

_____ Maxillary Sinus

_____ Sphenoidal Sinus

Figure 14.5 The paranasal sinuses.

Sagittal Section

Visualization of the sinuses and many other structures in the center of the skull is greatly facilitated if a sagittal section of the skull is available for study. Figure 14.6 is of such a section. Note, first of all, how the frontal bone is hollowed out in the forehead region to form the **frontal sinus.** The **sphenoidal sinus** (pink bone) is revealed as a space of considerable size. Immediately above it is a distinct saddlelike structure, the **sella turcica.**

Table 14.1 Location and function of skull foramina.

FORAMEN	LOCATION	PASSAGEWAY FOR
Anterior palatine foramen	maxillae	nasopalatine nerves and descending palatine vessels
Carotid canal	temporal bone	internal carotid artery
Foramen lacerum	between sphenoid and temporal bone	internal carotid artery and plexus of sympathetic nerves
Foramen magnum	occipital bone	brain stem
Foramen ovale	sphenoid bone	mandibular nerve and accessory meningeal artery
Foramen spinosum	sphenoid bone	branch of mandibular nerve
Greater palatine foramen	palatine bone	anterior palatine nerve and descending palatine vessels
Hypoglossal canal	occipital bone	12th cranial nerve (hypoglossal)
Jugular foramen	between temporal and occipital bones	inferior petrosal sinus, vagus n., glossophyaryngeal n., accessory n., etc.
Mandibular foramen	mandible	inferior alveolar nerve (branch of 5th cranial n.) and blood vessels
Mental foramen	mandible	mental nerve and blood vessels

This section also shows best the relationship of the ethmoid bone to the vomer. The **ethmoid** is the pale yellow bone in the nasal region. The **vomer** is the uncolored bone that is shaped somewhat like a plowshare. Extending upward from the superior margin of the vomer is the **perpendicular plate of the ethmoid.** These two bones, combined, form a bony septum in the nasal cavity. The uppermost projecting structure of the ethmoid is the **crista galli.** The air cells of the ethmoid are not revealed in this section because they are not on the median line.

Note also the relationship of the bones of the palate to the vomer. The anterior portion of the hard palate is the **palatine process of the maxilla.** Posterior to it is the **palatine bone** (orange color). The vomer is fused to these two bones of the hard palate.

Assignment:
Label figure 14.6.

The Mandible

The only bone of the skull that is not fused as an integral part of the skull is the lower jaw, or mandible. Figure 14.7 reveals the anatomical details of this bone.

The mandible consists of a horizontal portion, the **body,** and two vertical portions, the **rami.** Embryologically, the mandible forms from two centers of ossification, one on each side of the face. As the bone develops toward the median line, the two halves finally meet and fuse to form a solid ridge. This point of fusion on the midline is called the **symphysis** (label 10, figure 14.3). On each side of the symphysis are two depressions, the **incisive fossae.**

The superior portion of each ramus has a condyle, a coronoid process, and a notch. The **mandibular condyle** occupies the posterior superior terminus of the ramus. The process on the superior

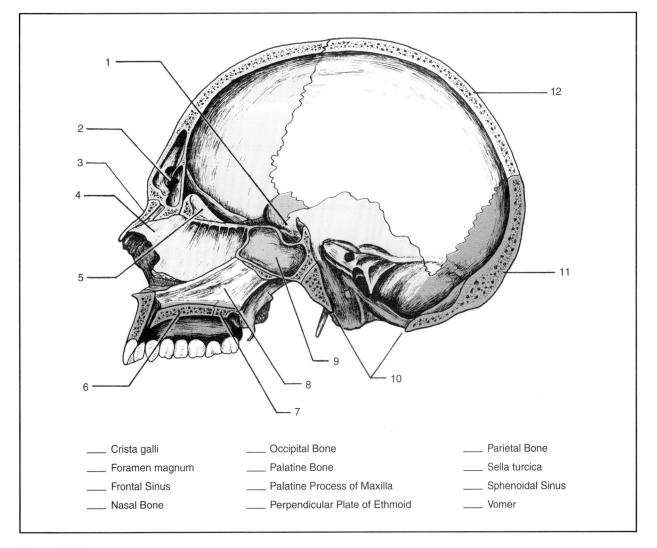

_____ Crista galli	_____ Occipital Bone	_____ Parietal Bone
_____ Foramen magnum	_____ Palatine Bone	_____ Sella turcica
_____ Frontal Sinus	_____ Palatine Process of Maxilla	_____ Sphenoidal Sinus
_____ Nasal Bone	_____ Perpendicular Plate of Ethmoid	_____ Vomer

Figure 14.6 Sagittal section of skull.

anterior portion of the ramus is the **coronoid process.** This tuberosity provides attachment for the *temporalis* muscle (see illustration A, figure 36.1). Between the mandibular condyle and the coronoid process is the **mandibular notch.** At the posterior inferior corners of the mandible, where the body and rami meet, are two protuberances, the **angles.** The angles provide attachment for the *masseter* and *internal pterygoid* muscles (figure 36.1).

A ridge of bone, the **oblique line,** extends at an angle from the ramus down the lateral surface of the body to a point near the mental foramen. This bony elevation is strong and prominent in its upper part but gradually flattens out and disappears, as a rule, just below the first molar.

On the internal (medial) surface of the mandible is another diagonal line, the **mylohyoid line.** It extends from the ramus down to the body. To this crest is attached a muscle, the *mylohyoid,* which forms the floor of the oral cavity. The bony portion of the body that exists above this line makes up a portion of the sides of the oral cavity proper.

Each tooth lies in a socket of bone called an *alveolus.* As in the case of the maxilla, the portion of this bone that contains the teeth is called the **alveolar process.** The alveolar process consists of two compact tissue bony plates, the *external* and *internal alveolar plates.* These two plates of bone are joined by partitions, or *septae,* that lie between the teeth and make up the transverse walls of the alveoli.

On the medial surfaces of the rami are two foramina, the **mandibular foramina.** On the external surface of the body are two prominent openings, the **mental foramina** (*mental:* chin).

Assignment:
Label figure 14.7 and the parts of the mandible in figure 14.3.

The Fetal Skull

The human skull at birth is incompletely ossified. Figure 14.8 reveals its structure. These unossified membranous areas, called *fontanels,* facilitate compression of the skull at childbirth. During labor the bones of the skull are able to lap over each other as the infant passes down the birth canal without causing injury to the brain.

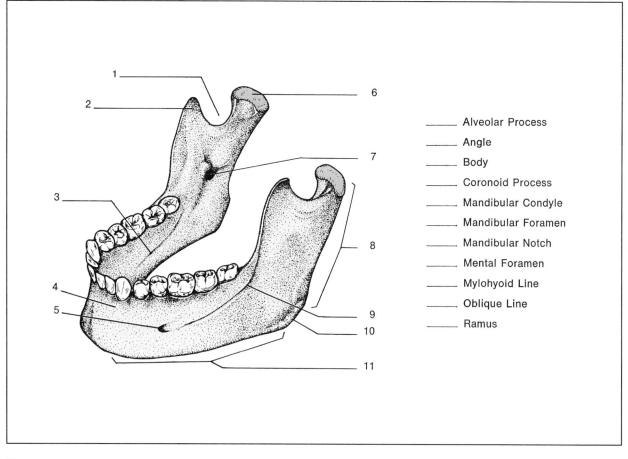

_____	Alveolar Process
_____	Angle
_____	Body
_____	Coronoid Process
_____	Mandibular Condyle
_____	Mandibular Foramen
_____	Mandibular Notch
_____	Mental Foramen
_____	Mylohyoid Line
_____	Oblique Line
_____	Ramus

Figure 14.7   **The mandible.**

There are six fontanels joined by five areas where future sutures of the skull will form. The largest fontanel is the **anterior fontanel,** a somewhat diamond-shaped membrane that lies on the median line at the juncture of the frontal and parietal bones. The **posterior fontanel** is somewhat smaller and lies on the median line at the juncture of the parietal and occipital bones. Between these two fontanels on the median line is a membranous area where the **future saggital suture** of the skull will form.

On each side of the skull, where the frontal, parietal, sphenoid, and temporal bones come together behind the eye orbit, is an **anterolateral fontanel.** Between the anterior and anterolateral fontanels can be seen a membranous line that is the area where the **future coronal suture** will develop.

The most posterior fontanel on the side of the skull is the **posterolateral fontanel,** which lies at the juncture of the parietal, temporal, and occipital bones. Between the anterolateral and posterolateral fontanels is a membranous line that will develop into the **future squamosal suture.** Ossification of these fontanels and membranous future sutures is usually completed in the two-year-old child.

Assignment:
Label figure 14.8.

Examine a fetal skull to identify all of the structures shown in figure 14.8.

Complete the Laboratory Report for this exercise.

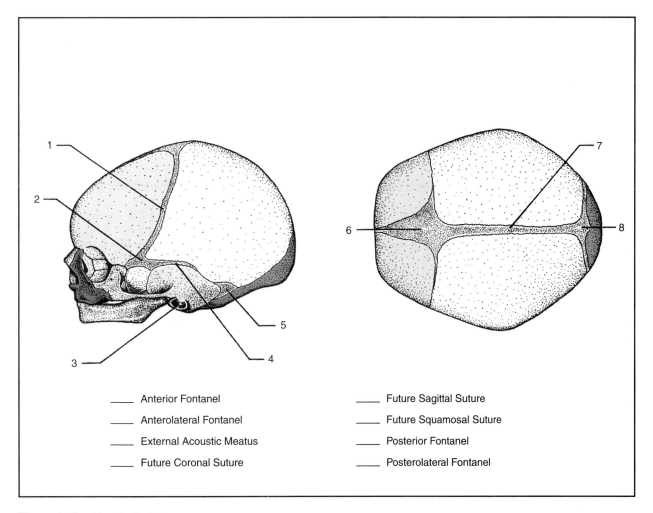

_____ Anterior Fontanel

_____ Anterolateral Fontanel

_____ External Acoustic Meatus

_____ Future Coronal Suture

_____ Future Sagittal Suture

_____ Future Squamosal Suture

_____ Posterior Fontanel

_____ Posterolateral Fontanel

Figure 14.8   The fetal skull.

15

The Vertebral Column and Thorax

The skeletal structure of the trunk of the body will be studied in this exercise. The vertebral column, ribs, sternum, and hyoid bone make up this part of the skeleton.

Materials:
    skeleton, articulated
    skeleton, disarticulated
    vertebral column, mounted

The Vertebral Column

The vertebral column consists of thirty-three bones, twenty-four of which are individual movable vertebrae. Figure 15.1 illustrates its structure. Note that the individual vertebrae are numbered from the top.

The Vertebrae

Although the vertebrae in different regions of the vertebral column vary considerably in size and configuration, they do have certain features in common. Each one has a structural mass, the **body,** which is the principal load-bearing contact area between adjacent vertebrae. The space between the surfaces of adjacent vertebral bodies is occupied by a fibrocartilaginous **intervertebral disk.** The collective action of these twenty-four disks imparts a vital cushion effect to the spinal column.

Note that in addition to the bodies, adjacent vertebrae contact each other on articular surfaces, which are located on the **transverse processes** (label 20). The **superior articular surfaces** (label 21) of each vertebra contact the **inferior articular surfaces** of the vertebra above it. At these three points adjacent vertebrae are secured together by ligaments, uniting the entire column into a single functioning unit.

In the center of each vertebra is an opening, the **spinal** (vertebral) **foramen,** which contains the spinal cord. Note that this foramen gets smaller as one progresses down the vertebral column.

In addition to the two transverse processes, each vertebra has a **spinous process** that projects out on its posterior surface. These spinous processes provide points of attachment for various muscles of the neck and back.

Extending backward from the body of each vertebra are two processes, the **pedicles,** which form a portion of the bony arch around the spinal foramen (see illustrations C and D). The openings between the pedicles of adjacent vertebrae allow spinal nerves to exit from the spinal cord. These openings are the **intervertebral foramina** (label 29) of the vertebral column. Between the transverse processes and the spinous process are two broad plates, called **laminae** (label 24). The two pedicles and two laminae constitute the *neural arch* of each vertebra.

Cervical Vertebrae   The upper seven bones are the *cervical vertebrae* of the neck. The first of these seven is the **atlas** (illustration A). Note that the spinal foramen is much larger here to accommodate a short portion of the brain stem, which extends down into the vertebral column.

Note in illustrations A, B, and C that the cervical vertebrae have a small **transverse foramen** in each transverse process. Collectively, these foramina form a passageway on each side of the spinal column for the vertebral artery and vein.

Observe, also, that the second cervical vertebra, or **axis** (illustration B), is unique in that it has a vertical protrusion, the **odontoid process** (dens), which provides a pivot for the rotation of the atlas. When the head is turned from side to side, movement occurs between the axis and atlas around this process.

Thoracic Vertebrae   Below the seven cervical vertebrae are twelve *thoracic vertebrae.* Illustration D reveals the structure of a typical thoracic vertebra. Note that these bones are larger and thicker than ones in the neck. A distinguishing feature of these vertebrae is that all twelve of them have **articular facets for ribs** on their transverse processes.

Lumbar Vertebrae   Inferior to the thoracic vertebrae lie five *lumbar vertebrae.* The bodies of these bones are much thicker than those of the other

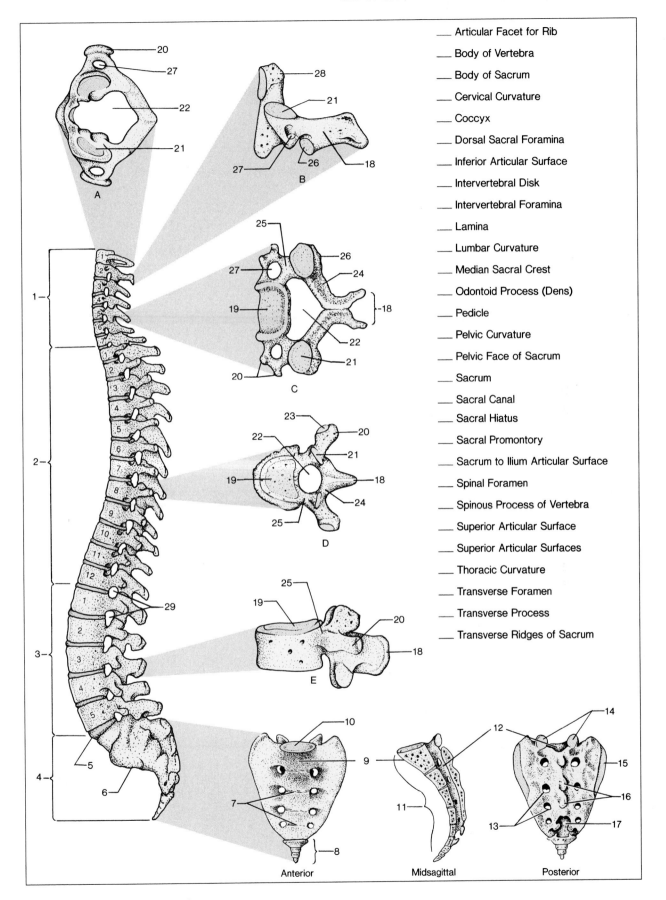

Figure 15.1 The vertebral column.

___ Articular Facet for Rib

___ Body of Vertebra

___ Body of Sacrum

___ Cervical Curvature

___ Coccyx

___ Dorsal Sacral Foramina

___ Inferior Articular Surface

___ Intervertebral Disk

___ Intervertebral Foramina

___ Lamina

___ Lumbar Curvature

___ Median Sacral Crest

___ Odontoid Process (Dens)

___ Pedicle

___ Pelvic Curvature

___ Pelvic Face of Sacrum

___ Sacrum

___ Sacral Canal

___ Sacral Hiatus

___ Sacral Promontory

___ Sacrum to Ilium Articular Surface

___ Spinal Foramen

___ Spinous Process of Vertebra

___ Superior Articular Surface

___ Superior Articular Surfaces

___ Thoracic Curvature

___ Transverse Foramen

___ Transverse Process

___ Transverse Ridges of Sacrum

vertebrae due to the greater stress that occurs in this region of the vertebral column. Illustration E is of a typical lumbar vertebra.

The Sacrum

Inferior to the fifth lumbar vertebra lies the *sacrum.* It consists of five fused vertebrae. Note that there are two oval **superior articular surfaces** (facets) that articulate with the inferior articular surfaces of the fifth lumbar vertebra. On each lateral surface of the sacrum is a **sacrum to ilium articular surface.** Note that the part of the sacrum that articulates with the intervertebral disk superior to it is called the **body of the sacrum.**

Observe that the anterior surface of the sacrum, or **pelvic face,** is curved backward and that the body of the first sacral vertebra forms a protrusion called the **sacral promontory.** Note also that four **transverse ridges** can be seen on the pelvic face that reveal where the five vertebrae are fused together.

Identify the **median sacral crest** (label 16) and the **dorsal sacral foramina** on the posterior surface. The neural arches of the fused sacral vertebrae form the **sacral canal,** which exits at the lower end as the **sacral hiatus.**

The Coccyx

The "tailbone" of the vertebral column is the **coccyx.** It consists of four or five rudimentary vertebrae. It is triangular in shape and is attached to the sacrum by ligaments.

Spinal Curvatures

Four curvatures of the vertebral column, together with the intervertebral disks, impart considerable springiness along its vertical axis. Three of them are identified by the type of vertebrae in each region: the **cervical, thoracic,** and **lumbar curves.** The fourth curvature, which is formed by the sacrum and coccyx, is the **pelvic curve.**

Assignment:
Label figure 15.1

The Thorax

The sternum, ribs, costal cartilages, and thoracic vertebrae form a cone-shaped enclosure, the *thorax.* Its components are illustrated in figures 15.2 and 15.3.

The Sternum   The sternum, or breastbone, consists of three separate bones, an upper **manubrium,** a middle **body** *(gladiolus),* and a lower **xiphoid** *(ensiform)* **process.** A **sternal angle** is formed where the inferior border of the manubrium articulates with the body.

On both sides of the sternum are notches (facets) where the sternal ends of the costal cartilages are attached. Note that the second rib fits into a pair of *demifacets (demi,* half) at the sternal angle.

The Ribs   There are twelve pairs of ribs. The first seven pairs attach directly to the sternum by costal cartilages and are called **vertebrosternal,** or **true, ribs.** The remaining pairs are called **false ribs.** The

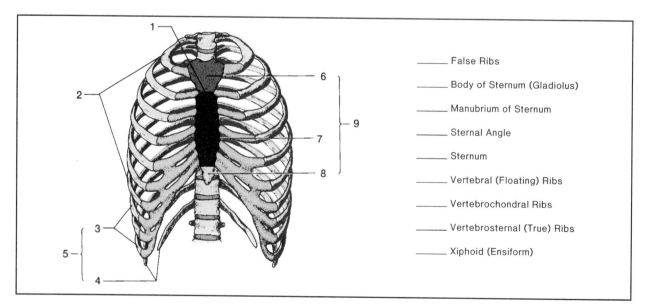

_____ False Ribs

_____ Body of Sternum (Gladiolus)

_____ Manubrium of Sternum

_____ Sternal Angle

_____ Sternum

_____ Vertebral (Floating) Ribs

_____ Vertebrochondral Ribs

_____ Vertebrosternal (True) Ribs

_____ Xiphoid (Ensiform)

Figure 15.2   The thorax.

upper three pairs of false ribs, the **vertebrochondral ribs,** have cartilaginous attachments on their anterior ends but do not attach directly to the sternum. The lowest false ribs, the **vertebral,** or **floating ribs,** are unattached anteriorly.

Figure 15.3 illustrates the structure of a central rib, a lateral view of a thoracic vertebra, and articulation details. Although considerable variability exists in size and configuration of the various ribs, the central rib reveals structures common to most ribs.

The principal parts of each rib are a head, neck, tubercle, and body. The **head** (label 15) is the enlarged end of the rib that articulates with the vertebral column. The **tubercle** (label 17) consists of two portions: an **articular portion** and a **nonarticular**

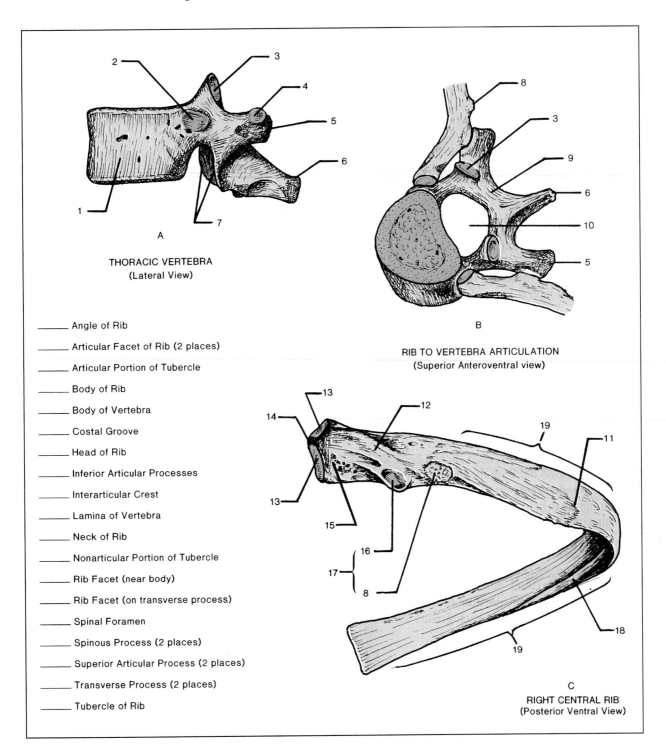

A

THORACIC VERTEBRA
(Lateral View)

B

RIB TO VERTEBRA ARTICULATION
(Superior Anteroventral view)

C

RIGHT CENTRAL RIB
(Posterior Ventral View)

_____ Angle of Rib

_____ Articular Facet of Rib (2 places)

_____ Articular Portion of Tubercle

_____ Body of Rib

_____ Body of Vertebra

_____ Costal Groove

_____ Head of Rib

_____ Inferior Articular Processes

_____ Interarticular Crest

_____ Lamina of Vertebra

_____ Neck of Rib

_____ Nonarticular Portion of Tubercle

_____ Rib Facet (near body)

_____ Rib Facet (on transverse process)

_____ Spinal Foramen

_____ Spinous Process (2 places)

_____ Superior Articular Process (2 places)

_____ Transverse Process (2 places)

_____ Tubercle of Rib

Figure 15.3   Rib anatomy and its articulation.

18

The Nerve Cells

The nervous system is composed of two kinds of cells: neurons and neuroglia. **Neurons** are functional units of the nervous system in that they transmit nerve impulses. **Neuroglial cells,** or **glia,** are connective, supportive, and nutritive; excitability is not characteristic. Although our prime concern here is with neurons, both kinds of nerve tissue will be studied.

The first portion of this exercise is descriptive and should be studied before examining prepared microscope slides in the laboratory. Figure 18.1 should be labeled and the questions on the Laboratory Report should be answered prior to laboratory studies.

As you study the neurons illustrated in figure 18.1, keep in mind that not all the structures shown in the illustration will be visible with your laboratory microscope. However, once you start studying microscope slides of the spinal cord and brain tissues, you will find that most of the cellular structures are visible to some extent.

Neurons

Because neurons perform different roles in different parts of the nervous system, they exist in a variety of sizes and configurations. Despite this diversity, all neurons have much in common. Figure 18.1 illustrates most structures that all neurons have in common.

The Cell Body and Its Processes

The large peripheral neurons illustrated in figure 18.1 are the principal neurons associated with skeletal muscle physiology. The cross section through the spinal cord illustrates the relationship of these neurons to the spinal cord and to each other. Three kinds of neurons are shown: sensory, motor and interneuron. These three neurons, together with the receptor and effector, make up a **reflex arc,** which is described more in detail in Exercise 41.

The cell body of a neuron is called a **perikaryon.** It usually contains a rather large nucleus surrounded by Nissl bodies and neurofibrils in the cytoplasm. **Nissl bodies** are dense aggregations of rough endoplasmic reticulum. They stain readily with basic aniline dyes such as toluidine blue or cresyl violet. They represent sites of protein synthesis. Some Nissl bodies are shown in the perikaryon of the motor neuron in figure 18.1.

Neurofibrils are slender protein filaments that extend throughout the cytoplasm parallel to the long axis. They resemble roadways that circumvent the Nissl bodies. It is significant that these minute fibers converge into the axon to form the core of the nerve fiber.

The most remarkable features of neurons are their cytoplasmic processes: the axons and dendrites. Traditional terminology once categorized these processes purely from a morphological sense; however, numerous exceptions require more accurate and universal definitions that emphasize their functional attributes.

A **dendrite,** or **dendritic zone,** functions in receiving signals from receptors or other neurons and plays an important part in integration of information.

An **axon,** on the other hand, is a single elongated extension of the cytoplasm that has the specialized function of transmitting impulses away from the dendritic zone.

The perikaryon can be located at any position along the conduction pathway. In the motor neuron of figure 18.1, the dendrites are the short branching processes of the parikaryon; the axon is the long single process. Although a neuron will usually have several dendrites, there will be only one axon. Some neurons, however, such as the *amacrine cells* of the retina, have no axon at all.

Fiber Construction

The neural fibers of all peripheral neurons are enclosed by a covering of **neurolemmocytes** (also called *Schwann cells*). This covering extends from the perikaryon to the peripheral termination of the fiber. In the larger peripheral neurons, such as those seen in figure 18.1, the neurolemmocytes are wrapped around the axon in a unique manner to form a myelin sheath of lipoprotein. The outer portion of the neurolemmocytes that contain nuclei and

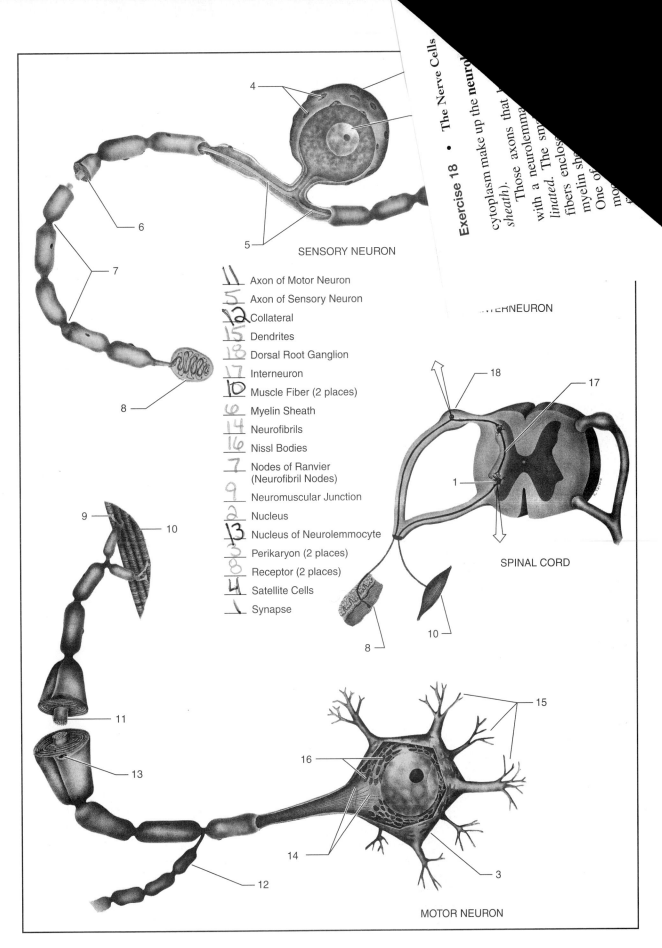

cytoplasm make up the **neuro**...
sheath). ...

Those axons that ...
with a neurolemma...
linated. The sm...
fibers enclose...
myelin she...
One of ...
mo...

SENSORY NEURON

11	Axon of Motor Neuron
5	Axon of Sensory Neuron
12	Collateral
15	Dendrites
18	Dorsal Root Ganglion
17	Interneuron
10	Muscle Fiber (2 places)
6	Myelin Sheath
14	Neurofibrils
16	Nissl Bodies
7	Nodes of Ranvier (Neurofibril Nodes)
9	Neuromuscular Junction
2	Nucleus
13	Nucleus of Neurolemmocyte
8	Perikaryon (2 places)
4	Receptor (2 places)
3	Satellite Cells
1	Synapse

INTERNEURON

SPINAL CORD

MOTOR NEURON

Figure 18.1 Sensory and motor neurons.

...lemma *(neurolemmal*

...have a thick myelin sheath ...are designated as being *mye-* ...aller peripheral neurons that have ...ed in neurolemmocytes, but lack the ...eath, are said to have *unmyelinated* fibers. ...f the most important functions of neurolem- ...cytes is to facilitate the regeneration of damaged fibers. When a fiber is damaged, regeneration occurs along the pathway formed by the neurolemmocytes.

The myelin sheath and neurolemma are formed during embryological development by the folding of the neurolemmocytes around the axons. The axon of the motor neuron in figure 18.1 reveals an enlarged cross section of the lipoprotein sheath and neurolemma.

Note in figure 18.1 that the myelin sheath and neurolemma are indented at regular intervals by **nodes of Ranvier** *(neurofibril nodes),* which are points of discontinuity between successive neu-rolemmocytes. Each internode consists of a single neurolemmocyte, and each node of Ranvier repre-sents a place where the exposed axon core lacks the myelin and neurolemma. The nodes play a signifi-cant role in facilitating the speed of nerve impulse transmission by *saltatory conduction,* a phenome-non to be studied in Exercise 44.

Types of Neurons

The variability in size, shape, and arborization of neurons is considerable. Neurons may be multipo-lar, bipolar, unipolar, or pseudounipolar. Some of them have perikarya as small as 4 μm, others may be as large as 150 μm. Some neurons have perikarya that are encapsulated with satellite cells; others are not encapsulated. Many have myelinated fibers; others do not. More detailed structural fea-tures of spinal cord and brain neurons follow.

Neurons of the Spinal Cord

As stated above, the principal neurons of the spinal cord are motor, sensory, and interneuron. Each type has unique characteristics.

Motor Neurons   As indicated in figure 18.1, these multipolar neurons have their perikarya locat-ed in the gray matter of the central nervous system (CNS). They are referred to as *efferent neurons* because they carry nerve impulses away from the CNS. The axons of these neurons are myelinated and terminate in skeletal muscle fibers. They func-tion in the maintenance of voluntary control over skeletal muscles.

The perikarya of these neurons are somewhat angular or star-shaped rather than oval. The cyto-plasm in a properly stained perikaryon reveals a nucleus, many Nissl bodies, and neurofibrils. Many short branching dendrites are present that make synaptic connections with axons of interneu-rons or sensory neurons. Illustrations C and D of figure HA-13 in the Histology Atlas reveal the appearance of these cells when observed under oil immersion optics.

The myelinated axons of these neurons often have **collaterals** emanating at right angles from a node of Ranvier. Collaterals are branches that pro-vide innervation to additional muscle fibers, other neurons, and even the same neuron in some instances.

Sensory Neurons   These neurons are also called *ganglion cells* because their perikarya are located in dorsal root ganglia outside of the CNS (see label 18, figure 18.1). Since they carry nerve impulses from a receptor toward the CNS, they are also often referred to as *afferent neurons.*

As illustrated in figure 18.1, the perikaryon of a sensory neuron is ovoid and is covered with a capsule of **satellite cells.** These cells have the same embryological origin and function as neurolemmo-cytes. Structurally, satellite cells resemble certain neuroglial cells (oligodendrocytes); however, the two types of cells do not have the same embryolog-ical origin. Illustration E, Figure HA-13 of the Histology Atlas reveals what satellite cells look like in a sectioned sensory ganglion.

Although sensory neurons appear to be unipo-lar, histologists classify them as *pseudounipolar* on the basis of their embryological development. Note that the entire fiber is myelinated from the receptor in the skin to the synaptic connection (synapse) in the spinal cord. The entire myelinat-ed fiber is designated as an **axon** even though part of it functions like a dendrite in carrying nerve impulses toward the cell body. A **synapse** (label 1, figure 18.1) is the point where the terminus of an axon of one neuron adjoins the dendritic zone of another neuron.

Interneurons   *(Associated* or *Internuncial)* These multipolar unmyelinated neurons exist in large numbers in the gray matter of the CNS. They are essentially the integrator cells between sensory and motor neurons. Most reflexes involve one or more interneurons between the sensory and motor neu-rons; however, some automatic responses, such as the knee-jerk reflex, are monosynaptic in that they lack interneurons.

Before going on to the study of the brain cells, label figure 18.1.

Neurons of the Brain

Four different types of neurons that are found in various parts of the brain are shown in figure 18.2. They are pyramidal, Purkinje, stellate, and basket cells.

Pyramidal Cells These multipolar neurons are the principal cells of the cortex of the cerebrum. The upper neuron in figure 18.2 is of this type. Figure HA-14 of the Histology Atlas has three photomicrographs of these cells that were made from three different slide preparations. The perikarya of these neurons are somewhat triangular to impart a pyramidlike shape.

A unique characteristic of these cells is that they have two separate sets of dendrites: (1) one long trunklike structure with many branches that ascends vertically in the cortex and (2) several basal branching dendrites that extend outward from the basal portion of the perikaryon. The single axon is distinguishable from the basal dendrites by its smoother surface.

Purkinje Cells The large multipolar neuron in the lower left-hand corner of figure 18.2 is a typical Purkinje cell. The perikarya of these neurons are located deep within the molecular layer of the cerebellum (see Figure HA-15, Histology Atlas). The large branching dendritic processes of these cells fill up most of the space in the molecular layer of the cerebellum. The axons of these neurons extend inward into the granule cell layer of the cerebellum.

Stellate Cells *(Golgi Type II)* These small neurons are found in both the cerebrum and cerebellum, where they act as linkage (association) cells for the pyramidal cells of the cerebrum and the Purkinje cells of the cerebellum. Both multipolar and bipolar types exist. See figures 18.2, HA-14, and HA-15.

Basket Cells These multipolar neurons are a variant of large stellate cells that are located deep in the molecular layer of the cerebellum in close association with the Purkinje cells (figures 18.2 and HA-15). They have short, thick, branching dendrites and a long axon. The axons of these neurons usually have five or six collaterals that make synaptic connections with dendrites of the Purkinje cells. Like the stellate cells, these cells function as association neurons.

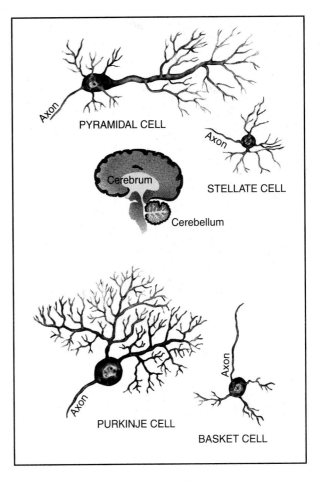

PYRAMIDAL CELL

Cerebrum

STELLATE CELL

Cerebellum

PURKINJE CELL

BASKET CELL

Figure 18.2 Neurons of the brain.

Neuroglia

Numerically, the neuroglial cells of the nervous system exceed neurons by a ratio of approximately 5:1. The term *neuroglia* (nerve glue) is applied to the ependyma, which lines the ventricles of the brain and spinal cord, and to the neuroglial cells that intermesh with neurons of the nervous system and retina.

The satellite cells and neurolemmocytes, which function much like neuroglial cells but have a different origin, are often referred to as *peripheral neuroglia* to set them apart.

Although neuroglial cells do not carry nerve impulses, they do perform many important functions within the CNS. First, during embryological development, they provide a framework in the CNS for young developing neurocytes. As the nervous system develops, the neuroglial cells also seem to acts as barriers to the formation of synaptic connections so that they control the development of functional reflex patterns in the CNS. They also appear to play a role as mediators for the normal metabolism of neurons, and they are involved in the degeneration and regeneration of nerve fibers.

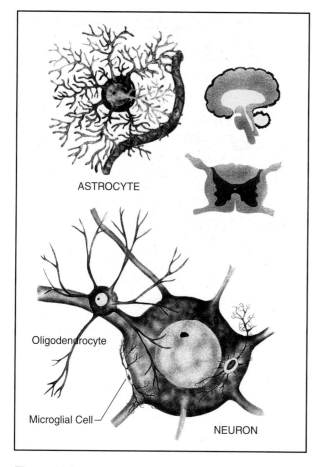

ASTROCYTE

Oligodendrocyte

Microglial Cell

NEURON

Figure 18.3 Neuroglial cells.

The neuroglia are divided into three distinct morphological categories: astrocytes, oligodendrocytes, and microglia. Oligodendrocytes and astrocytes, together, are often referred to as *macroglia.* Representatives of each group are illustrated in Figure 18.3.

Astrocytes Of the three types, these neuroglial cells have the largest cell bodies. There are two types: protoplasmic and fibrous astrocytes. The one shown in Figure 18.3 is of the protoplasmic variety.

Astrocytes may be attached to blood vessels, as shown in figure 18.3, or to the pia mater, or neurons, much like satellite cells. One known function of astrocytes is to sop up excess neurotransmitter molecules around neurons, thereby protecting neurons from receiving too much stimulation.

Oligodendrocytes Glial cells that have cell bodies somewhat smaller than astrocytes and have a close relationship to neurolemmocytes are called *oligodendrocytes.* These cells develop in rows along nerve fibers, where they produce myelin for myelinated nerve fibers. In tissue culture, these cells exhibit rhythmic pulsatile movements, which are not well understood.

Microglia These cells are much smaller than macroglial cells and exist in several different shapes. Two of them are seen attached to the perikaryon of a neuron in figure 18.3. The cytoplasmic processes on these cells are more twisted than those on macroglia. Microglia are scattered throughout the brain and spinal cord and have different functions at different stages in the development of the human body.

Before birth, the growing fetus generates many more neurons and glial cells than it needs. As unneeded cells die, young undifferentiated microglial cells, acting as phagocytes, remove the dead matter.

Once the sculpting of the nervous system is completed, microglia are transformed into resting stage cells. In this stage it appears that the cells produce growth factors needed by adjacent mature neurons and other glial cells.

While in the resting stage microglia have the capability to respond within minutes to damaging disturbances in their microenvironment and prepare to surround damaged neurons or other cells. If injury to neurons is severe, resulting in neuron death, the microglia begin to function as full-fledged phagocytes. When neuron damage is very mild, microglial cells tend to revert to the resting stage.

These immunological characteristics of microglia are very significant when you consider that leukocytes of the blood play no role in immunity for the CNS; in fact, leukocytes are considered to be harmful to cells of the CNS. Microscopic studies of healthy brain and spinal cord tissue rarely reveal the presence of white blood cells.

If microglia become infected or injured they can become damaging to surrounding nerve tissue and cause serious neurological disorders. Two examples of diseases that may be associated with injured microglia are AIDS and Alzheimer's disease.

Although the human immunodeficiency virus that causes AIDS does not affect neurons, it does invade microglia, causing them to produce inflammatory cytokines (cellular hormones) and other molecules that are toxic to neurons.

In Alzheimer's disease senile plaques form in brain tissue that contain beta amyloid (a protein fragment), microglia, astrocytes, and injured neurons. Some neurologists believe that the culprit is beta amyloid, which causes microglia to release substances harmful to neurons.

Microscopic Studies

After answering the questions on the Laboratory Report that pertain to cells of the nervous system examine prepared slides of the nervous system using oil immersion optics whenever necessary. The following procedure will involve most of the cells discussed in this exercise.

Materials:
prepared slides of:
  x.s. spinal cord (H10.341 or H10.342)
  spinal cord smear (H4.11)
  spinal ganglion (H4.25)
  cerebral cortex (H10.11 or H10.12)
  cerebellum (H10.23)

Spinal Cord   Examine the cross section of a rat spinal cord under low power and identify the structures labeled in figure HA-13. Study several of the large motor neurons in the gray matter with the high-dry objective.

Also, study a smear of ox spinal cord tissue and compare the cells with the photomicrograph in illustration D, figure HA-13.

Ganglion Cells   Study a slide of a section through a spinal ganglion and look for *satellite cells* that surround the sensory neurons. Refer to illustration E, figure HA-13.

Cerebral Cortex   Examine a slide of the cerebral cortex. Look for *pyramidal* and *stellate cells.* Refer to figure HA-14. Try to differentiate dendrites from axons on the pyramidal cells.

Note the presence of spiny processes called "gemmules" that are seen on the branches of the dendrites. These processes greatly increase the effective surface area of the dendrites, allowing large neurons to receive as many as 100,000 separate axon terminals, or synapses.

Cerebellum   Examine a slide of the cerebellum and look for *Purkinje cells, stellate cells,* and *basket cells.* Refer to figure HA-15.

Laboratory Report

Complete the first portion of Laboratory Report 18,19.

19

The Muscle Cells

Excitability in muscle cells manifests itself in **contractility.** This property is made possible by the presence of contractile protein fibers within the cells. The shortening of muscle cells, due to this property, is responsible for the various movements of appendages and organs of the body. Because of their elongated structure, muscle cells are often referred to as **muscle fibers.**

On the basis of cytoplasmic differences, there are two categories of muscle fibers: striated and smooth. **Striated muscle** fibers are characterized by regularly spaced transverse bands called *striae.* **Smooth muscle** fibers lack these transverse bands and have other distinctions to set them apart.

Striated muscle cells are further subdivided into skeletal and cardiac muscle. **Skeletal muscle** fibers are multinucleated, or *syncytial.* Attached primarily to the skeleton, they are responsible for voluntary body movements. **Cardiac muscle** cells, which make up the muscle of the heart wall, are not syncytial and have less-prominent striations than skeletal muscle.

Follow the same procedure for this exercise as for the last one: read over the text material and answer the questions on the Laboratory Report before entering the laboratory to study the various kinds of muscle tissue on prepared slides.

Smooth Muscle

Smooth muscle fibers are long, spindle-shaped cells as illustrated in figure 19.1. Each cell has a single elongated nucleus that usually has two or more nucleoli. When studied with electron microscopy, the cytoplasm reveals the presence of fine linear filaments of the same protein composition that exists in striated muscle. The fact that both smooth and striated fibers contain actin, myosin, troponin, and tropomyosin indicates that the chemistry of contraction is probably similar or identical in all types of muscle fibers.

Smooth muscle cells are found in the walls of blood vessels, walls of organs of the digestive tract, the urinary bladder, and other internal organs. They are innervated by the autonomic nervous system and, thus, are involuntarily controlled.

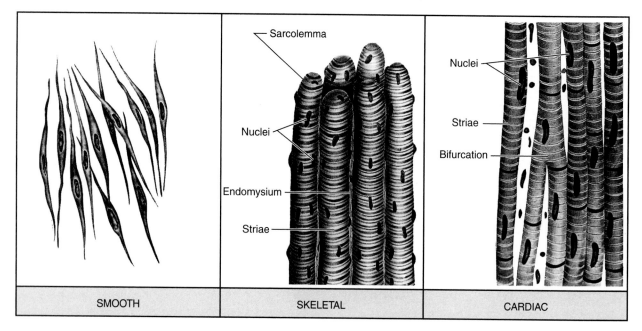

SMOOTH | SKELETAL | CARDIAC

Figure 19.1   Types of muscle tissue.

Skeletal Muscle

Skeletal muscle fibers are long, cylindrical, and multinucleated. Large numbers of these fibers are grouped together into bundles called **fasciculi.** The individual cells within each fascicle are separated from each other by a thin layer of connective tissue called **endomysium.** This separateness of muscle fibers enables each cell to respond independently to nerve stimuli.

Examination of the cytoplasm, or **sarcoplasm,** with a light microscope reveals a series of distinct dark and light **striae** (cross-stripes). The detailed structure of these bands will be explored when we study the chemistry of muscle contraction in Exercise 27. Surrounding each cell is a cell membrane called the **sarcolemma.** Note in figure 19.1 that the nuclei of each cell lie in close association with the sarcolemma.

Although striated muscle is referred to as being "skeletal" and "voluntary," there are certain striated muscles that are not attached to bone or subject to voluntary control. Two examples: (1) striated muscles of the tongue and external anal sphincter have no skeletal attachment, and (2) striated muscles of the pharynx and upper esophagus have no skeletal attachment and are not voluntarily controlled. Despite these differences, these aberrational examples are morphologically indistinguishable from other skeletal muscle.

Cardiac Muscle

Cardiac muscle cells are found in the wall of the heart and in the walls of the venae cavae, where they enter the right atrium of the heart. A unique characteristic of cardiac muscle is its ability to contract rhythmically and continuously as a result of intrinsic cellular activity.

Morphologically, cardiac muscle is readily distinguished from skeletal and smooth fibers, yet it shares some characteristics of each type. The principal distinguishing characteristics of cardiac muscle fibers are as follows:

- Instead of being anatomically syncytial, cardiac muscle is made up of separate cellular units that are separated from each other by **intercalated disks** (cell membranes). These disks stain somewhat darker than transverse striae.
- Some fibers of cardiac cells are bifurcated to form a branched three-dimensional network instead of forming only long straight cylinders.

- The elongated nuclei of each cellular unit lie deep within the cells instead of near the surface as in skeletal muscle.
- Cardiac tissue is not normally subject to voluntary control.

Laboratory Assignment

After answering the questions on the Laboratory Report that pertain to the characteristics of the various types of muscle tissue, examine prepared slides to identify the various kinds of cells.

Materials:
prepared slides of:

    striated muscle, teased (H3.11)
    striated muscle, x.s. and l.s. (H3.13)
    smooth muscle, teased (H3.21)
    muscle to tendon (H7.13)
    smooth muscle, sectioned (H3.22)
    cardiac muscle, teased (H3.31)
    cardiac muscle, l.s. (H3.32 or H3.33)

Skeletal Muscle   Scan a slide of striated muscle tissue with low power to find a muscle fiber with the most pronounced striae. Increase the magnification to 1000× (oil immersion) and see if you can differentiate the A and I bands.

Refer to illustration A, figure HA-7. Note the position of the nuclei in these fibers. Also examine a cross section of this tissue to identify the endomysium.

If a slide of muscle to tendon (H7.13) is available examine it under high-dry magnification and refer to illustration C, figure HA-7.

Cardiac Muscle   Scan a slide of this tissue with low power, looking for pronounced striae and intercalated disks. Increase the magnification to 1000× to observe the nature of the striae and disks. Can you find a place on the slide where bifurcation of the fibers is seen? Consult illustration A, figure HA-8, of the Histology Atlas.

Smooth Muscle   Study slides of this type of muscle tissue in the same manner as above. Finding individual fibers, separated from others, is not always as easy as it might seem. Compare your observations with illustrations B and C in figure HA-8.

Laboratory Report

Complete the last portion of combined Laboratory Report 18,19.

mechanism appears to be a compound, most likely a protein, that is embedded in the presynaptic membrane. The protein may act as a valve, enabling ACh to pass through when the action potential reaches the area.

This raises the question of the role of the synaptic vesicles. According to Dunant and Israel, it appears that these vesicles do two things: (1) they release stored ACh only when the cytoplasmic pool of the neurotransmitter has been exhausted by sustained nervous activity, and (2) they accumulate calcium ions that assist in the termination of the release of ACh.

Once the action potential is initiated in the sarcolemma of the muscle fiber, the following events occur within the muscle fiber:

- The action potential moves inward to the **T-tubule system** (label 5).
- Calcium ions, which permit contraction of the myofibrils, are released by the **sarcoplasmic reticulum** (color code: blue outside, green inside).
- ACh is inactivated by the enzyme *acetylcholinesterase* within 1/500 of a second.
- All energy for muscle fiber contraction is supplied by ATP from mitochondria in the area.

The action of acetylcholinesterase on ACh is to break it down into acetate and choline. This enzyme is found on the surfaces of the presynaptic and post-synaptic membranes as well as in the ground substance of the synaptic cleft.

Once the choline and acetate diffuse back into the axon terminal they are resynthesized back into ACh. The enzyme that catalyzes the synthesis of ACh is *choline acetyltransferase*. This enzyme is produced in the perikaryon of the neuron and travels down microtubules of the axon to the axon terminal, where it is stored in the cytoplasm.

These events, that depict how the nerve impulse bridges the gap from the axon terminal to a skeletal muscle fiber, resemble nerve to nerve transmission, and nerve to secretory cell, as well. One thing to keep in mind about neuron to neuron transmission in the brain, however, is that there are many neurotransmitters other than acetylcholine that bridge the synaptic clefts between different kinds of neurons.

Reference:

Dunant, Y., and M. Israel. The release of acetylcholine. *Scientific American* 252(4) (April 1985).

Laboratory Assignment

After labeling figure 20.1 examine a microscope slide of motor nerve endings (H4.31) under low-power and high-dry magnification. Consult illustration D, figure HA-7, for reference. Complete the Laboratory Report for this exercise.

PART 5 Instrumentation

The study of physiological phenomena, such as muscle contraction, heart action, blood pressure, and cerebral activity, requires the use of electronic instrumentation. An understanding of the nature and function of this type of equipment is essential in the laboratory. This unit on instrumentation is intended as an introduction to the principal types of equipment that are available. It is anticipated that the student will complete the Laboratory Report for the exercises in this unit before performing any of the experiments in Part 6.

The illustration on this page reveals the arrangement of equipment that might be used in a typical instrumentation setup. First, one must use a device, or combination of devices, to pick up the biological signal that is being studied. This will be an **electrode** if the signal is electrical, as in neural activity, or an **input transducer** if the signal is nonelectrical, as in temperature changes. Since biological phenomena often produce weak electrical signals, they are then fed into an **amplifier** for amplification. Finally, the amplified signals are converted to some form of visual or audio display that can be observed and quantified by the experimenter. The **output transducer** that provides this display may be an electrical meter, oscilloscope, loudspeaker, chart recorder, or computer monitor.

This unit consists of six exercises. Each exercise pertains to a particular type of equipment that will be used in different experiments. It is not anticipated that all the information on these twenty-two pages will be assimilated in one reading; however, the material should be studied carefully so that questions on the combined Laboratory Report 21–26 can be answered. The questions on this brief report are more or less of a general nature. They do not pertain to all the specific minutiae of all types of equipment. The principal value of this unit will be as a source of reference for the equipment that is encountered in various experiments.

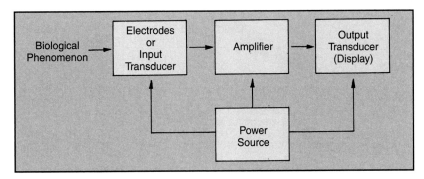

21 Electrodes and Input Transducers

Because electrodes and input transducers serve the same function in an experimental setup, we will study them here together. As indicated on the previous page, **electrodes** are used to pick up electrical signals, and **input transducers** are used to convert nonelectrical stimuli, such as heat, light, sound, or force, to an electrical signal that can be managed by the electronic equipment. Another type of electrode, the **stimulating electrode,** will also be studied here.

Before we get into the details of these three electrical components, let's review some of the characteristics of an electrical current. First of all, electricity flowing through a wire is the movement of free electrons from atom to atom of the conductor. The direction of movement is from the negative end of the wire, where there is a surplus of electrons, to the positive end. The terminal of a battery that supplies these electrons is the negative terminal and is called the *cathode.* The other terminal, which is positive, is the *anode.*

From the very early days of electrical science, electrical current has been considered to flow from the positive terminal of a power source to the negative terminal. However, since we know now that electrons move from the negative terminal to the positive one, the flow of electrons is opposite to the conventional concept of current direction; thus, it becomes necessary to distinguish one direction from the other by name. It is, therefore, customary to speak of *"electron flow" as taking place from negative to positive, and "current flow" from positive to negative.*

Another concept to keep in mind is the difference between direct current (DC) and alternating current (AC). In DC the electrons flow in only one direction (fixed polarity). This type of electrical energy is preferred for stimulating muscle and nerve preparations. With AC current the electrons flow in one direction for a brief interval and then reverse their direction in the next interval. These reversals of direction take place many times per second. The undulating nature of AC produces undesirable effects such as heating and accommodation. A problem that can arise with DC current is that prolonged stimulation of tissues can result in tissue damage due to hydrolysis. A more in-depth study of AC and DC currents will be encountered in Exercise 25, where the electronic stimulator is discussed.

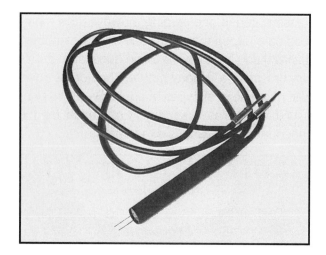

Figure 21.1 This type of electrode is suitable for stimulating muscles or nerves with electrical pulses to elicit desired responses.

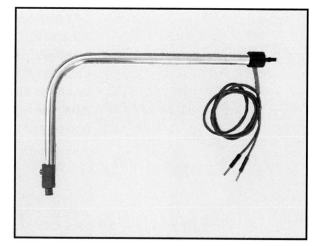

Figure 21.2 This type of electrode, called a Comb-electrode, is used for stimulating small portions of muscle tissue in certain types of experiments.

Stimulating Electrodes

When it is desirable to initiate a nerve impulse in a nerve fiber, or to cause a muscle to contract by direct stimulation, one must use a *stimulating electrode*. Many different kinds of electrodes are available for stimulating nerves and muscles. Two types are shown in figures 21.1 and 21.2. In addition, there are "needle," "pin," and "sleeve" electrodes, which have different applications in nerve and muscle physiology experiments.

The type of electrode used will depend on such factors as electrical conductivity, toxicity to tissues, size, chemical properties, ease of handling, and cost. Metals such as platinum, palladium, gold, and nickel-silver alloys are most frequently used because of their relative inertness and good conductivity. All stimulating electrodes have two contacts: a negative cathode and a positive anode. The wire ends of these electrodes usually have phone jack terminals that are easily connected to an electronic stimulator.

Pickup Type Electrodes

Electrodes that pick up electrical signals to be amplified and displayed exist in many configurations. Figures 21.3 and 21.4 reveal how diverse they may be. Note how small an EEG disk electrode (figure 21.3) is compared with the large plate-type electrode that is used for electrocardiograms (figure 21.4).

The greatest problem with electrodes that are used for picking up electrical signals through the skin is achieving maximum skin contact. The resistance of the cornified outer layer of the skin to low-voltage electrical transmission may be considerable. To overcome this resistance, it is often necessary to rub off some of the outer dead skin cells with an abrasive material and place a high-conductance solution between the skin and electrode. Such a solution may be 0.05N saline or a commercially prepared paste or jelly.

Input Transducers

As previously stated, *input transducers* are used to convert nonelectrical stimuli such as light, heat, pressure, sound, or movement to electrical signals that can be amplified and studied by an electronic setup. Examples of three transducers are illustrated in figures 21.5, 21.7, and 21.8.

Force Transducers

The conversion of muscular movements to electrical signals can be accomplished with various kinds of electromechanical (force) transducers. The Trans-Med/Biocom Model 1030 force transducer shown in figure 21.5 is the type that is used with Gilson chart recorders as illustrated in figure 29.1.

The Model 1030 is essentially a strain gage consisting of five steel leaf springs. It can measure muscle contraction or similar forces from 10 mg to 10 kg, which gives it a sensitivity range of an incredible 1:1,000,000. When small forces are to be measured, only the fixed single leaf is used. For loads approaching 10 kg, all five leaves are used; obviously, the number of leaves used depends on the size of the load within this range.

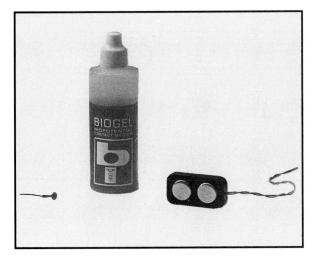

Figure 21.3 The small disk-type electrode on the left is for EEG measurements. The one on the right is a skin conduction electrode used for measuring galvanic skin response. The electrode jelly is used to reduce skin electrical resistance.

Figure 21.4 Two different types of electrodes that can be used for ECG recording. The one on the left is a suction type that establishes good skin contact on the chest. The plate type on the right is held in place with rubber straps.

This type of transducer functions through a series of resistors arranged as a *Wheatstone bridge* (figure 21.6). In a Wheatstone bridge there are four resistors (R_1, R_2, R_3, and R_4) connected to an electrical source (E) and a galvanometer (G). Electrical current flows through two parallel circuits or half bridges ($R_1 + R_2$ and $R_3 + R_4$). If the voltage (potential) at A and B are equal, no current will flow through the galvanometer and the bridge is said to be *balanced*. This condition can be expressed as

$$\frac{R_1}{R_2} = \frac{R_3}{R_4}$$

If all the resistors are constant except for R_1, which is variable, a change in R_1 will upset the balance and allow current to flow through the galvanometer.

In this type of strain gage transducer a silicon element acts as a variable resistor (R_1). It is a tubular component connected to the spring leaves. As deformation occurs in the leaves due to loading, proportional bending also occurs in the silicon resistor. This induced bending increases the electrical resistance in the element, causing the Wheatstone bridge to become unbalanced. When the bridge is unbalanced, electrical current flows from the transducer to produce a signal. The signal will be proportional to the stress applied. Although there are other factors, such as temperature adjustments, that are considered in the design and manufacture of these transducers, we will omit the embellishments here for reasons of clarity.

When a strain gage transducer is used it is usually necessary to balance the transducer first by making certain control adjustments on the output transducer (chart recorder). Specific instructions for balancing are outlined in Appendix B.

Pulse Transducers

Several different kinds of transducers have been developed for monitoring the human pulse. Figure 21.7 illustrates, diagrammatically, how a photoelectric plethysmograph functions. Note that a small light source produces a beam of light that is directed upward into the tissues of the finger. A few millimeters away from the light source is a photosensitive resistor (photoresistor). When a surge of blood passes through the pad of the finger, the light beam is disrupted and scattered, causing rays to be reflected to the photoresistor. Light striking the photoresistor causes a signal to be produced, which is fed to the amplifier and output transducer.

In addition to determining the pulse rate, this setup can be used to detect pulse pressure differences. Since this type of transducer lacks a Wheatstone bridge, no balancing of the transducer is required.

Other Types

There are several other types of transducers available for physiological work. For example, figure 21.8 illustrates a transducer that is able to convert air pressure differentials into electrical signals. Transducers are also available to detect fluid pressure differentials.

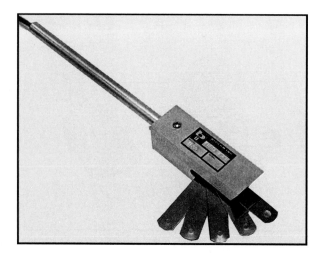

Figure 21.5 A strain gage force transducer, such as this one, utilizes a tubular silicon element that functions as the R_1 component in the Wheatstone bridge shown to the right in figure 21.6. Bending changes its electrical resistance.

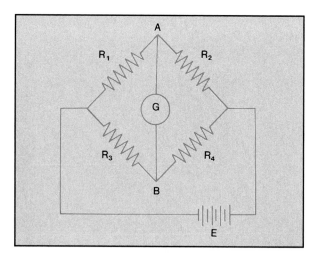

Figure 21.6 When this Wheatstone bridge is balanced, the electrical potentials at points A and B are equal and no current flows through the circuit. Bending in the leaves at R_1 causes unbalancing, and an electrical signal is produced.

A transducer that monitors temperature changes is called a **thermistor.** The fact that the flow of electricity through a wire changes with temperature changes makes it possible for thermistors to function.

A transducer that can convert sounds to an electrical signal is known to all: the **microphone.** Several of these various types will be encountered in different experiments in this manual.

Laboratory Report

Answer the questions on Laboratory Report 21–26 that pertain to electrodes.

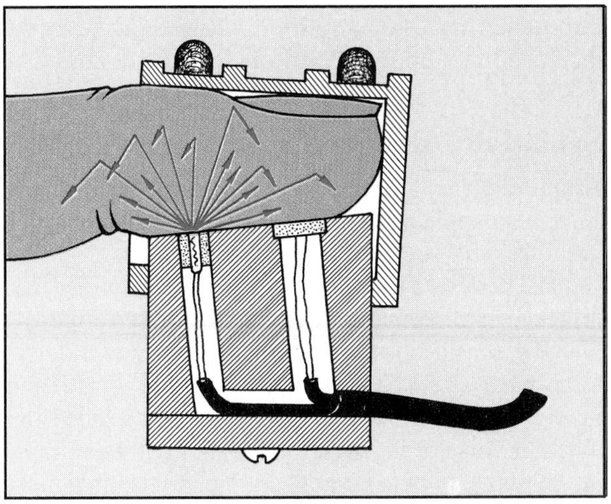

Figure 21.7 A pulse transducer uses light rays and a photoresistor to produce a signal.

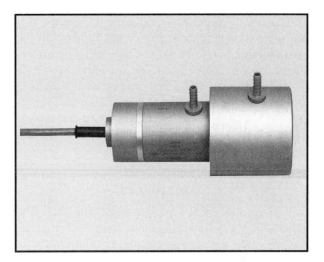

Figure 21.8 This transducer can produce electrical signals from air pressure differentials.

22 Amplifiers, Panel Meters, and Oscilloscopes

This exercise pertains to three items that are not used as independent components in any of the experiments in this book, yet they must be understood in the event that they are encountered in your laboratory work. No attempt will be made here to outline procedural protocol, however. If any of these components are used in an experiment, your instructor will provide supplemental instructions.

Amplifiers

Unless the weak electrical signals of biological systems are greatly increased many times (often a million or more), it is impossible to monitor them. Electronic amplifiers accomplish this task. Although none of the experiments in this book utilize separate amplifier units, such as the ones shown in figure 22.1, all chart recorders that will be used have amplifier units incorporated into them.

Amplifiers are usually designed to function within a limited range. The signal received by the unit (input signal) must have sufficient intensity to overcome the "electrical noise" of the amplifier's electronic circuitry, yet not be so great that it overloads and distorts the amplifier output. It is very important that the amplifier allow signals to be increased in amplitude only, and not be distorted.

Sensitivity Range of Amplifiers

Amplifiers vary in the degree of *sensitivity* to match particular applications. Whereas an ECG amplifier has to be very sensitive to pick up the weak electrical signals generated by cardiac activity, an amplifier that drives a chart pen need not be nearly as sensitive.

The sensitivity range of an amplifier will determine what size of input signal it can accept. If the input signal is so weak that it does not exceed the inherent noise of its circuitry, it is not within its minimum range. On the other hand, the input signal must not be so great that it distorts the amplifier output. The range of frequency in which an amplifier handles input without distortion is called its *linearity*. This information is important to the operator because it determines the fidelity of readout in the final analysis.

Power Supply

Amplifiers need direct current (DC) to function. Since their external source of electricity is usually conventional 110–120V AC, they must have a

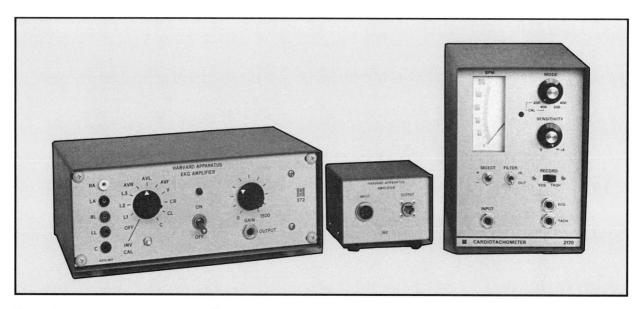

Figure 22.1 Amplifiers and amplifier-meter components.

power supply unit that converts the AC to DC. In addition to making this conversion, the power supply unit either increases or decreases the voltage required by the amplifier. Since present-day amplifiers are completely transistorized, the voltage requirements are quite low. This results in physically smaller power supply units that are usually an integral part of the amplifier unit.

Controls

The function of an amplifier will determine the types of controls that are needed. If the incoming signal remains constant, it may lack controls and be a very simple unit, such as the central unit in figure 22.1. Most amplifiers, however, will have "mode" and "gain" controls. If the amplifier is to be used in conjunction with a chart recorder, it may also have "zero offset" (centering) and calibration controls.

Mode A mode switch on an amplifier allows the instrument to accommodate either AC or DC input signals. When set at the AC position, a capacitor in the circuitry prevents DC signals from entering the amplifier; only AC signals pass through. In the DC position, both alternating and direct current signals are amplified. If a ground position is present, it is used to ground the input signal during adjustments of the connections at the input terminals. Grounding the amplifier input prevents severe deflection of recording instruments during these manipulations.

Gain A gain control is necessary to adjust the sensitivity of the amplifier to the strength of the incoming signal. If the input signal is strong, the dial should be turned to a low value. Weak signals require greater values. When the intensity, or *level*, of the input signal is unknown, it is best to set the gain control at the least sensitive position. The final sensitivity setting will be determined by advancing the control until a usable output signal is obtained.

Zero Offset A zero offset, or *centering*, control is used to move the zero line on the chart in either a positive or negative direction. This control permits the expansion of phenomena that are predominantly all positive or all negative by moving the pen away from the limits of its travel (see figure 22.2).

Electrical Interference

Electrical currents flowing through power cables, electrical appliances, and lighting systems generate magnetic and electrostatic fields that can cause unwanted interference in amplifier units. To prevent this type of electrical interference in electronic sys-

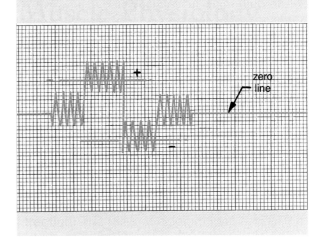

Figure 22.2 Using zero offset, or centering, control causes a shift in position of tracing.

tems, all interconnecting cables are *shielded*. This shielding usually consists of braided metallic coverings between the outer and inner insulation coatings, which are grounded to carry away unwanted currents or voltages. Additional methods for minimizing electrical interference will be outlined in various experiments where the problem becomes more acute.

Panel Meters

Electrical meters that are used to monitor signals are output transducers based on the principle of the *galvonometer* (figure 22.3). Such a meter is essentially a coil of wire that is mounted on a pair of pivots within the field of a permanent magnet. When a current passes through the coil, a magnetic field is generated by the coil. Since like poles repel and unlike poles attract, the coil is deflected against a spring in proportion to the current flowing through it. A pointer needle that is attached to the moving coil records

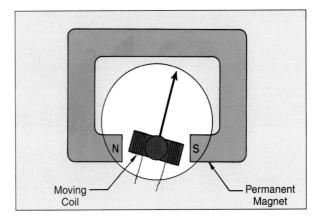

Figure 22.3 Current flowing through the magnet of the coil causes deflection of the needle.

the degree of angular displacement on a calibrated dial card. The meter on the cardiotachometer (right-hand unit, figure 22.1) is based on this principle.

This type of meter is readily converted to serve as an ammeter or voltmeter. In the case of an **ammeter,** which measures currents in amperes, an electrical bypass, or shunt, is added to diminish the amount of current that passes through the coil. In the **voltmeter,** which measures potential differences in volts, a resistor is added in series with the coil to allow only a small current to pass through the coil.

The Oscilloscope

The cathode ray oscilloscope is an output transducer that visually displays voltage signals on a fluorescent screen. A Tektronix dual-beam oscilloscope is shown in figure 22.5. The schematic diagram of an oscilloscope (figure 22.4), reveals how one of these instruments works.

Note that the electrical potential difference (voltage) of the input signal is expressed vertically and the time base is shown horizontally on the screen. This is achieved by the generation of a stream of electrons by an electron gun *(cathode)* at the back end of the tube. The electrons pass through a hole in an *anode* on their way to the fluorescent screen, where they produce a tiny luminous spot.

The position of the spot on the screen is determined by the electrostatic influence of two vertical (X) plates and two horizontal (Y) plates that lie alongside the beam pathway. Vertical movement of the beam (and spot) is controlled by the Y

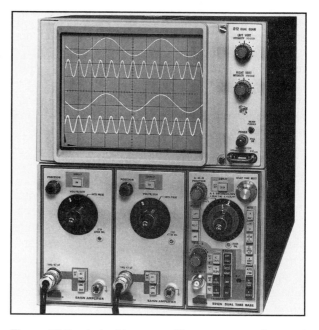

Figure 22.5 A dual-beam oscilloscope, manufactured by Tektronix, works well for two-channel experiments.

plates. These plates are energized by the input signal and represent the **voltage** of the signal. Horizontal movement, or **sweep,** of the beam is controlled by the X plates to establish the time base.

Laboratory Report

Answer the questions on Laboratory Report 21–26 that pertain to amplifiers, panel meters, and oscilloscopes.

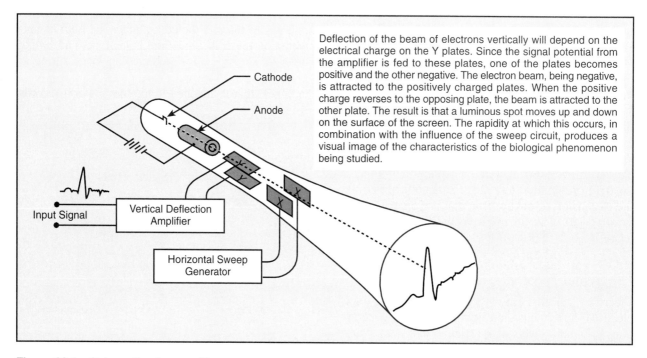

Deflection of the beam of electrons vertically will depend on the electrical charge on the Y plates. Since the signal potential from the amplifier is fed to these plates, one of the plates becomes positive and the other negative. The electron beam, being negative, is attracted to the positively charged plates. When the positive charge reverses to the opposing plate, the beam is attracted to the other plate. The result is that a luminous spot moves up and down on the surface of the screen. The rapidity at which this occurs, in combination with the influence of the sweep circuit, produces a visual image of the characteristics of the biological phenomenon being studied.

Cathode

Anode

Input Signal

Vertical Deflection Amplifier

Horizontal Sweep Generator

Figure 22.4 Schematic of an oscilloscope.

Gilson Chart Recorders

Of all the output transducers that are available in the physiology laboratory, the chart recorder is by far the one most widely used. Gilson Medical Electronics of Middleton, Wisconsin, manufactures and distributes three recorders that will be described here. They are the Unigraph, Duograph, and $\overline{5}/\underline{6}$H polygraph.

The Unigraph

Most single-event physiological experiments performed in this manual utilize the Gilson Unigraph. This unit is extensively used here because it is compact, rugged, versatile, and relatively easy to learn how to use.

The Unigraph can be used for studying skeletal muscle contraction, electrocardiograms, electroencephalograms, blood pressure variations, and many other activities. The physiological event is record-ed on calibrated heat-sensitive paper by a heated stylus; the paper moves along at either 2.5 mm or 25 mm per second.

Controls

Before attempting to use one of these instruments, one should be familiar with the following controls.

Mode Selection Control This control knob can be set at one of the six following settings: EEG, ECG, CC–Cal, DC–Cal, DC, and Trans. Refer to figure 23.2 to locate this knob.

If one is to make an electrocardiogram, the ECG setting will be used; for an electroencephalogram, the EEG setting is selected. The Trans setting is used when one uses a transducer. The Cal settings are used when it is necessary to calibrate a recording, as is true for doing ECGs.

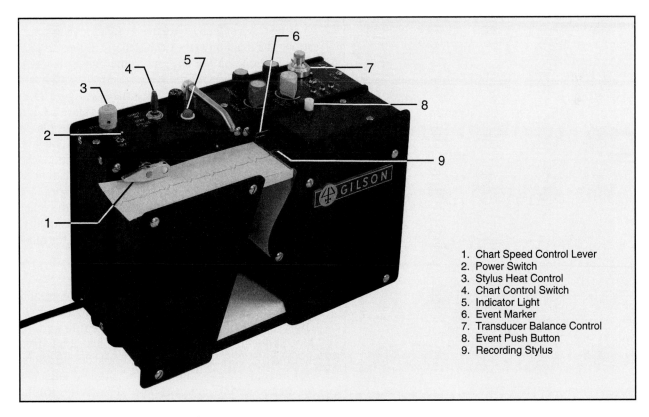

1. Chart Speed Control Lever
2. Power Switch
3. Stylus Heat Control
4. Chart Control Switch
5. Indicator Light
6. Event Marker
7. Transducer Balance Control
8. Event Push Button
9. Recording Stylus

Figure 23.1 The Gilson Unigraph.

Styluses The Unigraph has two styluses: a **recording stylus** and an **event marker stylus.** Labels 9 and 6, respectively, in figure 23.1 show where they are located. Figure 23.2 illustrates them more clearly.

Since stylus temperature is critical to produce a desired "trace" on the chart, there is a **stylus heat control** (label 3, figure 23.1) to regulate it. This control is usually set at the *two o'clock position* at the start of a recording. If the recording stylus is not warm enough, the tracing will be too faint. Turning the knob clockwise increases the temperature; counterclockwise movement will reduce it.

Whereas the recording stylus is used to record the amplified input signal, the event marker is used to record when an event begins, or when some new pertinent influence affects the recorded event. The event may be recorded manually by depressing the white event push button (label 8, figure 23.1) at the beginning of the event, or it may be recorded automatically if an event synchronization cable is used between the electronic stimulator and the Unigraph.

Centering Control *(Zero Offset Control)* This control, labeled CENTERING, is seen in figure 23.2 on the left side of the panel. It is used for moving the baseline of the recording stylus to the most desirable position. It should be kept in mind that the most desirable location of the baseline is not always the center of the paper; it may be near one of the edges.

Chart Speed The speed at which the paper moves along is controlled by the **speed control lever** (label 1, figure 23.1). When pointing to the left, as in figure 23.1, it is set at the faster speed of 25 mm per second. When positioned in the opposite direction (rotated backward 180°) it is set at the slower speed of 2.5 mm per second. For the chart to move at all, however, the chart control switch has to be in the Chart On position.

Toggle Switches The Unigraph has four toggle switches, two of which are extensively used in its operation. The main **power switch** (label 2, figure 23.1) is the small one on the left end of the unit. It has two settings: ON and OFF.

The large gray plastic switch (label 4, figure 23.1) is the **chart control switch.** As seen in figure 23.2, it has three settings: STBY, Chart On, and Stylus On. In the Stylus On position, the stylus heats up, but the chart does not move. At the Chart On position, the chart moves, and the recording stylus is activated (heated).

The other two small toggle switches at the other end of the unit come into play only occasionally. They should be set at the NORM positions for most experiments.

Sensitivity Controls Sensitivity of the Unigraph is controlled by the gain control and the sensitivity knob. The **gain control** is calibrated in MV/CM (millivolts per centimeter) and it has five settings: 2, 1, 0.5, 0.2, and 0.1. Look for this control in figure 23.2; note that it is situated adjacent to the centering control. The **sensitivity knob,** which is adjacent to the mode control, increases the sensitivity when it is turned clockwise.

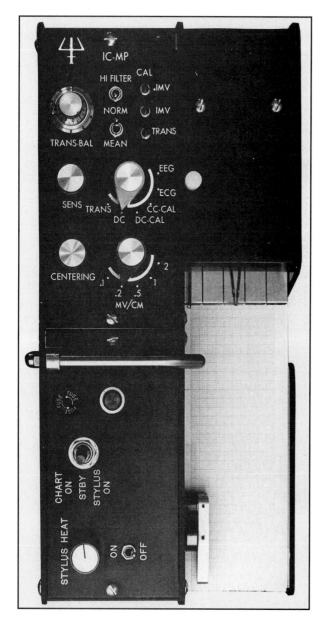

Figure 23.2 Unigraph controls.

Unigraph Operational Procedures

When setting up the Unigraph in an experiment the following points should be kept in mind.

Input   The cable end of electrodes or the transducer must be inserted into an input receptacle in the end of the Unigraph. The cable end must be compatible with the Unigraph. For some experiments it will be necessary to use an adapter unit between the cable end and the Unigraph.

Synchronization Cable   If an electronic stimulator is to be used, it is desirable to have an event synchronization cable between the stimulator and the Unigraph.

Power Source   The end of the power cord on this instrument has three prongs. This indicates that it must be used only in a 110V *grounded* outlet. Before plugging it in, however, the power switch should be OFF and the chart control switch at STBY.

Stylus Heat   Once the unit is plugged in, the power should be turned ON and the stylus heat control set at the two o'clock position.

After allowing a few minutes for the stylus to heat up, the trace should be tested by putting the chart control switch in the Chart On mode. If the trace line on the chart is too light, the darkness can be increased by turning the stylus heat control knob clockwise. This testing should be done with the chart speed set at the slow rate.

Balancing   If a force transducer is going to be used it will be necessary to balance the Wheatstone bridge in the circuitry, using the procedure outlined in Appendix B.

Calibration   When monitoring EEG, ECG, or EMG experiments it will be necessary to follow calibration instructions that are provided in Appendix B.

Mode Control   Before running the experiment, be sure to select the proper mode.

Sensitivity Control   Start all experiments at the lowest sensitivity setting, unless instructed otherwise. Sensitivity level can be increased gradually to arrive at the best level.

End of Experiment   When finished with the experiment always return all controls to their least sensitive settings, or to the OFF position. The chart control switch should be left at STBY.

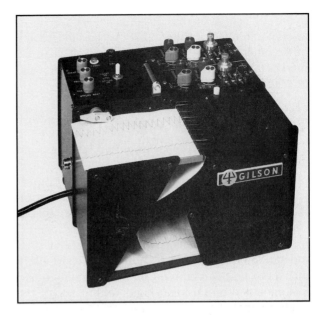

Figure 23.3   The Duograph.

The Duograph

The Gilson Duograph, figure 23.3, is a two-channel recorder that consists of two multipurpose amplifier modules. The modules (IC-MPs) are identical to the unit used in the Unigraph. This chart recorder enables one to monitor the interaction of two parameters simultaneously. Although it is heavier than the Unigraph (25 vs. 18 pounds), it is still extremely portable and can be carried with one hand. The chart is 127 mm wide and has two speeds (2.5 and 25 mm/sec) on the standard unit. An option of ten speeds is also available. Since the controls on the recorder are essentially the same as those on the Unigraph, learning to use one makes it easy to use the other.

The Gilson Polygraph

Recorders that can monitor three or more events simultaneously are called *polygraphs*. Gilson's polygraph is the $\overline{5/6}$H5 which is illustrated in figure 23.4. To facilitate accessibility, the five amplifier modules are mounted on a cantilevered unit above the chart. Each amplifier module can be easily removed to be replaced with a different one. This feature greatly enhances the versatility of the recorder.

The chart on the $\overline{5/6}$H is 300 mm wide and can be moved at ten different speeds: 0.1, 0.25, 0.5, 1.0, 2.5, 5, 10, 25, 50, and 100 mm per second. Five heated styluses provide individual tracings, and each one can be deactivated with a stylus lifter. A sixth stylus is also available if a servo unit is installed in the sixth space on the control panel. The stylus of a servo unit traces a pattern that overlaps the tracing of the other styluses. This polygraph also has a time marker that produces marks on the chart at one- and ten-second intervals.

Types of Amplifier Modules

There are six galvo-type amplifier modules and three servo modules available for the Gilson 5/6 H polygraph. Each **galvo module** has a rapid-response stylus that has a two-inch excursion on the chart. These units occupy the five spaces shown in figure 23.4 on the cantilevered control panel. The **servo channels** have a stylus that moves slowly over the full width of the chart. Only one servo unit can be used on the instrument, and it is mounted on the right-hand sixth space of the control panel. There is no servo unit shown in figure 23.4. A brief description of most of these modules follows.

IC-MP Multipurpose Module This is the same amplifier module that is utilized in the Unigraph and Duograph. It can be used for monitoring electrocardiograms (ECG), electroencephalograms (EEG), electromyograms (EMG), strain gage measurements, temperature changes, respiration, pulse, blood pressure, and so on.

IC-UM Multipurpose Module This unit is similar in applications to the IC-MP but differs in that it has a greatly increased sensitivity range. This module is considered to be a much more sophisticated unit than the IC-MP and is a finer research tool, particularly for recording EEGs.

IC-CC Electrocardiograph Module This module is intended primarily for electrocardiograms and electroencephalograms. It is equipped with a conventional ten-position lead switch that enables one to monitor ECGs from a standard five-lead patient cable.

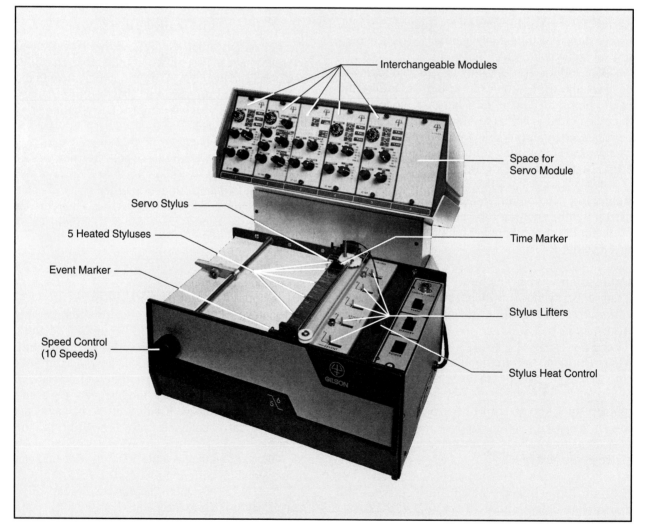

Figure 23.4 The Gilson polygraph.

IC-CT Cardiotachometer Module   This amplifier module is used to monitor the heart or pulse rate with a photoelectric finger pulse pickup like the one in figure 21.7.

IC-EMG Electromyograph Module   This unit is designed specifically to monitor bioelectrical signals produced by muscle activity. It is used with a three-lead patient cable that has two red leads and a single black one. The red leads are attached to electrodes on the belly of the muscle, and the black lead is attached to an electrode that acts as a ground. The electrode on the ground lead is attached to a point some distance away from the two red leads.

IC-S DC Servo Channel   This servo module records relatively slowly changing phenomena, such as temperature (via a thermistor probe), skin resistance, and other low-frequency DC data. It can also be used with pH, CO_2 and O_2 analyzers. Stylus excursion is 200 mm.

BG-5 Blood Gas Servo Channel   This unit records pCO_2, pO_2, pH, or H_2 directly from conventional electrodes that require very small samples. Stylus excursion is 200 mm.

Operation   To set up a multichannel experiment on the polygraph, one simply attaches the electrodes or transducers to the subject and connects the cables to the proper modules. For certain types of measurements, it is necessary to calibrate. Before any polygraph experiments are performed in this manual, however, you will have had considerable experience working with the Unigraph. The calibration procedures followed in using the Unigraph will be the same for the IC-MP module on the polygraph.

Laboratory Report

Answer the questions on Laboratory Report 21–26 that pertain to this exercise.

24

The Narco Physiograph®

If your laboratory is equipped with Narco Physiographs® instead of Gilson recorders, you will need to focus your attention on this exercise instead of Exercise 23. The Narco Physiograph®, which is manufactured by Narco Bio-Systems of Houston, Texas, is a four-channel polygraph with five ink-fed pens. Like the Gilson polygraph, it has interchangeable modules of various types.

Components

Refer to figure 24.1 to identify the following parts of the Narco Physiograph®.

Input Couplers

Note that on the upper portion of the sloping control panel are four units called *input couplers* (label 3). It is into these units that the transducers or electrodes feed the input signals that are to be monitored. Couplers are, essentially, preamplifier units that modify the signals before they enter the amplifier units (label 2).

Couplers of various types are available from the manufacturer. At the present time Narco produces thirteen different kinds of couplers. The Universal Coupler is probably the most widely used type. It is equivalent to the Gilson IC-MP module that is used in the Unigraph, Duograph, and polygraph. It can be used for monitoring ECG, EEG, EMG, and other DC and AC potentials. Other couplers that are available are the Strain Gage Coupler, GSR Coupler, Transducer Coupler, and Temperature Coupler.

Amplifier Controls

Note in figure 24.1 that each input coupler has its individual cluster of amplifier controls located below it. There are two knobs and one push button on each of these clusters.

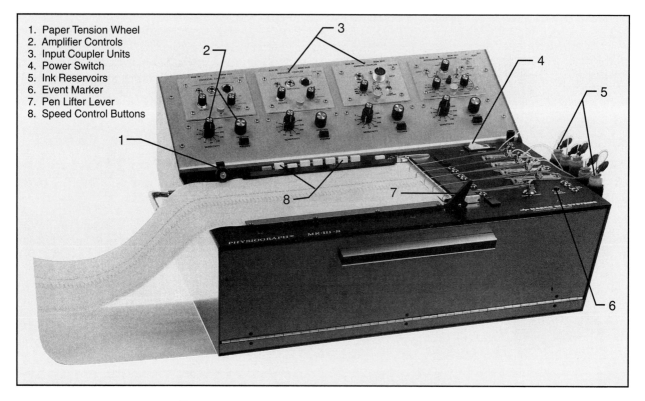

1. Paper Tension Wheel
2. Amplifier Controls
3. Input Coupler Units
4. Power Switch
5. Ink Reservoirs
6. Event Marker
7. Pen Lifter Lever
8. Speed Control Buttons

Figure 24.1　The Physiograph® Mark IIIS.

The control on the left, with ten settings, adjusts the sensitivity of the channel. It has an outer portion that enables one to select the desired millivolt setting (1, 2, 5, 10, 20, 50, 100, 200, 500, or 1000 millivolts). At setting 1, it takes only one millivolt to produce one centimeter pen movement; when set at 1000, it takes 1000 millivolts to produce the same amount of pen travel. Thus, we see that the 1 millivolt setting is the most sensitive setting available here. To adjust the sensitivity between any of these settings, one rotates the inner portion of this control. When the inner knob is completely clockwise, the sensitivity is identical to the reading on the dial. Counterclockwise rotation of the inner knob *decreases* the sensitivity.

The knob on the right (with POSITION under it) is used for repositioning the pen to locate the baseline of your recording in the most desirable spot. The location of the baseline will be determined by the degree of pen travel. For example, if the pen deflection is to be as much as 6 cm, it will be desirable to set the baseline substantially below the center of the arc of the pen. The total range of deflection of the pen is approximately 8 cm.

The push button below the positioning knob is used for starting the recording process. When depressed, this button is in the ON mode, which activates the pen. To stop the recording process one merely presses the button again, causing it to come up to a higher level, which is the OFF position.

Ink Pens

Note that the Mark IIIS has five ink pens: one for each of the four couplers and one to record one-second time intervals at the bottom of the chart. Each pen is supplied by ink from five separate **ink reservoirs** (label 5). These pens can be raised from the paper with the **pen lifter lever** (label 7). This lever actuates a metal bar that lifts all the pens simultaneously off the paper. *The pens should always be kept in the raised position when recording is not taking place.*

Ink reaches each pen through a small tube that connects with its ink reservoir. To get the ink to flow to the pen it is necessary to first raise the reservoir 2–3 cm and then squeeze the rubber bulb at the top of the reservoir as the index finger is held over the small hole at the tip of the rubber bulb. Once the tube is full of ink the fingertip is released before the bulb is released. When it becomes necessary to increase or decrease the flow of ink during recording, one simply raises or lowers the reservoir slightly.

To prevent clogging of pens it is necessary to remove all ink from the pens at the end of each laboratory period. With the pens in the raised position, the ink is drawn back into the reservoir by squeezing the reservoir bulb, placing the index finger over the hole on the bulb, and then releasing pressure on the bulb. The vacuum thus created will draw the ink back into the bulb.

Chart Speed

The Mark IIIS has seven push buttons (figure 24.2) that allow the following chart speeds: 0.05, 0.1, 0.25, 0.50, 1.0, 2.5, and 5.0 cm/sec. It also has an OFF push button on the left and a TIMER push button on the right end of the series.

To set the desired speed, one depresses the TIME button first (the up position is OFF), then the speed selection button is depressed, and finally, the OFF button is depressed again, putting the button in the ON position.

To change the speed, all one has to do is push a different speed button. To stop the paper movement, the OFF button is pressed. As soon as the paper stops moving, the pens should be raised with the pen lifter lever. At the end of the period it is important that both the OFF and TIME buttons be in the OFF position.

Operational Procedures

Although experiments performed on the Physiograph® will differ in certain respects, there are some procedures that should always be followed. These general procedures are outlined here.

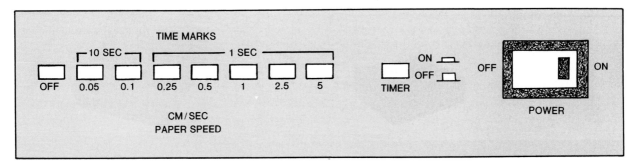

Figure 24.2 Push-button control panel of the Mark IIIS.

Dustcover

The dustcover should be removed from the unit, neatly folded, and placed somewhere where it will not be in the way or damaged. Because laboratory dust can affect some electronic components over a period of time, it is important that the unit be covered during storage.

Recording Paper

If there is no recording paper in the paper compartment, place a stack of paper in the compartment, which is accessed through the hinged front panel of the instrument. Feed the paper up through the opening provided in the top right margin of the paper compartment and then slide the paper under the paper guides on each side.

Draw the paper across the top of the recording surface and slide it under the paper tension wheel (label 1, figure 24.1) after lifting the wheel slightly by pulling the black lever straight up. After the paper is in place, release the lever. The paper should now be in firm contact with the paper drive mechanism and will move when the proper controls are activated.

Each small block on the paper measures 0.5 cm by 0.5 cm. Knowing this, one can determine the exact speed of the paper by the time marks.

Power Cord

The power cord is a loose component that should be inserted first into the right side of the Physiograph®, and then into a grounded electrical outlet.

Power Switch

Turn on the white power switch (label 4). Note that a small lighted indicator in the switch reveals that the unit has been turned on. This switch should be left in the ON position during the entire laboratory period and turned OFF at the end of the period.

Ink Pens

Fill the ink pens, observing the procedure outlined on the preceding page.

Paper Speed

Set the speed of the paper by first setting the TIMER push button in the ON mode and depressing the speed push button called for in the experiment. Do not press the OFF button until you are ready to start recording.

Coupler Attachment

Insert the cable from the experimental setup into the receptacle of the proper input coupler.

Amplifier Setting

Set the Sensitivity control at the sensitivity level recommended in the experiment. In most experiments a low sensitivity will be used first.

Transducer Balancing

If a transducer is to be used it will have to be balanced as per Appendix B.

Calibration

In certain experiments, such as in Exercise 29, it will have to be balanced as per Appendix B.

Monitoring

Once all the hookups have been completed, lower the pens to the paper and press the OFF button, which will start the paper moving, and then press the RECORD buttons on each amplifier cluster that is involved in the experiment. The events occurring in the experiment will now record.

Laboratory Report

Answer the questions on combined Laboratory Report 21–26 that pertain to the Narco Physiograph®.

25

The Electronic Stimulator

Stimulation of nerve and muscle cells in the laboratory is most readily accomplished with an electronic stimulator. Although we will focus our attention on the Narco and Grass stimulators, it is the general principles of stimulator operation that we will be primarily concerned with here.

The electronic stimulator accomplishes several things that are essential to controlled experiments. First of all, it converts alternating current (AC) to direct current (DC). As stated previously, DC moves continuously in one direction only. AC, on the other hand, reverses its direction repeatedly and continuously.

The advantage of DC stimulation is that there is an abrupt rise of voltage from zero to maximum and then a sudden return to zero when a pulse of electricity is applied to a nerve or muscle. This type of voltage control produces a **square wave** instead of a sine wave. The differences are seen in figure 25.1.

In addition to converting AC to DC, the stimulator, through a variety of controls and switches, can vary the voltage, duration, and frequency of pulses that are fed to a specimen. With the increase of **voltage,** the **amplitude,** or height, of the wave form is increased. The length of time from the start to end of a single pulse of electrical stimulus is the **duration,** and the number of pulses delivered per second is called the **frequency.** Some stimulators, such as the Grass SD9, can also vary the time between twin pulses; this time interval is designated as the **delay.** Figure 25.2 illustrates these various characteristics.

Control Functions

Figures 25.4 and 25.5 are of the Grass SD9 and Narco SM-1 electronic stimulators. Although the control panels of the two instruments differ in configuration and terminology, they both accomplish essentially the same function.

Power Switch

The power switch for the SD9 is located at the bottom of the front panel near the middle. Immediately above the switch is a red indicator light that lights up when the power is turned on.

The power switch for the SM-1 is located on the back panel. Like the SD9, the SM-1 has a red indicator light on the horizontal front panel.

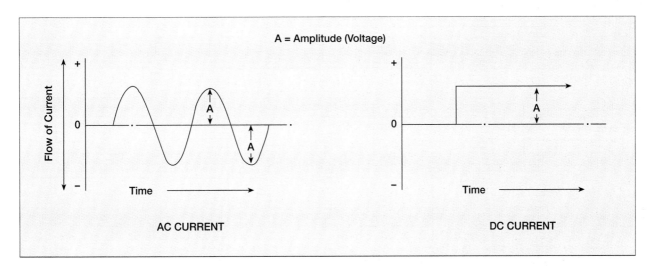

Figure 25.1   Comparison of AC and DC current.

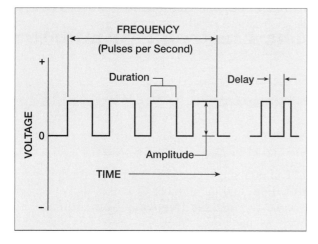

Figure 25.2 Square wave characteristics.

Voltage Settings

The intensity of any stimulus is varied by increasing or decreasing the voltage. Note that voltage on the SD9 is determined by two settings: a round knob in the upper right-hand corner and a decade (multiplier) switch below it. The voltage range here is from 0.1 to 100 volts.

Voltage adjustments on the SM-1 are made with the VOLTAGE RANGE switch and the VARIABLE control knob. The voltage range switch has three settings: 0–10, 0–100, and OFF. When the voltage range switch is set at the 0–100 setting, each digit on the variable dial is multiplied by 10; thus, the range on this scale is from 0 to 100 volts. When the voltage range switch is set on 0–10, the voltage output is between 0 and 10 volts. In the OFF mode no

electrical stimulus can occur. The lowest voltage output for this stimulator is 50 millivolts.

Duration Settings

The duration of each electrical pulse is measured in milliseconds (msec). On the SD9 this stimulus characteristic is controlled by the DURATION knob, which has a four-position decade switch below it. The combination of these two controls yields a range of 0.02 to 200 milliseconds (msec). The duration setting should not exceed 50% of the interval between pulses.

On the SM-1 duration is regulated by the WIDTH control, which is the left-hand knob on the horizontal panel. Its range is from 0.1 to 2.0 msec.

Mode Settings

Since stimulators can produce single stimuli, continuous pulses, and even twin pulses, there has to be a way of selecting the mode preferred for a particular experiment.

The two mode selection switches on the SD9 are located under the STIMULUS label. The right-hand MODE switch has three positions: repeat, single, and off. In the REPEAT position repetitive impulses are produced according to the setting on the Frequency control. To produce single pulses, the switch lever is depressed manually to the SINGLE position and released. One pulse is administered for each downward press of the lever; a spring returns the switch lever to the OFF position.

The other three-position switch under the stimulus label on the SD9 is used for twin pulses and hooking

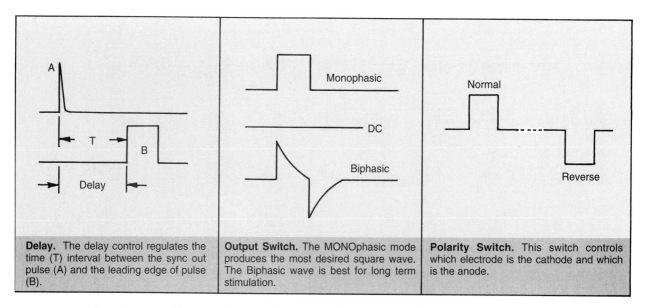

Delay. The delay control regulates the time (T) interval between the sync out pulse (A) and the leading edge of pulse (B).

Output Switch. The MONOphasic mode produces the most desired square wave. The Biphasic wave is best for long term stimulation.

Polarity Switch. This switch controls which electrode is the cathode and which is the anode.

Figure 25.3 Wave form modifications (Grass SD9 stimulator).

up with another stimulator. The MOD position is used when another stimulator is used in tandem with this one to modulate its output. When it is desirable to have the stimulator produce two pulses close together, the TWIN PULSES setting is used. When twin pulses or a second stimulator are not used, the setting should be on REGULAR. For most of our experiments this switch should be set at the regular setting.

The MODE switch on the SM-1 has three positions: continuous, single, and external triggering. When in the CONT position the stimulator will produce repetitive pulses according to the setting on the frequency control. When the switch lever is pressed to SINGLE, a single pulse will be delivered. The EXT TRIG position is used when an external triggering device is used. Such a device is plugged into a receptacle in the back of the stimulator.

Frequency Settings

The frequency control on a stimulator is set only when repetitive pulses are administered. In other words, the mode must be set at REPEAT on the SD9, or CONT on the SM-1.

On the SD9 the frequency control with its decade switch has a range of 0.20 to 200 pulses per second (PPS).

On the SM-1 the frequency range is from 0.1 Hz to 100 Hz. A RATE switch, which is calibrated ×1, ×0.1, and ×0.01, must be used in conjunction with

the round knob to get the proper setting. Incidentally, the Hz (Hertz) unit means essentially the same thing as PPS. One Hz represents one pulse per second.

Delay Settings

Stimulators that can produce twin pulses have a DELAY control to regulate the time interval between the pairs of pulses. The delay on the SD9 can be varied from 0.02 to 200 milliseconds, utilizing the control knob and four-place decade switch. The delay control can be used in either a single or repetitive mode. When the REPEAT mode is used, the delay time should not exceed 50% of the period between pulses to avoid overlap.

Since the Narco SM-1 stimulator does not produce twin pulses, there is no need for a delay control.

Output Switches on SD9

In addition to the aforementioned controls the Grass SD9 has two switches positioned under the OUTPUT label that can modify the wave form. Note that the three-position slide switch on the right is for selecting monophasic (MONO), biphasic (BI), or DC. Figure 25.3, middle illustration, shows the differences between these three types of current. The proper setting for most experiments will be in the monophasic position. However, when tissues are stimulated for a long period of time with the monophasic wave form,

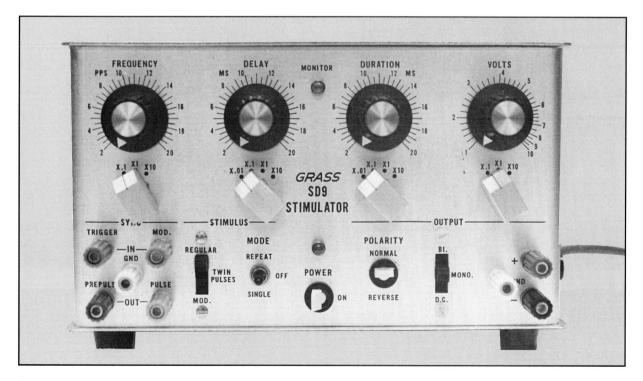

Figure 25.4 Control panel of the Grass SD9 electronic stimulator.

hydrolysis occurs, causing gas bubbles to form around the electrodes. For the longer duration experiments, the biphasic wave form will be preferable.

The DC position is seldom used due to the formation of electrolytes and gas bubbles in the preparation. Activation of the DC position overrides all other controls except voltage and mode.

The POLARITY switch to the left of the other switch is used for reversing direction of electron flow. In the NORMAL position the red output binding post is positive (+) with respect to the black binding post. In the REVERSE position the red binding post becomes negative (−) with respect to the black binding post. See the right-hand illustration in figure 25.3.

Operational Procedures

Actual operational procedures will be provided in each experiment as it is set up. For now, our primary concern is that one have a clear understanding of how the various controls are used on the electronic stimulator.

When the stimulator is first used in an experiment, you will want to review the functions of the various stimulator controls provided in this exercise.

Laboratory Report

Answer the questions on combined Laboratory Report 21–26 that pertain to electronic stimulators.

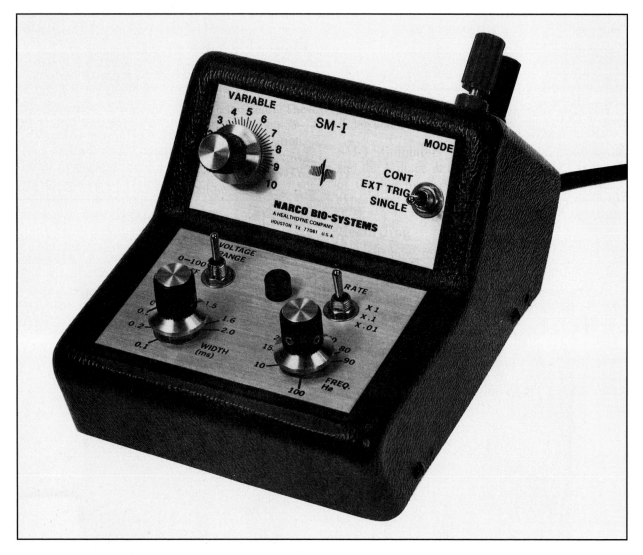

Figure 25.5 The Narco Bio-Systems SM-1 stimulator.

26

The Intelitool System

Four experiments in this laboratory manual utilize transducers with a computer to monitor physiological activities. The hardware and software for performing these experiments are manufactured and marketed by Intelitool, Inc., of Batavia, Illinois. The four exercises that utilize Intelitool equipment are Exercise 30 (the *Physiogrip*™ for muscle physiology), Exercise 43 (the *Flexicomp*™ for reflex monitoring), Exercise 63, (the *Cardiocomp*™ for ECG monitoring), and Exercise 77 (the *Spirocomp*™ for spirometry measurements).

An Intelitool data acquisition/analysis system is an excellent way to perform certain experiments if a compatible computer and printer are available. Since the equipment is designed to be used only on the human body (with safety), no animals are needed or sacrificed.

Although student computer knowledge is helpful in performing these experiments, it is not absolutely required. The reason is that the software instructions carefully guide the experimenter along, one step at a time, to the completion of the experiment. A great advantage of these systems over the use of chart recorders is that as the experiment is performed, data is accumulated automatically, and calculations are instantly made by the software. Another great advantage is that with a printer that has graphic capability, the results of the experiment can be printed out to produce a permanent record.

There is one limitation to these systems, however: they are all single channel systems. If one wishes to monitor two physiological activities simultaneously, such as ECG and respiration, one must use a Duograph or polygraph.

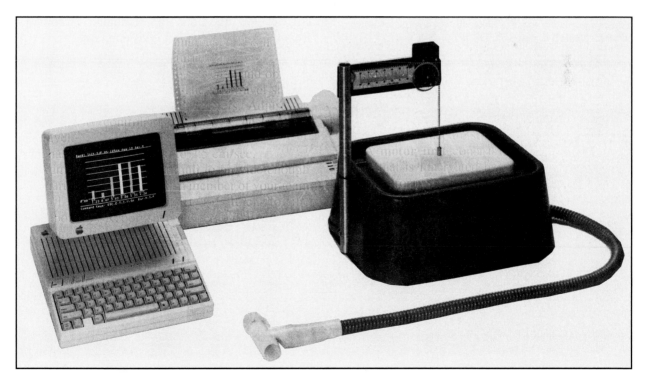

Figure 26.1 Intelitool setup for spirometry experiments.

All four Intelitool systems can be run on the following computer platforms*:

- IBM or compatible PC (MS-DOS 3.3 or higher)
- Macintosh (System 6.0.8 or higher)
- Apple II (ProDOS)

At this time only the Spirocomp can be run on Windows 95.

The instructions for each experiment are written with the assumption that you have a basic knowledge of the computer platform that you will be using. To accommodate the different platforms, separate instructions are provided in each exercise at each step when the computer is used.

To see how an Intelitool experiment is set up refer to figure 26.1. It illustrates the equipment that is used to perform breathing experiments (spirometry). A wet spirometer with hose attached, mouthpieces, computer, printer, and software are all the equipment that is needed. The Spirocomp transducer is an interface box that is bolted to the scale arm that has a pulley on it. As the subject breathes into the spirometry hose, respiratory volumes are recorded in memory and shown on the monitor as the air chambered bell moves up and down. Mathematical computations are automatically graphed by the software to determine pertinent information such as tidal volume, expiratory reserve volume, and vital capacity.

Platform refers to a specific combination of computer hardware and the operating system software.

Although setup instructions are included in each experiment, the actual setup process will probably be performed for you before you enter the laboratory. Even the software will be loaded and ready to use. Assembling the various components for each experiment consists primarily of connecting the Intelitool transducer unit to the computer with the proper cable. Only one system, the Physiogrip™, uses an electronic stimulator that is coupled to the subject's arm via a fitting on the front of the Physiogrip™ (see figure 30.1). All other systems have a single cable that leads to the computer.

No attempt will be made here to identify the manner in which one accesses the procedures for doing an Intelitool experiment. Once one gets into the program, prompts on the monitor screen and instructions in this manual will guide you through the experiment.

In addition to the instructions in this manual there are two manuals produced by Intelitool that will also be available to you. For each experiment, there is an Intelitool *Lab Manual* and a *User Manual*. Each manual is limited to the specific experiment being performed. There is no single manual that includes instructions for all four of the Intelitool systems. If some of the instructions in this manual seem confusing, you are encouraged to refer to the two Intelitool manuals for clarification.

Laboratory Report

Answer the questions on combined Laboratory Report 21–26 that apply to this exercise.

PART 6

6

Skeletal Muscle Physiology

The six exercises in this unit are concerned with an in-depth study of most aspects of skeletal muscle contraction. In Exercise 27 the ultrastructure of myofibrils will be studied in relation to what is actually observed under the light microscope during muscle contraction. The roles of calcium ions and ATP in muscle fiber contraction are also determined.

Exercise 28 is a simple exercise that accomplishes two things: first, it acquaints you with the workings of an electronic stimulator, and second, it prepares you for skills required to perform the experiments in Exercise 30.

Exercises 29 and 30 are classic experiments in which the simple twitch, recruitment, and wave summation are studied. In Exercise 29 a frog's muscle is electronically stimulated to observe the speed and strength of skeletal muscle contraction. Exercise 30 differs from Exercise 29 in that the former utilizes a computer with Intelitool hardware and software to observe the same phenomena on human tissue instead of frog tissue.

In Exercise 31 the effects of epinephrine, EDTA, and acetylcholine on muscle contraction will be studied. This experiment is open-ended in that a multiplicity of drugs can be tested here. Exercise 32 pertains to electromyography.

Histology Self-Quiz No. 1

If at this point you have completed all the histology assignments in the first 20 exercises in this manual you might wish to test your knowledge of histology by taking the self-quiz on page 483. It consists of thirty photomicrographs and seventy-five questions. After recording your answers on a sheet of paper, check your answers against the key, which you will find on page 414.

27

The Physicochemical Nature of Muscle Contraction

The microscopic changes that occur during skeletal muscle contraction will be observed here in a controlled experimental setup. The physicochemical nature of muscle contraction will be reviewed first for clarification of this complex phenomenon.

Skeletal muscle contraction involves many components within the muscle cell. The entire process is initiated by an action potential that reaches the neuromuscular junction. As we learned in Exercise 20, acetylcholine initiates depolarization of the muscle fiber. The depolarization proceeds along the surface (sarcolemma) of the muscle fiber and deep into the sarcoplasm by way of the T-tubule system of the sarcoplasmic reticulum.

Figure 27.2 illustrates the ultrastructure of a single muscle fiber. Note that the sarcoplasm of the fiber consists of many long tubular **myofibrils,** the **sarcoplasmic reticulum,** and **mitochondria. Nuclei** can also be seen beneath the **sarcolemma.** The abundance of mitochondria and their close association to the myofibrils indicates an availability of ATP for contraction.

Illustrations C and D in figure 27.2 reveal the molecular structure of the contractile **myofilaments** of the myofibrils. Myofibrils can be seen with a good light microscope, but myofilaments are visible only with electron microscopy.

Note that there are two types of myofilaments: actin and myosin. **Myosin** filaments are the thickest filaments and are the principal constituents of the dark **A band.** They are slightly thicker in the middle and taper toward both ends. The thinner **actin** filaments extend about 1 μm in either direction from the **Z line** and make up the lighter **I band.**

The shortening of myofibrils to achieve muscle fiber contraction is due to the interaction of actin and myosin in the presence of calcium, magnesium, and ATP. Two other protein molecules plus troponin and tropomyosin are also involved.

As the depolarization wave (action potential) moves along the sarcolemma and T-tubule system, large quantities of calcium ions are released by the sarcoplasmic reticulum. As soon as these ions are

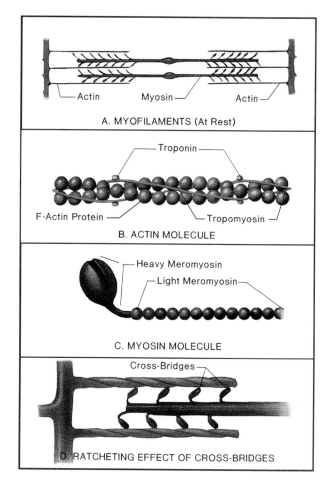

Figure 27.1 Myofilament structure.

released, attractive forces develop between the actin and myosin filaments that causes the ends of the actin fibers to be pulled toward each other, as shown in illustration D, figure 27.2. The exact mechanism that causes this shortening of the myofibrils is unknown, but the following explanation is the most acceptable theory we have at the present time.

It is in the nature of the actin and myosin molecules that we seek an answer to the attractive forces. The myosin molecule is made up of **heavy meromyosin** and **light meromyosin** (see figure 27.1). The heavy meromyosin forms a head that is flexible and able to pivot. The actin filament consists of three components: a double-stranded helix of **F-actin protein, tropomyosin,** and **troponin.** The tropomyosin is a long protein strand that lies within the groove of the F-actin protein. The third protein, troponin, occurs at various active sites along the helix. The cross-bridges of heavy meromyosin are able to bridge the gap between the myosin and actin molecules.

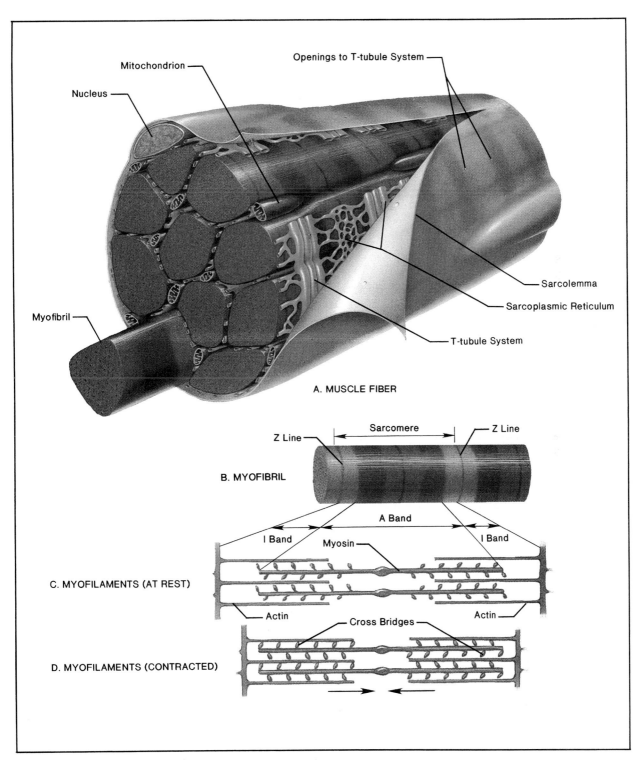

Figure 27.2   The ultrastructure of a muscle fiber.

The *ratchet theory of contraction* proposes that the following series of events takes place:

1. Calcium ions combine with troponin.
2. Calcium-bound troponin causes the tropomyosin to be pulled deeper into the groove of the two actin strands, exposing the active sites on the actin filament.
3. The heads of the cross-bridges form a bond with the active sites and pull on the actin filament with a ratchetlike power stroke. The energy for the first power stroke is derived from the meromyosin molecule that was previously activated by ATP.
4. After the first power stroke, ATP combines with the head of the cross-bridge, causing the head to separate from the active site to which it has been attached.
5. The ATP on the head is split by ATPase, which is contained in the heavy meromyosin. The energy released by this cleavage is used for cocking the head again, attaching the head to another active site, and performing another power stroke.

Microscopic Examination

To observe the nature of muscle contraction at the microscopic level, we will induce contraction in glycerinated rabbit muscle fibers with ATP and ions of magnesium and potassium. Measurements of the muscle fibers will be made before and after contraction to determine the percentage of contraction that occurs. Microscopic comparisons will be made of fibers in both the relaxed and contracted state to determine the changes that occur.

Materials:

test tube of glycerinated rabbit psoas muscle
(muscle is tied to a small stick)
dropping bottle of ATP in triple distilled water
dropping bottle of 0.25%ATP, 0.05M KCl, and
0.001M $MgCl_2$
3 microscope slides and cover glasses
sharp scissors
forceps
dissecting needle
petri dish
plastic ruler, metric scale
microscopes, dissecting and compound

1. Remove the stick of glycerinated muscle tissue from the test tube and pour some of the glycerol from the tube into a petri dish.
2. With scissors, cut the muscle bundle into segments about 2 cm long, allowing the pieces to fall into the petri dish of glycerol.
3. With forceps and dissecting needle, tease one of the segments into very thin groups of muscle fibers; single fibers, if obtained, will demonstrate the greatest amount of contraction. Strands of muscle fibers exceeding 0.2 mm in cross-sectional diameter are too thick and should not be used.
4. Place one of the strands on a clean microscope slide and cover with a cover glass. Examine under a high-dry or oil immersion objective. Sketch appearance of striae on the Laboratory Report.
5. Transfer three or more of the thinnest strands to a minimal amount of glycerol on a second microscope slide. Orient the strands straight and parallel to each other.
6. Measure the lengths of the fibers in millimeters by placing a plastic ruler underneath the slide. Use a dissecting microscope, if available. Record lengths on the Laboratory Report.
7. With the pipette from the dropping bottle, cover the fibers with a solution of ATP and ions of potassium and magnesium. Observe the reaction.
8. After 30 seconds or more, remeasure the fibers and calculate the percentage of contraction. Has the width of the fibers changed? Record the results on the Laboratory Report.
9. Remove one of the contracted strands to another slide. Examine it under a compound microscope and compare the fibers with those seen in step 4. Record the differences on the Laboratory Report.
10. Place some fresh fibers on another clean microscope slide and cover them with a solution of ATP and distilled water (no potassium or magnesium ions). Does contraction occur?

Laboratory Report

Record all data on the Laboratory Report and answer all the questions.

28

Mapping of Motor Points

The sensitivity of a muscle fiber to electrical stimulation through the skin surface is a function of the proximity of a neuromuscular junction, the thickness of the skin, localized variability in skin conductivity, and other factors. An electrical stimulus that causes a muscle twitch at one spot may have no effect on muscle fibers that are only a centimeter away. Specific spots on the skin where minimal electrical stimuli cause muscle twitching are called **motor points.** An attempt will be made here to identify specific spots on the skin where these motor points are located.

If the electrical stimulus is close enough to naked nerve endings in the skin, one may experience a slight pricking sensation as the stimulus is applied. Since the skin has many of these pain receptors, this sensation is often experienced; however, a muscle twitch may or may not be produced simultaneously with the pricking sensation. Our intent in this experiment will be to ignore this phase of the experiment. Only the motor points will be recorded.

In addition to learning about motor points, this simple experiment will provide you with a good opportunity to become more familiar with the electronic stimulator. The next few experiments in this manual rely heavily on the proper use of this instrument. By applying electrical stimuli to your own skin, you will acquire a greater appreciation of the flexibility and safety of the stimulator. Before performing this experiment, review Exercise 25, which pertains to the stimulator.

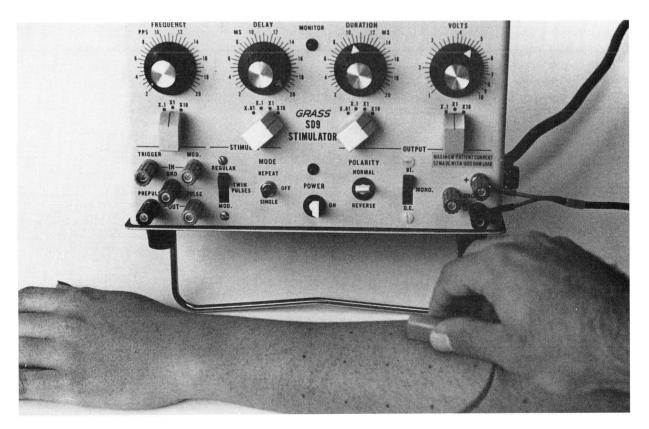

Figure 28.1   Experimental setup.

Procedure

In this experiment students will work in teams made up of a minimum of four students. One member will be the Subject, another will be the Prober, a third will be the Stimulator Operator, and the fourth member will be the Recorder. If teams need to be larger owing to the lack of equipment, some members can act as assistants to the Stimulator or Prober. If ample time is available, it may be desirable to allow two or more members of the group to act as subjects and permit different members to operate the stimulator. In this way, all members of each team will get a chance to develop skills with the stimulator.

Materials:

electronic stimulator (grounded electrical outlet is mandatory)
stimulator probe (Katech #105)
ruler
fine-point felt pen
electrode paste
alcohol swabs
abrasive soap
toothpicks

Stimulator Settings

The electrical stimuli to be applied to the skin will be minimal, or **threshold**—just sufficient to elicit a response. The threshold stimulus is governed by the *duration* (in milliseconds) and *voltage*. If the duration is too short, more voltage is required. The procedure is to start with a prolonged duration to find the minimum voltage and then gradually decrease the duration to a point at which the minimum voltage will still elicit a response.

Set the stimulator controls as follows to begin the stimulation procedure: Frequency at 2 pulses per second (pps), Delay* at 200 msec, Duration at 100 msec, volts at 1.0, Stimulus Mode switches at Regular Pulses and Repeat, Power Switch ON, Polarity at Normal, and Output switch at BIphasic. Connect the terminal jacks of the probe to the output posts of the stimulator.

Subject Preparation

The Subject will scrub the back of one forearm with a bar of abrasive soap to dislodge superficial layers of the *stratum corneum* and to dissolve excess skin oil.

*Delay setting does not really come into play in this experiment.

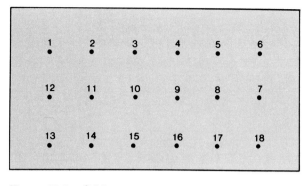

Figure 28.2 Grid pattern.

After the skin is completely dried, the Prober will lay out on the surface of the arm a grid of eighteen inked dots that are one inch apart. A fine-point felt tip pen works well. (The dots will vanish quickly with soap and water at the end of the experiment.) Figure 28.2 illustrates the nature of the grid. Numbers are shown here to indicate the sequence of stimulation. *Do not put numbers on the skin.*

On each pen mark, place a small amount of electrode paste. Position the Subject's arm in a relaxed manner on the table, with palm down. The Subject should look away from the Prober and Stimulator Operator while stimuli are applied to prevent the Subject from being influenced by the behavior of Prober or Stimulator Operator.

The Subject should use hand signals with the free hand to indicate sensations:

- **No sensation:** no signal.
- **Tingle, pricking sensation, or vague uncertain feeling:** hold up one finger.
- **Distinct pain:** hold up two fingers.

The Recorder will record voltages and spots where muscle twitches are observed. Your team is now ready to test each spot for motor points.

Testing with Voltage Increases

Now that the stimulator settings have been set for 1 volt, 100 msec duration, and twin pulses, we are ready to start the experiment. Proceed as follows:

1. Apply the stimulator probe to spot 1 and direct the Stimulator Operator to place the Mode switch on REPEAT. You are now applying 2 pulses per second of 1 volt each with 100 msec duration. *There should be no sensation.*
2. Test each of the 18 stations with this 1 volt stimulation, recording on the Laboratory Report any reaction that occurs at a particular station.

3. Clean the probe contacts by removing excess electrode paste from the space between the inner and outer contacts with a toothpick.
4. Now increase the stimulus to 5 volts and test each station with a 3-second stimulation. *Try to maintain the same amount of pressure on the probe.*
5. Repeat step 3.
6. After testing all 18 stations at 5 volts, increase the voltage again by another 5 volt increment and start all over again.
7. Continue this process of testing at increased increments of 5 volts until a twitch is observed by the Prober and Recorder or the Subject signals a sensation of pain. Be sure to repeat step 3 before each new test at an increased voltage.

   Important: Do not stimulate those stations a second time where a reaction was previously recorded.
8. Continue testing all nonreactive stations until 100 volts is reached. **At all times, be sure to prevent the Subject from observing stimulator settings.**

Determining Minimum Duration

In the above series of tests the duration was arbitrarily set at 100 msec and kept constant as the voltage was increased. We will now try to determine the minimum duration that will elicit a response at a few sensitive spots.

1. At the first spot where twitching was encountered, place the probe with the voltage set at the predetermined level.
2. Change the duration to 1 msec (large knob on 10, decade switch on 0.1×). Stimulate for 3 seconds and record the reaction.
3. Move on to five other stations where a contraction had previously been noted, skipping those that gave no response.
4. Now, increase the duration to 10 msec (move decade switch to 1×) and stimulate each one of the five same stations again.
5. Continue increasing the duration by increments of 10 msec until 150 msec has been reached. Do not exceed 150 msec duration.
6. Record results on the Laboratory Report.

Reversing Polarity

To observe if reversing the polarity has any effect on sensitivity to twitching, place the Polarity switch on REVERSE and administer a threshold stimulus of minimum duration to the last reactive spot tested. Record your observations on the Laboratory Report.

Laboratory Report

After recording all data on the Laboratory Report, answer all the questions.

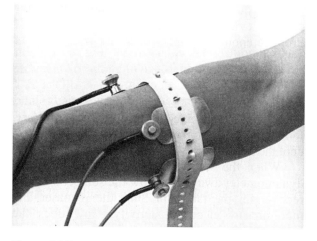

Figure 32.7 An alternative method of electrode application for an EMG demonstration.

Agonist vs. Antagonist Muscles

Opposing movements of muscle pairs are caused by *agonist* or *antagonist* muscles. For example, flexion of the forearm against the upper arm is achieved by contraction of the *biceps brachii*, the agonist. Extension occurs when the *triceps brachii*, its antagonist, contracts. The action of these two muscles must be coordinated so that when one contracts the other relaxes.

Activity in flexors cannot be accompanied by simultaneous activity in extensors, or rigidity will occur instead of flexion. Coordination may be demonstrated by recording the activity of both muscles simultaneously with a Duograph, polygraph, or dual-beam oscilloscope.

To demonstrate coordination, hook up three electrodes to each muscle of one arm through separate channels of whatever type of equipment is available. Have the subject execute the following maneuvers.

1. Flexion of arm against resistance. Assistant restrains subject's fist while flexion is attempted.
2. Extension against resistance.
3. Isometric tension (contraction) of both agonist and antagonist, simultaneously.

Laboratory Report

Complete the last portion of combined Laboratory Report 31, 32.

PART 7
The Major Skeletal Muscles

The eight exercises of this unit include around 105 skeletal muscles of the human body. Although this is not a complete listing of all the muscles, it does include practically all the surface muscles and most of the deeper ones. If time does not permit the study of all these muscles, the instructor will indicate which ones the student should know.

This is the first unit in which the cat will be used for dissection. Before studying cat muscles, it is recommended that the human musculature be studied first by labeling the illustrations. Although cat musculature differs to some extent from the human, the similarities far outweigh the differences.

One thing to keep in mind about the cat dissection: we are *not* concerned about the cat muscles from a comparative standpoint. The fact that the cat may have certain muscles that are lacking in the human, or that the origins or insertions might be slightly different in the cat, is unimportant. Our principal concern is to reinforce our knowledge of the human muscles by studying the similarities rather than the differences.

33

Muscle Structure

The movement of limbs and other structures by skeletal muscles is a function of the contractility of muscle fibers and the manner of attachment of the muscles to the skeleton. In this exercise our concern will be with the detailed structure of whole muscles and their mode of attachment to the skeleton.

Figure 33.1 reveals a portion of the arm and shoulder girdle with only two muscles shown. Several other muscles of the upper arm have been omitted that would obscure the points of attachment of these muscles. The longer muscle that is attached to the scapula and humerus at its upper end is the **triceps brachii.** The shorter muscle on the anterior aspect of the humerus is the **brachialis.**

Muscle Attachment

Each muscle of the body is said to have an origin and an insertion. The **origin** of the muscle is the immovable end. The **insertion** is the other end, which moves during contraction. Contraction of the muscle results in shortening of the distance between the origin and the insertion, causing movement of the insertion end.

Muscles may be attached to bone in three different ways: (1) directly to the periosteum, (2) by means of a tendon, or (3) with an aponeurosis. The upper end of the brachialis (the origin) is attached by the first method, i.e., directly to the periosteum. The insertion end of this muscle, however, does have a short tendon for attachment.

A **tendon** is a band or cord of white fibrous tissue that provides a durable connection to the skeleton. The tendon of the triceps insertion is much longer and larger due to the fact that the muscle exerts a greater force in moving the forearm.

An example of the aponeurosis type of attachment is seen in figure 39.1, page 182. Label 5 in that figure is the aponeurosis of the external oblique muscle. An **aponeurosis** is a broad flat sheet of glistening pearly-white fibrous connective tissue that attaches a muscle to the skeleton or another muscle.

Although most muscles attach directly to the skeleton, there are many that attach to other muscles or soft structures (skin) such as the lips and eyelids. In Exercises 36 through 40 the precise origins and insertions of various muscles will be studied.

Microscopic Structure

Illustrations B and C in figure 33.1 reveal the microscopic structure of a portion of the brachialis muscle. Note that the muscle is made up of bundles, or **fasciculi,** of skeletal muscle cells. A single fasciculus is shown, enlarged, in illustration C.

Protruding upward from the upper cut surface of the fasciculus is a portion of an individual muscle cell, or **muscle fiber.** Note that between the individual muscle fibers of each fascicle is a thin layer of connective tissue, the **endomysium.** This connective tissue enables each muscle fiber to react independently of the others when stimulated by nerve impulses. Surrounding each fascicle is a tough sheath of fibrous connective tissue, the **perimysium.**

Surrounding the fasciculi of the muscle is another layer of coarser connective tissue that is called the **epimysium.** Exterior to the epimysium is the **deep fascia,** which covers the entire muscle. Since this muscle lies close to the skin, the deep fascia of this muscle would have fibers that intermesh with the **superficial fascia** (not shown) that lies between the muscle and the skin. The connective tissue fibers of the deep fascia are also continuous with the tissue in the tendons, ligaments, and periosteum.

Illustration A reveals the continuity of the connective tissue of the endomysium, perimysium, and epimysium. A fasciculus of muscle fibers is shown bracketed in this illustration.

Assignment:
Label figure 33.1.

Examine some muscles on a freshly killed or embalmed animal, identifying as many of the aforementioned structures as possible.

Laboratory Report

Complete the first portion of Laboratory Report 33, 34.

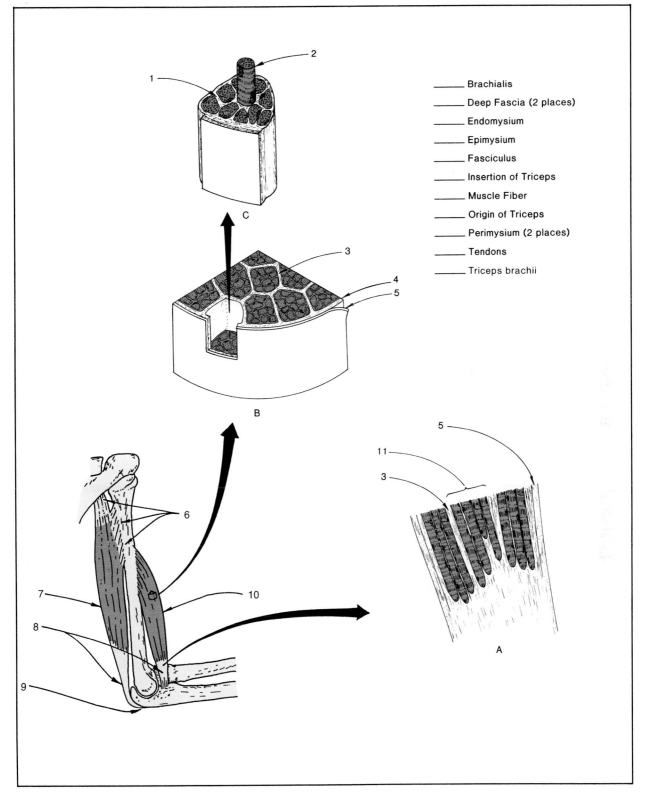

_____ Brachialis

_____ Deep Fascia (2 places)

_____ Endomysium

_____ Epimysium

_____ Fasciculus

_____ Insertion of Triceps

_____ Muscle Fiber

_____ Origin of Triceps

_____ Perimysium (2 places)

_____ Tendons

_____ Triceps brachii

Figure 33.1 Muscle anatomy and attachments.

34

Body Movements

The action of muscles through diarthrotic joints results in a variety of types of movement. The nature of movement is dependent on the construction of the individual joint and the position of the muscle. Muscles working through hinge joints result in movements that are primarily in one plane. Ball and socket joints, on the other hand, will have many axes through which movement can occur; thus, movement through these joints is in many planes. The different types of movement follow.

Flexion   When the angle between two parts of a limb is decreased the limb is said to be "flexed." Flexion of the arm takes place at the elbow when the antebrachium is moved toward the brachium. The term *flexion* may also be applied to movement of the head against the chest and of the thigh against the abdomen.

When the foot is flexed upward, the flexion is designated as being **dorsiflexion.** Movement of the sole of the foot downward, or flexion of the toes, is called **plantar flexion.**

Extension   Increasing the angle between two portions of a limb or two parts of the body is *extension.* Straightening the arm from a flexed position is an example of extension. Extension of the foot at the ankle is essentially the same as plantar flexion. When extension goes beyond the normal posture, as in leaning backward, the term **hyperextension** is used.

Abduction and Adduction   The movement of a limb away from the median line of the body is called *abduction.* When the limb moves toward the median line, *adduction* occurs. These terms may also be applied to parts of a limb, such as the fingers and toes, by using the longitudinal axes of the limbs as points of reference.

Rotation and Circumduction   The movement of a bone or limb around its longitudinal axis without lateral displacement is *rotation.* Rotation of the arm occurs through the shoulder joint. The head can be rotated through movement in the cervical vertebrae. If the rotational movement of a limb through a freely movable joint describes a circle at its termi-

nus, the movement is called *circumduction.* The arm, leg, and fingers can be circumducted.

Supination and Pronation   Rotation of the antebrachium at the elbow affects the position of the palm. When the palm is raised upward from a downward-facing position, *supination* occurs. When the rotation is reversed so that the palm is returned to a downward position, *pronation* takes place.

Inversion and Eversion   These two terms apply to foot movements. When the sole of the foot is turned inward or toward the median line, *inversion* occurs. When the sole is turned outward, *eversion* takes place.

Sphincter Action   Circular muscles such as those around the lips (*orbicularis oris*) and the eye (*orbicularis oculi*) are called *sphincter muscles.* When the fibers of these muscles contract, the openings they encircle are closed.

Muscle Grouping

For reasons of simplicity, muscles are often studied individually, as in figure 34.1. It should be kept in mind, however, that seldom, if ever, do they act alone. Rather, muscles are arranged in groups with specific functions to perform: e.g., flexion and extension, abduction and adduction, supination and pronation.

The flexors are the prime movers, or **agonists.** The opposing muscles, or **antagonists,** contribute to smooth movements by their power to maintain tone and give way to movement by the flexor group. Variance in the tension of flexor muscles results in a reverse reaction in the extensor muscles.

Muscles that assist the agonists to reduce undesired action or unnecessary movement are called **synergists.** Other groups of muscles that hold structures in position for action are called **fixation muscles.**

Laboratory Report

Identify the types of movement in figure 34.1 and complete the last portion of Laboratory Report 33,34.

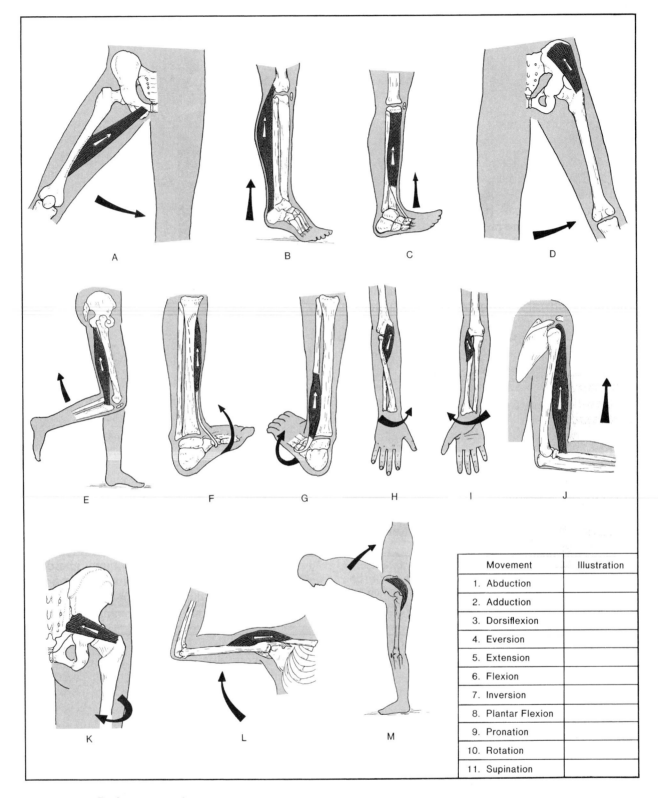

Movement	Illustration
1. Abduction	
2. Adduction	
3. Dorsiflexion	
4. Eversion	
5. Extension	
6. Flexion	
7. Inversion	
8. Plantar Flexion	
9. Pronation	
10. Rotation	
11. Supination	

Figure 34.1   Body movements.

Cat Dissection

Thorax and Shoulder Muscles (Superficial)

Cat muscles of the thorax fall into three categories: the pectoralis, trapezius, and deltoideus groups. Use figures 37.3, 37.4, and 37.5 for reference.

The Pectoralis Group

Pectoantebrachialis This narrow ribbonlike muscle is about 1 cm wide and extends from the upper sternum to the forearm fascia. Since there is no homolog in humans, *transect* and remove this muscle.

Pectoralis Major This muscle is about 5 cm wide and lies deep to the pectoantebrachialis. With a blunt probe separate it from the clavodeltoid and pectoralis minor. *Dissect.*

Pectoralis Minor Note that this muscle is larger than the pectoralis major. It arises from the sternum, passes diagonally craniad and laterad beneath the pectoralis major to insert on both the scapula and humerus. Note that in humans, this muscle inserts only on the scapula. *Transect* and *reflect.*

Xiphihumeralis The most caudal muscle of the pectoral group, its fibers run parallel to the pectoralis minor and eventually pass deep to the pectoralis minor to insert on the humerus. There is no human homolog. Remove this muscle.

The Trapezius Group

The trapezius muscle in humans is represented by three separate muscles in the cat. Use figures 37.4 and 37.5 to identify them on your specimen.

Clavotrapezius The most cranial trapezius muscle on the cat; covers the dorsal surface of the neck. It inserts on the clavicle and unites with the clavodeltoid to form a single muscle, the **brachiocephalicus.** This muscle is homologous to the portion of the human trapezius that inserts on the clavicle.

Acromiotrapezius This muscle lies caudad to the clavotrapezius. It is comparable to the part of the human trapezius that inserts on the acromion. *Dissect, transect* by cutting through the tendon of origin on the mid-dorsal line, and *reflect.*

Spinotrapezius This muscle is caudad to the acromiotrapezius and superficial to the cranial border of the latissimus dorsi. That part of the human trapezius that inserts on the spine of the scapula is comparable to this muscle. *Dissect, transect,* and *reflect.*

The Deltoid Group

The cat has three deltoid muscles that are homologs of the single human deltoid. Locate each of the following muscles.

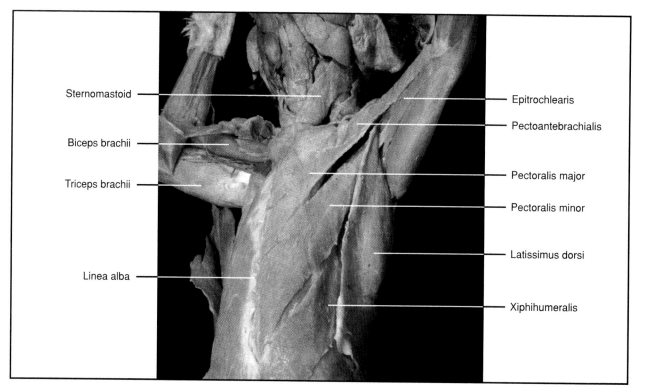

Figure 37.3 Ventrolateral aspect of superficial thoracic muscles of the cat.

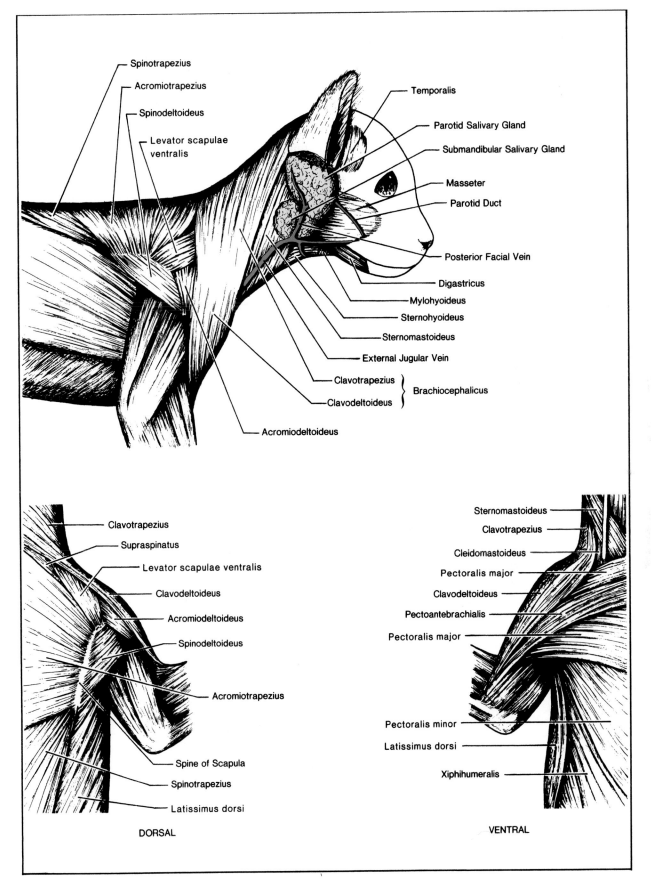

Figure 37.4 Trunk and shoulder muscles (superficial).

Clavodeltoideus *(Clavobrachialis)* The position of this muscle is best seen on the ventral view, figure 37.4, and on figure 37.5. This muscle is an extension of the clavotrapezius. It originates on the clavicle and inserts, primarily, on the proximal end of the ulna; some of its fibers, however, merge with the pectoantebrachialis to insert on the fascia of the forearm. *Dissect.*

Acromiodeltoideus This muscle is comparable to the acromial portion of the human deltoid. *Dissect.* Although freeing up this muscle is difficult, at least the cranial and caudal borders need to be clearly separated from adjacent muscles.

Spinodeltoideus This thick muscle lies caudad to the acromiodeltoid and arises from the spine of the scapula. Use the dorsal view, figure 37.4, and figure 37.5 for reference. *Dissect, transect,* and *reflect.*

Other Superficial Thoracic Muscles

In addition to the above three groups there are two more superficial muscles of importance:

Levator Scapulae Ventralis The location of this muscle is shown in the lateral and dorsal views, figure 37.4. Note that this muscle is caudad to the clavotrapezius and passes beneath the clavotrape-zius. It joins the craniolateral edge of the acromio-trapezius and inserts on the acromial part of the scapular spine. It arises from the transverse process of the atlas and the occipital bone. Since it has no human homolog, identify it and move on to the next muscle.

Latissimus Dorsi This broad, flat, triangular muscle has an extensive origin on the spines of the last six thoracic vertebrae, the lumbar vertebrae, the sacrum, and the iliac crest. The lumbodorsal fascia, an aponeurosis, contains the fibers of this origin. The muscle inserts on the proximal portion of the humerus. *Dissect, transect,* and *reflect.*

Deep Thoracic Muscles

Examine the following deeper muscles of the thorax using figures 37.6, 37.7, and 37.8 for reference.

Serratus Ventralis Pull the scapula away from the body wall and rotate it dorsad. Clean off the fascia from the body wall. Note the serrate margin of this muscle due to the origin by digitations from the upper eight or nine ribs.

The cranial portion of this muscle is homologous to the human *levator scapulae,* and the caudal portion is homologous to the *serratus anterior.* *Dissect.*

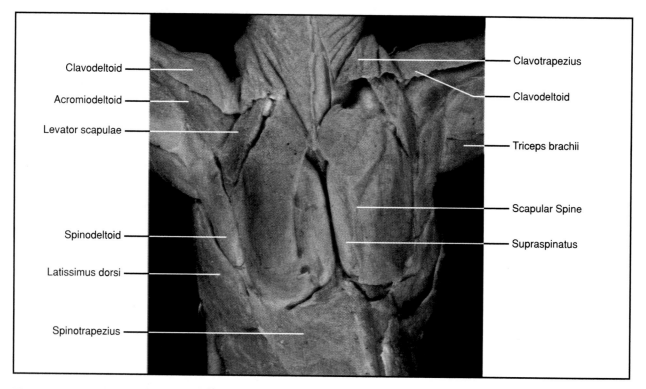

Clavodeltoid

Acromiodeltoid

Levator scapulae

Spinodeltoid

Latissimus dorsi

Spinotrapezius

Clavotrapezius

Clavodeltoid

Triceps brachii

Scapular Spine

Supraspinatus

Figure 37.5 Dorsal aspect of superficial thoracic muscles.

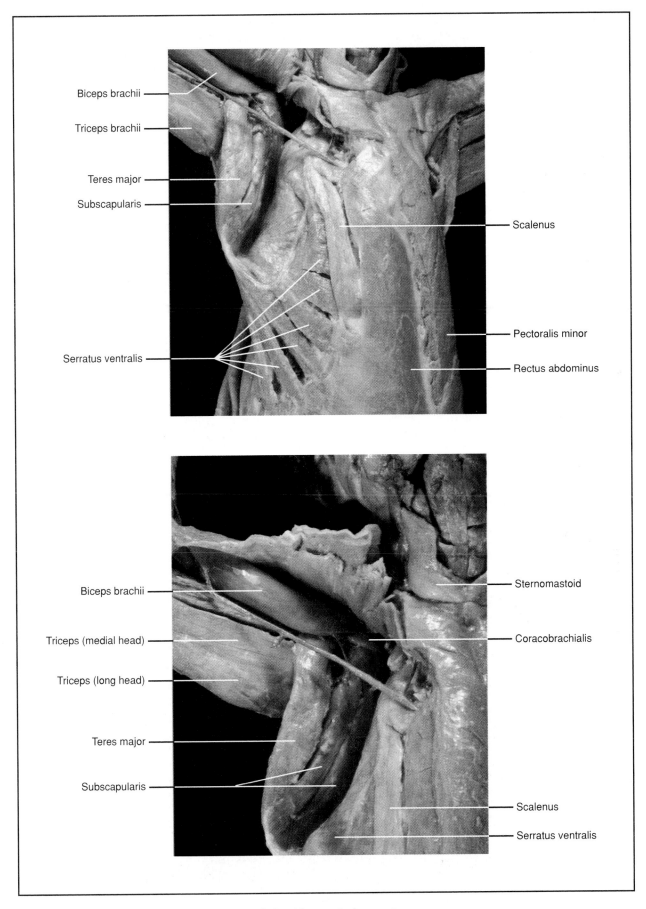

Figure 37.6 Deep muscles of cat thorax and shoulder, ventral aspect.

Rhomboideus This muscle (figure 37.7) lies deep to the acromiotrapezius and spinotrapezius, which should have been transected previously. Pull the two forelegs together on the ventral side. This will abduct the scapulae so that you can see the full extent of the rhomboids.

The rhomboideus extends from the spines of the upper thoracic vertebrae to the vertebral border of the scapula. The cranial (and larger) part of this muscle is homologous to the *rhomboideus minor* muscle in humans, and the caudal (smaller) part is homologous to the *rhomboideus major.* The rhomboideus major is larger than the rhomboideus minor in humans.

Levator Scapulae This is not a separate muscle but a cranial continuation of the serratus ventralis. Its origin is from the transverse processes of the caudal five cervical vertebrae, and it inserts on the vertebral border of the scapula ventral to the insertion of the rhomboideus. In the human, the insertion is comparable to that of the cat, but the origin is roughly comparable to the origin of the cat's levator scapulae ventralis.

Supraspinatus With the acromiotrapezius fully reflected, this muscle can be seen filling the supraspinous fossa of the scapula. From its origin in the supraspinous fossa it passes to the greater tubercle of the humerus.

Infraspinatus Reflect the transected spinodeltoid. The infraspinatus can now be observed extending from its origin in the infraspinous fossa of the scapula to the greater tubercle of the humerus.

Teres Major This thick band of muscle is located caudad to the infraspinatus muscle and covers the axillary border of the scapula. It extends forward to insert on the proximal end of the humerus. *Dissect.* Reflect the latissimus dorsi to expose the full extent of this muscle.

Teres Minor This small muscle is positioned between the infraspinatus and the long head of the biceps brachii. Since it is not labeled in any of the illustrations in this manual, you will have to look for it by reflecting the spinodeltoid. The muscle extends from the axillary border of the scapula to the humerus.

Laboratory Report

Complete the Laboratory Report for this exercise.

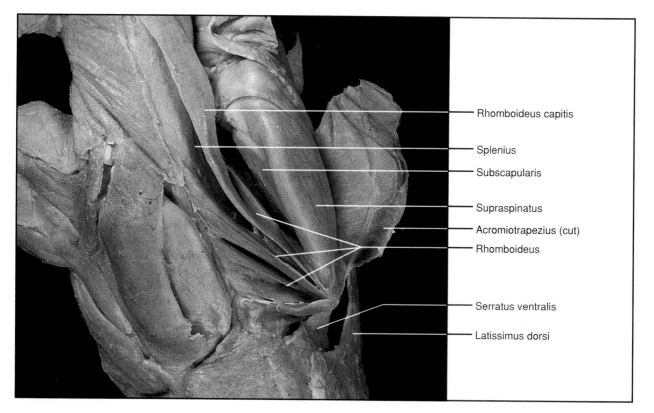

Rhomboideus capitis

Splenius

Subscapularis

Supraspinatus

Acromiotrapezius (cut)

Rhomboideus

Serratus ventralis

Latissimus dorsi

Figure 37.7 Deep muscles of cat's shoulder and back, dorsal aspect.

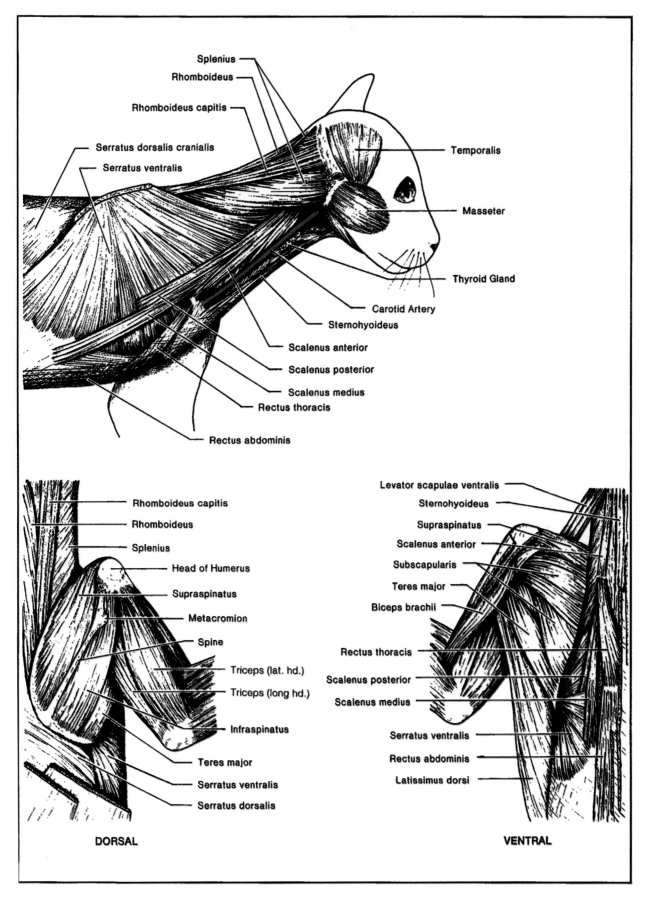

Figure 37.8 Deep trunk, shoulder, and neck muscles.

38

Upper Extremity Muscles

Muscles that move the upper arm are primarily muscles of the shoulder and trunk, which were studied in the last exercise. In this exercise the muscles of the brachium that control movements of the forearm will be studied. One muscle, however that was not mentioned in the last exercise that does move the upper arm is the *coracobrachialis.* It is a muscle of the upper arm and is illustrated in figure 38.1.

Coracobrachialis This muscle covers a portion of the upper medial surface of the humerus. It takes its *origin* on the apex of the coracoid process of the scapula. Its *insertion* is in the middle of the medial surface of the humerus.

> *Action:* Carries the arm forward in flexion and adducts the arm.

Forearm Movements

The principal movers of the forearm are the *biceps brachii, brachialis, brachioradialis,* and *triceps brachii.* These four muscles are also illustrated in figure 38.1. Note that incorporated into all the names of these upper arm muscles is the Latin term *brachium,* for the upper arm.

Biceps Brachii This is the large muscle on the anterior surface of the upper arm that bulges when the forearm is flexed. Its *origin* consists of two tendinous heads: a medial tendon, which is attached to the coracoid process, and a lateral tendon, which fits into the intertubercular groove on the humerus. The latter tendon is attached to the supraglenoid tubercle of the scapula. At the lower end of the muscle the two heads unite to form a single tendinous *insertion* on the radial tuberosity of the radius.

> *Action:* Flexion of the forearm; also, moves the radius outward to suppinate the hand.

Brachialis Immediately under the biceps brachii on the distal anterior portion of the humerus lies the brachialis. Its origin occupies the lower half of the humerus. Its insertion is attached to the front surface of the coronoid process of the ulna.

> *Action:* Flexion of the forearm.

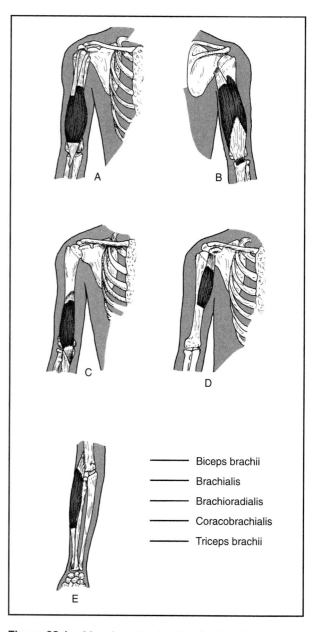

Figure 38.1 Muscles attached to the brachium.

—— Biceps brachii

—— Brachialis

—— Brachioradialis

—— Coracobrachialis

—— Triceps brachii

Brachioradialis This muscle is the most superficial muscle on the lateral (radial) side of the forearm. It originates above the lateral epicondyle of the humerus and inserts on the lateral surface of the radius slightly above the styloid process.

> *Action:* Flexion of the forearm.

Triceps Brachii The entire back surface of the brachium is covered by this muscle. It has three heads of *origin:* a long head arises from the scapula, a lateral head from the posterior surface of the humerus, and a medial head from the surface below the radial groove. The tendinous *insertion* of the muscle is attached to the olecranon process of the ulna.

Action: Extension of the forearm; antagonist of the brachialis.

Assignment:
Identify the muscles in figure 38.1 by placing the correct letter in front of each name in the legend.

Hand Movements

Muscles of the arm that cause hand movements are illustrated in figures 38.2 and 38.3. All illustrations are of the right arm; thus, if the thumb points to the left side of the page the anterior aspect is being viewed; when pointing to the right, the posterior view is observed. Keeping this in mind will facilitate understanding the descriptions that follow.

Supination and Pronation

Two muscles can cause supination: the biceps brachii and the supinator.

The **supinator** is a short muscle near the elbow that *arises* from the lateral epicondyle of the humerus and the ridge of the ulna. It curves around the upper portion of the radius and *inserts* on the lateral edge of the radial tuberosity and the oblique line of the radius.

Pronation is achieved by two pronator muscles shown in illustration B, figure 38.2. The upper muscle is the **pronator teres,** which *arises* on the medial epicondyle of the humerus and inserts on the upper lateral surface of the radius. The lower muscle, which is located in the wrist region, is the **pronator quadratus.** It *originates* on the distal portion of the ulna and *inserts* on the distal lateral portion of the radius.

Flexion of the Hand

Two muscles that flex the hand are shown in illustration C, figure 38.2. The muscle that extends diagonally across the forearm is the **flexor carpi radialis.** It *arises* on the medial epicondyle of the humerus and *inserts* on the proximal portions of the second and third metacarpals. The other flexor of the hand is the **flexor carpi ulnaris.** Its *origin* is on the medial epicondyle of the humerus and the posterior surface (olecranon process) of the ulna. Its *insertion* consists of a tendon that attaches to the base of the fifth metacarpal.

Action: Both muscles flex the hand; the radialis muscle causes abduction and the ulnaris muscle causes adduction.

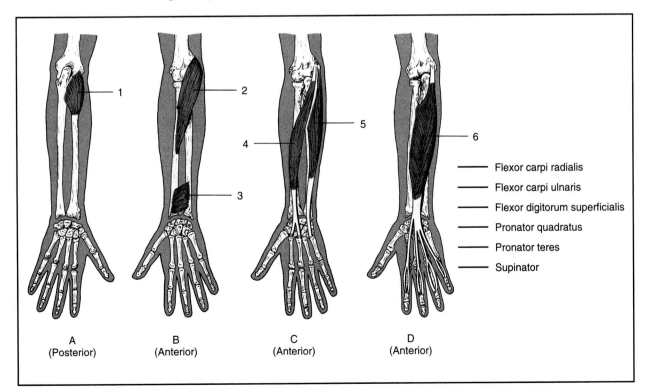

A
(Posterior)

B
(Anterior)

C
(Anterior)

D
(Anterior)

Flexor carpi radialis
Flexor carpi ulnaris
Flexor digitorum superficialis
Pronator quadratus
Pronator teres
Supinator

Figure 38.2 Forearm muscles (right arm shown).

Flexion of Fingers of the Hand

Three muscles flex the fingers. The **flexor digitorum superficialis** in illustration D, figure 38.2 flexes all fingers except the thumb. It *arises* on the humerus, ulna, and radius. Its *insertion* consists of tendons that are attached to the middle phalanges of the second, third, fourth, and fifth fingers.

The second flexor of the fingers is the **flexor digitorum profundus.** It is the larger muscle in illustration A, figure 38.3. It lies directly under the flexor digitorum superficialis. It *originates* on the ulna and the interosseous membrane between the radius and ulna. It *inserts* with four tendons on the distal phalanges of the second, third, fourth, and fifth fingers. This muscle flexes the distal portions of the fingers.

The third flexor is the **flexor pollicis longus** (Latin: *pollex,* thumb). It is the smaller muscle shown in illustration A, figure 38.3. Note that it *arises* on the radius, ulna, and interosseous membrane between these two bones. Its *insertion* consists of a tendon that is anchored to the distal phalanx of the thumb. It flexes only the thumb.

Extension of Wrist and Hand

Three muscles extend the hand at the wrist. Illustrations B and D, figure 38.3, reveal them.

The **extensor carpi radialis longus** and **brevis** are shown in illustration B. The brevis muscle is medial to the longus. Both muscles *originate* on the humerus, with the longus taking a more proximal position. The longus *inserts* on the second metacarpal; the brevis on the middle metacarpal.

The **extensor carpi ulnaris** is the third muscle involved in extending the hand. It is the muscle on the medial edge of the arm in illustration D. It *arises* on the lateral epicondyle of the humerus and part of the ulna. It *inserts* on the fifth metacarpal.

Extension and Abduction of Fingers

Three muscles cause extension of the fingers; one abducts the thumb. The **extensor digitorum communis** lies alongside the extensor carpi ulnaris; it is shown in illustration D, figure 38.3. It *arises* from the lateral epicondyle of the humerus and inserts on the distal phalanges of fingers two through five. It extends all fingers except the thumb.

The **extensor pollicis longus, extensor pollicis brevis,** and **abductor pollicis** move the thumb. The longus muscle is the lower one in illustration C. It *arises* on both the ulna and radius. It *inserts* on the distal phalanx of the thumb. The brevis muscle, which lies superior to it, is not shown in illustration C. It inserts on the proximal phalanx of the thumb and assists the longus in extending the thumb.

The abductor pollicis is the superior muscle shown in illustration C. It takes its *origin* on the interosseous membrane and it *inserts* on the lateral portion of the first metacarpal and trapezium. Its only action is to abduct the thumb.

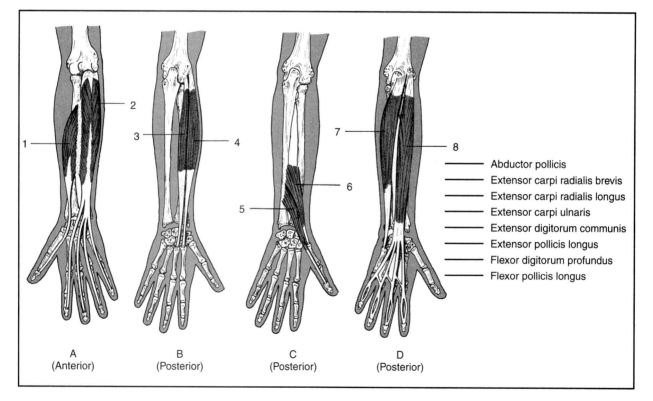

	Abductor pollicis
	Extensor carpi radialis brevis
	Extensor carpi radialis longus
	Extensor carpi ulnaris
	Extensor digitorum communis
	Extensor pollicis longus
	Flexor digitorum profundus
	Flexor pollicis longus

A (Anterior)   B (Posterior)   C (Posterior)   D (Posterior)

Figure 38.3 Forearm muscles (right arm shown).

Assignment:
Label figures 38.2 and 38.3, and complete the Laboratory Report for this exercise.

Cat Dissection

Forelimb

Muscles of the Brachium

Six homologs of the human brachium are seen in the cat. Except for the anconeus, all the human muscles were identified on pages 176 and 177.

One muscle of the cat's brachium, the *epitrochlearis,* has no homolog in the human and should be removed before studying the six homologs. It is a flat, thin, superficial muscle on the medial side of the brachium and can be seen in figure 38.5. It arises from the latissimus dorsi and is continuous, distally, with the fascia of the forearm. After removing this muscle, proceed to identify and free up with a probe the six homologs in the following order.

Triceps Brachii This muscle has three heads of origin, which merge to insert by a common tendon on the olecranon process of the ulna. The long head, located on the caudal surface of the arm, is the largest. The lateral head covers much of the lateral surface of the arm. *Dissect.* To reveal the medial head, which is between the long and lateral heads, *transect* and *reflect* the lateral head (see figure 38.6). Only the long head arises from the scapula.

Biceps Brachii In the cat this muscle lies deep in the pectoral muscles, just medial to their humoral insertions. This muscle is, therefore, located on the

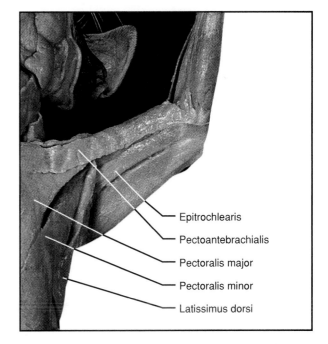

Epitrochlearis
Pectoantebrachialis
Pectoralis major
Pectoralis minor
Latissimus dorsi

Figure 38.5 Superficial forelimb muscles of the cat.

ventromedial surface of the arm (see figure 38.4). It inserts on the radius. *Dissect.*

Brachialis This muscle is located on the ventro-lateral surface of the arm, craniad to the lateral head of the triceps brachii. Refer to figure 38.6.

The insertion of the pectoralis major separates the biceps brachii and brachialis. Since this muscle arises from the lateral surface of the humerus, you will not be able to run a probe underneath it from its origin to its insertion on the ulna.

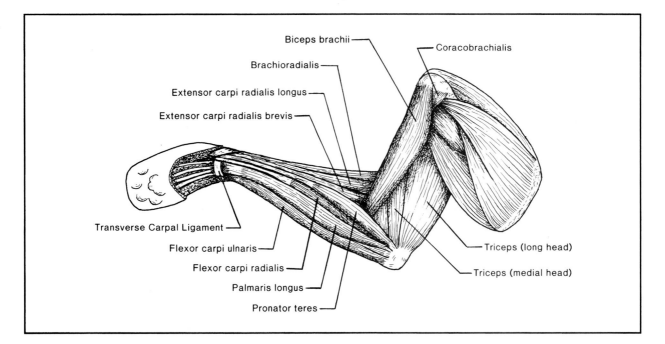

Biceps brachii
Brachioradialis
Extensor carpi radialis longus
Extensor carpi radialis brevis
Coracobrachialis
Transverse Carpal Ligament
Flexor carpi ulnaris
Flexor carpi radialis
Palmaris longus
Pronator teres
Triceps (long head)
Triceps (medial head)

Figure 38.4 Medial aspect of cat forelimb muscles.

Brachioradialis In the cat this muscle (figures 38.4 and 38.8) is a narrow ribbon, which may have been removed with the superficial fascia. It arises on the lateral side of the humerus and passes along the radial side of the forearm to insert on the styloid process of the radius. *Dissect.*

Coracobrachialis Use figures 38.4 and 38.7 to locate this small muscle. Note that it is on the medial side of the scapula caudad of the proximal (origin) end of the biceps brachii. This muscle is almost impossible to locate if the pectoralis major and minor muscles have not been previously transected and reflected.

Anconeus This small muscle is shown in figure 38.6. It consists of short superficial fibers that originate near the lateral epicondyle of the humerus, and it inserts near the olecranon. It assists the triceps in both the cat and human to extend the forearm.

Muscles of the Antebrachium

Eight muscles of the cat's forearm will be identified. As on the brachium homologs, free up each muscle with a probe. They are as follows:

Pronator Teres This muscle extends from the medial epicondyle of the humerus to the middle third of the radius. See figure 38.4. *Dissect.*

Palmaris Longus Note in figure 38.4 that this muscle is a large, flat, broad muscle near the center of the medial surface of the forearm. It arises from the medial epicondyle of the humerus and inserts by

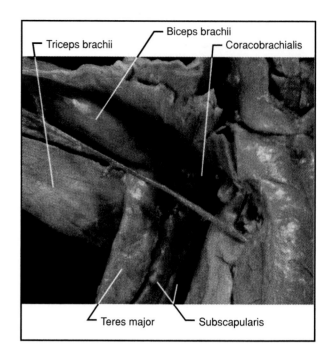

Figure 38.7 Brachium and shoulder muscles of cat.

means of four tendons on the digits. Note that at its distal end the four insertion tendons pass under a *transverse carpal ligament*. The palmaris longus is not found in about 10% of humans. *Dissect.*

Flexor Carpi Radialis This long, spindle-shaped muscle lies between the palmaris longus and the pronator teres. It arises from the medial epicondyle of the humerus and inserts at the base of the second and third metacarpals.

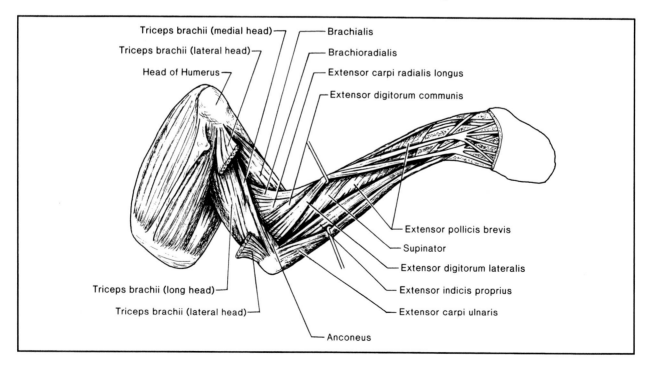

Figure 38.6 Cat forelimb muscles, lateral aspect.

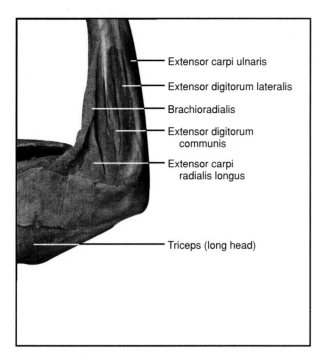

Figure 38.8 Superficial muscles of cat's forearm, lateral aspect.

Flexor Carpi Ulnaris Figure 38.4 shows the location of this muscle. What appears to be two muscles is actually two heads of origin: one originates from the medial epicondyle of the humerus and the other from the olecranon. About the middle of the ulna, the two heads join to form a single band of muscle fibers that inserts on the pisiform and hamate bones. *Dissect.*

Extensor Carpi Ulnaris This muscle, shown in figures 38.6 and 38.8, arises from the lateral epicondyle of the humerus and inserts on the base of the fifth metacarpal. *Dissect.*

Extensor Digitorum Communis Use figure 38.6 to locate this muscle. It arises from the lateral epicondyle of the humerus and extends to the wrist, where it divides into four tendons that insert on the second phalanges of digits 2–5. Note that these four tendons lie superficial to the three tendons of the extensor digitorum lateralis. *Dissect.*

Extensor Carpi Radialis Longus Use figure 38.8 to identify this muscle, which lies adjacent to the brachioradialis. Its origin is on the lateral supracondylar ridge of the humerus, and it inserts at the base of the second metacarpal. *Dissect.*

Supinator This muscle can be exposed by reflecting the extensor digitorum lateralis as shown in figure 38.6. This muscle originates from ligaments on the elbow. It inserts on the proximal end of the radius and acts as a lateral rotator of the radius (supination).

Complete exposure of the supinator would require transection and reflection of the overlying muscles. To avoid irreversible mutilation of your cat, do not do this.

Laboratory Report

Complete the Laboratory Report for this exercise.

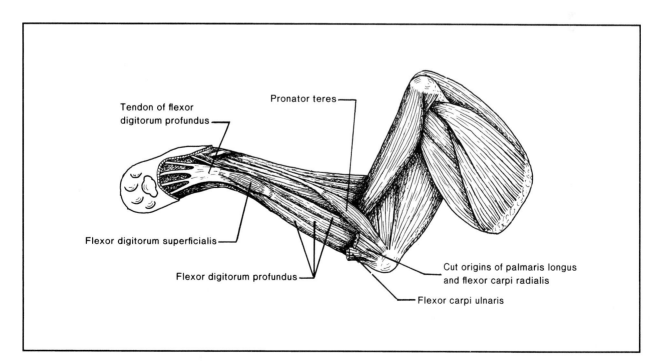

Figure 38.9 Deep muscles of cat's forearm, medial aspect.

39 Abdominal, Pelvic, and Intercostal Muscles

The muscles of the abdominal wall and pelvis will be studied in this exercise. Cat dissection will be utilized for this muscle study.

Abdominal Muscles

The abdominal wall consists of four pairs of thin muscles. The human illustration at the bottom of figure 39.1 has portions of the right abdominal wall removed to reveal the nature of the laminations. Identify the following muscles in this illustration.

Obliquus Externus *(External Oblique)* This muscle is the most superficial layer of the abdominal wall. It is ensheathed by the **aponeurosis of the obliquus externus,** which terminates at the linea alba and inguinal ligament.

The muscle takes its *origin* on the external surfaces of the lower eight ribs. Although it appears to insert on the linea semilunaris, its actual *insertion* is the **linea alba** (white line), where fibers of the left and right aponeuroses interlace on the midline of the abdomen.

The lower border of each aponeurosis forms the **inguinal ligament,** which extends from the anterior spine of the ilium to the pubic tubercle. Label 5, figure 39.3 is of this ligament.

Obliquus Internus *(Internal Oblique)* This muscle lies immediately under the external oblique, i.e., between the external oblique and the transversus abdominis.

Its *origin* is on the lateral half of the inguinal ligament, the anterior two-thirds of the iliac crest, and the thoracolumbar fascia. Its *insertion* is on the costal cartilages of the lower three ribs, the linea alba, and the crest of the pubis.

Transversus Abdominis This is the innermost muscle of the abdominal wall. It *arises* on the inguinal ligament, the iliac crest, the costal cartilages of the lower six ribs, and the thoracolumbar fascia. It *inserts* on the linea alba and the crest of the pubis.

Rectus Abdominis The right rectus abdominis is the long, narrow, segmented muscle running from the rib cage to the pubic bone. It is enclosed in a

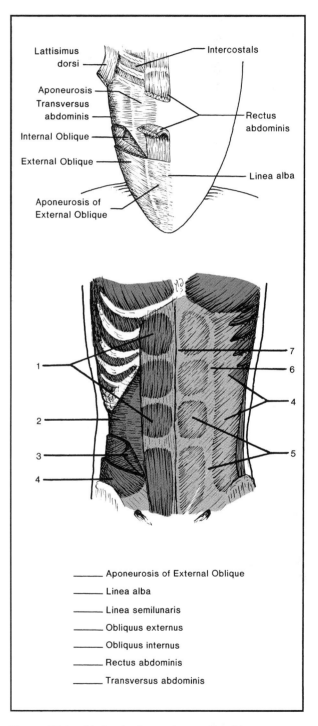

Lattisimus dorsi

Intercostals

Aponeurosis Transversus abdominis

Internal Oblique

External Oblique

Rectus abdominis

Linea alba

Aponeurosis of External Oblique

_____ Aponeurosis of External Oblique

_____ Linea alba

_____ Linea semilunaris

_____ Obliquus externus

_____ Obliquus internus

_____ Rectus abdominis

_____ Transversus abdominis

Figure 39.1 Abdominal muscles, cat and human.

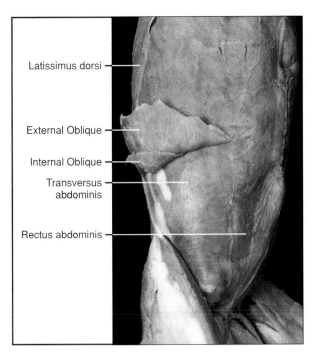

Latissimus dorsi

External Oblique

Internal Oblique

Transversus abdominis

Rectus abdominis

Figure 39.2 Abdominal muscles of the cat.

fibrous sheath formed by the aponeuroses of the above three muscles.

Its *origin* is on the pubic bone. Its *insertion* is on the cartilages of the fifth, sixth, and seventh ribs. The linea alba lies between the pair of muscles. Contraction of the rectus abdominis muscles aids in flexion of the spine in the lumbar region.

Collective Action

The above four abdominal muscles keep the abdominal organs compressed and assist in maintaining intraabdominal pressure.

They act as antagonists to the diaphragm. When the latter contracts, they relax. When the diaphragm relaxes, they contract to effect expiration of air from the lungs. They also assist in defecation, urination, vomiting, and parturition (childbirth delivery). Flexion of the body at the lumbar region is also achieved by these muscles.

Assignment:
Label figure 39.1.

Cat Dissection

Abdominal Muscles

As shown in the upper illustration of figure 39.1 and figure 39.2, the abdominal wall of the cat has the same four muscles as seen in humans. As you separate each muscle layer, pay particular attention to the direction of the fibers. Body wall strength is

greatly enhanced by the different directions of the fibers in each layer.

External Oblique This outer muscle of the body wall arises from the external surface of the lower nine ribs. Note how the cranial portion of this muscle interdigitates with the origin of the serratus ventralis. Insertion is on the iliac crest and by an aponeurosis that fuses with that of the opposite side to form the *linea alba.*

Loosen the cranial border of the muscle with a blunt probe and slide the probe underneath the muscle in a caudal direction. *Transect* alongside of the probe and *reflect.* (Remember, when you transect a muscle, your cut should be through the mid-belly half way between the origin and insertion.)

Internal Oblique To expose this muscle it will be necessary to make a shallow incision on the right side through the external oblique from the ribs to the pelvis. *Reflect* the external oblique. Since the external oblique is very thin, take care not to cut too deep. Note that the fibers of this muscle layer course at approximately right angles to those of the external oblique. Note, also, how the ends of the muscle fibers are continuous with the aponeurosis.

Transversus Abdominis This innermost muscle layer of the ventrolateral abdominal wall is very difficult to separate from the internal oblique; therefore, only a small "window" should be cut to expose the transversus abdominis.

Try to slide the tip of a blunt probe either underneath a small area of the aponeurosis or under the muscle fibers of the internal oblique. Make a cut to create a small opening and note the general transverse direction of the fibers of this muscle.

Rectus Abdominis This muscle extends from the pubis to the sternum on each side of the linea alba. It extends farther craniad in the cat than in the human. The cranial two-thirds of the aponeurosis of this muscle splits at the lateral border of the rectus abdominis, creating a dorsal leaf that passes dorsal, and a ventral leaf that passes ventral to the rectus abdominis. The dorsal and ventral leaves unite at the medial border of the rectus abdominis to help form the linea alba. Because the rectus abdominis is "wrapped" in the aponeurosis in this manner, it is difficult to pass a probe under this muscle from its origin to the insertion.

Intercostal Muscles

Although the human intercostal muscles were studied in Exercise 37, page 168, the same muscles in the cat were not studied at that time. Using figure 39.1, examine them on your specimen.

These muscles of respiration in the cat are essentially the same as in humans except that there are twelve sets in the cat and eleven sets in humans. Proceed to dissect these muscles as follows:

External Intercostals Reflect the external oblique to expose the lateral aspect of the rib cage. The external intercostal muscles can be seen occupying the intercostal spaces. Note that the direction of the fibers is forward and downward from the caudal border of one rib to the cranial border of the next rib caudad. Note, also, that the external intercostals do not reach the sternum and that the internal intercostals can be seen in the interval.

Internal Intercostals These muscles are immediately deep to the external intercostals. Cut a small window through an external intercostal between two ribs to expose one of these muscles. Note that the fibers course at approximately right angles to the fibers of the external muscle as they pass from the cranial border of the rib to the caudal border of the next rib craniad.

Pelvic Muscles

The principal muscles of the pelvic region are shown in figure 39.3. They are the quadratus lumborum, psoas major and iliacus. Muscles in this region of the cat will be dissected in the next exercise.

Quadratus Lumborum The muscle that extends from the iliac crest of the os coxa to the lowest rib in figure 39.3 is this muscle. Only the left muscle in figure 39.3 is completely exposed.

Its *origin* is on the iliac crest and the iliolumbar ligament that borders the iliac crest. Its *insertion* is on the lower border of the last rib and the apices of the transverse processes of the upper four lumbar vertebrae.

Occasionally, there is a second portion of this muscle positioned just in front of it. This part takes its *origin* on the upper borders of the transverse processes of the lower three or four lumbar vertebrae and *inserts* on the lower margin of the last rib. It does not extend down to the iliac crest.

Taken as a whole, when both portions of this muscle are present, the transverse processes of some of the lumbar vertebrae act as both origin and insertion—a truly unique situation.

Acting together, the right and left quadratus lumborums extend the spine at the lumbar verte-

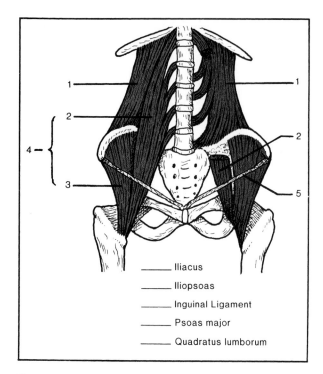

Figure 39.3 Pelvic muscles.

_____ Iliacus

_____ Iliopsoas

_____ Inguinal Ligament

_____ Psoas major

_____ Quadratus lumborum

brae. Lateral flexion or abduction results when one acts independently of the other.

Psoas Major This muscle is the long one shown in figure 39.3 on the right side of the body. A portion of the left psoas major has been cut away to reveal the full extent of the quadratus lumborum.

The psoas major *arises* from the sides of the bodies and transverse processes of the lumbar vertebrae. It *inserts* with the iliacus on the lesser trochanter of the femur.

The two psoas majors work synergistically with the rectus abdominis muscles to flex the lumbar region of the vertebral column.

Iliacus This muscle extends from the iliac crest to the proximal end of the femur. Its *origin* is the whole iliac fossa. Its *insertion* is the lesser trochanter of the femur. The psoas major and iliacus are jointly referred to as the *iliopsoas* because of their intimate relationship at their insertion.

The iliacus works synergistically with the psoas major to flex the femur on the trunk.

Laboratory Report

After labeling figure 39.3, answer the questions on combined Laboratory Report 39,40 that pertain to the muscles studied in this exercise.

40 Lower Extremity Muscles

Twenty-seven muscles of the leg will be studied in this exercise. As in the case of the upper extremity muscles, they are grouped according to type of movement.

Thigh Movements

Seven muscles that move the femur are shown in figure 40.1. All originate on a part of the pelvis.

Gluteus Maximus This muscle of the buttock region is covered by a deep fascia, the *fascia lata,* that completely invests the thigh muscles. Emerging downward from the fascia lata is a broad tendon, the **iliotibial tract,** which is attached to the tibia.

This muscle *arises* on the ilium, sacrum, and coccyx; it *inserts* on the iliotibial tract and the posterior part of the femur.

Action: Extension and outward rotation of the femur.

Gluteus Medius This muscle lies immediately under the gluteus maximus, covering a good portion of the ilium. Its *origin* is on the ilium and its *insertion* is on the lateral part of the greater trochanter.

Action: Abduction and medial rotation of the femur.

Gluteus Minimus This is the smallest of the three gluteal muscles and is located immediately under the gluteus medius. It, too, *arises* on the ilium; it *inserts* on the anterior border of the greater trochanter.

Action: Abduction, inward rotation, and slight flexing of the femur.

Piriformis This small muscle takes its *origin* on the anterior surface of the sacrum and *inserts* on the upper border of the greater trochanter.

Action: Outward rotation, some abduction, and extension of the femur.

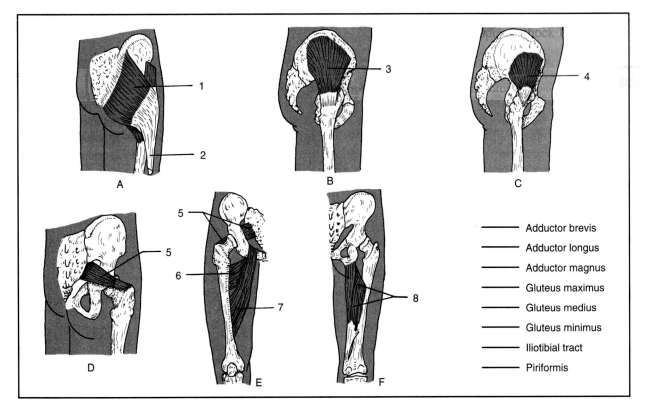

Adductor brevis
Adductor longus
Adductor magnus
Gluteus maximus
Gluteus medius
Gluteus minimus
Iliotibial tract
Piriformis

Figure 40.1   Muscles that move the femur.

185

Adductor Muscles   Three adductors are shown in illustrations E and F of figure 40.1. The **adductor magnus** (illustration F) is the largest one. It *arises* on the inferior portions of the pubis and ischium and it *inserts* on the linea aspera of the femur.

The **adductor longus** (illustration E) *originates* on the front of the pubis and *inserts* on the linea aspera. The **adductor brevis** (illustration E) *arises* on the posterior side of the pubis and *inserts* on the femur above the longus muscle.

Action: In addition to adduction, these muscles flex and rotate the femur medially.

Assignment:
Label figure 40.1

Thigh and Lower Leg Movements

The muscles illustrated in figure 40.2 are primarily concerned with flexion and extension of the lower part of the leg.

Hamstrings   The three muscles shown in illustration A, figure 40.2, are collectively known as the "hamstrings." They are as follows:

Biceps femoris This muscle occupies the most lateral position of the three hamstrings. It has two heads: one long, and the other short. The long head, which obscures the short one, *arises* on the ischial tuberosity. The short head *originates* on the linea aspera of the femur. *Insertion* of the muscle is on the head of the fibula and the lateral condyle of the tibia.

Semitendinosus Medial to the biceps femoris is this muscle, which *arises* on the ischial tuberosity and *inserts* on the upper end of the shaft of the tibia.

Semimembranosus This muscle occupies the most medial position of the three hamstrings. Its origin consists of a thick semimembranous tendon attached to the ischial tuberosity. Its insertion is primarily on the posterior medial part of the medial condyle of the tibia.

Collective Action: All of these muscles flex the calf upon the thigh. They also extend and rotate the thigh. Rotation by the biceps is outward; the other two muscles cause inward rotation.

Quadriceps Femoris   The large muscle that covers the anterior portion of the thigh is this muscle. It consists of four parts, which are shown in illustrations C and D, figure 40.2. Note that all of them unite to form a common tendon that passes over the patella to *insert* on the tibia.

Rectus femoris This portion of the quadriceps occupies a superficial central position. It *arises* by two tendons: one from the anterior inferior iliac spine and the other from a groove just above the acetabulum. The lower portion of the muscle is a broad aponeurosis that terminates in the tendon of insertion.

Vastus lateralis This lateral portion of the quadriceps is the largest of the four muscles. It *arises* from the lateral lip of the linea aspera.

Vastus medialis The most medial portion of the quadriceps is this muscle. It *arises* from the linea aspera.

Vastus intermedius Since this muscle is deep to the above three, it is shown separately in illustration D. Its *origin* is on the front and lateral surfaces of the femur.

Collective Action: The entire quadriceps femoris extends (straightens) the leg. Flexion of the thigh, however, is achieved by the rectus femoris.

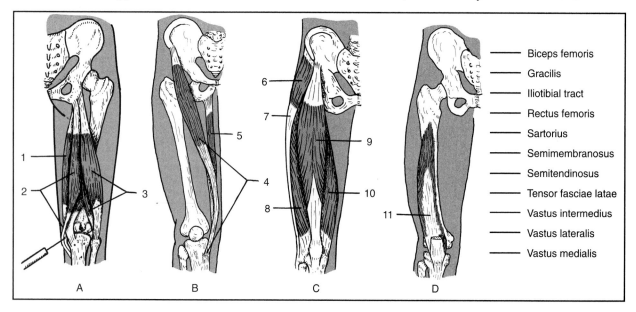

Biceps femoris
Gracilis
Iliotibial tract
Rectus femoris
Sartorius
Semimembranosus
Semitendinosus
Tensor fasciae latae
Vastus intermedius
Vastus lateralis
Vastus medialis

Figure 40.2   Muscles that move the tibia and fibula.

Sartorius This is the longest muscle shown in illustration B. It arises from the anterior superior spine of the ilium and inserts on the medial surface of the tibia.

> *Action:* Flexes the calf on the thigh, the thigh upon the pelvis, and rotates the leg laterally.

Tensor Fasciae Latae This muscle is shown in illustration C, figure 40.2. It *arises* from the anterior outer lip of the iliac crest, the anterior superior spine, and from the deep surface of the fascia lata. It is *inserted* between the two layers of the iliotibial tract at about the junction of the middle and upper thirds of the thigh.

> *Action:* Flexes the thigh and rotates it slightly medially.

Gracilis This muscle is located on the medial surface of the thigh. It *arises* from the lower margin of the pubic bone and *inserts* on the medial surface of the tibia near the insertion of the sartorius.

> *Action:* Adducts, flexes, and rotates the thigh medially.

Assignments:
Label figure 40.2

Lower Leg and Foot Movements

Figures 40.3 and 40.4 illustrate most of the muscles of the lower leg that cause flexion and foot movements.

Triceps Surae The large superficial muscle that covers the calf of the leg is the triceps surae. It consists of two parts that are shown in illustrations A

and B, figure 40.3. They are united by a common **tendon of Achilles** that *inserts* on the calcaneous of the foot.

> **Gastrocnemius** This outer portion of the triceps surae has two heads that *arise* from the posterior surfaces of the medial and lateral condyles of the femur. It can flex the calf on the thigh as well as cause plantar flexion.

> **Soleus** This muscle *arises* on the heads of the fibula and tibia. Plantar flexion is its only action.

Tibialis Anterior The anterior lateral portion of the tibia is covered by this muscle. It *arises* from the lateral condyle and upper two-thirds of the tibia. Its distal end is shaped into a long tendon that passes over the tarsus and *inserts* on the inferior surface of the first cuneiform and first metatarsal bones.

> *Action:* Dorsiflexion and inversion of the foot.

Tibialis Posterior This muscle lies on the posterior surfaces of the tibia and fibula. It *arises* from both these bones and the interosseous membrane. It *inserts* on the inferior surfaces of the navicular, cuneiforms, cuboid, and second through fourth metatarsals.

> *Action:* Plantar flexion and inversion of the foot; assists in maintenance of the longitudinal and transverse arches of the foot.

The Peroneus Muscles (Latin: *peroneus*, fibula) Three peroneus muscles are shown in illustrations A and B, figure 40.4. Note that all three of them *originate* on the fibula. The **peroneus longus** is the

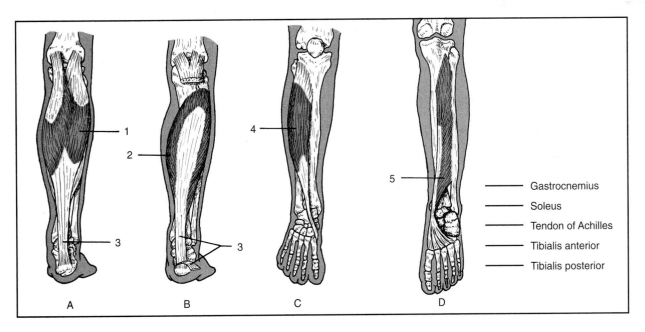

	Gastrocnemius
	Soleus
	Tendon of Achilles
	Tibialis anterior
	Tibialis posterior

Figure 40.3 **Lower leg muscles.**

longest one and takes its *origin* on the head and upper two-thirds of the fibula. The **peroneus tertius** is the smallest one, and the **peroneus brevis** is the one of in-between size. Note that the longus *inserts* on the first metatarsal and second cuneiform bones. The brevis and tertius muscles *insert* at different points on the fifth metatarsal bone.

Action: The longus and brevis muscles cause plantar flexion and eversion of the foot. The peroneus tertius causes dorsiflexion and eversion of the foot.

Flexor Muscles   Two flexor muscles of the foot are shown in illustration C, figure 40.4. The **flexor hallucis longus** (Latin: *hallux,* big toe) has its *origin* on the fibula and intermuscular septa. Its long distal tendon *inserts* on the distal phalanx of the great toe. It flexes the great toe.

The **flexor digitorum longus** is the longer muscle shown in illustration C that *originates* on the tibia and the fascia that covers the tibialis posterior. Its distal end divides into four tendons that *insert* on the bases of the distal phalanges of the second, third, fourth, and fifth toes. It flexes the distal phalanges of the four smaller toes.

Extensor Muscles   Two extensors of the foot are shown in illustration D, figure 40.4. The **extensor digitorum longus** is the longer one that takes its *origin* on the lateral condyle of the tibia, part of the fibula, and part of the interosseous membrane. Its tendon of insertion divides into four parts that *insert* on the superior surfaces of the second and third phalanges of the four smaller toes. It extends the prox-

imal phalanges of the four smaller toes. It also flexes and inverts the foot.

The **extensor hallucis longus** is the other extensor in illustration D. It takes its *origin* on the fibula and interosseous membrane. Its distal tendon *inserts* at the base of the distal phalanx of the great toe. This muscle extends the proximal phalanx of the great toe and aids in dorsiflexion of the foot.

Assignment:
Label figures 40.3 and 40.4.

Cat Dissection
Hip Muscles

Remove the superficial fat and fascia that encase the muscles of the hip and thigh so that muscle fiber direction can be determined. *Take care not to remove the fascia lata and iliotibial tract.* The latter is a white, tendinous band on the lateral side of the thigh. Although the sartorius muscle is usually considered a muscle of the thigh, it will be dissected here first to expose the full extent of the tensor fasciae latae.

Sartorius   This broad straplike muscle occupies the cranial half of the medial side of the thigh. Refer to figures 40.5 and 40.8 for reference. *Dissect, transect,* and *reflect.*

Tensor Fasciae Latae   This fan-shaped muscle is located on the anterolateral side of the hip region somewhat ventral to the other muscles in the group.

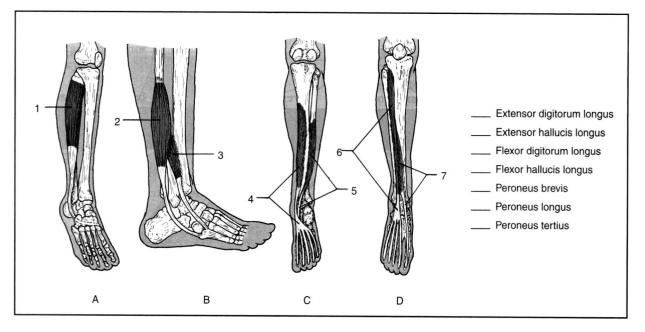

_____ Extensor digitorum longus

_____ Extensor hallucis longus

_____ Flexor digitorum longus

_____ Flexor hallucis longus

_____ Peroneus brevis

_____ Peroneus longus

_____ Peroneus tertius

Figure 40.4   Lower leg muscles.

Refer to figures 40.5 and 40.6. It arises from the lateral ilium and *inserts* in the iliotibial tract portion of the fascia lata. The fascia lata is a tough layer of connective tissue that can be seen covering the vastus lateralis muscle.

Slide a blunt probe underneath the fascia lata to separate it from the vastus lateralis and *transect* the fascia lata so that you can completely *reflect* the tensor fasciae latae. (The tensor muscle usually fuses somewhat with the gluteus maximus. This connection should be cut.)

Gluteus Medius   With the tensor fasciae latae fully reflected, you can see the full extent of the gluteus medius. This thick muscle arises from the lateral surface of the ilium and inserts on the greater trochanter. In the cat this muscle is much larger than the gluteus maximus and will be found craniad of the latter.

Gluteus Maximus   This small hip muscle lies just caudad to the gluteus medius. It arises from the last sacral and first caudal vertebrae and inserts on the greater trochanter; therefore, the origin and insertion in the cat are far less extensive than in the human. *Dissect.*

Caudofemoralis   This small muscle lies just caudad to the gluteus maximus. See figure 40.6. It is difficult to see because it is "hiding" below the cranial border of the biceps femoris. It arises on the

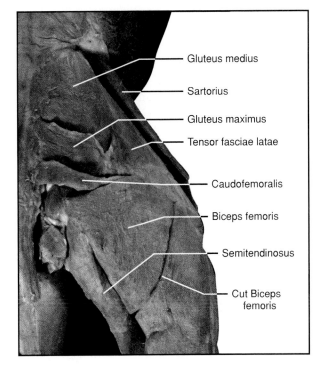

Figure 40.6   Hip muscles of the cat.

caudal vertebrae and inserts by a very thin tendon on the patella.

This muscle is not present in humans; however, in the cat the caudofemoralis and the gluteus maximus together are more comparable to the human

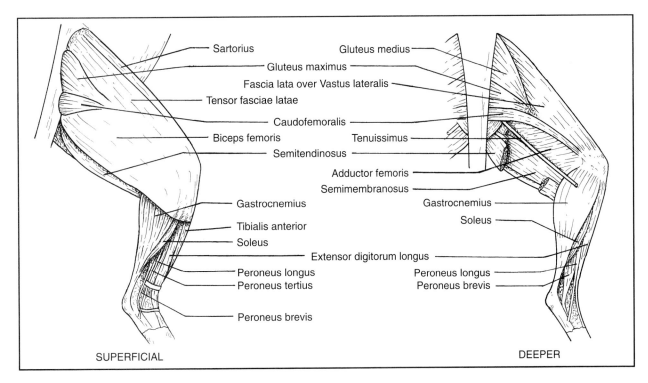

Figure 40.5   Lateral aspect of hindlimb.

gluteus maximus than is the cat's gluteus maximus alone. *Dissect.*

Thigh Muscles

Thirteen muscles make up the musculature of the cat's thigh. Identify them in the following order:

Sartorius   Previously dissected (see p. 188).

Gracilis   A wide, flat muscle covering most of the caudal portion of the medial side of the thigh. Refer to figures 40.7 and 40.9. At the insertion the fibers end in a very thin flat tendon, part of which becomes continuous with the fascia covering the distal portion of the leg. *Dissect, transect,* and *reflect.*

The Hamstrings   The cat has the same group of three muscles that are designated as "hamstrings" in humans.

> **Semimembranosus** This is a large, thick muscle seen just beneath the gracilis; therefore, the latter must be reflected to see its full extent. Dorsal to this muscle is seen the adductor femoris and ventral to it is the semitendinosus.
>
> *Dissect* by separating this muscle from the adductor femoris and semitendinosus. Whereas the human homolog inserts only on the tibia, this muscle in the cat inserts on both the femur and tibia.
>
> **Semitendinosus** This large band of muscle is on the ventral border of the thigh between the semimembranosus and the biceps femoris. It inserts on the tibia. *Dissect.*

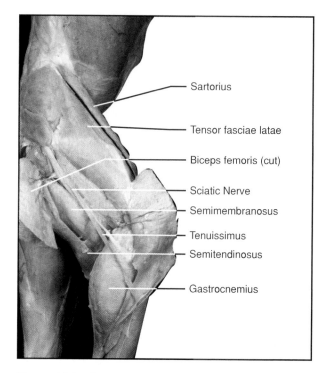

Figure 40.8   Deep muscles of thigh, lateral aspect.

Biceps Femoris This large broad muscle covers most of the lateral surface of the thigh. *Dissect, transect,* and *reflect.* When the muscle is transected, be careful not to cut the caudofemoralis and sciatic nerve. The nerve, shown in figure 40.8, appears as a white cord just beneath the biceps femoris.

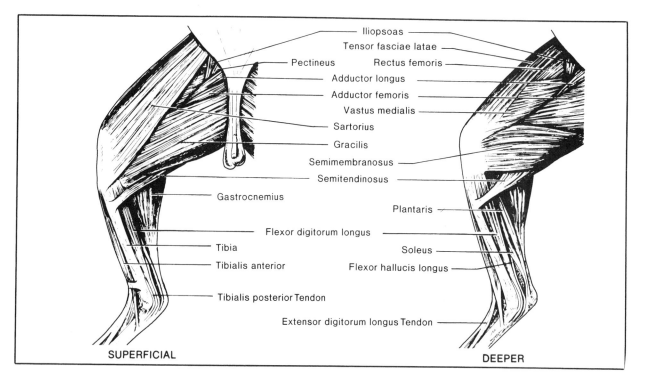

Figure 40.7   Medial aspect of hindlimb.

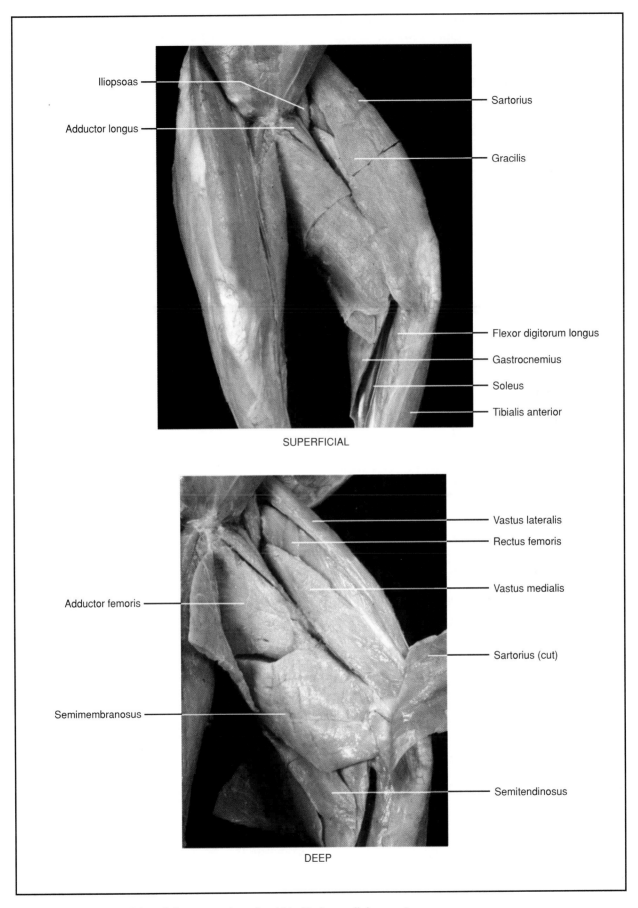

SUPERFICIAL

Iliopsoas

Adductor longus

Sartorius

Gracilis

Flexor digitorum longus

Gastrocnemius

Soleus

Tibialis anterior

Vastus lateralis

Rectus femoris

Vastus medialis

Adductor femoris

Sartorius (cut)

Semimembranosus

Semitendinosus

DEEP

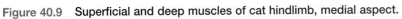

Figure 40.9 Superficial and deep muscles of cat hindlimb, medial aspect.

Adductor Femoris This muscle lies deep to the gracilis and dorsal to the semimembranosus. It originates on the hipbone and inserts on the femur. The adductor femoris corresponds to the adductor magnus and adductor brevis of the human. *Dissect.*

Adductor Longus This thin muscle lies along the cranial border of the adductor femoris and extends from the hipbone to the femur.

Pectineus A very small triangular muscle (see figure 40.7) that lies cranial to the adductor longus and extends from the hipbone to the proximal end of the femur.

Iliopsoas As in humans, this name is applied to a pair of muscles, the *iliacus* and the *psoas major.* Because they have a common insertion on the lesser trochanter, they are grouped together. Only the distal (insertion) end will be observed at this time. Note that the muscle fibers of the iliopsoas are oriented almost perpendicular to the fibers of the pectineus and adductor longus.

Quadriceps Femoris As in humans, the cat has a quadriceps femoris that consists of four muscles. This great extensor muscle of the knee joint has a common insertion by a large tendon that extends to the patella, attaches around the patella, and passes to the tibial tuberosity to insert.

> **Vastus lateralis** Large muscle deep to the tensor fasciae latae that occupies the craniolateral surface of the thigh.
>
> **Vastus medialis** Large muscle deep to the sartorius that occupies the craniomedial surface of the thigh.
>
> **Rectus femoris** A cigar-shaped muscle bordered laterally and medially by the vastus medialis. *Dissect, transect,* and *reflect.*
>
> **Vastus intermedius** A flat muscle deep to the rectus femoris and attached to the anterior surface of the femur. The rectus femoris must be reflected to see the full extent of this muscle. Try to separate the vastus intermedius from the vastus lateralis and vastus medialis.

Leg and Foot Muscles

Five muscles of the leg and foot of the cat will be identified here.

Gastrocnemius This muscle on the dorsal side of the leg has two heads of origin: a lateral head from the lateral epicondyle of the femur and a medial head from the meidal epicondyle of the femur. The muscle inserts by way of the calcaneous (Achilles) tendon onto the calcaneus.

In the cat a large *plantaris* (figure 40.7) muscle can be seen between the heads of the gastrocnemius. In humans the plantaris is a small fusiform muscle of little importance. *Dissect.*

Soleus Deep to the lateral head of the gastrocnemius lies the soleus muscle. Note that, like in humans, this muscle inserts with the gastrocnemius on the calcaneous bone by way of the calcaneous tendon.

Tibialis Anterior This tapered band of muscle is situated on the anterolateral (ventrolateral) aspect of the tibia. Follow the tendon of insertion of this muscle as it crosses the ankle obliquely to reach the medial surface of the foot.

Extensor Digitorum Longus This muscle is covered throughout most of its length by the tibialis anterior. Separate these two muscles and follow the tendon of insertion as it crosses the ankle ventrally and divides into four tendons that are distributed to the digits. *Dissect.*

Peroneus Muscles As do humans, the cat has three peroneus muscles: *peroneus longus, peroneus brevis,* and *peroneus tertius.* These muscles are found on the lateral side of the leg between the extensor digitorum longus and the soleus.

The positions of the tendons of insertion in relation to the lateral malleolus in the cat differ from those in humans; also, the insertion points are somewhat different. Therefore, it is best to identify these muscles simply as a "group" and not try to distinguish which muscle is which.

Surface Muscles Review

Figure 40.10 has been included in this exercise to summarize our study of the muscles of the body as a whole. Only the surface muscles are shown. To determine your present understanding of the human musculature, attempt to label these diagrams first *by not referring back to previous illustrations.* This type of self-testing will determine what additional study is needed.

Laboratory Report

Complete the last portion of combined Laboratory Report 39,40.

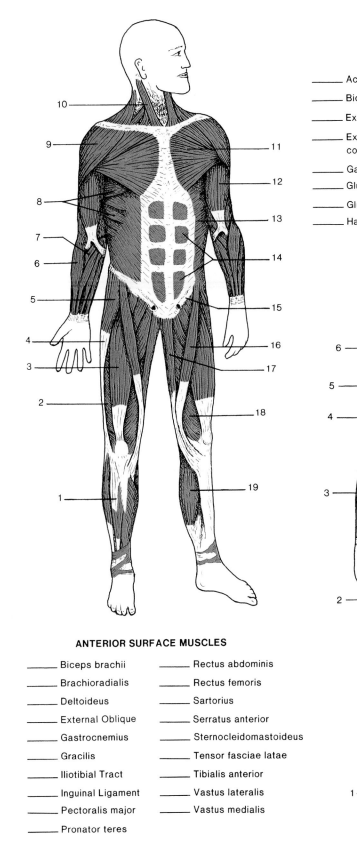

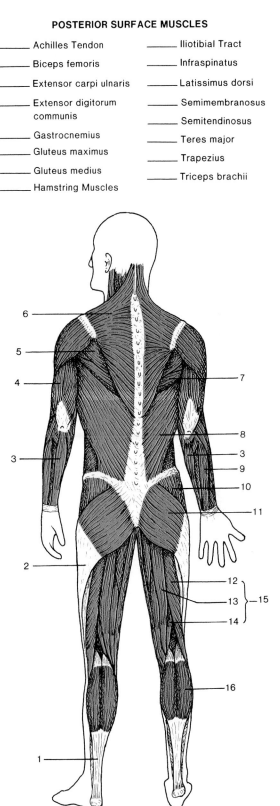

POSTERIOR SURFACE MUSCLES

_____ Achilles Tendon
_____ Biceps femoris
_____ Extensor carpi ulnaris
_____ Extensor digitorum communis
_____ Gastrocnemius
_____ Gluteus maximus
_____ Gluteus medius
_____ Hamstring Muscles

_____ Iliotibial Tract
_____ Infraspinatus
_____ Latissimus dorsi
_____ Semimembranosus
_____ Semitendinosus
_____ Teres major
_____ Trapezius
_____ Triceps brachii

ANTERIOR SURFACE MUSCLES

_____ Biceps brachii
_____ Brachioradialis
_____ Deltoideus
_____ External Oblique
_____ Gastrocnemius
_____ Gracilis
_____ Iliotibial Tract
_____ Inguinal Ligament
_____ Pectoralis major
_____ Pronator teres

_____ Rectus abdominis
_____ Rectus femoris
_____ Sartorius
_____ Serratus anterior
_____ Sternocleidomastoideus
_____ Tensor fasciae latae
_____ Tibialis anterior
_____ Vastus lateralis
_____ Vastus medialis

Figure 40.10 Major surface muscles of the body.

PART 8

The Nervous System

The nervous system consists of an intricate maze of neurons and receptors that functions to coordinate various physiological activities of the body. It includes the brain and spinal cord of the *central nervous system* (CNS) and is divided functionally into the *somatic* (voluntary) and *autonomic* (involuntary) systems.

This unit is divided into seven exercises. The first three exercises are devoted to the anatomy and physiology of reflexes. In Exercise 41 the anatomy of the spinal cord and its nerves is studied in relationship to the reflex arc. Exercise 42 has two parts: frog experimentation and a study of some of the more common diagnostic human reflexes. In Exercise 43 the patellar reflex is analyzed using the Intelitool Flexicomp.™

Exercise 44 is an experimental attempt to measure the approximate speed of the nerve impulse in the ulnar nerve of the arm by utilizing an electronic stimulator and chart recorder.

Exercises 45 and 46 pertain to an in-depth study of the anatomy of the brain. Although preserved human brains will be studied in these laboratory sessions, only sheep brains will be used for dissection. In the last experiment of this unit, brain waves will be monitored to produce an electroencephalogram (EEG) of a subject in the laboratory.

41

The Spinal Cord, Spinal Nerves, and Reflex Arcs

Automatic stereotyped responses to various types of stimuli enable animals to adjust quickly to adverse environmental changes. These automatic responses, generated by the nervous system, are called **reflexes.**

Reflexes that result in automatic regulation of body function can be either somatic or visceral. **Somatic reflexes** involve skeletal muscle responses, and **visceral reflexes** involve the adjustments of smooth muscle, cardiac muscle, and glands to stimuli.

The neural pathway utilized in performing a reflex is called a **reflex arc.** This pathway involves receptors, spinal nerves, the spinal cord, and effectors. It is the purpose of this exercise to study each of the components that are involved in both types of reflexes.

The Spinal Cord

The spinal cord is a downward extension of the medulla oblongata of the brain. Figure 41.1 reveals its gross structure, as seen posteriorly. To expose it, the posterior portions of the vertebrae and sacrum have been removed.

The spinal cord starts at the upper border of the atlas and terminates as the **conus medullaris** at the lower border of the first lumbar vertebra.

In fetal life the spinal cord occupies the entire length of the vertebral canal (spinal cavity), but as the vertebral column elongates during growth, the spinal cord fails to lengthen with it; thus, the vertebral canal extends downward beyond the end of the spinal cord.

Extending downward from the conus medullaris is an aggregate of fibers called the **cauda equina** (horse's tail), which fills the lower vertebral canal. The innermost fiber of the cauda equina, which is located on the median line, is called the **filum terminale internus.** This delicate prolongation of the conus medullaris becomes the **filum terminale externa** after it passes through the **dural sac** (dura mater), which lines the vertebral canal. The dural sac, incidentally, is continuous with the dura mater that surrounds the brain. Only a portion of it (label 25) is shown in figure 41.1. The remainder of it has been left out of this illustration for clarity.

The Spinal Nerves

The spinal nerves emerge from the spinal cord in pairs through the intervertebral foramina on each side of the spinal cavity. There are eight cervical, twelve thoracic, five lumbar, and five sacral pairs. These thirty pairs of nerves, plus one pair of coccygeal nerves, make a total of thirty-one pairs of spinal nerves.

References to the sectional view of the spinal cord in figure 41.1 reveals that each spinal nerve is connected to the spinal cord by structures called **anterior** and **posterior roots.** Note that the posterior root (label 21) has an enlargement, the **spinal ganglion,** which contains cell bodies of sensory neurons. Neuronal fibers from these roots pass through the outer **white matter** of the spinal cord into the inner **gray matter** where neuronal connections *(synapses)* are made.

Cervical Nerves The first cervical nerve (C_1) emerges from the spinal cord in the space between the base of the skull and the atlas. The eighth cervical nerve (C_8) emerges between the seventh cervical and first thoracic vertebrae. Note that the first four cervical nerves are united in the neck region to form a network called the **cervical plexus.** The remaining cervical nerves (C_5, C_6, C_7, and C_8) and the first thoracic nerve unite to form another network called the **brachial plexus** in the shoulder region.

Thoracic *(Intercostal)* **Nerves** The first thoracic nerve (T_1) emerges through the intervertebral foramen between the first and second thoracic vertebrae. The twelfth thoracic nerve (T_{12}) emerges through the foramen between the twelfth thoracic and first lumbar vertebrae. Each of these nerves lies adjacent to the lower margin of the rib above it.

Lumbar Nerves The first lumbar nerve (L_1) emerges from the intervertebral foramen between the first and second lumbar vertebrae. Close to where the nerve emerges from the spinal cavity, it divides to form an upper **iliohypogastric** nerve and a lower **ilioinguinal** nerve.

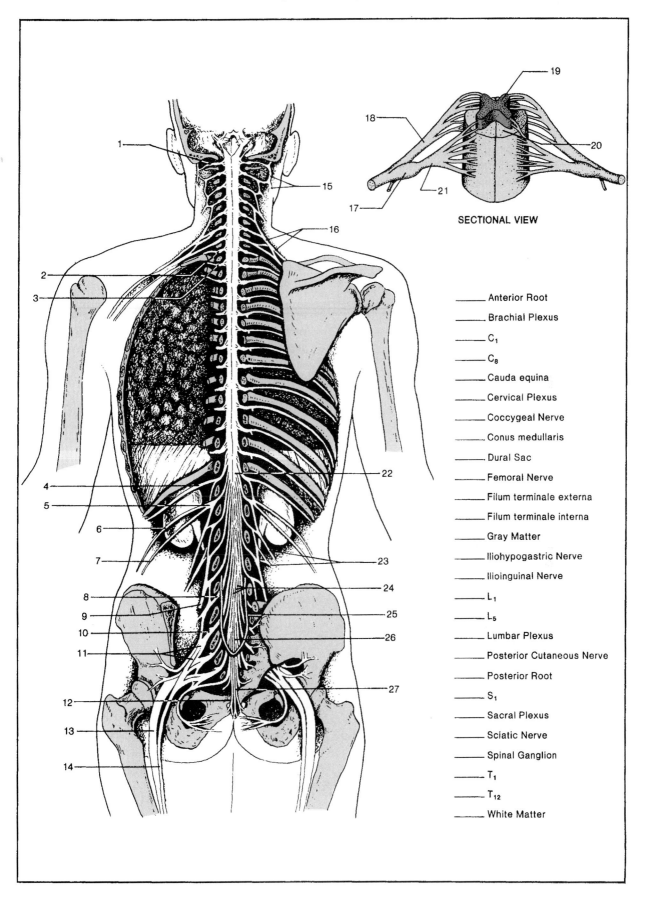

SECTIONAL VIEW

_____ Anterior Root

_____ Brachial Plexus

_____ C_1

_____ C_8

_____ Cauda equina

_____ Cervical Plexus

_____ Coccygeal Nerve

_____ Conus medullaris

_____ Dural Sac

_____ Femoral Nerve

_____ Filum terminale externa

_____ Filum terminale interna

_____ Gray Matter

_____ Iliohypogastric Nerve

_____ Ilioinguinal Nerve

_____ L_1

_____ L_5

_____ Lumbar Plexus

_____ Posterior Cutaneous Nerve

_____ Posterior Root

_____ S_1

_____ Sacral Plexus

_____ Sciatic Nerve

_____ Spinal Ganglion

_____ T_1

_____ T_{12}

_____ White Matter

Figure 41.1 The spinal cord and spinal nerves.

Note in figure 41.1 that L_1, L_2, L_3, and most of L_4 are united to form the **lumbar plexus** (label 23). The largest trunk emanating from this plexus is the **femoral** nerve, which innervates part of the leg. In reality, the femoral is formed by the union of L_2, L_3, and L_4.

Sacral and Coccygeal Nerves The remaining nerves of the cauda equina include five pairs of sacral and one pair of coccygeal nerves. A **sacral plexus** (label 11) is formed by the union of roots of L_4, L_5, and the first three sacral nerves (S_1, S_2, and S_3). The largest nerve that emerges from this plexus is the **sciatic** nerve, which passes down into the leg. The smaller nerve that parallels the sciatic is the **posterior cutaneous** nerve of the leg. The **coccygeal** nerve (label 12) contains nerve fibers from the fourth and fifth sacral nerves (S_4 and S_5).

Assignment:
Label figure 41.1.

Spinal Cord Sectional Details

Injury to the spinal cord is medically grave in that once the spinal cord is traumatized or severed, the damage may be permanent. Fortunately, the bony protection of the spinal cavity and the tissues around the spinal cord afford considerable protection. The presence of various kinds of connective tissue, meninges, and cerebrospinal fluid act as a complex shock absorber to protect the vulnerable cord. The cross-sectional view shown in figure 41.2 illustrates how the cord and spinal nerves are surrounded by these protective layers.

Cord Structure Before examining the surrounding tissues, identify the internal details of the spinal cord in figure 41.2. The most noticeable characteristic of the spinal cord is the pattern of the **gray matter,** which has some semblance to the configuration of the outstretched wings of a swallow-tailed butterfly. Surrounding the darker material is the **white matter,** which consists primarily of myelinated nerve fibers.

Along its median line the spinal cord is divided by a septum and a fissure. Note that the dorsal portion of the cord is bisected by a **posterior median septum** of neuroglial tissue that extends deep into the white matter. On the ventral surface of the cord is an **anterior median fissure.** In the gray matter on the median line lies a tiny **central canal,** evidence of the tubular nature of the spinal cord. This canal is continuous with the ventricles of the brain and extends into the filum terminale. It is lined with ciliated ependymal cells.

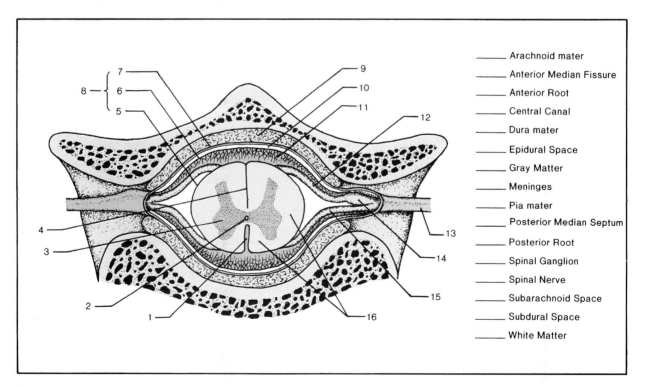

_____ Arachnoid mater
_____ Anterior Median Fissure
_____ Anterior Root
_____ Central Canal
_____ Dura mater
_____ Epidural Space
_____ Gray Matter
_____ Meninges
_____ Pia mater
_____ Posterior Median Septum
_____ Posterior Root
_____ Spinal Ganglion
_____ Spinal Nerve
_____ Subarachnoid Space
_____ Subdural Space
_____ White Matter

Figure 41.2 The spinal cord and meninges.

Spinal Nerves Note how the spinal nerves (label 13, figure 41.2) emerge from between the vertebrae on each side of the spinal column. Identify the **posterior root** with its **spinal ganglion** and the **anterior root.**

The Meninges Surrounding the spinal cord (and brain) are three meninges (*meninx,* singular). The outermost membrane is the **dura mater** (label 7), which also forms a covering for the spinal nerves. A cutaway portion of the dura on each side in figure 41.2 reveals the inner spinal nerve roots. The dura mater is the toughest of the three meninges, consisting of fibrous connective tissue. Between the dura and the bony vertebrae is an **epidural space,** which contains loose (areolar) connective tissue and blood vessels.

The innermost meninx is the **pia mater.** It is a thin delicate membrane that is actually the outer surface of the spinal cord. Between the pia mater and dura mater is the third meninx, the **arachnoid mater.** Note that the other surface of the arachnoid mater lies adjacent to the dura mater. The space between the dura and arachnoid mater is actually a potential cavity that is called the **subdural space.** The inner surface of the arachnoid mater has a delicate fibrous texture that forms a netlike support around the spinal cord. The space between the arachnoid mater and the pia mater is the **subarachnoid space.** This space is filled with cerebrospinal fluid that provides a protective fluid cushion around the spinal cord.

Assignment:
Label figure 41.2.

The Somatic Reflex Arc

A brief mention of the somatic reflex arc was made in Exercise 18. In that exercise our prime concern was with the structure of neuronal elements that make up the circuit. Figure 41.3 is very similar to figure 18.1, page 87, although figure 41.3 is simpler.

The principal components of this type of reflex arc are (1) a **receptor,** such as a neuromuscular spindle or a cutaneous end-organ that receives the stimulus; (2) a **sensory** (afferent) **neuron,** which carries impulses through a peripheral nerve and posterior root to the spinal cord; (3) and **interneuron** (association neuron), which forms synaptic connections between the sensory and motor neuron in the gray matter of the spinal cord; (4) a **motor** (efferent) **neuron,** which carries nerve impulses from the central nervous system through the anterior root to the effector via a peripheral nerve; and (5) an **effector** (muscle tissue), which responds to the stimulus by contracting.

Reflex arcs may have more than one interneuron or may be completely lacking in interneurons. If no interneuron is present, as is true of the knee jerk reflex, the reflex arc is said to be *monosynaptic.* When one or more interneurons exist, the reflex arc is classified as being *polysynaptic.*

The Visceral Reflex Arc

Muscular and glandular responses of the viscera to internal environmental changes are controlled by the **autonomic nervous system.** The reflexes of this system follow a different pathway in the nervous system. Figure 41.4 illustrates the components of a visceral reflex arc.

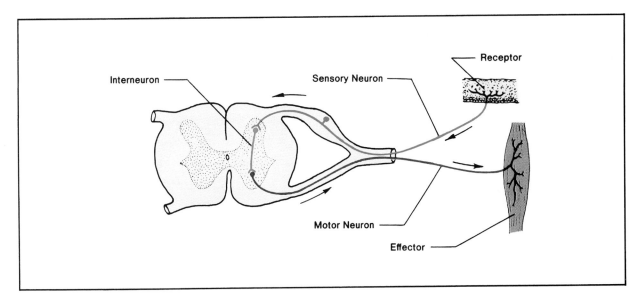

Figure 41.3 The somatic reflex arc.

The principal anatomical difference between somatic and visceral reflex arcs is that visceral reflex arcs have two efferent neurons instead of only one. These two efferent neurons make synaptic connection with each other outside of the central nervous system in an autonomic ganglion.

Two types of autonomic ganglia are shown in figure 41.4: vertebral and collateral. The **vertebral autonomic ganglia** are united to form a chain that lies along the vertebral column. The **collateral ganglia** are located farther away from the central nervous system.

A visceral reflex originates with a stimulus acting on a receptor in the viscera. Impulses pass along the dendrite of a **visceral afferent neuron.** Note that the cell bodies of these neurons are located in the **spinal ganglion.** The axon of the visceral afferent neuron forms a synapse with the **preganglionic efferent neuron** in the spinal cord. Impulses in this neuron are conveyed to the **postganglionic efferent neuron** at synaptic connections in either type of autonomic ganglion (note the short portions of cutoff postganglionic effer-

ent neurons in the vertebral autonomic ganglia). From these ganglia the postganglionic efferent neuron carries the impulses to the organ that is innervated.

In this brief mention of the autonomic nervous system, the student is reminded that the system consists of two parts: the sympathetic and parasympathetic divisions. The *sympathetic* (thoracolumbar) *division* includes the spinal nerves of the thoracic and lumbar regions. The *parasympathetic* (craniosacral) *division* incorporates the cranial and sacral nerves. Most viscera of the body are innervated by nerves from both of these divisions. This form of double innervation enables an organ to be stimulated by one system and inhibited by the other.

Assignment:
Label figure 41.4.

Laboratory Report

Complete the Laboratory Report for this exercise.

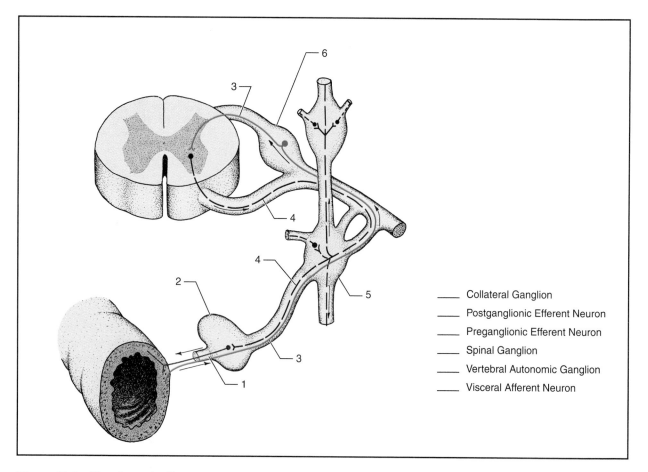

_____ Collateral Ganglion
_____ Postganglionic Efferent Neuron
_____ Preganglionic Efferent Neuron
_____ Spinal Ganglion
_____ Vertebral Autonomic Ganglion
_____ Visceral Afferent Neuron

Figure 41.4 The visceral reflex arc.

Somatic Reflexes

We learned in Exercise 41 that somatic reflexes are automatic responses that occur in skeletal muscles when the proper stimuli are applied to appropriate receptors. In this laboratory period we will study these somatic reflexes, first in the frog, then in humans.

Frog Experiments

Somatic reflexes, whether in amphibians or humans, have five features worthy of study: (1) function, (2) speed of reaction, (3) radiation, (4) inhibition, and (5) synaptic fatigue. Since the frog is a good laboratory specimen to use for studying reflexes, we will use a pithed frog to explore these various facets of muscular response.

Frog Preparation

"Single pith" a frog to destroy its brain, using the following procedure:

Materials:
    small frog
    sharp dissecting needle

1. Rinse the frog under cool tap water and grip it in your left hand as shown in illustration A, figure 42.1.
2. Use your left index finger to force the frog's nose downward so that the head makes a sharp angle with the trunk. Now, locate the transverse groove between the skull and the vertebral column by pressing down on the skin with your fingernail to mark its location.
3. Push a sharp dissecting needle into the crevice, forcing it forward into the center of the skull through the foramen magnum at the base of the skull. Twist the needle from side to side to destroy the brain. This process of "single pithing" produces what is referred to as a *spinal frog.*
4. After five or six minutes, test the effectiveness of the pithing by touching the cornea of each eye with a dissecting needle. If the lower eyelid on each eye does not rise to cover the eyeball, the pithing is complete.

    This test is valid *only* after approximately five minutes to allow recovery from a temporary state of neural (spinal) shock to the entire spinal cord.

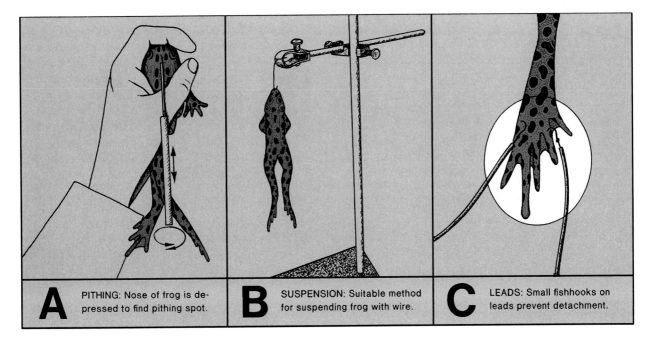

A PITHING: Nose of frog is depressed to find pithing spot.

B SUSPENSION: Suitable method for suspending frog with wire.

C LEADS: Small fishhooks on leads prevent detachment.

Figure 42.1 **Frog pithing and setups.**

5. Determine whether spinal shock has ended by pinching one of the toes with forceps. If a withdrawal reflex occurs, spinal shock has ended, and your specimen is ready to use.

Functional Nature of Reflexes

Muscle tone, movement, and coordinated action in a spinal frog would indicate that these physiological activities occur independently of cognition and are the result of reflex activity. Test for these phenomena as follows:

Materials:
>   spinal frog
>   ring stand and clamp
>   galvanized iron wire, pliers, forceps
>   filter paper squares (3 mm)
>   28% acetic acid
>   10% NaHCO$_3$ in squeeze bottle
>   squeeze bottle of tap water
>   beaker (250 ml size)

1. With the frog in a squatting position, gently draw out one of the hind legs. Is any pull exerted by the animal? When the foot is released, does it return to its previous position?
2. Squeeze the muscles of the hind legs. Do they feel flaccid or firm? Does muscle tone seem to exist?
3. Can you induce the animal to jump by prodding its posterior?
4. Place the animal in a sink basin filled with water. Does it attempt to swim? Does it float or sink? Report all answers to above questions on the Laboratory Report.
5. Suspend the frog with a wire hook through the lower jaw as shown in illustration B, figure 42.1.
6. With forceps, dip a small square of filter paper in acetic acid, shake off excess, and apply to the ventral surface of the thigh for about 10 seconds. Is the frog able to remove the irritant? Do you think that the frog feels the burning sensation caused by the acid? Explain.
7. Wash the acid off the leg with 10% NaHCO$_3$, spraying it on the affected area with a squeeze bottle. Allow the liquid to drain from the leg into an empty beaker.
8. Let stand for 1 or 2 minutes and rinse with water from squeeze bottle.
9. Place another piece of filter paper and acid on another portion of the body (leg or abdomen). Observe reaction. What happens if you restrain the limb that attempts removal? Does any other adaptive behavior take place?
10. Record all observations on the Laboratory Report.

Reaction Time of Reflexes

To determine whether the reaction time to a stimulus is influenced by the strength of the stimulus, proceed as follows:

Materials:
>   7 test tubes (20 mm dia) in test tube rack
>   3 beakers (150 ml size)
>   1 graduate (10 ml size)
>   1 squeeze bottle of tap water
>   1 squeeze bottle of 1% HCl
>   1 squeeze bottle of 1% NaHCO$_3$
>   Kimwipes

1. Dispense 30 ml of 1% NaHCO$_3$ into a beaker.
2. Label seven test tubes: 1.0, .5, .4, .3, .2, .1, and .05%.
3. Measure out 10 ml of 1% HCl from squeeze bottle and pour it into the first tube.
4. Into the other six tubes, measure out the correct amounts of 1% HCl and tap water to make 10 ml each of the correct percentages of HCl. (Examples: 5 ml of 1% HCl and 5 ml of water in the .5% tube; 4 ml of 1% HCl and 6 ml of water in the .4% tube, etc.)
5. Pour the contents from the .05% tube into a clean beaker and allow the long toe of one of the hind legs to be immersed in it for 90 seconds. Be sure to prevent other parts of the foot from touching the sides of the beaker.

   If the foot is withdrawn, record on the Laboratory Report the number of seconds that elapsed from the time of immersion to withdrawal.

   If there is no withdrawal, remove the beaker of .05% HCl at 90 seconds.
6. Bathe the foot in the beaker of 1% NaHCO$_3$ for 15 seconds and rinse the foot by spraying with water. Use the third unused beaker to catch any water sprayed on the foot. Dry the foot off with Kimwipes and let rest for 2 or 3 minutes.
7. Make two more tests with the 0.05% HCl and record the average for the three tests. Be sure to neutralize, wash, and dry between tests.
8. Pour the 0.05% HCl back into its test tube and empty the tube of .1% HCl into the beaker.
9. Place the long toe into this solution, as previously, and again record the withdrawal time in seconds.
10. After neutralization, bathing, drying, and resting the foot, repeat the test two times as above, with this solution. Be sure not to exceed 90 seconds.
11. Repeat these procedures for all other tubes of diluted HCl, progressively increasing the acid concentration from low to high.

12. Plot the reaction times against the concentration on the graph provided on the Laboratory Report.

Reflex Radiation

If the foot of a spinal frog is subjected to increasing intensity of electrical stimulation, a phenomenon called *reflex radiation* occurs. To demonstrate it, one applies a tetanic stimulus, first at a low level, and then gradually increases it. Proceed as follows to observe what takes place.

Materials:
electronic stimulator
2-wire leads with small fishhooks soldered to the leads

1. Connect the jacks of the 2-wire leads to the output posts of the stimulator. (These leads have very small fishhooks soldered to one end to enable easy attachment to the skin. Handle them gingerly. They are sharp!)
2. Attach the hooks to the skin of the left foot as shown in illustration C, figure 42.1.
3. Set the stimulator to produce a minimum tetanic stimulus as follows: Voltage at 0.1 volt, Duration at 50 msec, Frequency at 50 pps, Stimulus switch at REGULAR, Polarity at NORMAL, and Output switch on BIphasic.
4. Press the Mode switch to REPEAT and look for a response. If no response, sequentially increase the voltage by doubling each time (i.e., 0.2, 0.4, 0.8, etc.) until a response occurs.

    Describe the response on the Laboratory Report.
5. Increase the voltage in steps of 2 or 3 volts at a time and note any changes that occur in the reflex pattern.

Reflex Inhibition

Just as it is possible to stifle a sneeze or prevent a knee jerk, it should be possible to override the reflex to acid with electrical stimulation. With the stimulator still hooked up to the left foot from the previous experiment, proceed as follows:

1. Lower the toes of the right foot into a beaker of HC1 of a concentration that produced a moderately fast reflex response.
2. As the right foot is lowered into the acid, administer a moderate tetanizing stimulus to the left foot.
3. Adjust the voltage until the foot in the acid is inhibited.
4. Record on the Laboratory Report the acid concentration and voltage that inhibited the reflex action.

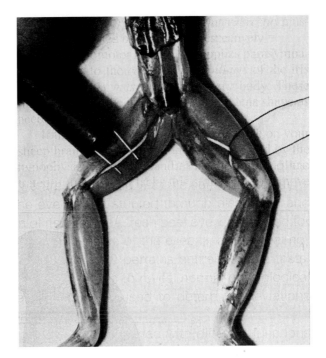

Figure 42.2 Setup for determining when synaptic fatigue occurs. Note string on nerve to prevent trauma when nerve is manipulated.

Synaptic Fatigue

If your specimen is still responding well to stimuli, it may be used for this last test. If it does not prove viable, replace it with a freshly pithed specimen for this experiment.

1. Remove the skin from both thighs (see figure 42.2).
2. Stimulate the sciatic nerve of the *left leg* and note that muscle contraction occurs in both legs.
3. Continue this stimulation until the muscles in the *right leg* fail to respond.
4. After 30 seconds rest, stimulate the left leg again to make sure that *no right leg muscle contraction occurs.*
5. Now, stimulate the sciatic nerve of the right leg instead of the left one.

    Does this cause muscles of the right leg to respond? Where did fatigue take place when the left sciatic nerve was overstimulated?

Reflexes in Medical Diagnosis

Reflex testing is a standard, useful clinical procedure employed by physicians in search of neurological pathology. Damage to intervertebral disks, tumors, polyneuritis, apoplexy, and many other conditions can be better understood with the aid of reflex studies. The diagnostic reflexes studied here are the ones most frequently employed by physicians.

Interpretation of reflex responses is often subjective and requires considerable experience on the part of the diagnostician. Our purpose in performing the tests here, obviously, is not to diagnose, but rather to test, observe, and understand why certain tests are preformed.

Two Types of Reflexes

Clinically, reflexes are categorized as being either deep or superficial. The **deep reflexes** include all reflexes that are elicited by a sharp tap on an appropriate tendon or muscle. They are also called *jerk, stretch,* or *myotatic reflexes.* The receptors (spindles) for these reflexes are located in the muscle, not in the tendon. When the tendon is tapped, the muscle is stretched; stretching the muscle, in turn, activates the muscle spindle, which triggers the reflex response.

Superficial reflexes are withdrawal reflexes elicited by noxious or tactile stimulation. They are also called *cutaneous reflexes.* Instead of percussion initiating these reflexes, the skin is stroked or scratched to induce response.

Reflex Aberrations

Abnormal responses to stimuli may be diminished (hyporeflexia), exaggerated (hyperreflexia), or pathological. **Hyporeflexia** may be due to malnutrition, neuronal lesions, aging, deliberate relaxation, etc. **Hyperreflexia** is often accompanied by marked muscle tone due to the loss of inhibitory control by the motor cortex. Among other things, it may also be caused by strychnine poisoning. **Pathological reflexes** are reflex responses that occur in one or more muscles other than the muscle where the stimulus originates.

These three deviations are what a physician looks for in performing these tests. The following scale is used for calculating each response:

$++++$ very brisk, hyperactive; often indicative of pathology; may be associated with clonus (spasms)
$+++$ brisker than average, may or may not be indicative of pathology
$++$ average; normal
$+$ somewhat diminished
0 no response

Test Procedure

Perform the seven reflex tests shown in figure 42.3 according to the procedures outlined below.

Materials:
reflex hammer

Biceps Reflex This reflex is a deep reflex that causes flexion of the arm. It is elicited by holding the subject's elbow with the thumb pressed over the tendon of the biceps brachii, as shown in illustration A, figure 42.3. To produce the desired response, strike a sharp blow to the first digit of the thumb with the reflex hammer.

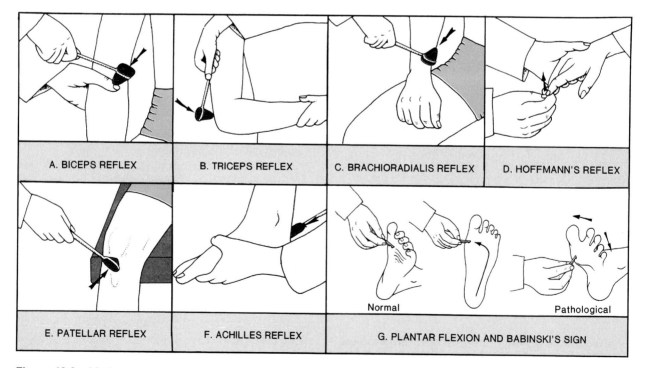

| A. BICEPS REFLEX | B. TRICEPS REFLEX | C. BRACHIORADIALIS REFLEX | D. HOFFMANN'S REFLEX |
| E. PATELLAR REFLEX | F. ACHILLES REFLEX | G. PLANTAR FLEXION AND BABINSKI'S SIGN | |

Figure 42.3 Methods used for producing seven types of somatic reflexes.

If the reflex is absent or diminished (0 or +), apply reinforcement by asking the subject to clench his/her teeth or squeeze his or her thigh with the other hand. Test both arms and record the degree of response on the Laboratory Report.

This reflex functions through C_5 and C_6 spinal nerves.

Triceps Reflex This deep reflex causes extension of the arm in normal individuals. To demonstrate this reflex, flex the arm at the elbow, holding the wrist as shown in illustration B, figure 42.3, with the palm facing the body.

Strike the triceps brachii tendon above the elbow with the pointed end of the reflex hammer. Use the same reinforcement techniques as above, if necessary. Test both arms and record the degree of response on the Laboratory Report.

This reflex functions through C_7 and C_8 spinal nerves.

Brachioradialis Reflex In normal individuals this reflex manifests itself in flexion and pronation of the forearm and flexion of the fingers.

To demonstrate this reflex, direct the subject to rest the hand on the thigh in the position shown in illustration C, figure 42.3. Strike the forearm with the wide end of the reflex hammer about one inch above the end of the radius. The illustration reveals the approximate spot. Use reinforcement techniques, if necessary. The tendon that is struck here is for the brachioradialis muscle of the arm.

This reflex functions through the same spinal nerves (C_5 and C_6) as the biceps brachii reflex.

Hoffmann's Reflex This reflex is an example of a deep reflex in which the response does not occur only in the muscle that is stretched. If the response to the stimulus here is broad, affecting several fingers, other than the finger stimulated, there is indication of pyramidal tract (spinal cord) damage.

To induce this reflex, flick the terminal phalanx of the index finger upward, as shown in illustration D, figure 42.3.

Pathology is indicated if the thumb is adducted and flexed and the other fingers exhibit twitchlike flexion. Such a broad field of reflex action that results from a brief stretch of the flexor muscles of the index finger indicates deep reflex hyperactivity and pathological characteristics.

Test both hands of the subject and record results on the Laboratory Report.

Patellar Reflex This monosynaptic reflex is variously referred to as the *knee reflex, knee jerk,* or *quadriceps reflex.*

To perform this test, the subject should be seated on the edge of a table with the leg suspended and somewhat flexed over the edge. To elicit the typical response, strike the patellar tendon just below the kneecap. The correct spot is shown in illustration E, figure 42.3.

If the response is negative, utilize the **Jendrassic maneuver** for facilitation. In this technique, the subject locks the fingers of both hands in front of the body and pulls the hands against each other isometrically.

A phenomenon called **clonus** is sometimes seen when inducing this reflex. Clonus is characterized by a succession of jerklike contractions (spasms) that follow the normal response and persist for a period of time. This condition is a manifestation of hyperreflexia and indicates damage within the central nervous system.

Test both legs, recording the results on the Laboratory Report. This reflex functions through L_2, L_3, and L_4 spinal nerves.

Achilles Reflex This reflex is also referred to as the *ankle jerk.* It is characterized by plantar flexion when the Achilles tendon is struck a sharp blow. Stretching this tendon affects the muscle spindles in the triceps surae, causing it to contract.

To perform this test, grip the foot with the left hand, as shown in illustration F, figure 42.3, forcing it upward somewhat. With the subject relaxed, strike the Achilles tendon as shown. Hyporeflexia here is often associated with hypothyroidism.

This reflex functions through S_1 and S_2 spinal nerves.

Plantar Flexion *(Babinski's sign)* This reflex is the only one in this series that is of a superficial nature. Note in illustration G, figure 42.3, that the normal reaction to stroking the sole of the foot in an adult is plantar flexion. If dorsiflexion occurs, starting in the great toe and spreading to the other toes (Babinski's sign), it may be assumed that there is myelin damage to fibers in the pyramidal tracts. Incidentally, dorsiflexion is normal in infants, especially if they are asleep. Babinski's sign disappears in infants once myelinization of nerve fibers is complete. This reflex functions through S_1 and S_2 spinal nerves.

Perform this test using a hard object such as a key. Follow the pattern shown in the middle illustration.

Laboratory Report

Complete the Laboratory Report for this exercise.

43

The Patellar Reflex:
Using the Intelitool Flexicomp™

In Exercise 42 we learned that physicians use reflex responses to detect neural pathology. In general, the judgment of the existence or absence of pathology is based on past experience and is necessarily quite subjective. It is possible, however, with the proper software and hardware to quantify certain reflex responses to confirm suspected damage to the nervous system.

The Intelitool Flexicomp™ system will be used here to remove some of the subjectivity in observing the patellar reflex. Since this same hardware can be used to analyze the biceps and triceps reflexes, you may wish to test them also. With the setup illustrated in figure 43.1, you will be able to accomplish the following:

- Quantify the stimulus response in terms of degree of deflection.
- Quantify time intervals, such as the latent period and time to maximal response.
- Demonstrate the effect of neural facilitation on the involuntary reflex response.

- Contrast the involuntary reflex with a voluntary response involving cognition.

It is assumed that you have a basic knowledge of the computer platform that you will be using. The platforms for which Flexicomp™ is available are:

- IBM or compatible PC (MS-DOS 3.3 or higher)
- Macintosh (System 6.0.8 or higher)
- Apple II (ProDOS)

All equipment will be set up before you enter the laboratory. The software will be loaded and ready to use. Figure 43.1 illustrates the setup. You will note as you progress with the experiment that separate instructions are provided to accommodate the software differences between the three platforms.

You will be performing these various experiments with a laboratory partner. One individual will be the subject, and the other will operate the computer and apply the mallet to the subject's knee. Proceed as follows to perform the various tests.

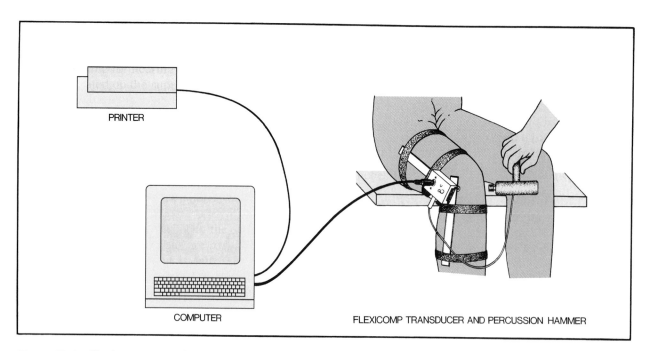

PRINTER

COMPUTER

FLEXICOMP TRANSDUCER AND PERCUSSION HAMMER

Figure 43.1 Flexicomp setup for monitoring the patellar reflex.

Materials:
> computer, monitor, and printer
> displacement transducer with Velcro straps
> percussion hammer (computer interfaced)
> transducer cable (game port connector)
> Flexicomp™ program diskette
> data disk (initialized)
> calibration template
> *Flexicomp™ Lab Manual*
> *Flexicomp™ User Manual*

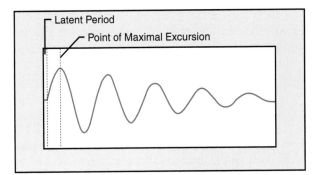

Figure 43.2   Response to stimulus application.

1. Calibrate the Transducer.

Before hooking up the subject, calibrate the Flexicomp™ transducer, following the instructions on the screen. It will be calibrated first at 90° and then at 150°. Adjustments are made by turning the aluminum knob on the transducer.

Important: The thumbscrew locking tab in the angle of the transducer **must be loosened** at this point. Failure to do so may cause damage to the transducer. The thumbscrew should always be kept loose whenever data acquisition is not in progress, and the transducer should never be forced if it has hit an internal stop. Be sure to loosen the thumbscrew first!

If calibration fails for any reason during the experiment, the software will detect the flaw and force you to redo the calibration procedure.

2. Attach Flexicomp™ to Subject's Leg.

With the subject seated on the edge of a table, as shown in figure 43.1, strap the transducer to the right thigh and lower leg.

Note that the transducer box is facing outward and that the hinge of the transducer is aligned with the knee joint. Try to minimize any twisting forces on the transducer.

3. Center the Trace on the Screen.

MS-DOS: Select "Experiment Menu" from the FLEXICOMP™ MENU. Then select "Real Time Experiment Mode." The data acquisition screen will appear.

With the **thumbscrew loose** press the button on the end of the mallet. As the trace moves across the screen, turn the adjustment knob until the trace is just below the center of the screen (it may take several tries to get it right).

When the trace has been vertically positioned, tighten the thumbscrew locking tab. Press the "C" key to clear the computer's memory and then "Y" to confirm. You are now ready to proceed to step 4.

Macintosh: Select **New** from the **File** menu to open a data acquisition window if one is not already open. With the **thumbscrew loose** press the button on the end of the mallet.

As the trace moves across the screen, turn the adjustment knob until the trace is just below the center of the screen (it may take several tries to get it right).

When the trace has been vertically positioned, tighten the thumbscrew locking tab. Press the "C" key to clear the computer's memory and then "Y" to confirm. You are now ready to proceed to step 4.

Apple II: Select "Experiment Mode" from the FLEXICOMP™ MAIN MENU, and then choose "Real Time Mode." With the **thumbscrew loose,** press the button on the end of the mallet.

As the trace moves across the screen, turn the adjustment knob until the trace is just below the center of the screen (it may take several tries to get it right).

When the trace has been vertically positioned, tighten the thumbscrew locking tab. Press the "C" key to clear the computer's memory and then "Y" to confirm. You are now ready to proceed to step 4.

4. Begin Data Collection.

Strike the subject's patellar ligament (just below the kneecap) hard enough to elicit a good response. The plot will automaticlly stop when it reaches the edge of the screen.

Repeat the application of the mallet several times. If your technique is good you should get similar results each time.

5. Review the Data.

MS-DOS: Press ESC to return to the main FLEXICOMP™ MENU and select "Review/Analyze Menu." Choose "Displacement/Time Analysis" and press the "S" key to stop the trace.

Then use the "Z", "J," and arrow (cursor) keys to position the data markers and measure the latent period, maximum excursion, and time to maximum excursion. Record the vaues you obtain:

Latent period _____

Maximum excursion _____

Time to max. excursion _____

Use the "F" or "B" key to move forward or backward through the data one frame at a time. Note that the left edge of the screen represents the instant the stimulus was applied.

Macintosh: Select **Displacement** from the **Analysis** menu. Click and drag in the plot area to position the data markers and measure the latent period, maximum excursion, and time to maximum excursion. Record the values.

Latent period _____

Maximum excursion _____

Time to max. excursion _____

To select a specific event, click on the event number at the top of the window. Note that the left edge of the screen represents the instant the stimulus was applied.

Apple II: Press ESC to return to the FLEXI-COMP™ MAIN MENU and select "Review/Analyze." Choose "Review-Time/Displacement Analysis" and press the "S" key to stop the trace.

Then use the "D," "J," and arrow (cursor) keys to position the data markers and measure the latent period, maximum excursion, and time to maximum excursion. Record the values here:

Latent period _____

Maximum excursion _____

Time to max. excursion _____

Use the "F" or "B" key to move forward or backward through the data one frame at a time. Note that the left edge of the screen represents the instant the stimulus was applied.

6. Save Data to Disk.

MS-DOS: Press ESC to return to the main FLEXI-COMP™ MENU and select "Data File/Disk Commands." Then choose SAVE File and give the file an appropriate name and press the Return (Enter) key.

Macintosh: Select **Save** from the **File** menu and name the file appropriately.

Apple II: From the MAIN MENU select "Save Current Data" and then select "Catalog/Save Data File." On single floppy drive systems, you will be prompted to insert a data disk. Give the file an appropriate name and press the Return key.

7. Demonstrate Facilitation Effects.

With the subject clasping his or her hands in front and **pulling hard,** repeat steps 3 through 6.

Obtain several frames of data in this manner. Note any differences in the measurements between this and the first set of data. What accounts for the differences caused by facilitation?

8. Evaluate Voluntary Reflex.

With the subject facing away from you, repeat steps 3 through 6, as you randomly tap the hammer gently on the table. The subject should respond as soon as he or she hears the tap by gently kicking.

Compare latent periods for this set of data with that of the involuntary knee jerk responses. Explain the differences you find.

9. Optional Experiments.

If you wish to continue with additional reflex experiments, consult the two Intelitool manuals. The following brief descriptions describe the general procedures.

Biceps Jerk   With the subject's arm relaxed on the desk, gently press his or her biceps tendon with your thumb. The biceps tendon can be felt in the center of the hollow of the elbow. Strike your thumb with the hammer to elicit the reflex response.

Triceps Jerk   Have the subject lie down on the table with the elbow bent so that his or her arm lies loosely across the abdomen. Support the arm being tested at the forearm balance point (near the elbow) with your supporting hand balled up into a fist. Strike the triceps tendon about 2 inches above the elbow.

Ankle Jerk   The subject should kneel on a chair, with back to the examiner and feet projecting over the edge of the chair (shoes and socks off).

Strike the Achilles tendon in its center (at the level of the ankle) and observe the plantar extension of the foot.

Laboratory Report

Since there is no Laboratory Report for this experiment, your instructor will indicate how to write up this experiment.

44

Action Potential Velocities

The velocity of nerve impulse conduction is not the same in all nerve fibers of the body. Unmyelinated fibers of small diameter may have velocities as low as 0.5 meter per second. Fibers of this type are seen in portions of the autonomic nervous system. Large diameter myelinated fibers of the somatic type, on the other hand, may carry nerve impulses at speeds as high as 130 meters per second. This high velocity in myelinated fibers is due to the flow of ions from one node of Ranvier to another along the fiber. Such ionic movement through the axoplasm and extracellular fluid from node to node is called *saltatory conduction.*

To determine the speed of the nerve impulse along a nerve, all that is necessary is to determine how long it takes the action potential to traverse a known distance. That distance in meters divided by the time (total time less latency) will yield the velocity in meters per second. These measurements can readily be determined with either an oscilloscope, polygraph, or Duograph. Although a Duograph is illustrated in the setup in figure 44.1, a polygraph or oscilloscope would work as well. Your instructor may wish to demonstrate the procedure first, using a dual-beam oscilloscope.

The Ulnar Nerve

The electrode hookup shown in figure 44.1 will be used to monitor action potentials in the ulnar nerve. This nerve has been selected for this study because of its simplicity and accessibility. The fact that it lies close to the surface of the arm makes it easy to stimulate through the skin with an electronic stimulator, such as the Grass SD9. An understanding of the exact location of this nerve is essential to know where to place the various electrodes.

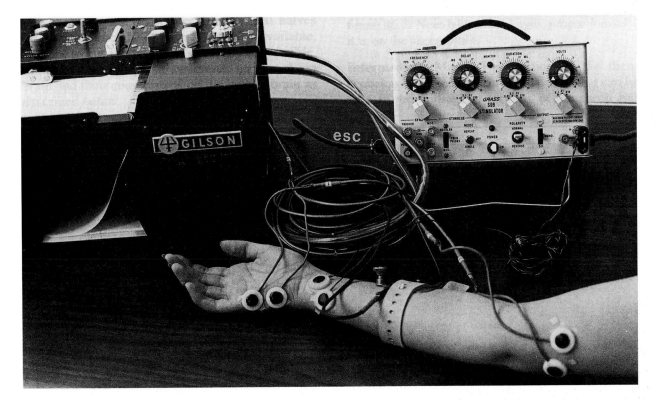

Figure 44.1 Experimental setup.

The ulnar nerve arises from the medial cord of the brachial plexus, deriving its fibers from the eighth cervical (C$_8$) and first thoracic (T$_1$) nerves. Figure 44.2 illustrates the path it follows through the arm. Note that in the upper arm the nerve lies on the medial surface, just under the skin. In the elbow region it rests in a groove posterior to the medial epicondyle of the humerus. In the forearm it lies, superficially, on the medial surface, and supplies branches to the *flexor carpi ulnaris* and the medial half of the *flexor digitorum profundus.*

In the hand the ulnar nerve divides to form deep and superficial branches. The superficial branch has branches that extend down both sides of the fifth finger and the medial side of the fourth finger. The deep branch innervates the *abductor digiti quinti* and two other muscles of the small finger. It is the latter muscle that will concern us in this experiment.

Electrode Placement

Figure 44.3 reveals a closeup of the arrangement of the electrodes on the arm. Note that the two stimulator electrodes (anode and cathode) are located on the wrist at point A. A pair of recording electrodes that are attached at point B on the side of the hand are located over the abductor digiti quinti muscle. These electrodes will transmit a signal to the Duograph when the muscle contracts.

The two electrodes above the elbow at point C pick up the action potential as it travels from point A through point C. The signal from these electrodes will be fed into the other channel on the Duograph.

The ECG electrode that is strapped in place on the arm serves as a common ground for the recording electrodes that are located at points B and C. Note that there are two alligator clamps attached to it.

The production of an action potential in a nerve by artificial electrical stimulation is referred to as an **evoked potential** as opposed to potentials produced endogenously during the voluntary contraction of muscles. When a nerve is stimulated somewhere along its course, evoked potentials move in opposite directions from the point of stimulation.

An impulse that moves along the fiber in the natural direction is called **orthodromic conduction.** In this experiment the evoked potential that moves from points A to B to stimulate the *abductor digiti quinti* is orthodromic since all fibers of the deep branch are motor. The other evoked action potential that moves in the opposite direction is called **antidromic conduction.**

The action potential that moves along motor fibers from point A to C is antidromic. Any nerve

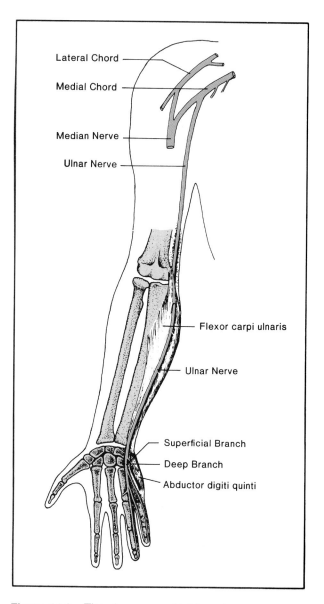

Figure 44.2 The ulnar nerve (right arm).

impulses carried by sensory fibers in the ulnar nerve from point A to C are orthodromic. If the stimulus is strong enough, orthodromic impulses in the sensory fibers will trigger a secondary response called a **reflex potential.**

Team Assignments

The class will be divided into teams of four members each. Since there may not be enough equipment available for all teams to work at the same time, it may be necessary for some groups to work on some other experiment until the equipment becomes available. Each team will consist of a subject, a director, a stimulator operator, and a Duograph operator. Responsibilities are as follows:

The Subject

The subject must be capable of keeping the experimental arm completely relaxed during the entire procedure. He or she should rest assured that the low amperage used in this experiment cannot cause injury, although some slight discomfort may be experienced occasionally.

Careful selection of the subject is crucial in achieving success in this experiment. A good subject is not squeamish, timid, coy, silly, or hyperexcitable; nor is a good subject brash, careless, or uninterested.

The Director

The director is responsible for coordinating the activities of the team. Ideally, this person should be bold, but not reckless, and careful, but not plodding. An important role of the director is to prepare the subject by locating the course of the ulnar nerve and placing the electrodes where they belong.

At the conclusion of the experiment, it is the director's responsibility to disconnect the subject, clean the skin and electrodes, and return the electrodes to their proper places. The director and the two equipment operators will work together to clean the work area at the end of the period.

The Duograph Operator

The Duograph operator is responsible for setting up and putting away this instrument. Selection of the proper chart speeds, notations of pertinent information on the chart, and making sensitivity adjustments are some of this individual's responsibilities.

Three events will be recorded simultaneously: (1) the onset and duration of the stimulus, (2) the evoked muscle potential, and (3) the evoked nerve potential.

The Stimulator Operator

A complete understanding of the functions of the various controls of the stimulator qualifies one as the stimulator operator. Stimuli will be manually delivered at predetermined intervals. Adjustments called for by other team members will be made as needed.

Procedure

As soon as all members of your team have agreed upon their responsibilities, proceed as follows to perform the experiment.

Materials:
    Gilson Duograph or polygraph
    electronic stimulator
    two 3-lead patient cables
    2-lead cable for stimulator
    event synchronization cable
    6 skin electrodes (Gilson self-adhering,
        #E1081K)
    ECG plate electrode
    electrode paste
    Scotchbrite pad (grade #7447)
    alcohol or alcohol swabs
    meterstick

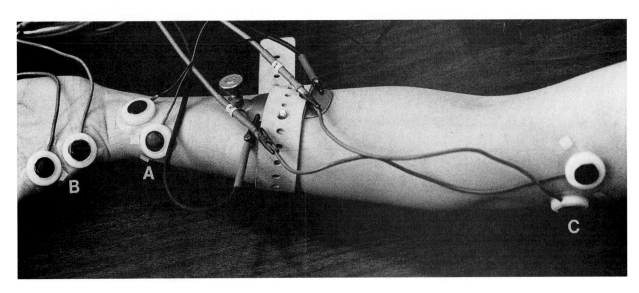

Figure 44.3 Arrangement of electrodes.

Equipment Setup

The Duograph Operator and Stimulator Operator should work together to set up the equipment as follows:

1. Plug in the power cords for the Duograph and stimulator.
2. Connect the stimulator to the Duograph with the event synchronization cable. This cable is labeled **esc** in figure 44.1.
3. Arrange the equipment in the manner shown in figure 44.1.
4. Attach two leads from the wrist electrodes to the output posts of the stimulator.
5. Plug the two 3-lead cables into the Duograph.

   One of these cables will receive signals from the hand. The other one will monitor nerve impulses in the arm. Make the leads of these cables accessible to the Director, who will make the electrode hookups.
6. Set the **stimulator controls** as follows: Duration at 1 msec, Voltage at 1 volt, Stimulus switch on REGULAR, Mode at OFF, Power on, Polarity switch on NORMAL, and the Output switch at MONO.
7. Set the **Duograph controls** as follows: Power switch on, c.c. switch at STBY, stylus heat control knob at the two o'clock position, chart speed at 25 mm per sec, Mode on both channels at EEG, and Gain at 2 MV/CM.

Preparation of Subject

The Director will prepare the subject in the following manner:

1. Position the subject comfortably so that the arm rests on the table as shown in figure 44.1.
2. If enough table space is available, locate the subject to the right of the stimulator instead of directly in front of it to allow room for the stimulator and Duograph operators.
3. Gently scrub the skin areas with a Scotch-brite pad and swab with alcohol.
4. Attach the electrodes to places on the arm as shown in figure 44.3.

   Refer to figures 32.3 through 32.6 on page 151 for the procedure to be used in placing the adhesive pads on the self-adhering electrodes.
5. Be sure to apply electrode paste to the ECG plate electrode before strapping it in place.

6. Connect the two electrode wires of the hand to the #1 and #2 leads of one 3-lead cable.
7. Connect the two electrode wires of the upper arm to the #1 and #2 leads of the other 3-lead cable.
8. Clamp the ground wires from each of these cables to the ECG plate electrode in the middle of the arm.
9. With a meterstick measure the distances between points A and B and A and C and record these measurements on the Laboratory Report.

Stimulation and Recording

Now that all electrodes are in place proceed, as follows, to determine the threshold stimulus that will just barely activate the *abductor digiti quinti* muscle:

1. Starting with a Duration of 1 msec and 1 volt, depress the Mode switch to SINGLE to see if the muscle reacts.
2. If no response occurs, increase the Duration in 1 msec increments until muscle reaction occurs.

   At some point below 10 msec it should be possible to establish the threshold voltage.
3. If there still is no response, try increasing the voltage in 1 volt increments at 5 msec Duration until there is a response.
4. After the threshold stimulus has been determined, return the voltage to a subthreshold level.
5. With the chart moving at 25 mm per second, depress the Mode switch on the stimulator to SINGLE at appoximately one-second intervals as you increase the voltage from subthreshold until maximal potentials are obtained. The voltage should be increased at uniform increments.
6. Be sure to label the tracing with pertinent information.
7. Disconnect the subject and shut off the instruments.
8. Clean up the work area and put away all equipment.

Laboratory Report

Calculate the velocities of the nerve impulse in the ulnar nerve according to the instructions provided on the Laboratory Report. Answer all questions on the report.

45

Brain Anatomy:
External

This study of the human brain will be done in conjunction with dissection of the sheep brain. The ample size and availability of sheep brains make them preferable to brains of most other animals; also, the anatomical similarities that exist between the human and sheep brains are considerably greater than their differences.

Preserved human brains will also be available for study in this laboratory period. Dissection, however, will be confined to the sheep brain. A discussion of the meninges will precede the sheep brain dissection.

The Meninges

It was observed in Exercise 41 that the spinal cord, as well as the brain, is surrounded by three membranes called the *meninges*. Due to the importance of these protective membranes it will be necessary to reemphasize certain anatomical characteristics about them.

Dura Mater   The outermost meninx is the *dura mater,* which is a thick tough membrane made up of fibrous connective tissue. The lateral view of the opened skull in figure 45.1 reveals the dura mater lifted away from the exposed brain. Note that on the median line of the skull it forms a large **sagittal sinus** (label 6) that collects blood from the surface of the brain. Observe, also, that many cerebral veins empty into this sinus.

Extending downward between the two halves of the cerebrum is an extension of the dura mater, the **falx cerebri.** This structure shows up only in the frontal section of figure 45.1. At its anterior end, the falx cerebri is attached to the crista galli of the ethmoid bone.

Arachnoid Mater   Inferior to the dura mater lies the second meninx, the *arachnoid mater* (label 8, figure 45.1). Between this delicate netlike membrane and the surface of the brain is the **subarachnoid space,** which contains cerebrospinal fluid. Note that the arachnoid mater has small projections of tissue, the **arachnoid granulations**, that extend up into the interior of the sagittal sinus. These granulations allow cerebrospinal fluid to diffuse from the subarachnoid space into the venous blood of the sagittal sinus.

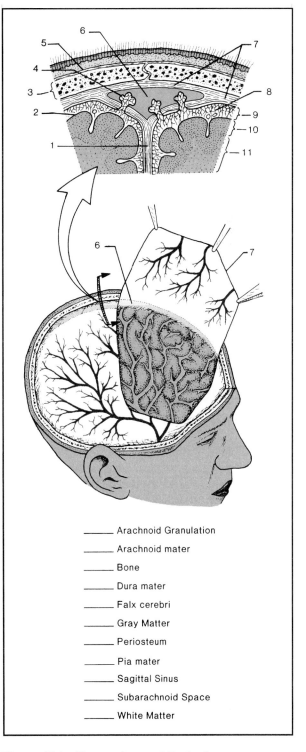

_____ Arachnoid Granulation

_____ Arachnoid mater

_____ Bone

_____ Dura mater

_____ Falx cerebri

_____ Gray Matter

_____ Periosteum

_____ Pia mater

_____ Sagittal Sinus

_____ Subarachnoid Space

_____ White Matter

Figure 45.1   The meninges of the brain.

213

Pia Mater   The surface of the brain is covered by the third meninx, the *pia mater*. This membrane is very thin. Note that the surface of the brain has grooves, or **sulci,** that increase the surface of the **gray matter** (label 10). As in the case of the spinal cord, the gray matter is darker than the inner **white matter** because it contains cell bodies of neurons. The white matter consists of neuron fibers that extend from the gray matter to other parts of the brain and spinal cord.

Assignment:
Label figure 45.1.

Sheep Brain Dissection

Two procedures are provided here for this sheep brain study: (1) an *in situ* study of fresh brain material and (2) the dissection of a preserved, formalin-hardened sheep brain. In the *in situ* dissection, the brain can be observed as it appears, undisturbed, in the skull. To get to the brain it is necessary to remove the top of the skull with an autopsy saw. This type of dissection enables one to see many things that cannot be seen in a preserved specimen that has already been removed from the skull.

The value of the second dissection is that the formalin-hardened tissue will reveal structures that are usually destroyed when trying to remove the fresh brain from the skull. If time restrictions prevent doing both dissections, the second dissection is preferred.

Sheep Head Dissection

To examine the sheep's brain within the cranium, it will be necessary to remove the top of the skull with an autopsy saw, as illustrated in figures 45.2 through 45.5. The blade of this type of saw cuts through bone with a reciprocating action. Since it does not have a rotary action, it is a relatively safe tool to use; however, care must be exercised to avoid accidents. For maximum safety, it will be necessary for one person to hold the sheep's head while another individual does the cutting. *At no time should the holder's hands be in the cutting path of the saw blade.*

If only one or two saws are available, it will not be possible for everyone to start cutting at the same time. Since it takes five to ten minutes to remove the top of the skull, some students will have to be working on other parts of this exercise until a saw is available. Proceed as follows:

Materials:
fresh sheep head
autopsy saw
dissecting kit and screwdriver (8″ long)
black felt pen

1. With a black felt pen, mark a line on the skull across the forehead between the eyes, around the sides of the skull, and over the occipital condyles. Refer to figure 45.2.
2. While firmly gripping the saw with both hands, cut around the skull following the marked line. Brace your hands against the table for support. Don't try to "free hand" it. Use the large cutting edge for straight cuts, the small cutting edge for sharp curves.

   Be sure that the head is held securely by your laboratory partner. Cut only deep enough to get through the bone. Try not to damage the brain.

   Rest your hands after cutting an inch or two. To avoid accidents, do not allow your wrists to get too tired. As mentioned above, watch out for your partner's hands!
3. After cutting all the way around the skull, place the end of a screwdriver into the saw cut (figure 45.4) and pry upward to loosen the bony plate. Cut free any remaining portions.
4. Gently lift the bone section off the top of the skull. Note how fragments of the **dura mater** prevent you from completely removing the cap. Use a sharp scalpel or scissors to cut the dura mater loose as the skull cap is lifted off.
5. Examine the inside surface of the removed bone and study the dura mater. Note how thick and tough it is. Its toughness is due to the presence of collagenous fibers.
6. Locate the **sagittal sinus.** Open it with a scalpel or scissors and see if you can see any evidence of **arachnoid granulations.**
7. Examine the surface of the brain. Observe that it is covered with convolutions called **gyri** (*gyrus,* singular). Between the gyri are grooves called **sulci.**
8. Locate the middle meninx, **or arachnoid mater,** on the surface of the brain. This meninx is a very thin, netlike membrane that appears to be closely attached to the brain. In actuality, there is a thin space, the **sub-arachnoid space,** between the arachnoid mater and the surface of the brain. Although this space contains **cerebrospinal fluid** in a live animal, its presence on your specimen will probably not be detectable.
9. Gently force your second, third, and fourth fingers under the anterior portion of the brain, lifting the brain from the floor of the cranial cavity. Take care not to damage the brain.

   Locate the **olfactory bulbs** (see figure 45.8). Note that they are attached to the forepart of the skull. From these bulbs pass olfactory nerves through the cribriform plate into the nasal cavity.

10. Free the olfactory bulbs from the skull with your scalpel.

11. Raise the brain further to expose the **optic chiasma** (see figure 45.8). Note how the optic nerves that pass up through the skull are continuous with this structure. It is in the optic chiasma that some of the neurons from the retina of each eye cross over.

12. Sever the optic nerves as close to the skull as possible with your scalpel and raise the brain a little further to expose the two **internal carotid arteries** that supply the brain with blood. Do you remember from your study of the skull what foramina in the skull allow these blood vessels to enter the cranium?

13. Sever the internal carotid arteries and raise the brain a little more to expose the **infundibulum,** which connects with the hypophysis. Figure 45.7 illustrates the relationship of the infundibulum to the hypophysis.

14. Carefully dissect the **hypophysis** out of the sella turcica without severing the infundibulum.

15. Continue lifting the brain to identify as many cranial nerves as you can. Refer to figures 45.7 and 45.8 to identify these nerves.

16. Remove the brain completely from the cranium. Section the brain, frontally, to expose the inner material. Note how distinct the **gray matter** is from the **white matter.**

17. Dispose of the remains in the proper waste receptacle and proceed to the next dissection, which will utilize perserved material.

Preserved Sheep Brain Dissection

In this portion of our brain study we will be concerned with the finer details of its external and internal structure. Constant comparisons will be made with the human brain. Once you have located a structure on the sheep brain, try to find a

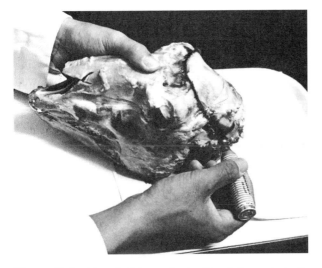

Figure 45.2 The skull is marked with a felt pen to indicate the line of cut.

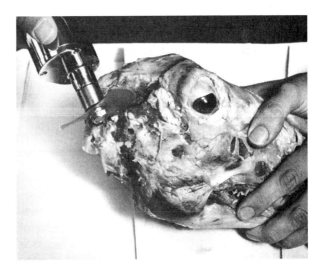

Figure 45.3 While the head is held securely by an assistant, the cut is made with an autopsy saw.

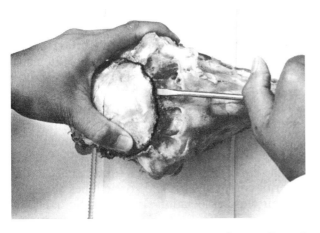

Figure 45.4 A large screwdriver is used to pry the cut section of the skull off the head.

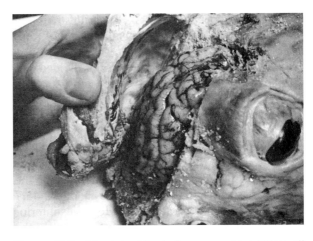

Figure 45.5 The intact brain is exposed by lifting off the cut portion of the skull.

the hypophysis without damaging other structures, gently lift it up with forceps and sever the infundibulum with a scalpel or a pair of scissors.

IV Trochlear Nerve This nerve provides muscle sense and motor stimulation of the *superior oblique* muscle of the eye. As in the case of the oculomotor, this nerve also emerges from the midbrain. This nerve is very small in diameter and one of the more difficult ones to locate.

V Trigeminal Nerve Just posterior to the trochlear nerve lies the largest cranial nerve, the trigeminal. Although it is a mixed nerve, its sensory functions are much more extensive than its motor functions. Because of its extensive innervation of the mouth and face, it is described in detail on pages 222–224.

VI Abducens Nerve This small nerve provides innervation of the *lateral rectus* muscle of the eye. It is a mixed nerve in that it provides muscle sense as well as muscular contraction. On the human brain it emerges from the lower part of the pons near the medulla.

VII Facial Nerve This nerve consists of motor and sensory divisions. Label 15, figure 45.9 reveals that this nerve on the human brain has two branches. It innervates muscles of the face, salivary glands, and taste buds of the anterior two-thirds of the tongue.

VIII Vestibulocochlear *(Auditory)* **Nerve** This nerve goes to the inner ear. It has two branches: the *vestibular division,* which innervates the semicircular canals to function in equilibrium, and the *cochlear division,* which innervates the cochlea to function in hearing. The vestibulocochlear nerve emerges just posterior to the facial nerve on the human brain.

IX Glossopharyngeal Nerve This mixed nerve emerges from the medulla posterior to the vestibulocochlear. It functions in reflexes of the heart, taste, and swallowing. Taste buds on the back of the tongue are innervated by some of its fibers. Efferent fibers innervate muscles of the pharynx (swallowing) and the parotid salivary glands (secretion).

X Vagus Nerve This cranial nerve, which originates on the medulla, exceeds all of the other cranial nerves in its extensive ramifications. In addition to supplying parts of the head and neck with nerves, it has branches that extend down into the chest and abdomen. It is a mixed nerve.

Sensory fibers in this nerve go to the heart, external acoustic meatus, pharynx, larynx, and thoracic and abdominal viscera.

Motor fibers pass to the pharynx, base of the tongue, larynx, and to the autonomic ganglia of the thoracic and abdominal viscera. Many references will be made to this nerve in subsequent discussions of the physiology of the lungs, heart, and digestive organs.

XI Accessory Nerve Since this nerve emerges from both the brain and spinal cord it is sometimes called the *spinal accessory* nerve. Note that on both the sheep and human brains it parallels the lower part of the medulla and the spinal cord with branches going into both structures. On the human brain it appears to occupy a more posterior position than the twelfth nerve. Afferent and efferent spinal components innervate the *sternocleidomastoid* and *trapezius* muscles. The cranial portion of the nerve innervates the pharynx, upper larynx, uvula, and palate.

XII Hypoglossal Nerve This nerve emerges from the medulla to innervate several muscles of the tongue. It is a mixed nerve. On the human brain in figure 45.9, it has the appearance of a cut-off tree trunk with roots extending into the medulla.

The Hypothalamus

That portion of the diencephalon that includes the mammillary bodies, infundibulum, and part of the hypophysis is collectively called the *hypothalamus.* The entire diencephalon, including the hyopthalamus, will be studied more in detail in the next exercise, when the internal parts of the brain are revealed by dissection. For now, let's consider the mammillary bodies and hypohysis.

Mammillary Bodies This part of the hypothalamus lies superior to the hypophysis. Although it appears as a single body on the sheep, it consists of a pair of rounded eminences just posterior to the infundibulum on the human. These bodies receive fibers from the olfactory areas and ascending pathways of the brain. Fibers also exit from these bodies to the thalamus and other brain nuclei.

Hypophysis This structure, also known as the *pituitary gland,* is attached to the base of the brain by a stalk, the **infundibulum.** It consists of two distinctly different parts: an anterior adenohypophysis and a posterior neurohypophysis. Only the neurohypophysis and infundibulum are considered to be part of the hypothalamus because they both have the same embryological origin as the brain.

The adenohypophysis originates as an outpouching of the ectodermal stomodeum (mouth cavity) and is quite different histologically from the neurohypophysis. A close functional relationship exists between the hypothalamus, hypophysis, and the autonomic nervous system.

Assignment:

> The remainder of this exercise pertains to the labeling of figures 45.10 and 45.11 and does not pertain to sheep brain dissection. To continue the sheep brain dissection, proceed to Exercise 46.

Label figure 45.9.

Functional Localization of the Cerebrum

Extensive experimental studies on monkeys, apes, and humans have resulted in considerable knowledge of the functional areas of the cerebrum. Figure 45.10 shows the locations of some of these centers. Before attempting to identify each of these areas be sure you know the boundaries (sulci and fissures) of the cerebral lobes (figure 45.7).

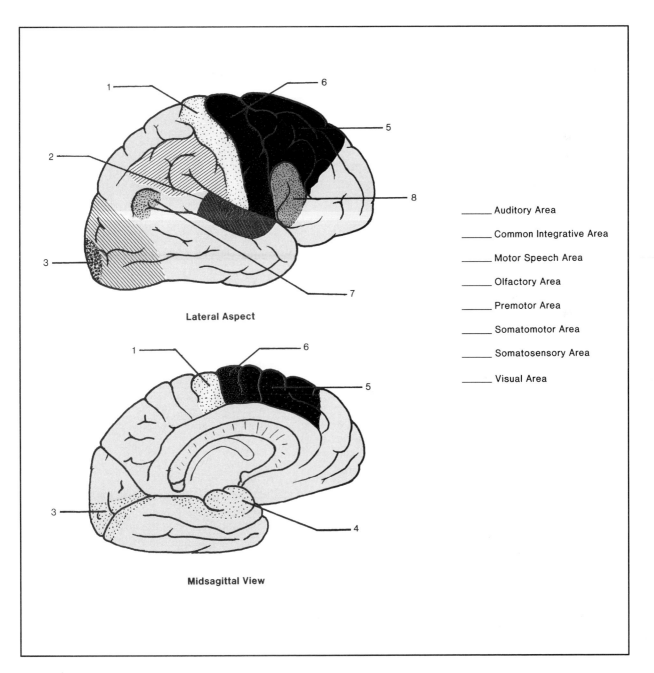

Lateral Aspect

Midsagittal View

_____ Auditory Area

_____ Common Integrative Area

_____ Motor Speech Area

_____ Olfactory Area

_____ Premotor Area

_____ Somatomotor Area

_____ Somatosensory Area

_____ Visual Area

Figure 45.10 Functional areas of the human brain.

Somatomotor Area This area occupies the surface of the precentral gyrus of the frontal lobe. Electrical stimulation of this portion of the cerebral cortex in a conscious human results in movement of specific muscular groups.

Premotor Area The large area anterior to the somatomotor area is the *premotor area*. It exerts control over the motor area.

Somatosensory Area This area is located on the postcentral gyrus of the parietal lobe. It functions to localize very precisely those points on the body where sensations of light touch and pressure originate. It also assists in determining organ position. Other sensations, such as aching pain, crude touch, warmth, and cold are localized by the thalamus rather than this area.

Motor Speech Area This area is located in the frontal lobe just above the lateral cerebral fissure, anterior to the somatomotor area. It exerts control over the muscles of the larynx and tongue that produce speech.

Visual Area This area is located on the occipital lobe. Note that only a small portion of it is seen on the lateral aspect; the greater portion of this area is on the medial surface of the cerebrum. On this surface it extends anteriorly to the parieto-occipital fissure, becoming narrower as it approaches the fissure. This area receives impulses from the retina via the thalamus. Destruction of this region causes blindness, although light and dark are still discernible.

Auditory Area This area receives nerve impulses from the cochlea of the inner ear via the thalamus. It is responsible for hearing and speech understanding. It is located on the superior temporal gyrus that borders on the lateral cerebral fissure.

Olfactory Area This area is located on the medial surface of the temporal lobe. The recognition of various odors occurs here. Tumors in this area cause individuals to experience nonexistent odors of various kinds, both pleasant and unpleasant.

Association Areas Adjacent to the somatosensory, visual, and auditory areas are *association areas* that lend meaning to what is felt, seen, or heard. These areas are lined regions in 45.10.

Common Integrative Area The integration of information from the above three association areas and the olfactory and taste centers is achieved by a small area that is called the *common integrative,* or *gnostic, area.* This area is located on the **angular gyrus,** which is positioned approximately midway between the three association areas.

Assignment:
Label figure 45.10.

The Trigeminal Nerve

Although a detailed study of all cranial nerves is precluded in an elementary anatomy course, the trigeminal nerve has been singled out for a thorough study here. This fifth cranial nerve is the largest one that innervates the head and is of particular medical and dental significance.

Figure 45.11 illustrates the distribution of this nerve. You will note that it has branches that supply the teeth, tongue, gums, forehead, eyes, nose, and lips. The following description identifies the various ganglia and branches.

On the left margin of the diagram is the severed end of the nerve at a point where it emerges from the brain. Between this cut end and the three main branches is an enlarged portion, the **Gasserian ganglion.** This ganglion contains the nerve cell bodies of the sensory fibers in the nerve. The three branches that extend from this ganglion are the *ophthalmic, maxillary,* and *mandibular* nerves. Locate the bony portion of the skull in figure 45.11 through which these three branches pass.

Ophthalmic Nerve The ophthalmic nerve is the upper branch that passes out of the cranium through the *superior orbital fissure.* It has three branches that innervate the lacrimal gland, the upper eyelid, and the skin of the nose, forehead, and scalp. One of its branches also supplies the cornea of the eye, the iris, and the ciliary body (a muscle attached to the lens of the eye).

Superior to the lower branch of the ophthalmic is the **ciliary ganglion,** which contains nerve cell bodies that are incorporated into reflexes controlling the ciliary body. Unlike the other two divisions of the trigeminal, the ophthalmic nerve and its branches are not involved in dental anesthesia.

Maxillary Nerve This nerve, also known as the *second division of the trigeminal nerve,* is a sensory nerve that provides innervation to the nose, upper lip, palate, maxillary sinus, and upper teeth. It is the branch that comes off just inferior to the ophthalmic nerve.

Note that the maxillary nerve has two short branches that extend downward to an oval body, the **sphenopalatine ganglion.** The major portion of the maxillary nerve becomes the **infraorbital nerve,** which passes through the *infraorbital nerve canal* of the maxilla. This canal is shown with dotted lines

exercise for

46

The nature of
studying sect
through it. In
several sectio
ties (ventricle
amus, fornix,

Materials:

preserved
preserved
human br
dissecting
long shar
dissecting

With the pres
down on a
meat-cutting
dinal cerebra
tissue, cuttin
on the midl
attempt to cu
Proceed to i
brain and the
described bel

Co

Fo

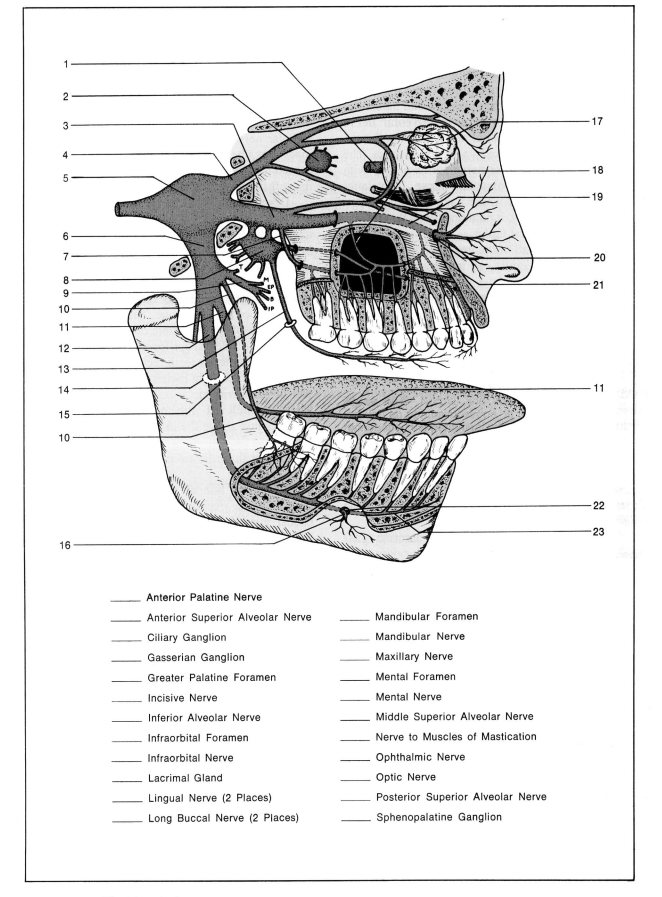

1
2
3
4
5
6
7
8
9
10
11
12
13
14
15
10
16

17
18
19
20
21
11
22
23

_____ Anterior Palatine Nerve

_____ Anterior Superior Alveolar Nerve   _____ Mandibular Foramen

_____ Ciliary Ganglion   _____ Mandibular Nerve

_____ Gasserian Ganglion   _____ Maxillary Nerve

_____ Greater Palatine Foramen   _____ Mental Foramen

_____ Incisive Nerve   _____ Mental Nerve

_____ Inferior Alveolar Nerve   _____ Middle Superior Alveolar Nerve

_____ Infraorbital Foramen   _____ Nerve to Muscles of Mastication

_____ Infraorbital Nerve   _____ Ophthalmic Nerve

_____ Lacrimal Gland   _____ Optic Nerve

_____ Lingual Nerve (2 Places)   _____ Posterior Superior Alveolar Nerve

_____ Long Buccal Nerve (2 Places)   _____ Sphenopalatine Ganglion

Figure 45.11   The trigeminal nerve.

Figure 46.1

Within the third ventricle is seen a **choroid plexus** (label 2, figure 46.2), which secretes cerebrospinal fluid (CSF). Although not shown in any of the human brain illustrations here, there are choroid plexuses in the lateral ventricles also. A choroid plexus in the fourth ventricle is shown in figure 46.2.

Cerebrospinal Fluid Pathway The CSF is constantly being produced by choroid plexuses of all the ventricles. Figure 46.3 indicates the pathway of the fluid in the brain.

CSF that originates in the lateral ventricles enters the third ventricle through the **interventricular foramen** (foramen of Monro). Note in figure 46.3 that this foramen lies anterior to the fornix and choroid plexus of the third ventricle.

From the third ventricle the CSF passes into the **cerebral aqueduct** (aqueduct of Sylvius). This duct conveys the CSF to the fourth ventricle. The **choroid plexus of the fourth ventricle** lies on the posterior wall of this ventricle.

From the fourth ventricle the CSF exits into the subarachnoid space that surrounds the cerebellum through three apertures: one on the median line and two lateral apertures.

In figure 46.3 the **median aperture** (foramen of Magendie) is the lower opening that has two arrows leading from it. That part of the subarachnoid space that it empties into is the **cisterna cerebellomedullaris.** Since the **lateral apertures** (foramina of Luschka) are located on the lateral walls of the fourth ventricle, only one is shown in figure 46.3.

From the cisterna cerebellomedullaris the CSF moves up into the **cisterna superior** (above the cerebellum) and finally into the subarachnoid space around the cerebral hemispheres. From the subarachnoid space this fluid escapes into the blood of the sagittal sinus through the **arachnoid granulations.** The CSF passes down the subarachnoid space of the spinal cord on the posterior side and up along the anterior surface.

Assignment:
After identifying all the structures of the sheep brain, label figure 46.2.

Label figure 46.3.

Frontal Sections

Illustrations A and B of figure 46.4 reveal frontal sections through two slightly different regions of the sheep brain. Once you study sections such as these, the human brain section in illustration C will be relatively easy to label.

Infundibular Section (Sheep Brain)

Place a whole sheep brain on a dissecting tray with the *dorsal side down.* Cut a frontal section with a long knife by starting at the point of the infundibulum. Cut through perpendicularly to the tray and to the longitudinal axis of the brain. Then, make a second cut parallel to the first cut ¼ inch further back. The latter cut should be through the center of the hypophysis.

Place this slice, *with the first cut upward,* on a piece of paper toweling for study. Illustration A, figure 46.4, is of this surface. Proceed to identify the following structures. You may need to refer to figure 46.1.

Note the pattern and thickness of the **gray matter** of the cerebral cortex. Force a probe down into a sulcus. What meninx do you break through as the probe moves inward?

Locate the two triangular **lateral ventricles.** Probe into one of the ventricles to see if you can locate the **choroid plexus.** Identify the **corpus callosum** and **fornix.**

Identify the **third ventricle,** which is situated on the median line under the fornix. Also, identify the **thalamus** and **intermediate mass.** Does the thalamus appear to consist of white or gray matter? What is the function of the white matter between the thalamus and the cerebral cortex?

Hypophyseal Section (Sheep Brain)

Prepare a section of the sheep brain by cutting another ¼ inch slice of the posterior portion of the brain. Place this slice, *anterior face upward* on paper toweling. This section through the hypophysis should look like illustration B, figure 46.4. Identify the structures in illustration B that are labeled.

Infundibular Section of Human Brain

Now that you have identified the parts of the sheep brain, identify the following structures in illustration C in figure 46.4:

Locate first the two triangular **lateral ventricles** and the **third ventricle.** Note how the **longitudinal cerebral fissure** extends downward to the **corpus callosum,** which forms the upper wall of the lateral ventricles. Although there is no label for the *septum pellucidum,* can you find it? Identify the two masses of gray matter just below the septum pellucidum that constitute the **fornix.**

Identify the areas on each side of the third ventricle that make up the nuclei of the **hypothalamus.** Approximately eight separate masses of gray matter are seen here. What physiological functions are regulated by these nuclei?

Locate the two large areas on the walls of the lateral ventricles that constitute the **thalamus.** Note, also, how these two portions of the thalamus are united by the **intermediate mass** that passes through the third ventricle.

Labels 12, 13, 14, and 15 are masses of gray matter in the cerebral hemispheres that are collectively known as the *basal ganglia.* The largest basal ganglia are the **putamen** and **globus pallidus** (labels 12 and 13). The latter is medial to the putamen. Together the putamen and globus pallidus form a triangular mass, the **lentiform nucleus.** These two ganglia exert a steadying effect on voluntary muscular movements.

The **caudate nuclei** are smaller basal ganglia located in the walls of the lateral ventricles superior

to the thalamus. These nuclei have something to do with muscular coordination, since surgically produced lesions in these centers can correct certain kinds of palsy.

Although some anatomists prefer to designate the thalamus and hypothalamus as basal ganglia, functionally, they differ considerably.

Assignment:
Label figure 46.4.

Laboratory Report

Complete the Laboratory Report for this exercise.

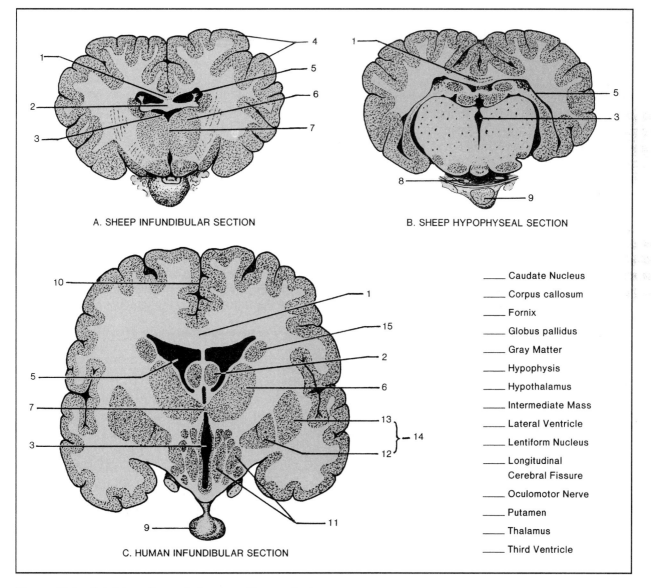

A. SHEEP INFUNDIBULAR SECTION

B. SHEEP HYPOPHYSEAL SECTION

C. HUMAN INFUNDIBULAR SECTION

_____ Caudate Nucleus
_____ Corpus callosum
_____ Fornix
_____ Globus pallidus
_____ Gray Matter
_____ Hypophysis
_____ Hypothalamus
_____ Intermediate Mass
_____ Lateral Ventricle
_____ Lentiform Nucleus
_____ Longitudinal Cerebral Fissure
_____ Oculomotor Nerve
_____ Putamen
_____ Thalamus
_____ Third Ventricle

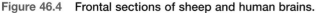

Figure 46.4 Frontal sections of sheep and human brains.

47 # Electroencephalography

The movement of nerve impulses from one neuron to another throughout the brain produces background electrical activity that is observable with the appropriate electronic equipment. The study of this electrical activity is called *electroencephalography.* The monitoring of this phenomenon may be achieved with needle electrodes inserted directly into the scalp or by placing surface electrodes on the scalp. The record produced by either method is called an **electroencephalogram,** or **EEG.**

Although considerable variation exists in the electrical activity of the brains of different individuals, normal patterns have been established. Electroencephalograms have many medical applications. They can be used for diagnosing organic malfunctions such as epilepsy, brain tumor, brain abscess, cerebral trauma, meningitis, encephalitis, and certain congenital conditions. They are particularly useful in establishing the existence of cerebral thrombosis and embolism. Despite its diagnostic value, however, a completely normal EEG is frequently seen even when considerable brain damage is present. In addition, normal EEGs are usually encountered in most individuals affected with nonorganic disorders.

Normal Wave Patterns

Most of the time, brain waves are irregular and exhibit no general pattern. At other times, however, distinct patterns can be detected. Although some of these patterns are characteristic of specific abnormalities such as epilepsy, there are distinctly normal patterns such as alpha, beta, theta, and delta waves as illustrated in figure 47.1.

Alpha Waves   When an individual is in a relaxed state of mind, with eyes closed and no particular thought in mind, *alpha waves* are most frequently seen. These waves occur at a rate of 8 to 13 cycles per second (cps).

When the eyes are opened and the subject's attention is drawn to a thought or visual object, the alpha rhythm usually changes to fast, irregular, shallow waves. This change of wave pattern is called *alpha block.*

Beta Waves   These waves occur at a frequency of 14 to 25 cps. There appear to be two types of beta waves: *beta I waves,* which occur with increased mental activity, and *beta II waves,* which are elicited by more intense mental activity.

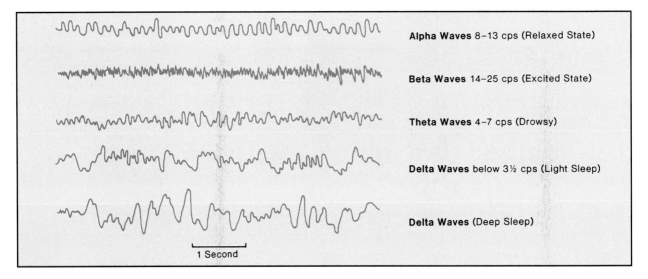

Figure 47.1   Normal electroencephalograms.

Theta Waves   The frequency range of these waves is between 4 and 7 cps. When seen in children, these waves are considered to be normal; in adults, however, it usually indicates disappointment and frustration. There are also certain brain disorders that exhibit these waves.

Delta Waves   At frequencies below 3½ cps the brain waves are designated as being *delta waves*. This brain wave pattern is normal for infants and occurs during sleep in adults.

The brain wave pattern of infants and children exhibits fast beta rhythms as well as the slower theta and delta rhythms. As children mature into adolescence, the rhythms develop into a characteristic alpha pattern of adulthood. Physiological conditions such as low body temperatures, low blood sugar levels, and high levels of carbon dioxide in the blood can slow down the alpha rhythm; the opposite conditions speed it up.

Pathological Evidence

The five EEGs shown in figure 47.2 are representative of what might be expected of individuals with brain damage. Note the absence of stability that is typical for the alpha rhythm of figure 47.1.

In evaluating EEGs, the neurologist looks for patterns that are considered *slow* or *fast,* as well as those that are paroxysmal. Records that are moderately slow or moderately fast are generally considered to be *mildly abnormal.* Those that are very slow or very fast are *definitely abnormal.* A very fast one would be the fourth pattern down in figure 47.2. In this case it represents an individual who suffered motor seizures and amnesia from barbiturate intoxication.

Considerable extremes in voltage, such as seen in the first tracing (epilepsy), are very significant. Epileptic patterns will vary with the type of seizure that occurs. Characteristically, epileptic seizures create large spikes that indicate considerable voltage differences.

Procedure

The clarity of wave patterns produced in EEG measurements is a function of the type of equipment and technique. A good oscilloscope produces excellent results. The procedures outlined here are for the Unigraph.

The electrical potentials emitted by the brain are of such low magnitude that the Unigraph will be operated at its maximum sensitivity (0.1 MV/CM). Ideally, one would use needle electrodes that are inserted into the scalp. For obvious legal reasons, such electrodes will not be used here; instead, we will use small disk electrodes that will be pressed tightly against the skin.

To overcome skin resistance, the skin will be gently scrubbed with a Scotchbrite pad before applying the electrodes. This scrubbing, plus the employment of an electrode jelly, will greatly enhance skin conductivity. An elastic headband will be used to exert constant pressure to the electrodes.

To record an EEG, students will work in groups of four. One member of the group will be the subject, another member will prepare the subject, and the remaining two will work with the equipment.

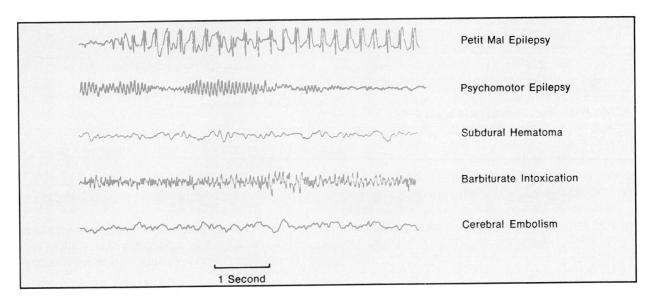

Petit Mal Epilepsy

Psychomotor Epilepsy

Subdural Hematoma

Barbiturate Intoxication

Cerebral Embolism

1 Second

Figure 47.2   Abnormal electroencephalograms.

Three electrodes will be attached to the subject: two on the forehead and one at the base of the skull in the occiput region. The electrodes will be attached to the three leads of a patient cable. The latter is plugged into the Unigraph.

The success of this experiment hinges considerably on environmental conditions as well as electrode placement. Ideally, the room should be darkened and quiet. A relaxed subject is essential, particularly when trying to record alpha waves. Subjects who are trained in Yoga or are capable of self-hypnosis can produce EEG patterns that are quite striking.

While the subject is being prepared, it will be necessary for the other two members to calibrate the Unigraph. Proceed as follows:

Materials:

Unigraph or polygraph
patient cable (3 leads)
3 EEG disk-type electrodes
3 circular Band-Aids
electrode jelly
alcohol swabs
Scotchbrite pad (grad #7447)
elastic wrapping

Calibration

For the EEG tracing to have any significant meaning, it is necessary to calibrate the Unigraph so that a one centimeter deflection of the stylus equals 0.1 millivolt; thus, each millimeter of vertical deflection will equal 0.01 millivolt. Proceed as follows to calibrate your instrument:

1. Turn the Mode selector control to CC/Cal (Capacitator Coupled Calibrate).
2. Turn on the power switch and turn the stylus heat control knob to the two o'clock position.
3. Set the speed selector lever to the slow position. The free end of the lever should be pointed toward the styluses.
4. Move the chart control switch (c.c.) to Chart On and observe the width of the tracing on the paper. Adjust the stylus heat control knob to produce the desired width of tracing.
5. Set the Gain control knob on 0.1 MV/CM.
6. Push down the 0.1 MV button to determine the amount of deflection. **Hold the button down a full two seconds before releasing.** If it is released too quickly the tracing will not be perfect.

   By turning the sensitivity knob, adjust the deflection so that it deflects exactly 1 cm in each direction. The unit is now properly calibrated.

7. Set the Mode selector control on EEG and return the c.c. switch to STBY. Place the speed selector lever in the fast position. The instrument is now ready for use.

Preparation of Subject

Two electrodes will be fixed to the forehead near the hairline and one will be placed in the occiput region of the head on the right side. The electrodes that are attached to the back of the head and right forehead will pick up the brain waves. The electrode on the left forehead acts as a ground. Proceed as follows:

1. With a Scotchbrite pad, *lightly* scrub the right and left forehead areas four or five times to remove the dead surface cells. Do not scrub so hard that you draw blood! Examine the surface of the pad for evidence of cleansing (white powder on pad).

   Refer to figure 47.3 to note the general area to locate the electrode.

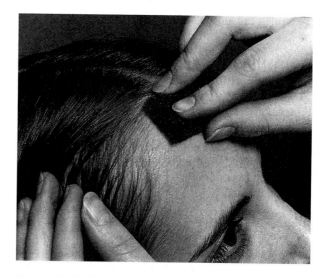

Figure 47.3 The skin surface is prepared by rubbing lightly with a Scotchbrite pad.

2. Disinfect each area with an alcohol swab.
3. Affix an electrode to a round Band-Aid or strip of adhesive tape, add a drop of electrode jelly, and attach to the forehead as shown in figures 47.4, 47.5, and 47.6.

   Be sure to position the electrode so that its flat surface will be against the skin. Follow this procedure for both electrodes that are attached to the forehead.
4. Position the third electrode on the back of the right side in the occiput region, using electrode jelly. Do not use a Band-Aid. To hold it in place,

put an elastic wrapping over it and the other two electrodes. Refer to figure 47.7.

5. Clamp the leads of the patient cable to the wires of the electrodes as follows:

   Lead 1—to the occiput electrode.
   Lead 2—to the right forehead electrode.
   Unnumbered Lead—to the left forehead electrode.

6. Have the subject lie down on a cot or laboratory tabletop. Make sure that all leads are free and not binding under the head or shoulders. Leads should not be disturbed by body movements during the test. Subject should lie perfectly still.

7. Plug the end of the 3-lead cable into the Unigraph. You are now ready to monitor the brain waves.

Monitoring

With the Unigraph set at 0.1 MV/CM in the EEG mode, place the c.c. switch at Chart On and observe the EEG pattern produced on the chart. The chart should be moving at the fast speed (25 mm per second).

In most cases the typical wave pattern is irregular. Be sure to keep the subject as quiet as possible. Even eye movements can alter the pattern. After about 2 minutes of recording, attempt to induce alpha waves using the suggestions that follow.

Alpha Rhythm   Although alpha waves are the most common wave form in normal adults, they are not always easy to demonstrate in a laboratory situation. Producing complete relaxation in the subject

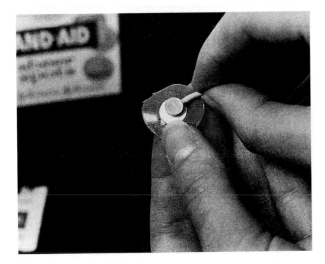

Figure 47.4   With flat surface upward, the electrode is placed on an adhesive patch.

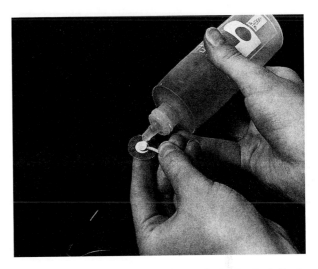

Figure 47.5   A very small drop of electrode jelly is placed on an electrode.

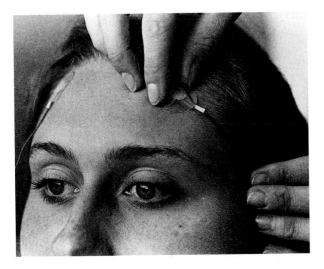

Figure 47.6   The forehead electrodes are positioned near the hairline.

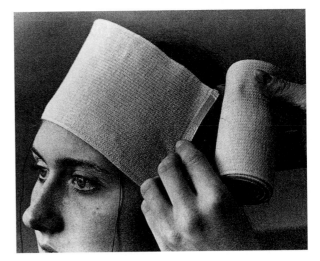

Figure 47.7   An elastic band is wrapped tightly over all three electrodes.

is very important. Extraneous disturbances, such as conversation, slamming doors, or furniture moving will prevent their formation. With the c.c. switch at Chart On, make the following type of statement to the subject in an attempt to induce relaxation:

> "With your eyes closed, relax all the muscles of your body as much as you can. Try not to concentrate on any one thought. Let your thoughts flit lightly from one thought to another. You have absolutely no problems. Life is beautiful. Everybody is kind, good, and your friend."

Allow the tracing to run intermittently for about 5 minutes; then, place the c.c. switch in the OFF position. Count the waves per second. Is it an alpha rhythm pattern (8 to 13 cps)? Determine the maximum number of microvolts seen in a specific wave.

Alpha Block   To demonstrate alpha block, place the c.c. switch on Chart On again and get the subject in the same mental state as previously. When the alpha rhythm is present, have the subject open his or her eyes (as you press the event marker button) and ask some questions that require concentration. Observe the pattern for evidence of alpha block. Continue to present the subject with questions and problems for a period of 4 or 5 minutes. Mark the paper with a pen or pencil where the questioning began.

Hyperventilation   Start the chart moving again. Have the subject relax with closed eyes to reestablish the alpha rhythm. Depress the event marker and ask the subject to take deep breaths at a rate of 40 to 50 per minute for a period of 2 or 3 minutes. This will hyperventilate the lungs, causing a lowering of the level of carbon dioxide in the blood. Observe the tracing and note the changes that occur.

Laboratory Report

Remove the chart from the Unigraph and review the results with the subject. Attach samples of the EEG to the Laboratory Report sheet in correct places, using tape or adhesive.

If time permits, repeat the experiment with other members of the group acting as subjects.

Answer all the questions on the Laboratory Report.

9

The Sense Organs

This unit comprises four exercises. The first two pertain to the eye and the last two are concerned with the ear. In Exercise 48, the anatomy of the eye will be studied through dissection and visual examination of its interior. A fresh beef eye will be used for dissection. To observe the inside of the eyeball, an ophthalmoscope will be used. Exercise 49 pertains entirely to eye tests. Some of these tests will be used to reveal normal physiological phenomena; other tests will be used to determine deviations from normal vision.

Exercises 50 and 51 are concerned with different functions of the ear. In Exercise 50 that part of the ear that functions in hearing will be studied. Different kinds of hearing tests will be employed here to detect deviations from normal. In Exercise 51 the part of the ear that functions in the maintenance of equilibrium will be studied. A variety of tests pertaining to equilibrium will be performed here.

Anatomy of the Eye

In preparation for the study of eye function in the next exercise, our focus here will be on the anatomy of the eye through dissection, ophthalmoscopy, and microscopy. Before performing these studies, however, label figures 48.1, 48.2, and 48.3, which pertain to the lacrimal system, internal anatomy, and the extrinsic muscles.

The Lacrimal Apparatus

Each eye lies protected in a bony recess of the skull and is covered externally by the **eyelids** *(palpebrae)*. Lining the eyelids and covering the exposed surface of the eyeball is a mucous membrane called the **conjunctiva.** The conjunctival surfaces are kept constantly moist by protective secretions of the lacrimal glands, which lie between the upper eyelid and the eyeball.

Figure 48.1 reveals the external anatomy of the right eye, with particular emphasis on the lacrimal apparatus. Note that the lacrimal gland consists of two portions: a large **superior lacrimal gland** and a smaller **inferior lacrimal gland.** Secretions from these glands pass through the conjunctiva of the eyelid via six to twelve small ducts.

The flow of fluid from these ducts moves across the eyeball to the *medial canthus* (angle) of the eye, where the fluid enters two small orifices, the **lacrimal puncta.** These openings lead into the **superior** and **inferior lacrimal ducts;** the superior one is in the upper eyelid and the inferior one is in the lower eyelid. These two ducts empty directly into an enlarged cavity, the **lacrimal sac,** which, in turn, is drained by the **nasolacrimal duct.** The tears finally exit into the nasal cavity through the end of the nasolacrimal duct.

Two other structures, the caruncula and the plica semilunaris, are seen near the puncta in the medial canthus of the eye. The **caruncula** is a small red conical body that consists of a mound of skin containing sebaceous and sweat glands and a few small hairs. It is this structure that produces a whitish secretion that constantly collects in this region. Lateral to the caruncula is a curved fold of conjunctiva, the **plica semilunaris.** This structure contains some smooth muscle fibers; in the cat and many other animals it is more highly developed than in humans and is often referred to as the "third eyelid."

Assignment:
Label figure 48.1

Internal Anatomy

Figure 48.2 is a horizontal section of the right eye as seen looking down upon it. Note that the wall of the

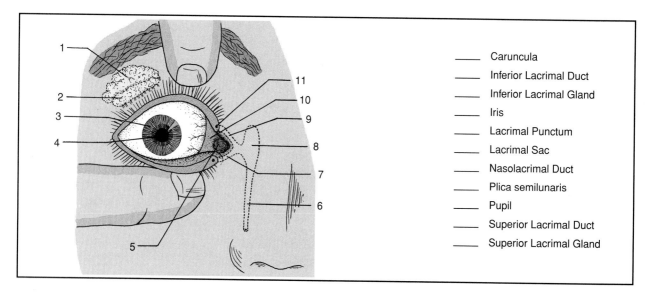

_____ Caruncula

_____ Inferior Lacrimal Duct

_____ Inferior Lacrimal Gland

_____ Iris

_____ Lacrimal Punctum

_____ Lacrimal Sac

_____ Nasolacrimal Duct

_____ Plica semilunaris

_____ Pupil

_____ Superior Lacrimal Duct

_____ Superior Lacrimal Gland

Figure 48.1   The lacrimal apparatus of the eye.

eyeball consists of three layers: an outer **scleroid coat,** a middle **choroid coat,** and an inner **retina.** The continuity of the surface of the retina is interrupted only by two structures, the optic nerve and the fovea centralis.

The **optic nerve** (label 3) contains fibers leading from the rods and cones of the retina to the brain. That point on the retina where the optic nerve makes its entrance is lacking in rods and cones. The absence of light receptors at this spot makes it insensitive to light; thus, it is called the **blind spot.** It is also known as the **optic disk.**

Note that the optic nerve is wrapped in a sheath, the **dura mater,** which is an extension of the dura mater that surrounds the brain. To the right of the blind spot is a pit, the **fovea centralis,** which is the center of a round yellow spot, the *macula lutea.* The macula, which is only a half a millimeter in diameter, is composed entirely of cones and is that part of the eye where all critical vision occurs.

Light entering the eye is focused on the retina by the **lens,** an elliptical crystalline clear structure suspended in the eye by a **suspensory ligament.** The substance of the lens is a completely transparent protein called *crystallin,* which is formed from cells during embryological development. The cells that produce this protein self-destruct (apoptosis) before birth.

Attached to the suspensosry ligament is the **ciliary body,** a circular smooth muscle that can change the shape of the lens to focus on objects at different distances.

In front of the lens is a cavity that contains a watery fluid, the **aqueous humor.** Between the lens and the retina is a larger cavity that contains a more viscous, jellylike substance, the **vitreous body.** Immediately in front of the lens lies the circular colored portion of the eye, the **iris.** It consists of circular and radiating muscle fibers that can change the size of the **pupil** of the eye. The iris regulates the amount of light that enters the eye through the pupil.

Covering the anterior portion of the eye is the **cornea,** a clear, transparent structure that is an extension of the scleroid coat. It acts as a window to the eye, allowing light to enter.

Aqueous humor is constantly being renewed in the eye. The enlarged inset of the lens-iris area in figure 48.2 illustrates where the aqueous humor is produced (point A) and where it is reabsorbed (point B).

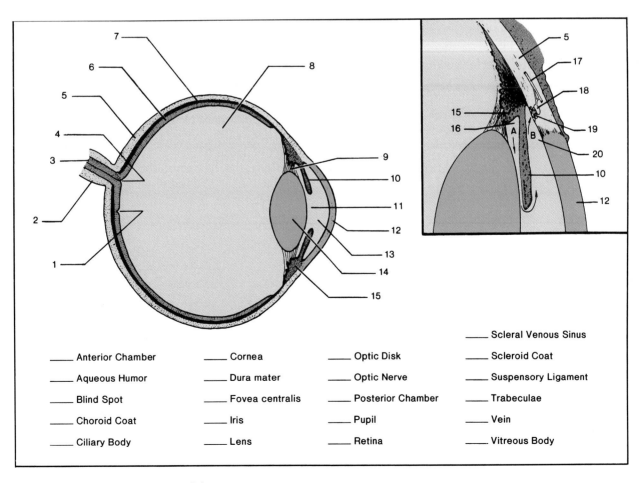

_____ Anterior Chamber	_____ Cornea	_____ Optic Disk	_____ Scleral Venous Sinus
_____ Aqueous Humor	_____ Dura mater	_____ Optic Nerve	_____ Scleroid Coat
_____ Blind Spot	_____ Fovea centralis	_____ Posterior Chamber	_____ Suspensory Ligament
_____ Choroid Coat	_____ Iris	_____ Pupil	_____ Trabeculae
_____ Ciliary Body	_____ Lens	_____ Retina	_____ Vein
			_____ Vitreous Body

Figure 48.2 Internal anatomy of the eye.

At point A, which is the **posterior chamber,** aqueous humor is produced by capillaries of the ciliary body. As this fluid is produced, it passes through the pupil into the **anterior chamber** and is reabsorbed into minute spaces called **trabeculae** at the juncture of the cornea and the iris (point B). From the trabeculae the fluid passes to the **scleral venous sinus** (canal of Schlemm), label 18. This venous sinus drains into the venous system, represented as a short **vein** in figure 48.2.

Assignment:
Label figure 48.2.

The Ocular Muscles

The ability of each eye to move in all directions within its orbit is controlled by six *extrinsic ocular muscles*. Movement of the upper eyelid is controlled by a single muscle, the **superior levator palpebrae.**

Figure 48.3, which is a lateral view of the right eye, reveals the insertions of all seven of these muscles. Their origins are on various bony formations at the back of the eye orbit.

The muscle on the side of the eyeball that has a portion of it removed is the **lateral rectus.** Opposing it on the medial side of the eye is the **medial rectus.** The muscle that inserts on top of the eye is the **superior rectus;** its antagonist on the bottom of the eye is the **inferior rectus.** Just above the stub of the lateral rectus is seen the insertion of the **superior oblique,** which passes through a cartilaginous loop, the **trochlea.** Opposing the superior oblique is the **inferior oblique,** of which only a portion of the insertion can be seen.

Assignment:
Label figure 48.3.

Beef Eye Dissection

The nature of the tissues of the eye can best be studied by actual dissection of an eye. A beef eye will be used for this study. Figure 48.4 illustrates various steps that will be followed.

Materials:
beef eye, preferably fresh
dissecting instruments
dissecting tray

1. Examine the external surfaces of the eyeball. Identify the **optic nerve** and look for remnants of **extrinsic muscles** on the sides of the eyeball.
2. Note the shape of the **pupil.** Is it round or elliptical? Identify the **cornea.**
3. Holding the eyeball as shown in illustration 1, figure 48.4, make an incision into the scleroid coat with a sharp scalpel about one-quarter of an inch away from the cornea. Note how difficult it is to penetrate.
4. Insert scissors into the incision and carefully cut all the way around the cornea, *taking care not to squeeze the fluid out of the eye.* Refer to illustration 2, figure 48.4.
5. Gently lift the front portion off the eye and place it on the tray with the inner surface upward. The lens usually remains attached to the vitreous body, as shown in illustration 3. If the eye is preserved (not fresh), the lens may remain in the anterior portion of the eye.
6. Examine the inner surface of the anterior portion and identify the thickened, black circular **ciliary body.** What function does the black pigment perform?
7. Study the **iris** carefully. Can you distinguish **circular** and **radial muscle fibers**?

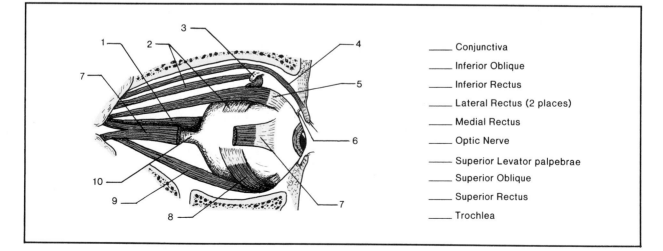

_____ Conjunctiva
_____ Inferior Oblique
_____ Inferior Rectus
_____ Lateral Rectus (2 places)
_____ Medial Rectus
_____ Optic Nerve
_____ Superior Levator palpebrae
_____ Superior Oblique
_____ Superior Rectus
_____ Trochlea

Figure 48.3   The extrinsic muscles of the eye.

8. Is there any **aqueous humor** left between the cornea and iris? Compare its consistency with the **vitreous body.**

9. Separate the lens from the vitreous body by working a dissecting needle around its perimeter as shown in illustration 3. If the lens is from a fresh eye (not preserved), hold the lens up by its edge with a forceps and look through it at some distant object.

What is unusual about the image? Next, place the lens on printed matter. Are the letters magnified?

Note: Lenses from preserved specimens are usually not transparent due to coagulation of the crystallin by preservatives.

10. Compare the consistency of the lens at its center and near its circumference by squeezing the lens between your thumb and forefinger. Do you detect any differences?

11. Locate the **retina** at the back of the eye. It is a thin colorless inner coat that separates easily from the pigmented **choroid coat.**

Now locate the **blind spot** *(optic disk).* This is the area where the optic nerve forms a juncture with the retina.

12. Note the irridescent nature of a portion of the choroid coat. This reflective surface is called the **tapetum lucidum.** It enables the eyes of animals to reflect light at night and appears to enhance night vision by reflecting some light back into the retina.

13. Answer all questions on the Laboratory Report that pertain to this dissection.

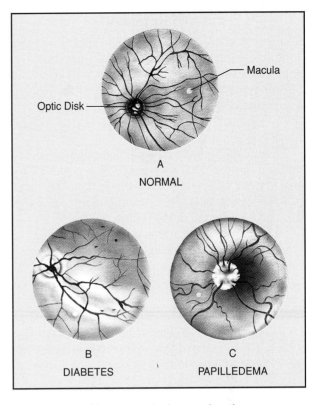

Figure 48.5 Normal and abnormal retinas as seen through an ophthalmoscope. In illustration B *(diabetic retinopathy)* the peripheral vessels are generally irregular, fewer in number, and often exhibit small hemorrhages. *Papilledema,* or "choked disk" (illustration C), may be evidence of hypertension, brain tumor, or meningitis. In this case the veins are considerably enlarged and often hemorrhagic.

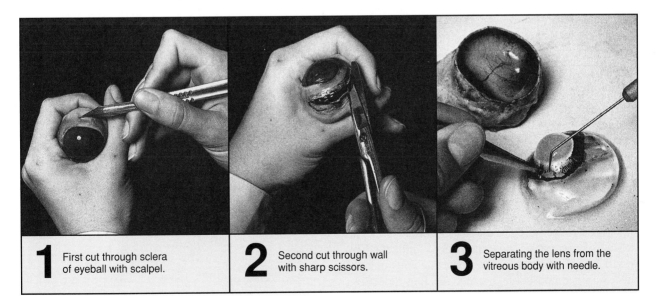

1 First cut through sclera of eyeball with scalpel.

2 Second cut through wall with sharp scissors.

3 Separating the lens from the vitreous body with needle.

Figure 48.4 Three steps in beef eye dissection.

Ophthalmoscopy

Routine physical examinations invariably involve an examination of the **fundus,** or interior of the eye. A careful examination of the fundus provides valid information about the general health of the patient. The retina of the eye is the only part of the body where relatively large blood vessels may be inspected without surgical intervention.

Figure 48.5 illustrates what one might expect to see in a normal eye (A), the eye of a diabetic (B), and the eye of one who exhibits a "choked disk" (C). These are only a few examples of pathology that can be detected in the fundus.

The instrument that one uses for examining the fundus is called an **ophthalmoscope.** In this exercise you will have an opportunity to examine the interior of the eyes of your laboratory partner. Your partner, in turn will examine your eyes. It will be necessary to make this examination in a darkened room. Before this examination can be made, however, it is essential that you understand the operation of the ophthalmoscope.

The Ophthalmoscope

Figure 48.6 reveals the construction of a Welch Allyn ophthalmoscope. Within its handle are two C-size batteries that power an illuminating bulb that is located in the viewing head. Note that at the top of the handle there is a **rheostat control** for regulating the intensity of the light source. Only after depressing the lock button can this control be rotated.

The head of the instrument has two rotatable notched disks: a large lens selection disk and a smaller aperture selection disk. A viewing aperture is located at the upper end of the head.

While looking through the viewing aperture, the examiner is able to select one of twenty-three small lenses by rotating the **lens selection disk** with the index finger held as shown in figure 48.7. Twelve of the lenses on this disk are positive (convex) and eleven of them are negative (concave). Numbers that appear in the **diopter window** indicate which lens is in place. The positive lenses are represented as black numbers, the negative lens numbers are red. A black "0" in the window indicates that no lens is in place.

If the eyes of both subject and examiner are normal *(emmetropic),* no lens is needed. If the eye of the subject or examiner is farsighted *(hypermetropic),* the positive lenses will be selected. Nearsighted *(myopic)* eyes require the use of negative lenses. The degree of myopia or hypermetropia is indicated by the magitude of the number. Since the eyes of both the examiner and subject affect the lens selection, the selection becomes entirely empirical.

The **aperture selection disk** enables the examiner to change the character of the light beam that is

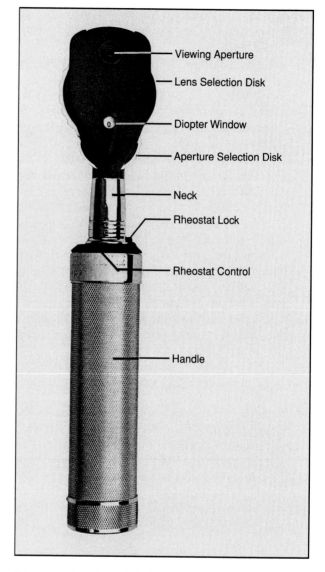

Figure 48.6   The ophthalmoscope.

Labels: Viewing Aperture, Lens Selection Disk, Diopter Window, Aperture Selection Disk, Neck, Rheostat Lock, Rheostat Control, Handle

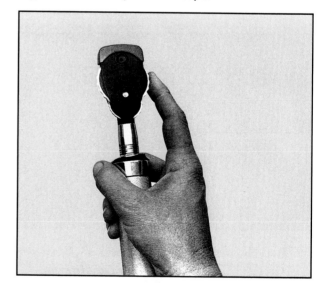

Figure 48.7   Viewing lenses of ophthalmoscope are rotated into position by moving the lens selection disk with the index finger.

projected into the eye. The light may be projected as a grid, straight line, white spot, or green spot. **The green spot is most frequently used** because it is less irritating to the eye of the subject, and it causes the blood vessels to show up most clearly.

Using an ophthalmoscope for the first time seems difficult for some, relatively easy for others, and nearly impossible for a few. If the examiner and subject have normal eyes, usually there is a minimum of difficulty. However, when the examiner's eyes are severely myopic, astigmatic, or hypermetropic, difficulties can occur. Even for examiners with normal vision there is the problem of reflection of light from the retina back into the examiner's eye. The best way to minimize this problem is to **direct the light beam toward the edge of the pupil** rather than through the center.

A precautionary statement must be heeded concerning light exposure. **Avoid subjecting an eye to more than one minute of light exposure.** After one minute exposure, allow several minutes of rest for recovery of the retina.

Procedure

Before using the ophthalmoscope in the darkened room, it will be necessary for you to become completely familiar with its mechanics.

Materials:
ophthalmoscope
metric tape or ruler

Desk Study of Ophthalmoscope

Turn on the light source by depressing the lock button and rotating the rheostat control. Observe how the light intensity can be varied from dim to very bright. Rotate the aperture selection disk as you hold the ophthalmoscope about two inches away from your desktop. How many different light patterns are there? Select the large white spot for the next phase of this study.

Rotate the lens selection disk until a black "5" appears in the diopter window. While peering through the viewing window at the print on this page, bring the letters into sharp focus. Have your lab partner measure with a metric tape the distance in millimeters between the ophthalmoscope and the printed page when the lettering is in sharp focus. Record this distance on the Laboratory Report. Do the same thing for the 10D, 20D, and 40D lenses, recording all measurements.

Now, set the lens selection disk on "0" and the light source on the green spot. The instrument is properly set now for the beginning of an eye examination. Read over the following procedure *in its entirety* before entering the darkened room. This will prepare you for what you will be doing.

The Examination

Figures 48.8 and 48.9 illustrate how the ophthalmoscope is held when viewing different eyes. Note that the examiner uses the right hand when examining the right eye and the left hand when examining the left eye. Proceed as follows:

1. With the "0" in the diopter window and the light turned on, grasp the ophthalmoscope, as shown in figure 48.8. Note that the index finger rests on the lens selection disk.
2. Place the viewing aperture in front of your right eye and steady the ophthalmoscope by resting

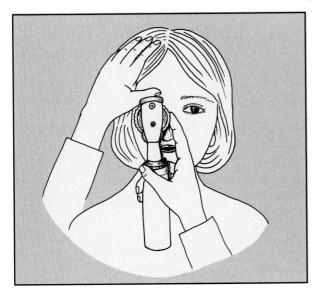

Figure 48.8 When examining the right eye, the ophthalmoscope is held with the right hand.

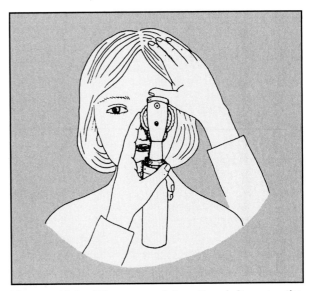

Figure 48.9 When examining the left eye, the ophthalmoscope is held with the left hand.

the top of it against your eyebrow. See figure 48.10.

Start viewing the subject's right eye at a distance of about 12 inches. Keep your subject on your right and instruct your subject to **look straight ahead at a fixed object at eye level.**

3. Direct the beam of light into the pupil and examine the lens and vitreous body. As you look through the pupil, a **red reflex** will be seen. If the image is not sharp, adjust the focus with the lens selection disk.

4. While keeping the pupil in focus, move in to within 2 inches of the subject, as illustrated in figure 48.11. The red reflex should become more pronounced. **Try to direct the light beam toward the edge of the pupil** rather than in the center to minimize reflection from the retina.

5. Search for the **optic disk** and adjust the focus, as necessary, to produce a sharp image. Observe how blood vessels radiate out from the optic disk. Follow one or more of them to the periphery.

Locate the **macula.** It is situated about two disk diameters *laterally* from the optic disk. Note that it lacks blood vessels. It is easiest to observe when the subject is asked to look directly into the light, a position that **must be limited to one second only.**

6. Examine the entire retinal surface. Look for any irregularities, depressions, or protrusions from its surface. It is not unusual to see a roundish elevation, not unlike a pimple on its surface.

To examine the periphery, instruct the subject to (1) **look up** for examination of the superior retina, (2) **look down** for examination of the inferior retina, and (3) **look laterally and medially** for those respective areas.

7. **Caution:** Remember to limit the right eye examination to **one minute!** After this time limit is reached, examine the left eye, using the left hand to hold the ophthalmoscope and reversing the entire procedure.

8. Switch roles with your laboratory partner. When you have examined each other's eyes examine the eyes of other students. Report any unusual conditions to your instructor.

Histological Studies

Most prepared slides available for laboratory study are made from either the rabbit or the monkey. Both animals are suitable. Proceed as follows to identify structures depicted in the photomicrographs of figures HA-16 and HA-17 of the Histology Atlas.

Materials:
prepared slides of:
    monkey eye, x.s. (H10.635, H10.64 or H10.76)
    rabbit eye, sagittal section (H10.61 or 10.62)

The Cornea   With the lowest powered objective on your microscope scan a cross section of the eye of a rabbit or monkey until you find the cornea. Refer to illustrations A and B, figure HA-16. Note that the cornea consists of three layers: the epithelium, stroma, and endothelium. For fine detail, as in illustration B, use high-dry.

The **corneal epitheium** is the outer portion that consists of stratified squamous. It is, essentially, the conjunctival portion of the cornea. Note that the outermost cells of this layer are flattened squamous cells and that the basal cells are columnar.

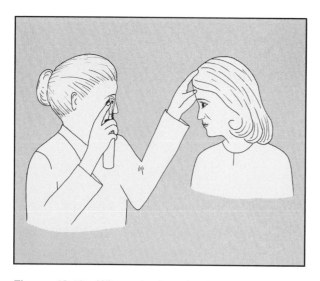

Figure 48.10 When viewing the lens and vitreous body, use this position, relative to the subject.

Figure 48.11 When examining the retinal surfaces move in close as illustrated here.

The **corneal endothelium** is the innermost layer of cells, which is adjacent to the aqueous humor. This thin layer consists of low cuboidal cells.

The **corneal stroma** *(substantia propria)* constitutes nine-tenths of the thickness of the cornea. It consists of collagenous fibrils, fibroblasts, and cementing substance. The fibrils are arranged in lamellae that run parallel to the corneal surface. The components of this stroma are held together by a mucopolysaccharide cement. The chemical structure and fibril arrangement contribute to the transparency of the cornea.

The Lens   Note (illustration C, Figure HA-16) that the lens is enclosed in a homogenous elastic **capsule** to which the **suspensory ligament** is attached. The lens is formed from epithelial cells that become elongated to a fibrous shape (lens fibers), losing most of their organelles and nuclei. All that remains in a mature lens fiber are a few microtubules and clumps of free ribosomes. The principal component of the lens is a protein called *crystallin,* which is formed during embryological development.

The Ciliary Muscle   This ring of smooth muscle tissue is a part of the body wall of the eye. Identify the **ciliary processes,** which form ridges on the ciliary muscle. The latter provide an anchor for the suspensory ligaments.

Body Wall of the Eye   Examine a section through the body wall at the back of the eye. Identify the sclera, choroid coat, and retina. Note that the **sclera** consists of closely packed collagenous fibers, elastic fibers, and fibroblasts.

Note, also, the large amount of pigment in the **choroid coat** that is produced by **melanocytes.** This is the layer that supplies nourishment to the retina and scleroid coat. Look for blood vessels.

The **retina** is the inner photosensitive layer of the body wall.

Retinal Layers   The retina is composed of five main classes of neurons: the *photoreceptors* (rods and cones), *bipolar cells, ganglion cells, horizontal cells,* and *amacrine cells.* The first three form a direct pathway from the retina to the brain. The horizontal and amacrine cells form laterally directed pathways that modify and control the message being passed along the direct pathway.

Study a section of the retina and identify the various layers. Refer to figure HA-17 for reference. Any slide of monkey eye section is suitable for retinal study.

At the base of the retina, where it meets the choroid layer, are the **rods** and **cones.** The nuclei for these photoreceptors are located in the **outer nuclear layer.** To stimulate these receptors light must pass through all the other cell layers.

Nuclei of the amacrine and bipolar neurons are located in the **inner nuclear layer.** Dendrites of these association neurons make synaptic connections with axons of the rods and cones in the **outer synaptic layer.**

The nuclei of ganglion cells are located closest to the exposed surface of the retina: in a layer designated as the **ganglion cell layer.** The dendrites of ganglion cells make synaptic connections with amacrine and bipolar cells in the **inner synaptic layer.** Nonmyelinated axons of the ganglion cells fill up the **nerve fiber layer;** they converge at the optic disk to form the optic nerve. Some neuroglial cell nuclei can be seen in the nerve fiber layer.

The Fovea Centralis   Study a slide that reveals the structure of the fovea. Compare your slide with the photomicrographs in illustrations B and C in figure HA-17.

Laboratory Report

Answer the questions on the first portion of combined Laboratory Report 48,49.

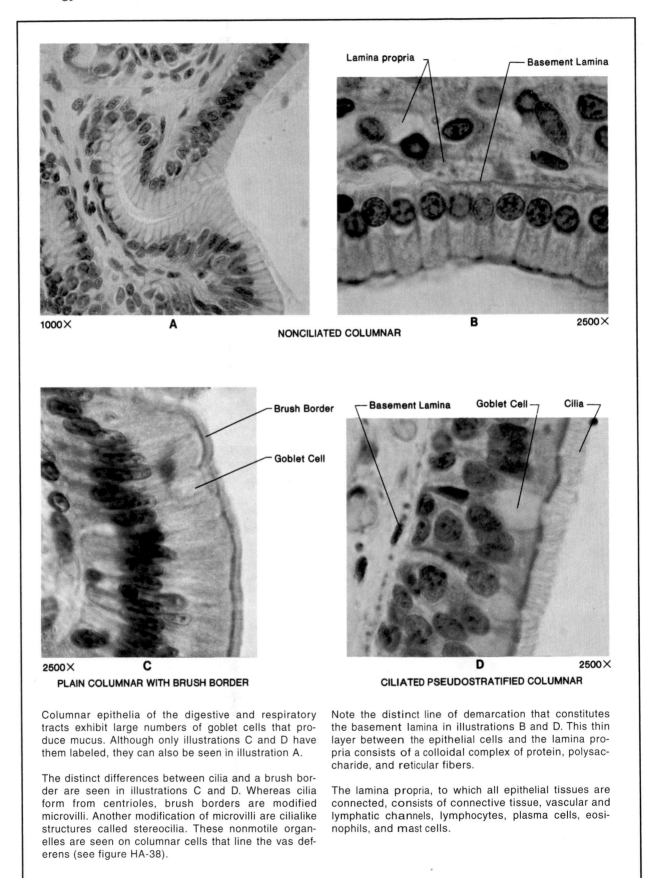

1000×   **A**

Lamina propria   Basement Lamina

B   2500×

NONCILIATED COLUMNAR

Brush Border

Goblet Cell

2500×   **C**

PLAIN COLUMNAR WITH BRUSH BORDER

Basement Lamina   Goblet Cell   Cilia

D   2500×

CILIATED PSEUDOSTRATIFIED COLUMNAR

Columnar epithelia of the digestive and respiratory tracts exhibit large numbers of goblet cells that produce mucus. Although only illustrations C and D have them labeled, they can also be seen in illustration A.

The distinct differences between cilia and a brush border are seen in illustrations C and D. Whereas cilia form from centrioles, brush borders are modified microvilli. Another modification of microvilli are cilialike structures called stereocilia. These nonmotile organelles are seen on columnar cells that line the vas deferens (see figure HA-38).

Note the distinct line of demarcation that constitutes the basement lamina in illustrations B and D. This thin layer between the epithelial cells and the lamina propria consists of a colloidal complex of protein, polysaccharide, and reticular fibers.

The lamina propria, to which all epithelial tissues are connected, consists of connective tissue, vascular and lymphatic channels, lymphocytes, plasma cells, eosinophils, and mast cells.

Figure HA-2   Columnar epithelium.

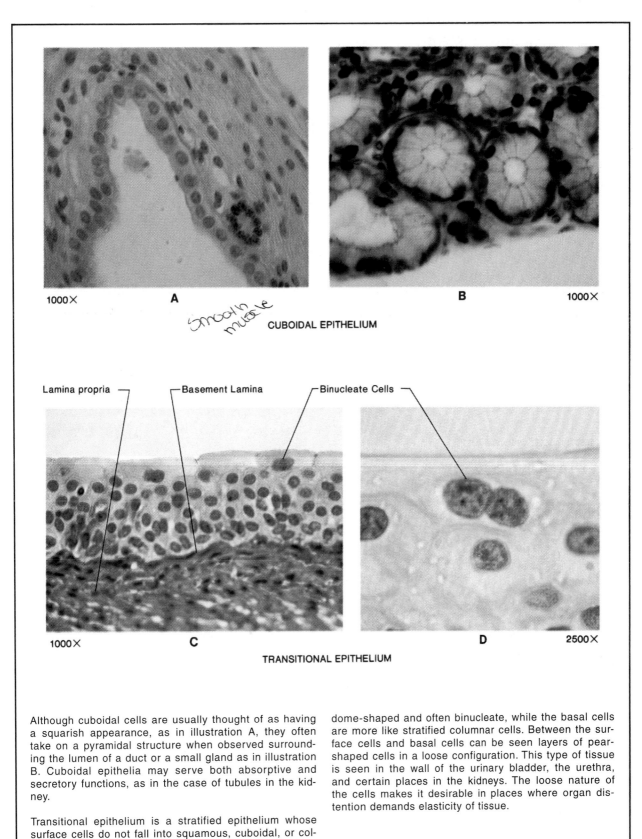

1000×   **A**

Smooth muscle (handwritten)

B   1000×

CUBOIDAL EPITHELIUM

Lamina propria   Basement Lamina   Binucleate Cells

1000×   **C**   **D**   2500×

TRANSITIONAL EPITHELIUM

Although cuboidal cells are usually thought of as having a squarish appearance, as in illustration A, they often take on a pyramidal structure when observed surrounding the lumen of a duct or a small gland as in illustration B. Cuboidal epithelia may serve both absorptive and secretory functions, as in the case of tubules in the kidney.

Transitional epithelium is a stratified epithelium whose surface cells do not fall into squamous, cuboidal, or columnar categories. Note that the surface cells are dome-shaped and often binucleate, while the basal cells are more like stratified columnar cells. Between the surface cells and basal cells can be seen layers of pear-shaped cells in a loose configuration. This type of tissue is seen in the wall of the urinary bladder, the urethra, and certain places in the kidneys. The loose nature of the cells makes it desirable in places where organ distention demands elasticity of tissue.

Figure HA-3   Cuboidal and transitional epithelium.

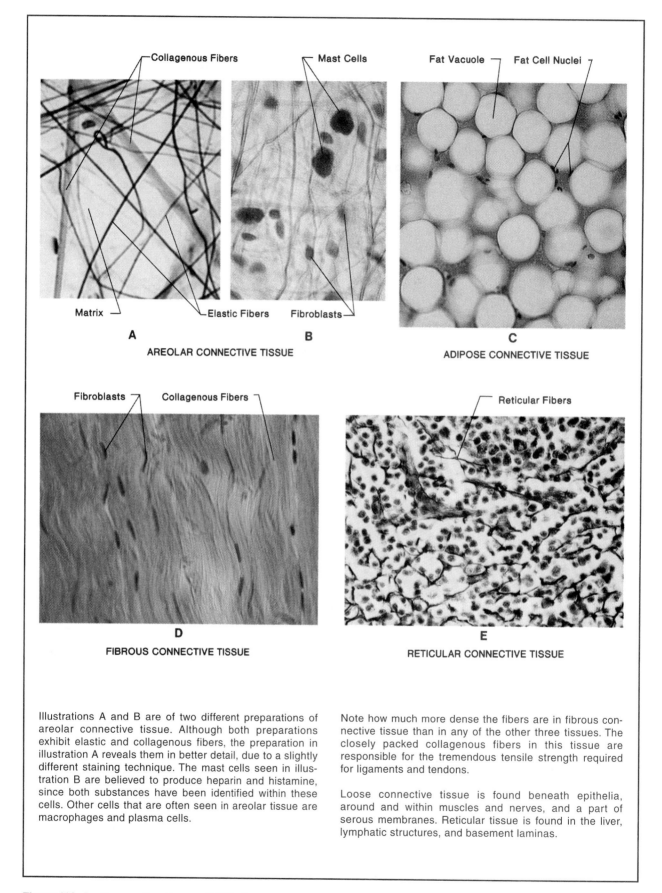

Collagenous Fibers · Mast Cells · Fat Vacuole · Fat Cell Nuclei

Matrix · Elastic Fibers · Fibroblasts

A · **B**

AREOLAR CONNECTIVE TISSUE

C

ADIPOSE CONNECTIVE TISSUE

Fibroblasts · Collagenous Fibers · Reticular Fibers

D

FIBROUS CONNECTIVE TISSUE

E

RETICULAR CONNECTIVE TISSUE

Illustrations A and B are of two different preparations of areolar connective tissue. Although both preparations exhibit elastic and collagenous fibers, the preparation in illustration A reveals them in better detail, due to a slightly different staining technique. The mast cells seen in illustration B are believed to produce heparin and histamine, since both substances have been identified within these cells. Other cells that are often seen in areolar tissue are macrophages and plasma cells.

Note how much more dense the fibers are in fibrous connective tissue than in any of the other three tissues. The closely packed collagenous fibers in this tissue are responsible for the tremendous tensile strength required for ligaments and tendons.

Loose connective tissue is found beneath epithelia, around and within muscles and nerves, and a part of serous membranes. Reticular tissue is found in the liver, lymphatic structures, and basement laminas.

Figure HA-4   Connective tissues (1000×).

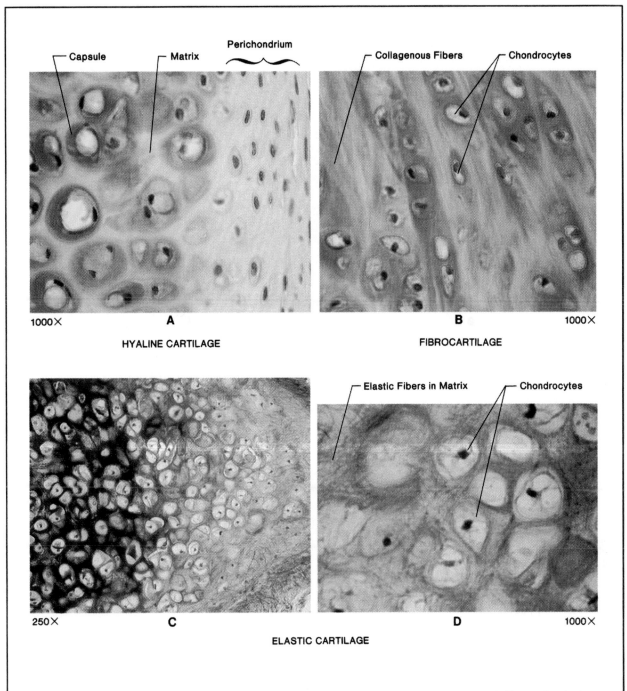

Figure HA-5 Types of cartilage.

All types of cartilage are similar to each other in that the chondrocytes are contained in smooth-walled spaces called lacunae. Observe that the walls of the lacunae are quite dense and stain darker than surrounding matrix. These walls are referred to as capsules. Note that the chondrocytes appear to have large vacuoles in them. This apparent vacuolation is an artifact of tissue preparation due to the leaching out of fat and glycogen by solvents used in the staining process. Note that a dense layer of connective tissue, the peri-chondrium, is shown in illustration A. Except for the free surfaces of articular cartilage, all other hyaline cartilage structures are invested with a perichondrium.

The firmness of the matrix in hyaline cartilage is due to the presence of condroitin sulfate and collagen. Fibrocartilage is much stronger than hyaline cartilage because of the preponderance of collagenous fibers in its matrix. The presence of elastic fibers as well as collagenous fibers in elastic cartilage makes this type of cartilage more flexible.

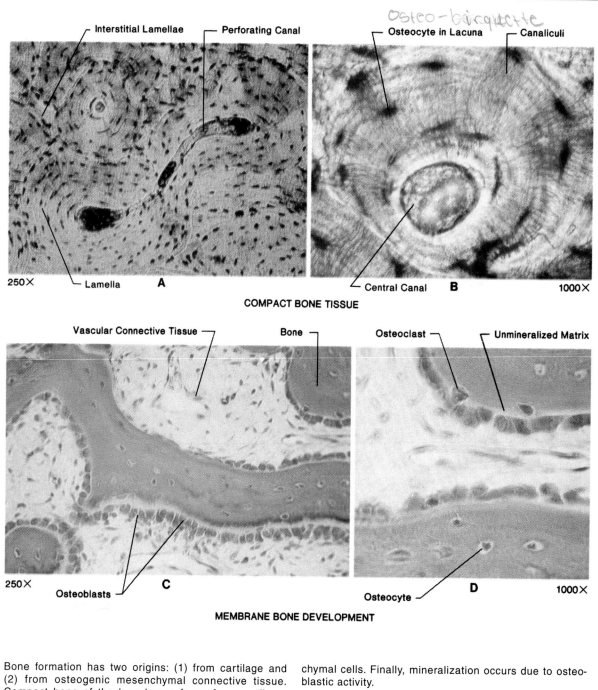

COMPACT BONE TISSUE

MEMBRANE BONE DEVELOPMENT

Bone formation has two origins: (1) from cartilage and (2) from osteogenic mesenchymal connective tissue. Compact bone of the long bones forms from cartilage; bones of the skull, on the other hand, develop as shown in illustrations C and D. The latter are often referred to as "membrane bones."

Note the loci of the different types of bone cells in the above illustrations. Illustrations C and D reveal that osteoblasts in membranous bone formation are seen on the leading edge of the forming bone. The first stage is the secretion of matrix by the osteoblasts. Reticular fibers are then added to the matrix by surrounding mesen-

chymal cells. Finally, mineralization occurs due to osteoblastic activity.

Osteoclasts are larger multinucleated cells that are involved in shaping bone structure by bone resorption. These cells are surrounded by a clear area (Howship's lacuna), evidence of mineral resorption. The osteoclast seen in illustration D is in an early stage of development; thus, Howship's lacuna is not very large. Once a bone cell becomes completely surrounded by bone, it is referred to as an osteocyte. Nourishment in mature compact bone reaches osteocytes through canaliculi.

Figure HA-6   Bone histology.

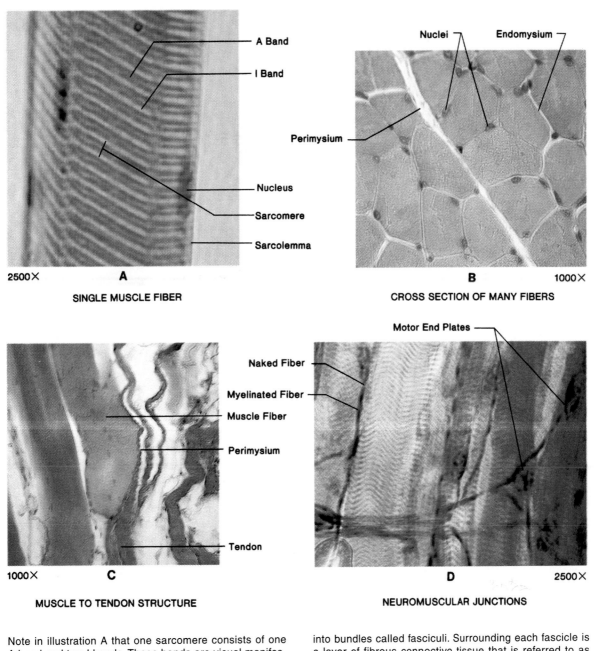

A Band
I Band
Nucleus
Sarcomere
Sarcolemma

2500× **A**
SINGLE MUSCLE FIBER

Nuclei **Endomysium**
Perimysium

B 1000×
CROSS SECTION OF MANY FIBERS

Naked Fiber
Myelinated Fiber
Muscle Fiber
Perimysium
Tendon

1000× **C**
MUSCLE TO TENDON STRUCTURE

Motor End Plates

D 2500×
NEUROMUSCULAR JUNCTIONS

Note in illustration A that one sarcomere consists of one A band and two I bands. These bands are visual manifestations of the intricate arrangement of actin and myosin myofilaments. Interaction between these filaments causes muscle contraction.

A significant characteristic of skeletal muscle cells is that they are multinucleated, or syncytial. Note that the nuclei are elongated and situated near the sarcolemma of the cell. The peripheral location of the nuclei shows up well in illustration B.

Illustration B reveals that muscle fibers are separated from each other by endomysium, and grouped together into bundles called fasciculi. Surrounding each fascicle is a layer of fibrous connective tissue that is referred to as the perimysium.

Illustration C illustrates how the fibrous connective tissue of the endomysium, perimysium, and epimysium is continuous with the tendons which attach muscles to bone. The connective tissue of tendons, in turn, is continuous with the periosteum of bone.

Note in illustration D how the myelinated motor nerve fibers lose their myelin sheaths and become "naked" where they join the motor end plates.

Figure HA-7 Skeletal muscle microstructure.

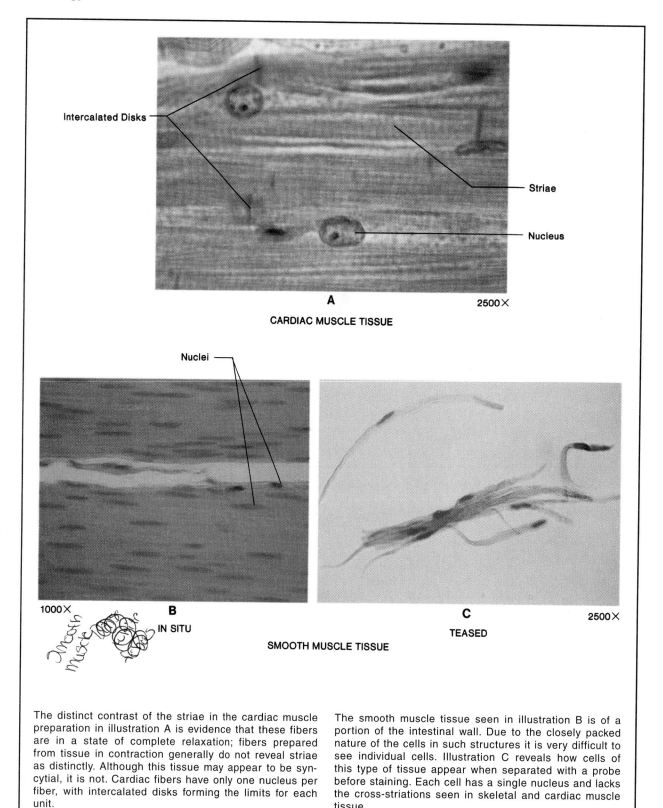

Intercalated Disks

Striae

Nucleus

A

2500×

CARDIAC MUSCLE TISSUE

Nuclei

1000×   **B**

IN SITU

C   2500×

TEASED

SMOOTH MUSCLE TISSUE

The distinct contrast of the striae in the cardiac muscle preparation in illustration A is evidence that these fibers are in a state of complete relaxation; fibers prepared from tissue in contraction generally do not reveal striae as distinctly. Although this tissue may appear to be syncytial, it is not. Cardiac fibers have only one nucleus per fiber, with intercalated disks forming the limits for each unit.

The smooth muscle tissue seen in illustration B is of a portion of the intestinal wall. Due to the closely packed nature of the cells in such structures it is very difficult to see individual cells. Illustration C reveals how cells of this type of tissue appear when separated with a probe before staining. Each cell has a single nucleus and lacks the cross-striations seen in skeletal and cardiac muscle tissue.

Figure HA-8   Cardiac and smooth muscle tissue.

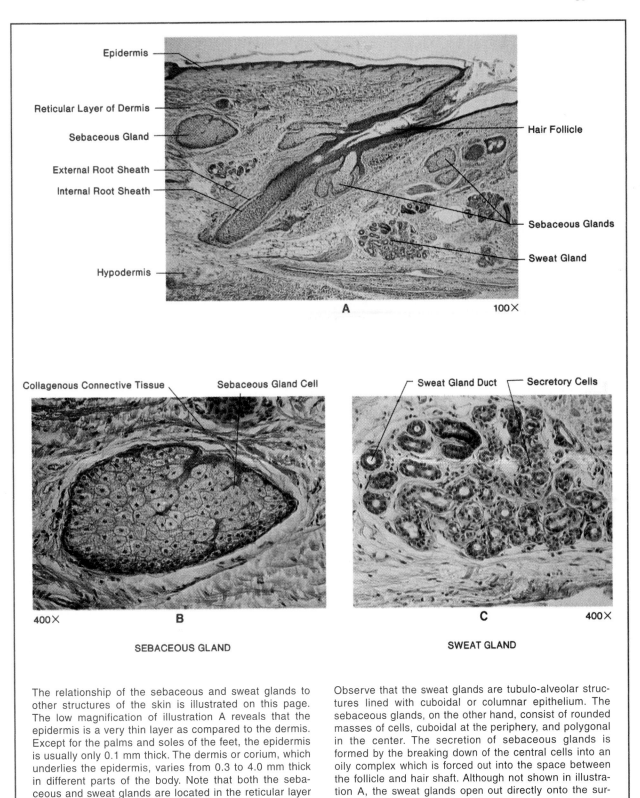

Epidermis

Reticular Layer of Dermis

Sebaceous Gland

External Root Sheath

Internal Root Sheath

Hypodermis

Hair Follicle

Sebaceous Glands

Sweat Gland

A   100×

Collagenous Connective Tissue   Sebaceous Gland Cell

Sweat Gland Duct   Secretory Cells

400×   B

SEBACEOUS GLAND

C   400×

SWEAT GLAND

The relationship of the sebaceous and sweat glands to other structures of the skin is illustrated on this page. The low magnification of illustration A reveals that the epidermis is a very thin layer as compared to the dermis. Except for the palms and soles of the feet, the epidermis is usually only 0.1 mm thick. The dermis or corium, which underlies the epidermis, varies from 0.3 to 4.0 mm thick in different parts of the body. Note that both the sebaceous and sweat glands are located in the reticular layer of the dermis.

Observe that the sweat glands are tubulo-alveolar structures lined with cuboidal or columnar epithelium. The sebaceous glands, on the other hand, consist of rounded masses of cells, cuboidal at the periphery, and polygonal in the center. The secretion of sebaceous glands is formed by the breaking down of the central cells into an oily complex which is forced out into the space between the follicle and hair shaft. Although not shown in illustration A, the sweat glands open out directly onto the surface of the skin.

Figure HA-9   Scalp histology.

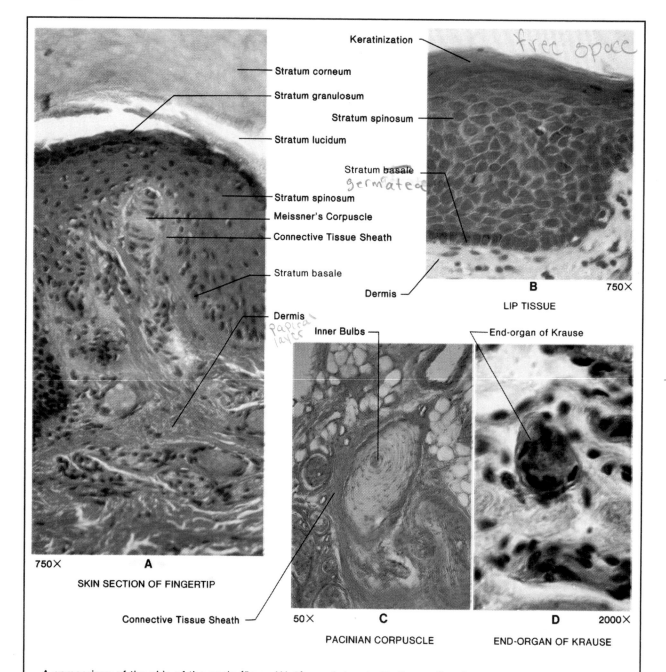

Keratinization

free space

Stratum corneum

Stratum granulosum

Stratum lucidum

Stratum spinosum

Stratum spinosum

Meissner's Corpuscle

Stratum basale
germated

Connective Tissue Sheath

Stratum basale

Dermis

Dermis

B 750×

LIP TISSUE

papiral layes

Inner Bulbs

End-organ of Krause

750×

A

SKIN SECTION OF FINGERTIP

Connective Tissue Sheath

50× **C**

PACINIAN CORPUSCLE

D 2000×

END-ORGAN OF KRAUSE

A comparison of the skin of the scalp (figure HA-9), a fingertip (illustration A), and the lip (illustration B) reveals that the integument differs in construction in different regions of the body. These differences are due to the presence or absence of hair and the degree of keratinization. The fingertips, palms, and soles of the feet have epidermal layers with a thick stratum corneum and no hair follicles. Skin on the lips has some keratinization, but to a much lesser degree than the fingertips.

Three receptors of the skin are illustrated here: Meissner's, Pacinian, and Krause corpuscles. Meissner's corpuscle, which is seen in illustration A,

is located in the papillary layer of the dermis. Note that it is encased in a sheath of connective tissue. These receptors are sensors of discriminative touch.

Pacinian corpuscles (illustration C) consist of laminated collagenous material and several inner bulbs. Pressure on the lamina triggers impulses in nerve endings in the bulbs. They are so large that they can be seen without magnification.

End-organs of Krause are small corpuscles that are sensitive to cold temperatures. They, too, are quite numerous in the skin.

Figure HA-10 Skin structure.

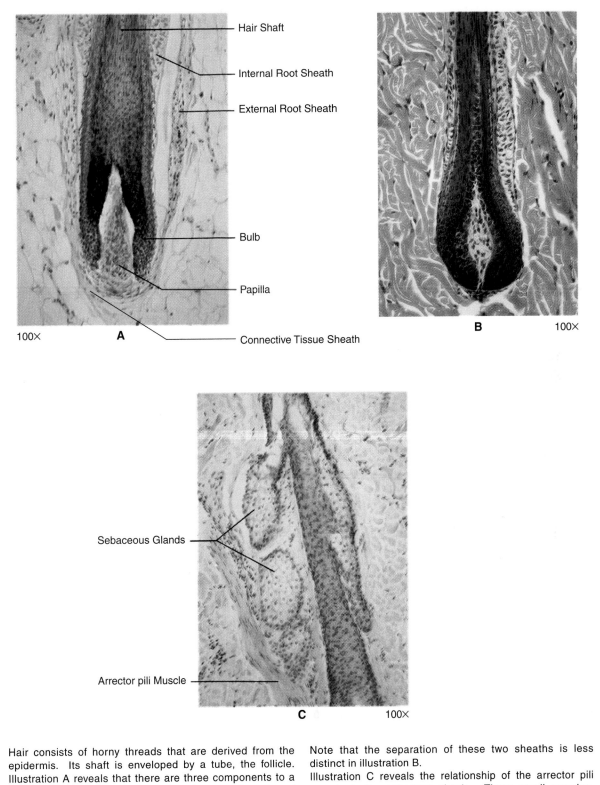

Hair Shaft

Internal Root Sheath

External Root Sheath

Bulb

Papilla

Connective Tissue Sheath

100×   **A**

B   100×

Sebaceous Glands

Arrector pili Muscle

C   100×

Hair consists of horny threads that are derived from the epidermis. Its shaft is enveloped by a tube, the follicle. Illustration A reveals that there are three components to a hair follicle: an internal root sheath, an external root sheath, and a connective tissue sheath. The two inner sheaths are derived from the epidermis; the outer connective tissue sheath is formed from the dermis.

Note that the separation of these two sheaths is less distinct in illustration B.

Illustration C reveals the relationship of the arrector pili muscle to the sebacous glands. These small muscles, which consist of smooth muscle tissue, play a role in forcing sebum out into the hair follicle. The result of their action produces what we call "goose bumps" on the skin.

Figure HA-11   **Hair structure.**

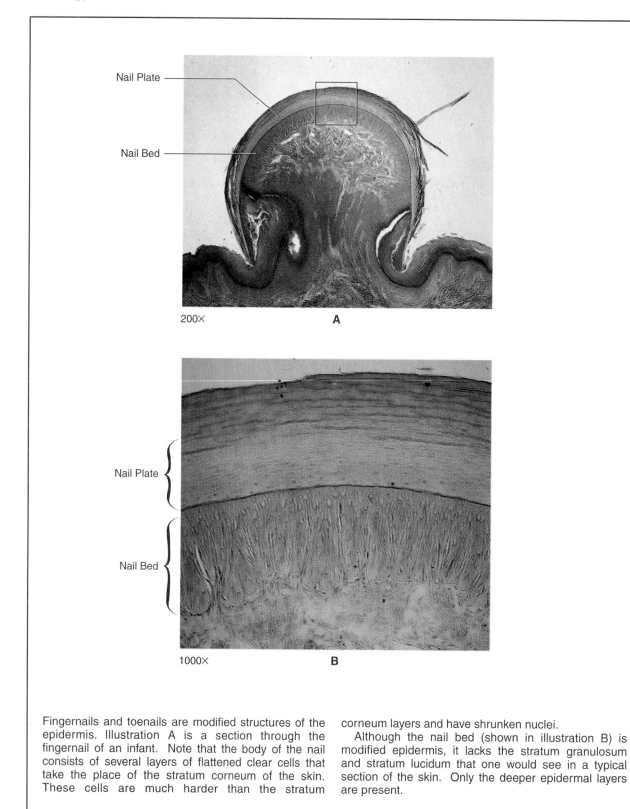

200×   **A**

Nail Plate

Nail Bed

Nail Plate

Nail Bed

1000×   **B**

Fingernails and toenails are modified structures of the epidermis. Illustration A is a section through the fingernail of an infant. Note that the body of the nail consists of several layers of flattened clear cells that take the place of the stratum corneum of the skin. These cells are much harder than the stratum corneum layers and have shrunken nuclei.

Although the nail bed (shown in illustration B) is modified epidermis, it lacks the stratum granulosum and stratum lucidum that one would see in a typical section of the skin. Only the deeper epidermal layers are present.

Figure HA-12   Fingernail structure.

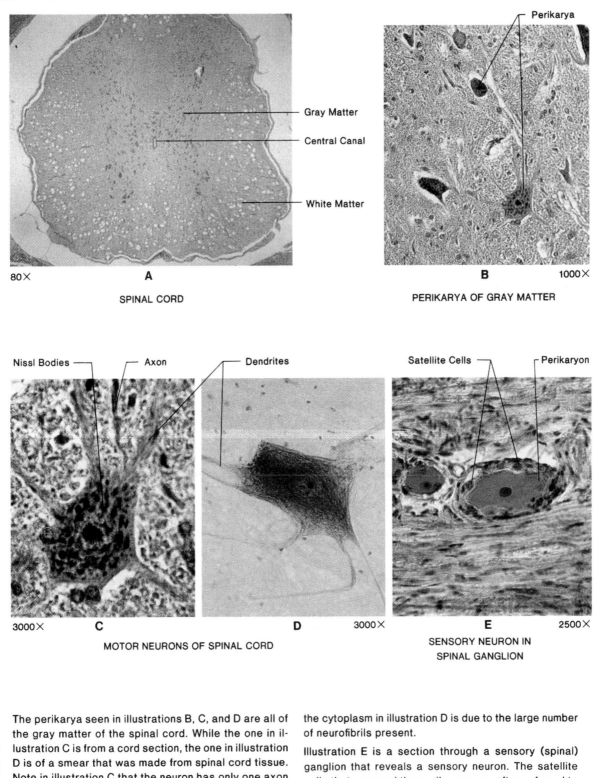

80× **A** SPINAL CORD

Gray Matter

Central Canal

White Matter

Perikarya

B 1000× PERIKARYA OF GRAY MATTER

Nissl Bodies — Axon — Dendrites Satellite Cells — Perikaryon

3000× **C** MOTOR NEURONS OF SPINAL CORD **D** 3000× **E** 2500× SENSORY NEURON IN SPINAL GANGLION

The perikarya seen in illustrations B, C, and D are all of the gray matter of the spinal cord. While the one in illustration C is from a cord section, the one in illustration D is of a smear that was made from spinal cord tissue. Note in illustration C that the neuron has only one axon and several dendrites. This neuron also exhibits a large number of Nissl bodies. The netlike appearance of the cytoplasm in illustration D is due to the large number of neurofibrils present.

Illustration E is a section through a sensory (spinal) ganglion that reveals a sensory neuron. The satellite cells that surround the perikaryon are often referred to as capsule cells.

Figure HA-13 Neurons of the spinal cord and spinal ganglia.

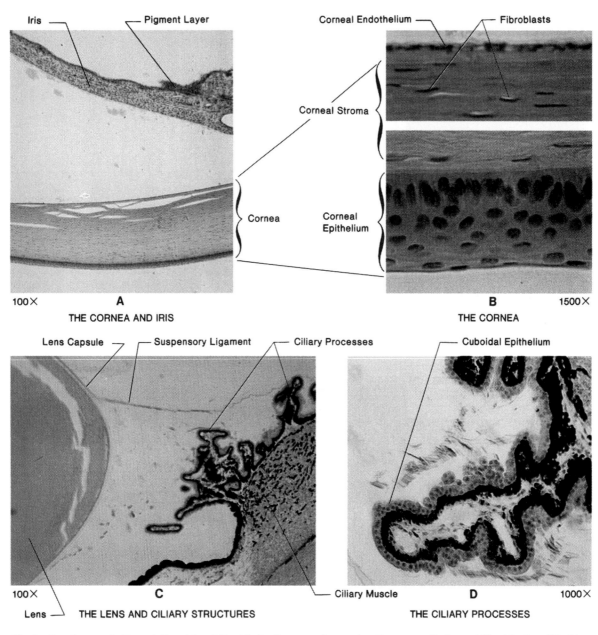

A 100× THE CORNEA AND IRIS

Iris — Pigment Layer

B 1500× THE CORNEA

Corneal Endothelium — Fibroblasts
Corneal Stroma
Corneal Epithelium

C 100× THE LENS AND CILIARY STRUCTURES

Lens Capsule — Suspensory Ligament — Ciliary Processes
Cornea
Lens — Ciliary Muscle

D 1000× THE CILIARY PROCESSES

Cuboidal Epithelium

Illustration A reveals the relationship of the iris to the cornea. If a portion of the lens were shown in this photomicrograph, it would be above the iris, since the iris lies between the lens and the cornea.

Illustration B is a compacted enlargement of the cornea, in which many of the middle layers have been removed so that all significant layers can be seen. Note that the outer corneal epithelium consists of stratified squamous epithelium, and that the inner corneal endothelium consists of a single layer of low cuboidal cells. Between these two layers is the corneal stroma (*substantia propria*), which makes up nine-tenths of the thickness of the cornea. This transparent stroma is made up of collagenous fibrils and fibroblasts, all held together with a mucopolysaccharide cement.

Note in illustration C that the lens is encased in an elastic capsule to which the suspensory ligament is attached. The lens consists of carcasses of epithelial cells that replaced most of their cytoplasm with the protein *crystallin* as they died during embryological development. This process of cell death, called *apoptosis*, is a common phenomenon in cellular physiology (consider keratinization). Note, also, that the ciliary processes shown in illustration D are covered with cuboidal epithelial cells. It is these cells that produce the aqueous humor that fills the space between the lens and the cornea.

Figure HA-16   The cornea, lens, iris, and ciliary structures.

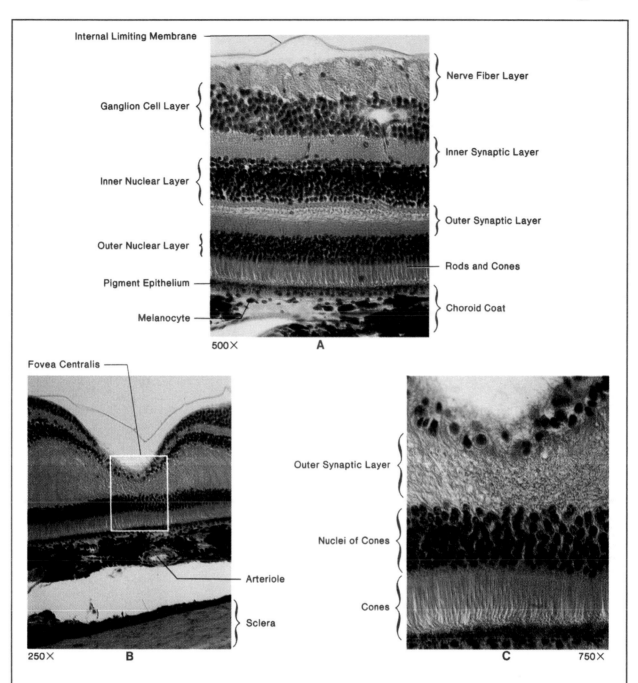

Illustration A reveals the various neuronal layers of the retina. Note that the photoreceptors (rods and cones) are in the deepest layer next to the choroid coat. Light striking the retina, must pass through all these layers to reach the receptors.

The nuclei seen in the outer nuclear layer are of the rods and cones. The outer synaptic (*plexiform*) layer is where the synapses are made between axons of the rods and cones and the dendrites of bipolar cells and the processes of horizontal cells. The inner nuclear layer contains nuclei of bipolar neurons and association neurons (horizontal and amacrine cells). The inner synaptic (*plexiform*) layer is the place where synaptic connections are made between the cells of the inner nuclear layer and the ganglion cells. The nerve fiber layer consists of nonmyelinated axons of the ganglion cells. These fibers converge on the optic disk to form the optic nerve. Illustrations B and C reveal the nature of the fovea centralis. Note that the receptors in the fovea consist only of cones and that the inner synaptic and nerve fiber layers are lacking.

Figure HA-17   The retina of the eye.

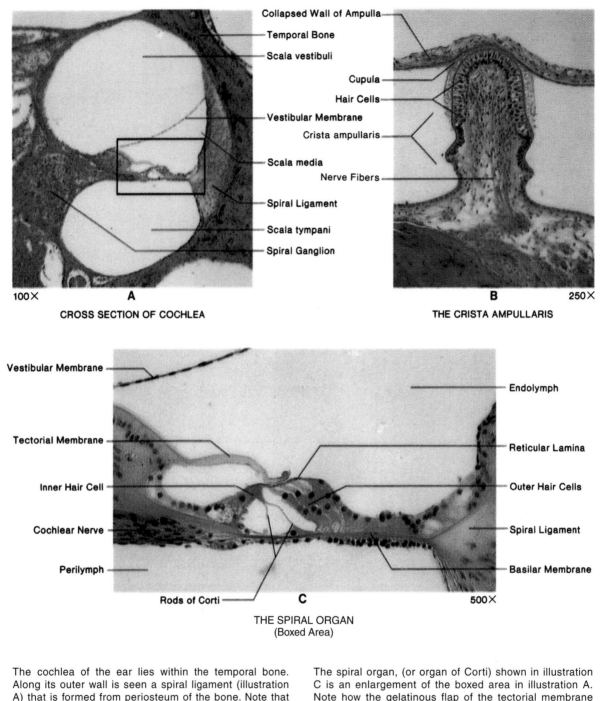

A 100× **CROSS SECTION OF COCHLEA**

Collapsed Wall of Ampulla
Temporal Bone
Scala vestibuli
Cupula
Hair Cells
Vestibular Membrane
Crista ampullaris
Scala media
Nerve Fibers
Spiral Ligament
Scala tympani
Spiral Ganglion

B 250× **THE CRISTA AMPULLARIS**

Vestibular Membrane
Endolymph
Tectorial Membrane
Reticular Lamina
Inner Hair Cell
Outer Hair Cells
Cochlear Nerve
Spiral Ligament
Perilymph
Basilar Membrane
Rods of Corti **C** 500×

THE SPIRAL ORGAN
(Boxed Area)

The cochlea of the ear lies within the temporal bone. Along its outer wall is seen a spiral ligament (illustration A) that is formed from periosteum of the bone. Note that fibers of the spiral ligament are continuous with the basilar membrane. Observe that the cochlea has three chambers: scala vestibuli, scala tympani, and scala media. The cochlear duct is the flexible portion of the cochlea that is bounded on one side by the vestibular membrane and on the other side by the basilar membrane. Endolymph is present in the cochlear duct; perilymph fills the scala vestibuli and scala tympani.

The spiral organ, (or organ of Corti) shown in illustration C is an enlargement of the boxed area in illustration A. Note how the gelatinous flap of the tectorial membrane lies in contact with hairs that emerge from the hair cells. Discrimination of sound frequencies results from nerve impulses sent to the brain via the cochlear nerve in conjunction with vibrations in the endolymph and basilar membrane.

The obliteration of the cupula in illustration B occurred during slide preparation. Other structures in the crista ampullaris are normal.

Figure HA-18   The cochlea and crista ampullaris.

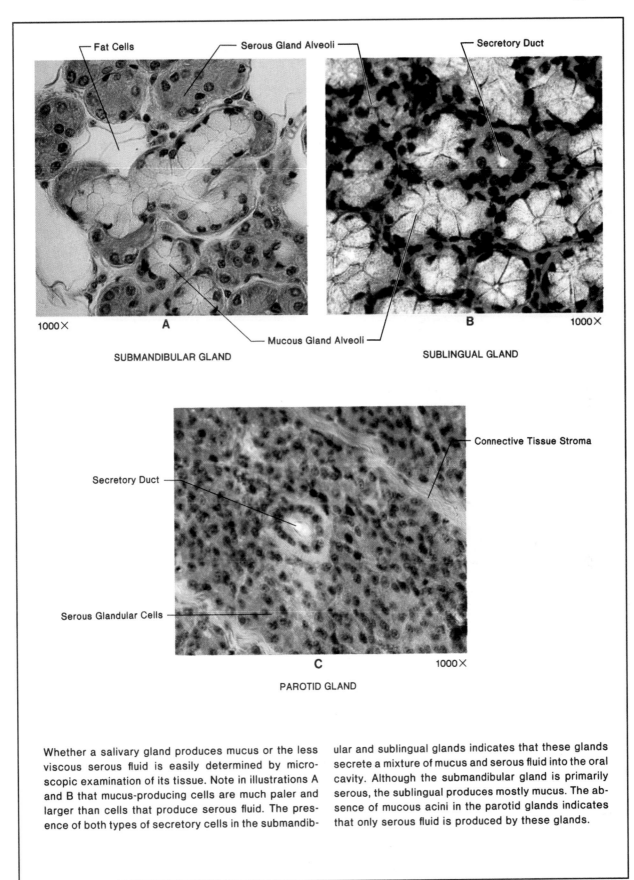

Fat Cells — Serous Gland Alveoli — Secretory Duct

1000× **A**

SUBMANDIBULAR GLAND

B 1000×

— Mucous Gland Alveoli —

SUBLINGUAL GLAND

Secretory Duct —

Connective Tissue Stroma

Serous Glandular Cells —

C 1000×

PAROTID GLAND

Whether a salivary gland produces mucus or the less viscous serous fluid is easily determined by microscopic examination of its tissue. Note in illustrations A and B that mucus-producing cells are much paler and larger than cells that produce serous fluid. The presence of both types of secretory cells in the submandibular and sublingual glands indicates that these glands secrete a mixture of mucus and serous fluid into the oral cavity. Although the submandibular gland is primarily serous, the sublingual produces mostly mucus. The absence of mucous acini in the parotid glands indicates that only serous fluid is produced by these glands.

Figure HA-19   Histology of major salivary glands.

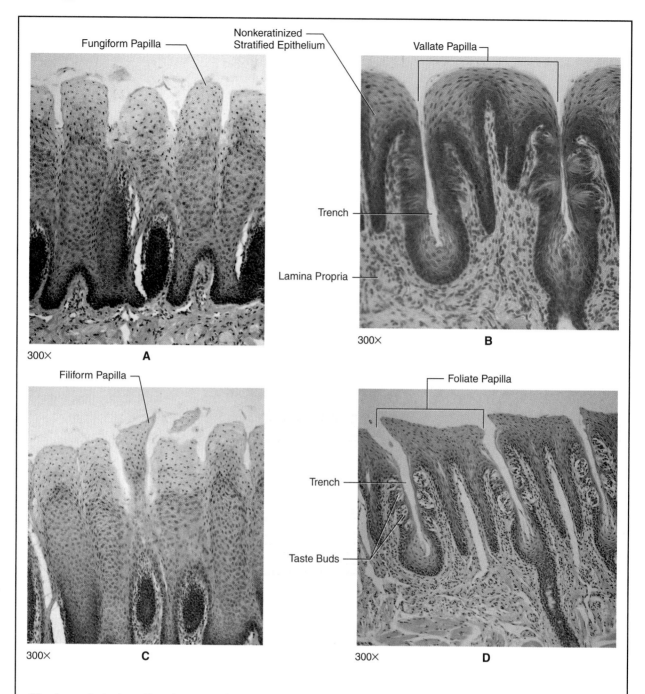

The four principal papillae that cover the dorsum of the tongue are the fungiform, filiform, vallate, and foliate papillae. All of them are projections of the dermis of the skin.

The fungiform papillae (illustration A) have rounded eminences and are deep red. Although they are found primarily on the apex and sides of the tongue, they are scattered irregularly over the dorsum.

Filiform papillae (illustration C) have a threadlike (filiform) appearance. At the back of the tongue they are arranged in lines parallel with the two rows of vallate papillae. On the apex of the tongue their arrangement is in lines extending from one side of the tongue to the other.

Vallate papillae (illustration B) are large projections located at the back of the tongue. They vary from eight to twelve in number. Note that each papilla is covered with stratified squamous epithelium and is surrounded by a deep trench. In the walls of these trenches are located hundreds of taste receptors (taste buds).

Foliate papillae (illustration D) are ridges seen on the sides of the tongue. They resemble vallate papillae in that they have taste buds on their trench walls.

Figure HA-20   Papillae of the tongue.

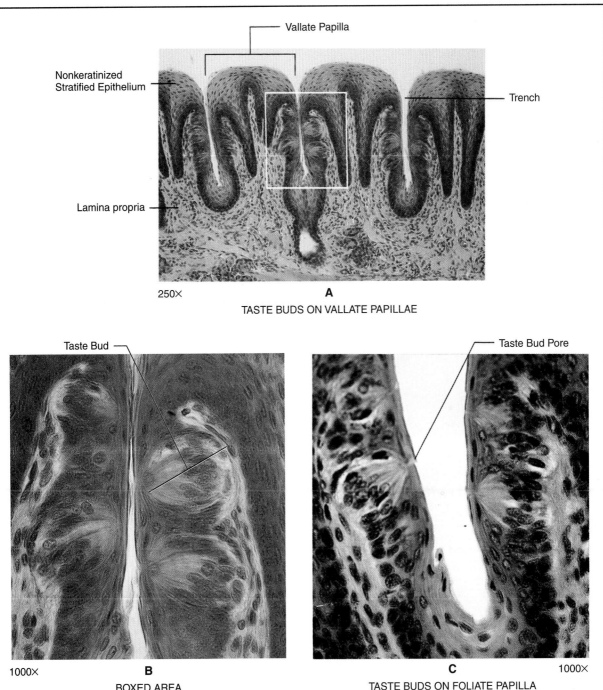

Vallate Papilla

Nonkeratinized
Stratified Epithelium

Trench

Lamina propria

250× **A**

TASTE BUDS ON VALLATE PAPILLAE

Taste Bud

Taste Bud Pore

1000× **B**
BOXED AREA

C 1000×
TASTE BUDS ON FOLIATE PAPILLA

The tongue may have as many as 20–30 vallate papillae. Each papilla is surrounded by a deep trench as seen in illustration A. Since the walls of each papilla will have taste buds covering the entire surface of the trench, it is conceivable that hundreds of taste receptors are present on these structures.

Note that each taste bud has a pore that opens out into the trench. At the base of each trench are duct openings from serous glands (glands of von Ebner) that keep the trench moist and assist in dissolving substances that stimulate the taste buds. To be able to see nerve fibers leading into the sensory cells of a taste bud it is necessary to stain the tissue with a special silver dye. No such dye was used on these slide preparations.

Foliate papillae (illustration C) are located on the sides of the tongue. The superior clarity of the taste buds in this photomicrograph is due to superb tissue preparation.

Figure HA-21   Taste buds.

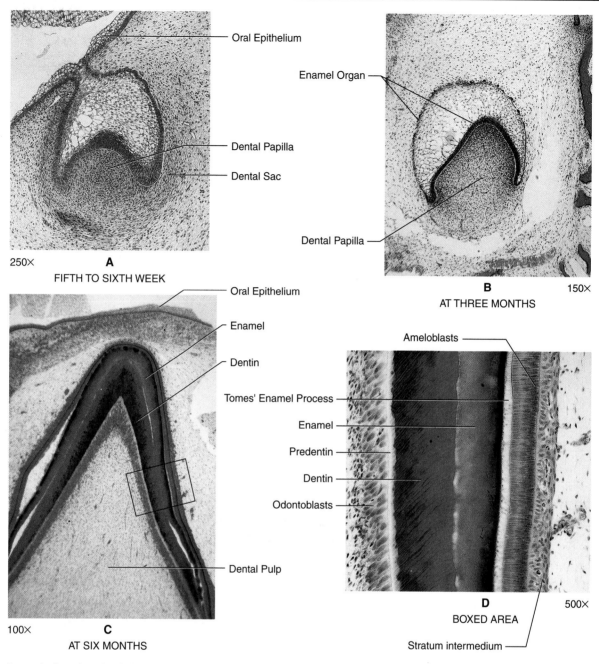

A
250×
FIFTH TO SIXTH WEEK

- Oral Epithelium
- Dental Papilla
- Dental Sac

B   150×
AT THREE MONTHS

Enamel Organ

Dental Papilla

C   100×
AT SIX MONTHS

- Oral Epithelium
- Enamel
- Dentin
- Tomes' Enamel Process
- Enamel
- Predentin
- Dentin
- Odontoblasts
- Dental Pulp

D   500×
BOXED AREA

Ameloblasts

Stratum intermedium

In each dental arch of the embryo ten tooth buds form by proliferation of cells in the oral epithelium. At first the buds are solid and rounded, but as the cells multiply they push inward (invaginate) as shown in illustration A. The oral epithelium (ectoderm) forms a two-layered structure called the enamel organ. Mesenchymal cells (mesoderm) in the distal portion of the bud push in to form the dental papilla. Illustration C reveals that by the fifth or sixth month the enamel organ produces the enamel of the tooth, and the cells of the dental papilla differentiate to form dentin.

Dentin is laid down just before the appearance of enamel. It is produced by a layer of elongated cells called odontoblasts that form from mesenchymal cells of the dental papilla. Note in illustration D that uncalcified dentin, or predentin, is formed first. Predentin is quickly converted to mature dentin, which consists of collagenous fibers and a calcified component called apatite.

Enamel is produced by cells called ameloblasts. Note in illustration D that these cells produce a clear layer called Tomes' enamel process, which becomes converted to mature enamel by calcification with mineral salts. As new enamel forms, the ameloblasts move outward away from the dentin and odontoblasts. When the tooth erupts, the ameloblasts are sloughed off.

Figure HA-22   Embryological stages in tooth development.

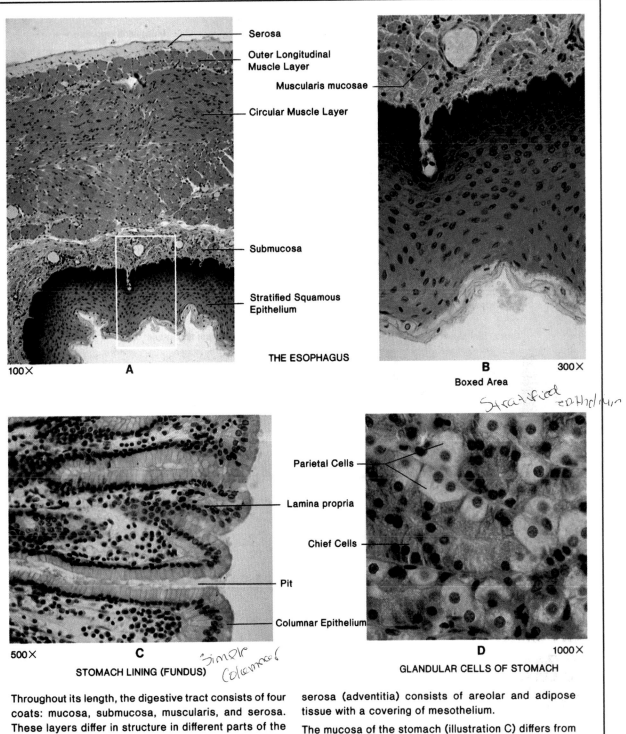

Serosa

Outer Longitudinal
Muscle Layer

Muscularis mucosae

Circular Muscle Layer

Submucosa

Stratified Squamous
Epithelium

THE ESOPHAGUS

100×   **A**

B   300×

Boxed Area

Stratified epithelium

Parietal Cells

Lamina propria

Chief Cells

Pit

Columnar Epithelium

500×   **C**

STOMACH LINING (FUNDUS)   *Simple Columnar*

D   1000×

GLANDULAR CELLS OF STOMACH

Throughout its length, the digestive tract consists of four coats: mucosa, submucosa, muscularis, and serosa. These layers differ in structure in different parts of the tract.

Note that the mucosa of the esophagus consists of stratified squamous epithelium, similar to what is seen in the oral cavity. The submucosa is a layer of areolar tissue that contains mucous glands, nerves, blood vessels, and some muscle tissue. The muscularis layer consists of longitudinal and circular fibers. The

serosa (adventitia) consists of areolar and adipose tissue with a covering of mesothelium.

The mucosa of the stomach (illustration C) differs from the esophagus in that it consists of simple columnar epithelial cells. The surface of the mucosa contains numerous pits. Deep in the pits lie glands that contain parietal and chief cells. The chief cells are serous cells that produce the precursor of pepsin; the parietal cells produce the antecedent of hydrochloric acid and the intrinsic factor gastrin.

Figure HA-23   The esophagus and stomach.

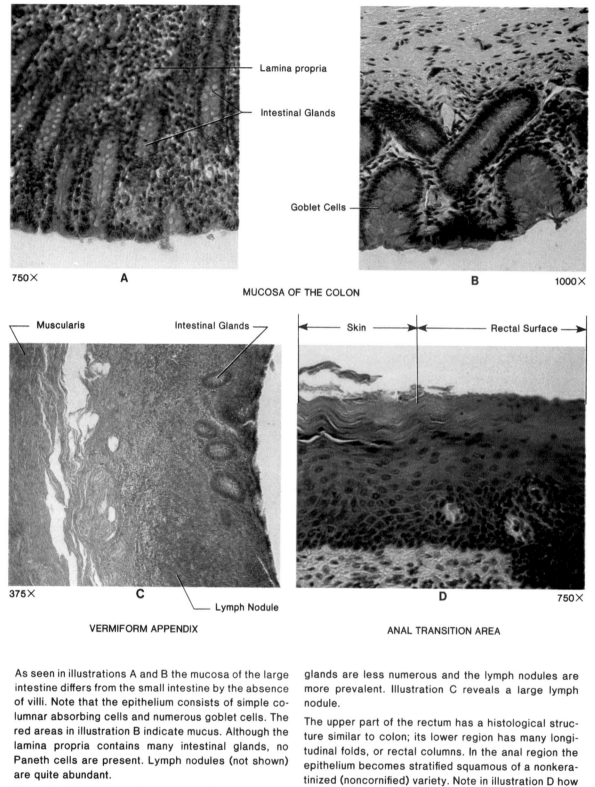

Lamina propria

Intestinal Glands

Goblet Cells

750×   A   B   1000×

MUCOSA OF THE COLON

Muscularis   Intestinal Glands   Skin   Rectal Surface

375×   C   750×   D

Lymph Nodule

VERMIFORM APPENDIX   ANAL TRANSITION AREA

As seen in illustrations A and B the mucosa of the large intestine differs from the small intestine by the absence of villi. Note that the epithelium consists of simple columnar absorbing cells and numerous goblet cells. The red areas in illustration B indicate mucus. Although the lamina propria contains many intestinal glands, no Paneth cells are present. Lymph nodules (not shown) are quite abundant.

The wall of the appendix (illustration C) has a histological structure quite similar to the colon, except that the glands are less numerous and the lymph nodules are more prevalent. Illustration C reveals a large lymph nodule.

The upper part of the rectum has a histological structure similar to colon; its lower region has many longitudinal folds, or rectal columns. In the anal region the epithelium becomes stratified squamous of a nonkeratinized (noncornified) variety. Note in illustration D how this epithelium changes to the characteristic keratinized epithelium of the skin on the outside of the body.

Figure HA-26   The lower digestive tract.

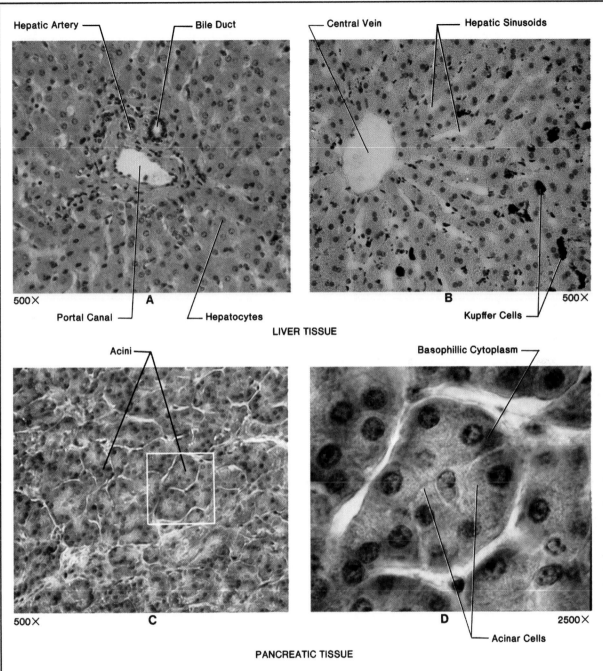

Hepatic Artery — — Bile Duct

Portal Canal — — Hepatocytes

500× A

LIVER TISSUE

— Central Vein — Hepatic Sinusoids

B 500×

Kupffer Cells

Acini —

500× C

Basophillic Cytoplasm —

D 2500×

— Acinar Cells

PANCREATIC TISSUE

The liver consists of many lobules that contain two kinds of cells: hepatocytes and Kupffer cells. The hepatocytes are arranged in plates that radiate outward around a central vein (illustration B) and have vascular channels, called sinusoids, between them. The hepatocytes synthesize bile components and several blood components such as albumin and blood clotting proteins. The secretions of these cells drain into the blood in the central vein of the lobule by way of the sinusoids. The central veins of each liver lobule, in turn, empty into a portal canal which lies between the lobules. The Kupffer cells are stationary phagocytic cells of the reticuloendothelial system that filter out of the blood any bacteria or other foreign material passing through the liver. The Kupffer cells seen in illustration B have been blackened by the uptake of carbon particles that were injected into the animal prior to sacrifice for tissue preparation.

Illustrations C and D illustrate the exocrine cells of the pancreas. Note how the pyramid-shaped cells are arranged in alveoli (acini). The basophilic portion of the cells is specialized for the production of zymogen, an inactive protein precursor.

Figure HA-27   Liver and pancreas histology.

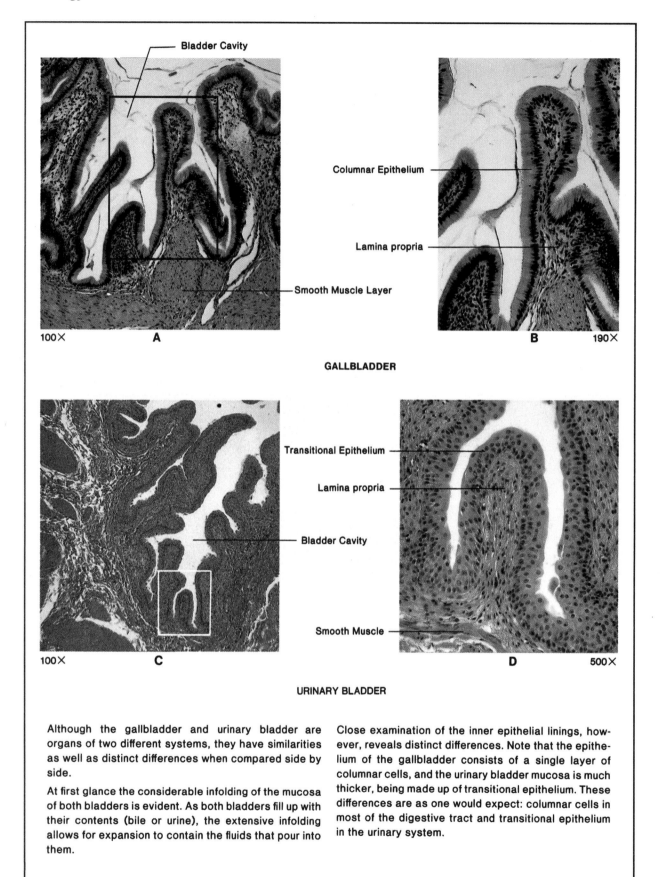

Labels in Figure A (100×):
- Bladder Cavity
- Smooth Muscle Layer

Labels in Figure B (190×):
- Columnar Epithelium
- Lamina propria

GALLBLADDER

Labels in Figure C (100×):
- Bladder Cavity

Labels in Figure D (500×):
- Transitional Epithelium
- Lamina propria
- Smooth Muscle

URINARY BLADDER

Although the gallbladder and urinary bladder are organs of two different systems, they have similarities as well as distinct differences when compared side by side.

At first glance the considerable infolding of the mucosa of both bladders is evident. As both bladders fill up with their contents (bile or urine), the extensive infolding allows for expansion to contain the fluids that pour into them.

Close examination of the inner epithelial linings, however, reveals distinct differences. Note that the epithelium of the gallbladder consists of a single layer of columnar cells, and the urinary bladder mucosa is much thicker, being made up of transitional epithelium. These differences are as one would expect: columnar cells in most of the digestive tract and transitional epithelium in the urinary system.

Figure HA-28 The gallbladder and urinary bladder.

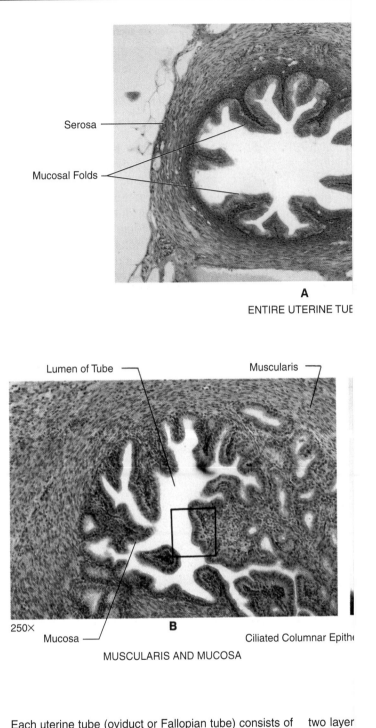

Figure HA-35   The uterine tube.

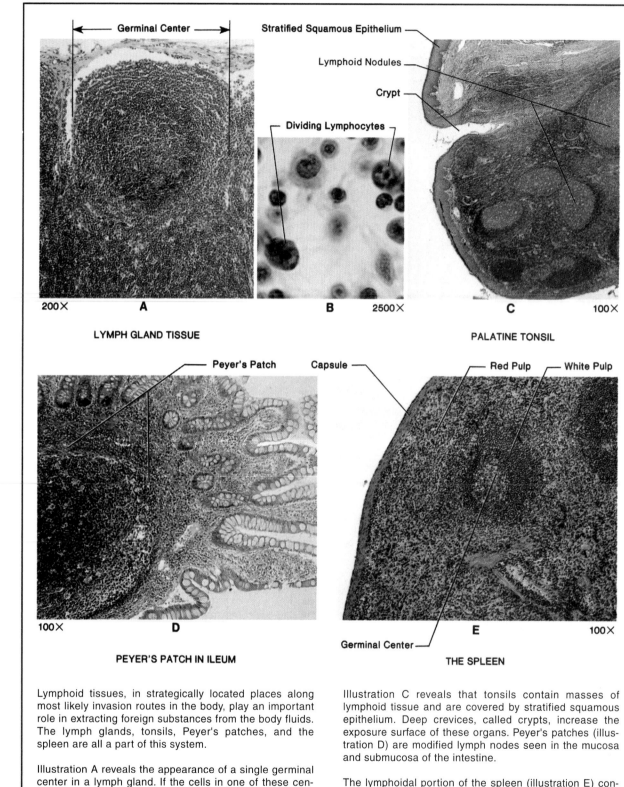

Figure HA-29   Lymphoid tissues in lymph gland, palatine tonsil, ileum, and spleen.

Each uterine tube (oviduct or Fallopian tube) consists of an inner mucosa, a middle muscularis, and an outer serosa. The epithelium of the mucosa consists of simple columnar cells, some of which are ciliated (see illustration C).
The muscularis portion of the oviduct consists of only

two layer
longitudi
the musc
This mu
function
ovulation

Lymphoid tissues, in strategically located places along most likely invasion routes in the body, play an important role in extracting foreign substances from the body fluids. The lymph glands, tonsils, Peyer's patches, and the spleen are all a part of this system.

Illustration A reveals the appearance of a single germinal center in a lymph gland. If the cells in one of these centers is observed with oil immersion optics (illustration B), multiplying cells can be seen.

Illustration C reveals that tonsils contain masses of lymphoid tissue and are covered by stratified squamous epithelium. Deep crevices, called crypts, increase the exposure surface of these organs. Peyer's patches (illustration D) are modified lymph nodes seen in the mucosa and submucosa of the intestine.

The lymphoidal portion of the spleen (illustration E) consists of the "white" pulp, which is white only in unstained tissue. Within the white pulp are diffuse germinal centers. The spleen also stores blood and recycles worn-out erythrocytes.

Perimetrium ⎯⎯

Smo

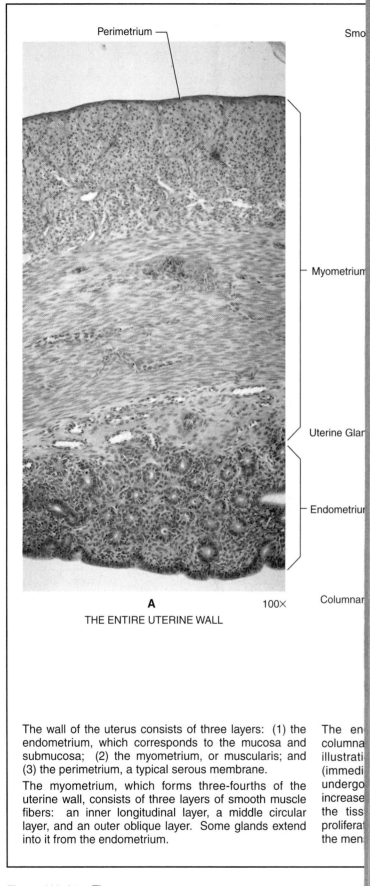

Myometrium

Uterine Glar

Endometriu

A　　　100×

Columnar

THE ENTIRE UTERINE WALL

The wall of the uterus consists of three layers: (1) the endometrium, which corresponds to the mucosa and submucosa; (2) the myometrium, or muscularis; and (3) the perimetrium, a typical serous membrane.

The myometrium, which forms three-fourths of the uterine wall, consists of three layers of smooth muscle fibers: an inner longitudinal layer, a middle circular layer, and an outer oblique layer. Some glands extend into it from the endometrium.

The en
columna
illustrati
(immedi
undergo
increase
the tiss
prolifera
the men

Figure HA-34　**The uterus.**

52

A White Blood Cell Study:
The Differential WBC Count

The blood is an opaque, rather viscous fluid that is bright red when oxygenated and dark red when depleted of oxygen. Its specific gravity is normally around 1.06, and its pH is slightly alkaline. When centrifuged, blood becomes separated into a dark red portion made up of **formed elements** and a clear straw-colored portion called **plasma.** The formed element content of blood is around 47% in men and 42% in women.

In this exercise we will study the leukocytic portion of the formed elements in blood. This study may be made from a prepared stained microscope slide or from a slide made from your own blood. Figures 52.1 and 52.2 will be used to identify the various kinds of leukocytes. If you are going to prepare a slide from your own blood, *be sure to review the sanitary precautions that are provided on the preceding page.*

The formed elements of blood consist of erythrocytes (red blood cells), leukocytes (white blood cells), and blood platelets. As illustrated in figure 52.1, the *erythrocytes* are non-nucleated cells with depressed centers that contain hemoglobin. They are the most numerous elements in blood—around 4.5 to 5.5 million cells per cubic millimeter.

The *leukocytes* are of two types: granulocytes and agranulocytes. The *granulocytes* are so named because they have conspicuous granules in their cytoplasm. **Neutrophils, eosinophils,** and **basophils** fall into this category. The *agranulocytes* lack pronounced granules in the cytoplasm; the **monocytes** and **lymphocytes** are of this type.

The *blood platelets* are noncellular elements in the blood that assist in the blood clotting process.

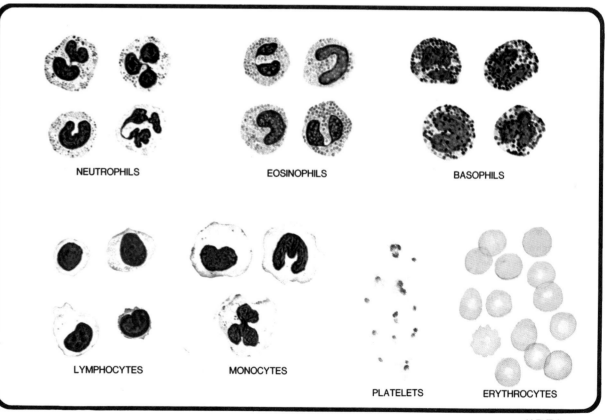

NEUTROPHILS　　　EOSINOPHILS　　　BASOPHILS

LYMPHOCYTES　　　MONOCYTES

PLATELETS　　　ERYTHROCYTES

K.P. Talaro

Figure 52.1　Formed elements of blood.

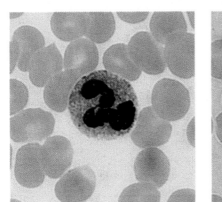

NEUTROPHIL

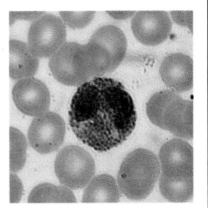

EOSINOPHIL

BASOPHIL

GRANULOCYTES

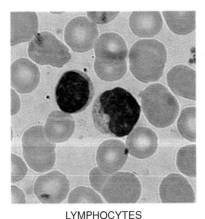

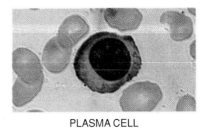

BLOOD PLATELETS

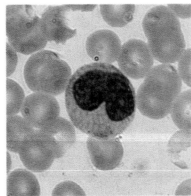

LYMPHOCYTES

MONOCYTE

AGRANULOCYTES

PLASMA CELL
(abnormal)

Neutrophils differ from the other two granulocytes in having smaller and paler granules in the cytoplasm. Nuclei of these leukocytes are characteristicallylobulated, with bridges between the lobules (seefigure 52.1). During infections, when many of these arebeing produced in the bone marrow, the nuclei arehorseshoe-shaped; these juvenile cells are often called "stab cells".

Eosinophils, generally, have bilobed nuclei as above;however, juveniles have nuclei that are horseshoe-shaped (figure 52.1). Note that the eosinophilicgranules in the cytoplasm are large and pronounced.

Basophils, which are few in number, are distinguished from the other granulocytes by the presence of dark basophilic granules in the cytoplasm. The nuclei of these cells are large and varied in shape. The fact that heparin has been identified in these cells indicates that they probably secrete this substance into the blood.

Two sizes of lymphocytes—large and small—are shown above and in figure 52.1. The small ones, which are more abundant than the large ones, have large, dense nuclei surrounded by a thin layer of basophilic cytoplasm. The large lymphocytes have indented nuclei and more cytoplasm than small lympocytes.

Monocytes are the largest cells found in normal blood. The nucleus may be ovoid, indented, or horse-shoe shaped, as illustrated in figure 52.1. Monocytes have considerably more cytoplasm than lymphocytes and are voracious phagocytes.

Blood platelets, which form from megakaryocytes in the bone marrow, function in the formation of blood clots.

Plasma cells are only rarely seen in the blood. The photomicrograph above was made from the blood of a patient with plasma cell leukemia. Although they resemble lymphocytes, the plasma cells have more cytoplasm that is very basophilic.

Figure 52.2   Photomicrographs of formed elements in the blood (5,000×).

In this study of the formed elements we will attempt to determine the percentage of each type of leukocyte on a slide. By scanning an entire slide and counting the various types, you will have an opportunity to encounter most, if not all, types. The erythrocytes and blood platelets will be ignored. The main purpose of this exercise will be to learn how to distinguish the various types of white blood cells, not to become proficient in performing differential WBC counts.

The normal percentage ranges for the various leukocytes are as follows: neutrophils 50–70%, lymphocytes 20–30%, monocytes 2–6%, eosinophils 1–5%, and basophils 0.5–1%. Because of the extremely low percentages of eosinophils and basophils, it is necessary to examine at least 100 leukocytes to increase the possibility of encountering one or two of them on a slide.

Deviations from the above percentages may indicate serious pathological conditions. High neutrophil counts, or **neutrophilia,** often signal localized infections such as appendicitis or abscesses in some part of the body. **Neutropenia,** a condition in which there is a marked decrease in the number of neutrophils, occurs in typhoid fever, undulant fever, and influenza. **Eosinophilia** (high eosinophil count)

may indicate allergic conditions or invasions by parasitic roundworms such as *Trichinella spiralis,* the pork worm. Counts of eosinophils may rise to as much as 50% of total leukocytes in cases of trichinosis. High lymphocyte counts, or **lymphocytosis,** are present in whooping cough and some viral infections. Increased numbers of monocytes, or **monocytosis,** may indicate the presence of the Epstein-Barr virus, which causes infectious mononucleosis.

A white cell count that determines the relative percentages of each type of leukocyte is called a **differential white blood cell count.** It is this type of count that will be performed here in this laboratory period. If a prepared slide is to be used, ignore the instructions below under the heading of Preparation of a Slide and proceed to the heading Performing the Cell Count.

CAUTION: See page 301.

Materials:
    prepared blood slide stained with Wright's or Giemsa's stains (H2.851 or H2.852)

for staining a blood smear:
    2 or 3 clean microscope slides (with polished edges)
    sterile disposable lancets

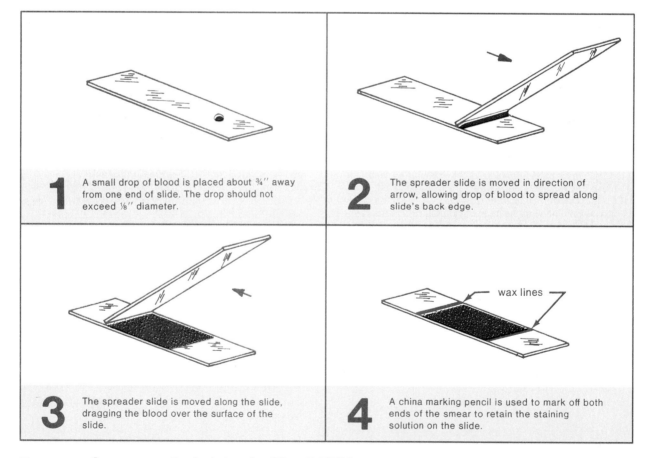

1 A small drop of blood is placed about ¾″ away from one end of slide. The drop should not exceed ⅛″ diameter.

2 The spreader slide is moved in direction of arrow, allowing drop of blood to spread along slide's back edge.

3 The spreader slide is moved along the slide, dragging the blood over the surface of the slide.

4 A china marking pencil is used to mark off both ends of the smear to retain the staining solution on the slide.

wax lines

Figure 52.3 Smear preparation technique for differential WBC count.

sterile absorbent cotton and 70% alcohol (or
    disposable alcohol swabs)
Wright's stain
distilled water in dropping bottle
wax pencil
bibulous paper
immersion oil

Preparation of a Slide

Figure 52.3 illustrates the procedure that will be
used to make a stained slide of a blood smear. The
most difficult step in making such a slide is getting
an ideal spread of the blood, which is thick at one
end and thin at the other end. If done properly, the
smear will have a gradient of cellular density that
will make it possible to choose an area that is ideal
for study. The angle at which the spreading slide is
held will determine the thickness of the smear. It
may be necessary for you to make more than one
smear to get one that is suitable.

1. Clean three or four slides with soap and water.
   Handle them with care to avoid getting their flat
   surfaces soiled. Although only two slides may
   be needed, it is often necessary to repeat the
   spreading process; thus, the extra slides.
2. Scrub the middle finger with 70% alcohol and
   perforate it with a lancet.
3. Put a drop of blood on the slide about ¾″ from
   one end and spread with another slide in the
   manner illustrated in figure 52.3. *Note that the
   blood is dragged over the slide, not pushed.*

   Do not pull the slide over the smear a second time. *If you don't get an even smear the first
   time, repeat the process on a fresh clean slide.*
   To get a smear that will be the proper thickness,
   hold the spreading slide at an angle somewhat
   greater than 45°.
4. Draw a line on each side of the smear with a wax
   pencil to confine the stain which is to be applied
   to the slide. (Note: This step is helpful for begin-
   ners and usually omitted by professionals.)
5. Cover the film with Wright's stain, counting
   the drops as you add them. Stain for **4 minutes**
   and then add the same number of drops of distilled water to the stain and let stand for anoth-

er **10 minutes.** Blow gently on the mixture
every few minutes to keep the solutions mixed.
6. Gently wash off the slide under running water
   for **30 seconds** and shake off the excess. Blot
   dry with bibulous paper.

Performing the Cell Count

Whether you are using a prepared slide or one that
you have just stained, the counting procedure is
essentially the same. Ideally, one should use an oil
immersion lens for this procedure; high-dry optics
can be used, but they are not as reliable for best
results. Proceed as follows:

1. Scan the slide with the low-power objective to
   find an area where cell distribution is best. A
   good area is one in which the cells are not
   jammed together or scattered too far apart.
2. Once an ideal area has been selected, place a
   drop of immersion oil on the slide near one edge
   and lower the oil immersion objective into the
   oil. If you are using a slide you have just prepared you needn't worry about oil being placed
   directly on the stained smear.

   If the high-dry objective is to be used, omit
   placing oil on the slide.
3. Systematically scan the slide, following the
   pathway indicated in figure 52.4. As each
   leukocyte is encountered identify it, using figures 52.1 and 52.2 for reference.

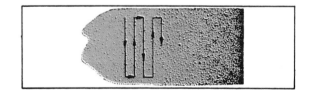

Figure 52.4   Examination path for a differential count.

4. Tabulate your count on the Laboratory Report
   sheet according to the instructions there. It is
   best to remove your Lab Report sheet from the
   back of the manual for this identification and
   tabulation procedure.

53

Total Blood Cell Counts

The number of leukocytes and erythrocytes present in a cubic millimeter of blood is routinely determined in diagnostic blood analyses. Although the differential WBC count performed in the last exercise is very useful, it alone cannot give us a complete picture of the health of the patient. One must also know the total number of white and red blood cells per unit of volume (a cubic millimeter).

In this exercise we will have an opportunity to determine accurate blood cell counts using a special cell-counting slide called a **hemacytometer.** Although using a device such as this for counting blood cells is very satisfactory, the procedure is tedious and has been supplanted in the modern medical laboratory by automated electronic counting devices that are much more rapid. Since no equipment of this type is available here, we will use the hemacytometer.

Total WBC Count

The number of white blood cells in a cubic millimeter of normal blood will range from 5000 to 9000 cells. The exact number of cells will vary with the time of day, exercise, and other factors. If a patient has a high WBC count, such as 17,000

leukocytes per cubic millimeter, and an abnormally high neutrophil count, as determined by a differential count, one can safely assume that an infection is present somewhere in the body.

To determine the number of leukocytes in a cubic millimeter of blood, one must dilute the blood and count the WBCs on a hemacytometer, using the low-power objective of a microscope. Figures 53.1 and 53.2 illustrate the various steps in diluting the blood and charging a hemacytometer for this count.

Note in figure 53.1 that blood is drawn up into a special pipette and then diluted in the pipette with a weak acid solution. After shaking the pipette to mix the acid and blood, a small amount of the diluted blood is allowed to flow under the cover glass of the hemacytometer as shown in illustration 1, figure 53.2.

When examined under the low-power objective, the diluted blood will reveal only white blood cells since all red blood cells have been destroyed by the acid in the dilution fluid. Illustration 2, figure 53.2, reveals the five areas on the hemacytometer that are tabulated for a WBC total count. Mathematical calculations are used to determine the number of leukocytes per cubic millimeter. Proceed as follows:

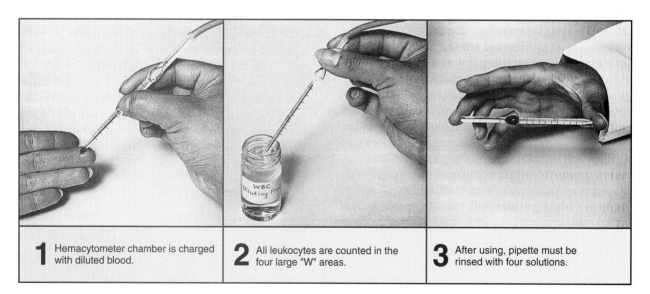

1 Hemacytometer chamber is charged with diluted blood.

2 All leukocytes are counted in the four large "W" areas.

3 After using, pipette must be rinsed with four solutions.

Figure 53.1 Dilution and mixing procedures for WBC blood count.

Preparation of Hemacytometer

Working with your laboratory partner, assist each other to prepare a "charged" hemacytometer.

Materials:

    hemacytometer and cover glass
    WBC diluting pipette and rubber tubing
    WBC diluting fluid
    mechanical hand counter
    pipette cleaning solutions
    cotton, alcohol, lancets, clean cloth

1. Wash the hemacytometer and cover glass with soap and water, rinse well, and dry with a clean cloth or Kimwipes.
2. Produce a free flow of blood, wipe away the first drop, and draw the blood up into the diluting pipette to the 0.5 mark. See illustration 1, figure 53.1. Any slight excess over the 0.5 mark can be withdrawn by placing the pipette tip on blotting paper.

    If the blood goes substantially past the 0.5 mark, discharge the blood from the pipette and flush it out with acid, water, alcohol, and acetone (in that order). See illustration 3, figure 53.2.

    The ideal way is to draw up the blood *exactly* to the mark on the first attempt.
3. As shown in illustration 2, figure 53.1, draw the WBC diluting fluid up into the pipette until it reaches the 11.0 mark.
4. Place your thumb over the tip of the pipette, slip off the tubing, and place your third finger over the other end (illustration 3, figure 53.1).

5. Mix the blood and diluting fluid in the pipette for **2–3 minutes** by holding it as shown in illustration 3, figure 53.1. The pipette should be held parallel to the tabletop and moved through a 90° arc, with the wrist held rigidly.
6. Discharge one-third of the bulb fluid from the pipette by allowing it to drop onto a piece of paper toweling.
7. While holding the pipette as shown in illustration 1, figure 53.2, deposit a **tiny drop** on the sloping polished surface of the counting chamber next to the edge of the cover glass. **Do not let the tip of the pipette touch the polished surface for more than an instant.** If it is there too long, the chamber will overfill.

    A properly filled chamber will have diluted blood filling only the space between the cover glass and counting chamber. No fluid should run down into the moat.
8. Charge the other side if the first side was overfilled.

Performing the Count

Place the hemacytometer on the microscope stage and bring the grid lines into focus under the **low-power** (10×) objective. Use the coarse adjustment knob and reduce the lighting somewhat to make both the cells and lines visible.

Locate one of the "W" (white) sections shown in illustration 2, figure 53.2. Since the diluting fluid contains acid, all erythrocytes have been destroyed (hemolyzed); only the leukocytes will show up as very small dots.

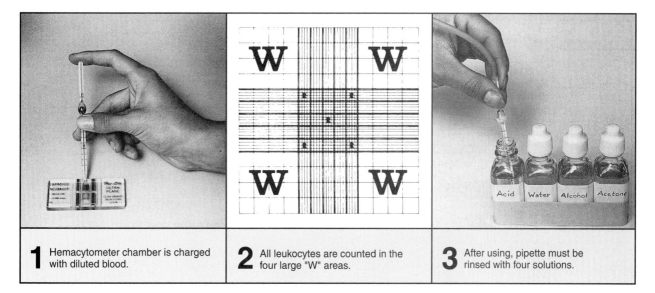

1 Hemacytometer chamber is charged with diluted blood.

2 All leukocytes are counted in the four large "W" areas.

3 After using, pipette must be rinsed with four solutions.

Figure 53.2   Charging the hemacytometer, counting areas, and pipette cleaning.

Do the cells seem to be evenly distributed? If not, charge the other half of the counting chamber after further mixing. If the other chamber had been previously charged unsuccessfully by overflooding, wash off the hemacytometer and cover glass, shake the pipette for 2–3 minutes, and recharge it.

Count all the cells in the four "W" areas, using a mechanical hand counter. To avoid overcounting of cells at the boundaries, **count the cells that touch the lines on the left and top sides only.** Cells that touch the boundary lines on the right and bottom sides should be ignored. This applies to the boundaries of each entire "W" area.

Discharge the contents of the pipette and rinse it out sequentially with acid, water, alcohol, and acetone (illustration 3, figure 53.2).

Calculations

To determine the number of leukocytes per cubic millimeter, multiply the total number of cells counted in the four "W" areas by 50. The factor of 50 is the product of the volume correction factor and dilution factor, or

$$2.5 \times 20 = 50$$

The volume correction factor of 2.5 is arrived at in this way: Each "W" area is eactly one square millimeter by 0.1 mm deep. Therefore, the volume of each "W" is 0.1 mm^3. Since four "W" sections are counted, the total amount of diluted blood that is examined is 0.4 mm^3. And since we are concerned with the number of cells in one cubic millimeter instead of 0.4 mm^3, we must multiply our count by 2.5, derived by dividing 1.0 by 0.4.

Record your results on the Laboratory Report.

Red Blood Cell Count

Normal RBC counts for adult males average around 5,400,000 ($\pm$600,000) cells per cubic millimeter. The normal average for women is 4,600,000 ($\pm$500,000) per cubic millimeter. The difference in sexes does not exist before puberty. At high altitudes, higher values will be normal for all individuals.

Low and high RBC counts result respectively in anemia or polycythemia. Although low RBC counts

do result in anemia, individuals can be anemic and still have normal RBC counts. If the cells are small, or if they lack sufficient hemoglobin, anemia will result. Thus, **anemia** may be defined, simply, as a condition in which the oxygen-carrying capacity of the blood is reduced.

Polycythemia, a condition characterized by above-normal RBC counts, may be due to living at high altitudes *(physiological polycythemia)* or red marrow malignancy *(polycythemia vera)*. In physiological polycythemia, counts may run as high as $8,000,000/\text{mm}^3$. In polycythemia vera, counts of 10–11 million are not uncommon.

Although the general procedures for the RBC count are essentially the same as for counting white blood cells, the diluting fluid, pipette, and mathematics are different. The RBC pipette differs from the WBC pipette in that the RBC pipette has 101 scribed above the bulb instead of 11. The RBC diluting fluid may be one of several isotonic solutions such as physiological saline or Hayem's solution. To perform a red blood count, follow this procedure:

Materials:
RBC diluting pipette
rubber tubing and mouthpiece
RBC diluting fluid
other supplies used for WBC count

1. Draw the blood up to the 0.5 mark and the diluting fluid up to the 101 mark.
2. Mix the blood and diluting fluid in the same manner as performed for WBC counts.
3. Count all the cells in the five "R" areas (see illustration 2, figure 53.2). Use the **high-power objective** (40 or 50$\times$) and observe the same rules for "line counts" as applied to WBC counts.
4. Multiply the total count of the five areas by 10,000 (dilution factor is 200, volume correction factor is 50).
5. Clean the pipette with acid, water, alcohol, and acetone at the end of the period.

Laboratory Report

Record your results in section B and table II (section E) of Laboratory Report 52–58.

Hemoglobin Percentage Measurement

Since hemoglobin is the essential oxygen-carrying ingredient of erythrocytes, its quantity in the blood determines whether a person is anemic. We learned in Exercise 53 that one may have a normal RBC count and still be anemic if the erythrocytes are smaller than normal. Such a condition, which is called *microcytic anemia,* results in one having an insufficient amount of hemoglobin. In other cases, a normal count in which the erythrocytes have a low hemoglobin percentage *(hypochromic cells)* may also result in anemia. Thus, it is apparent that the amount of hemoglobin present in a unit volume of blood is the critical factor in determining whether an individual is anemic.

Determination of the hemoglobin content of blood can be made by various methods. One of the oldest methods, and incidentally a very inaccurate one, is the Tallqvist scale. In this technique a piece of blotting paper that has been saturated with a drop of blood is compared with a color chart to determine the percentage of hemoglobin. Very few, if any, physicians utilize this method today. A rapid, accurate method is to insert a cuvette of blood into a photocolorimeter that is calibrated for blood samples. Such a method,

however, requires a rather expensive piece of electronic equipment. It also requires a considerable quantity of blood.

A relatively inexpensive instrument for hemoglobin determinations is the *hemoglobinometer.* The *Hb-Meter,* manufactured by American Optical, is such a device. Figures 54.1 through 54.4 illustrate the procedure to follow in using the Hb-Meter. It is this piece of equipment that we will use in this laboratory period to determine the hemoglobin content of a blood sample.

The hemoglobin content of blood is expressed in terms of grams per 100 ml of blood. Three different standards are used depending on the community where the test is performed. For our purposes we will use the 15.6 gm/100 ml as standard

For adult males, the normal on this scale is 14.9 ± 1.5 gm/100 ml; for adult females, the normal on this scale is 13.7 ± 1.5; for children at birth, 21.5 ± 3; for children at 4 years, 13 ± 1.5 gm/100 ml. A conversion scale exists on the side of the A/O Hb-Meter that allows one to determine hemoglobin percentages from grams Hb/100 ml. Proceed as follows to determine your own hemoglobin percentage:

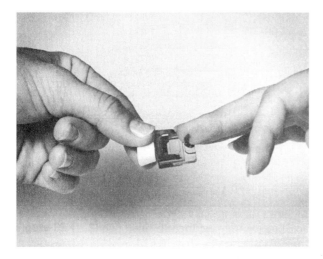

Figure 54.1 A fresh drop of blood is added to the moat plate of the blood chamber assembly. The blood should flow freely without excessive finger pressure.

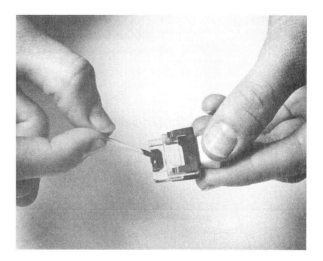

Figure 54.2 Blood sample is hemolyzed on moat plate with a wooden hemolysis applicator. Thirty-five to forty-five seconds are required for complete hemolysis.

Materials:
    American Optical Hb-Meter
    hemolysis applicators
    lancets
    cotton
    alcohol

1. Disassemble the blood chamber by pulling the two pieces of glass from the metal clip. Note that one piece of glass has an H-shaped moat cut into it. This piece will receive the blood. The other piece of glass has two flat surfaces and serves as a cover plate.
2. Clean both pieces of glass with alcohol and Kimwipes. Handle by edges to keep clean.
3. Reassemble the glass plates in the clip so that the grooves on the moat plate face the cover plate. The moat plate should be inserted only halfway to provide an exposed surface to receive the drop of blood. See figures 54.1 and 54.2.
4. Disinfect and puncture the finger with a disposable lancet.
5. Place a drop of blood on the exposed surface of the moat plate, as shown in figure 54.1.
6. Hemolyze the blood on the plate by mixing the blood with the pointed end of a hemolysis appli-

cator as shown in figure 54.2. It will take 35–45 seconds for all the red blood cells to rupture. Complete hemolysis has occurred when the blood loses its cloudy appearance and becomes a transparent red liquid.
7. Push the moat plate in flush with the cover plate and insert the sample into the side of the instrument, as shown in figure 54.3.
8. Place the eyepiece to your eye with the left hand in such a manner that the left thumb rests on the light switch button on the bottom of the Hb-Meter.
9. While pressing the light button with the left thumb, move the slide button on the side of the instrument back and forth with the right index finger until the two halves of the split field match. The index mark on the slide knob indicates the grams of hemoglobin per 100 ml of blood. Read the percent hemoglobin on the 15.6 scale.

Laboratory Report

Record your results on table II in section E of Laboratory Report 52–58.

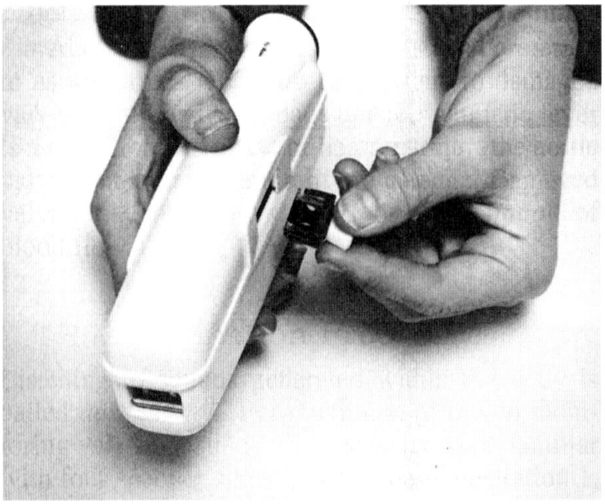

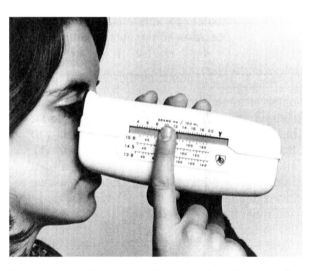

Figure 54.3 Charged blood chamber is inserted into slot of Hb-Meter. Before insertion, the unit should be tested to make certain the batteries are active.

Figure 54.4 Blood sample is analyzed by moving slide button with right index finger. When the two colors match in density, the gm/100 ml is read on the scale.

55

Packed Red Cell Volume

In Exercises 53 and 54 we employed two different tests for determining the presence or absence of anemia: (1) the RBC count and (2) the use of a hemoglobinometer. We have seen that anemia may be due to a dilution of the number of erythrocytes or to a deficiency of hemoglobin. It was observed that the most significant factor is the amount of hemoglobin that is present in a given volume of blood. A third method that can be used in anemia diagnosis is to determine the volume percentage of blood that is occupied by the red blood cells that have been packed by centrifugation. This method is called the **VPRC,** or volume of packed red cells.

The VPRC is determined by centrifuging a blood sample in a special centrifuge tube called a *hematocrit* or in a special size of heparinized capillary tubing. The centrifuge must be specifically designed to accommodate the hematocrit or capillary tubing.

In this exercise, we will utilize a micro method for determining the VPRC. As illustrated in figure 55.1, only a drop of blood is needed to perform the test. Capillary tubing, which has been heparinized, is used to collect the blood sample. It is centrifuged at high speed for only four minutes. (Macro methods that utilize hematocrits usually centrifuge the blood for 30 minutes at a much slower speed.) After centrifugation,

the percentage of total volume occupied by the packed red blood cells is read on a special tube reader (figure 55.6) or directly on the head of the centrifuge, depending on the type of centrifuge.

An interesting relationship exists between grams of hemoglobin per 100 ml and the VPRC: the VPRC is usually three times the gm Hb/100 ml. In men, the normal VPRC range is between 40% and 54%, with 47% as average. In women, the normal range is between 37% and 47%, with 42% as average. The principal advantages of this method over total RBC counts and Hb-Meter usage are: (1) simplicity, (2) speed, and (3) high degree of accuracy. Although this micro method is not as accurate as macro methods, it still functions in an accuracy range of ±2% for most blood samples. Proceed as follows to determine the VPRC of your own blood.

Materials:
    lancets
    cotton
    alcohol
    hematocrit centrifuge
    tube reader
    heparinized capillary tubes
    Clay-Adams Seal-Ease

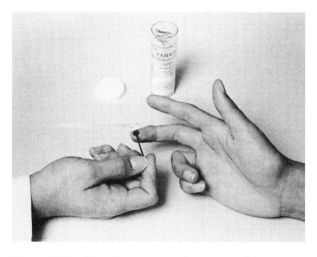

Figure 55.1   Blood is drawn up into a heparinized capillary tube for VPRC determination.

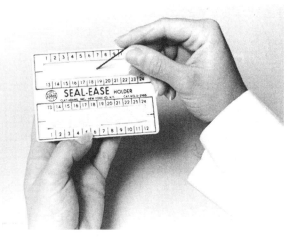

Figure 55.2   After filling with blood, the end of the capillary tube is sealed with clay.

1. Produce a free flow of blood on a finger. Wipe away the first drop of blood.
2. Place the red marked end of the capillary tube into the drop of blood and allow the blood to be drawn up about two-thirds of the way into the tube. Holding the open end of the tube downward from the blood source will cause the tube to fill more rapidly.
3. Seal the blood end of the tube with Seal-Ease. See figure 55.2.
4. Place the tube into the centrifuge with the sealed end against the ring of rubber at the circumference. Load the centrifuge with an even number of tubes (2, 4, 6, etc.) to properly balance the head.
5. Secure the inside cover with a wrench as illustrated in figure 55.4.

6. Lower the outside cover and securely latch it.
7. Turn on the centrifuge, setting the timer for 4 minutes.
8. Remove the centrifuged sample from the centrifuge and determine the VPRC by placing the tube in the mechanical tube reader. Instructions for reading are on the instrument.

Note: On some centrifuges the tubes can be read directly on the head without using a tube reader.

Laboratory Report

Record your results on table II in section E of Laboratory Report 52–58.

Figure 55.3 The capillary tubes are placed in the centrifuge with the sealed end toward the perimeter.

Figure 55.4 The inner safety lid is tightened in place with a lock wrench.

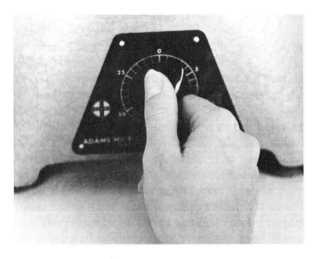

Figure 55.5 The timer on the centrifuge is set for four minutes by turning the dial clockwise.

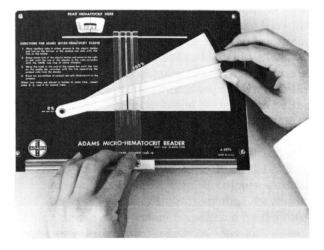

Figure 55.6 The capillary tube is placed in the tube reader to determine the VPRC.

56

Coagulation Time

The coagulation of blood is a complex phenomenon involving over thirty substances. The majority of these substances inhibit coagulation and are called *anticoagulants;* the remainder, which promote coagulation, are designed as *procoagulants.* Whether the blood will coagulate depends on which group predominates in a given situation. Normally, the anticoagulants predominate, but when a vessel is ruptured, the procoagulants in the affected area assume control, causing a clot *(fibrin)* to form in a relatively short period of time.

Of the various methods that have been devised to determine the rate of blood coagulation, the one outlined here is very easy to perform and very reliable. Figures 56.1 through 56.3 illustrate the general procedure. Proceed as follows:

Materials:
    lancets, cotton, alcohol
    capillary tubes (0.5 mm dia)
    3-cornered file

1. Puncture the finger to expose a free flow of blood. **Record the time.**
2. Place one end of the capillary tube into the drop of blood. Hold the tube so that the other end is **lower** than the drop of blood so that the force of gravity will aid the capillary action.
3. At **one-minute intervals,** break off small portions of the tubing by scratching the glass first with a file.
    Important: Separate the broken ends slowly and gently while looking for coagulation. Coagulation has occured when threads of fibrin span the gap between the broken ends as shown in figure 56.3.
4. **Record the time** as that from which the blood first appeared on that finger to the formation of fibrin.

Laboratory Report

Record your results on table II, section E, of Laboratory Report 52–58.

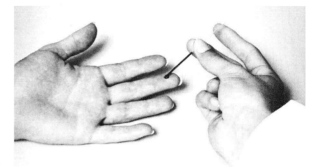

Figure 56.1 Blood is drawn up into a nonheparinized capillary tube.

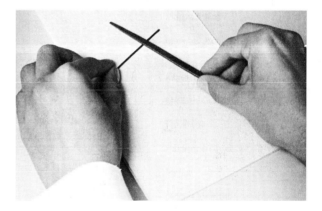

Figure 56.2 At one-minute intervals, sections of the tube are filed and broken off for the test.

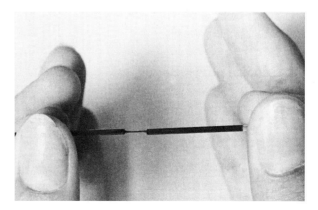

Figure 56.3 When the blood has coagulated, strands of fibrin will extend between broken ends of tube.

57

Sedimentation Rate

If citrated blood is allowed to stand vertically in a tube for a period of time, the red blood cells will fall to the bottom of the tube, leaving clear plasma in the upper region. The distance that the cells fall in one hour is called the **sedimentation rate** (more commonly, *sed rate*). With the Landau micro method, the sed rate in normal adults is 0–6 mm; in children the normal rate is 0–8 mm. During menstruation and pregnancy, the sed rate invariably exceeds 8 mm in women. More significantly, the sed rate in all acute general infections and malignancies is accelereated; pulmonary tuberculosis, rheumatic carditis, rheumatoid arthritis, and Hodgkin's disease are examples of situations that result in high sedimentation rates.

The most influential factor to affect the sedimentation rate is the size of sedimenting particles in the blood. *Rouleaux formation,* which results from the adherence of erythrocytes to form stacks of cells like poker chip piles, is the principal cause of high sedimentation rates.

The vertical stacking of the cells in the blood sample increases the particle mass without greatly increasing surface area resistance to settling. When inflammation exists in the body, the negative surface charge of erythrocytes is reduced, causing the cells to lack essential repulsive forces. The result is rouleaux formation. The negative charge of erythrocytes is a function of sialic acid groups in the plasma membrane. These acid groups, in turn, are attenuated by the presence of macromolecules such as fibrinogen and gamma globulin in plasma. At present it is the general consensus that high levels of fibrinogen play an important role in causing high sedimentation rates by promoting rouleaux formation.

The importance of the sedimentation rate in diagnostic work cannot be overemphasized. It is a routine blood test that is as significant to the physician as the pulse, blood pressure, temperature, and urine analysis.

In performing the following test, take care to produce an ample puncture. Most first-time attempts at performing this test fail due to an insufficient flow of blood from the punctured finger.

Materials:
  lancets
  cotton
  alcohol
  pipette cleaning solutions (4-bottle kit)
  Landau sed-rate pipette
  Landau rack
  mechanical suction device
  5% sodium citrate
  clean microscope slide (optional)

1. Draw sodium citrate solution up to the first line that completely encircles the pipette. Use the mechanical suction device shown in figure 57.1.
2. Produce a free flow of blood, wipe off the first drop with cotton, and draw it up into the pipette until the blood reaches the second line. Take care not to get any air bubbles into the blood. *Note:* If air bubbles do occur in the blood, carefully expel the mixture onto a clean microscope slide and draw it up again.
3. Mix the blood and citrate by drawing the two fluids up into the bulb and expelling them into the lumen of the tube. **Do this six times.** If any bubbles occur in the tube lumen at this time, use the slide technique described above.
4. Adjust the top level of the blood as close to zero as possible. It is practically impossible to get it to stop exactly on zero.
5. Remove the suction device by placing the lower end of the pipette on the index finger of the left hand before pulling the device off the other end. If the lower end is not completely sealed, this step will cause the blood to be pulled out of the pipette.
6. Place the lower end of the pipette on the base of the pipette rack and the other end at the top of the rack. Be sure that the tube is exactly perpendicular. Record the time at which it is put in the rack.
7. After one hour, measure the distance from the top of the plasma to the top of the red blood cells with a plastic millimeter scale.
8. Rinse the pipette with acid, water, alcohol, and acetone from the pipette cleaning kit.

Laboratory Report

Record your results on table II, section E, of Laboratory Report 52–58.

Figure 57.1 Anticoagulant sodium citrate is drawn up into the pipette to the first line.

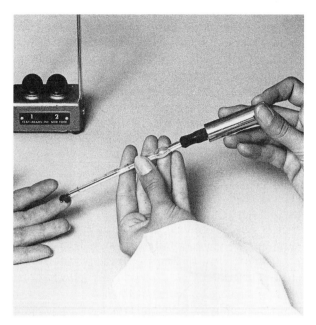

Figure 57.2 Blood is drawn up into the pipette and mixed with sodium citrate in the pipette.

Figure 57.3 After the blood and anticoagulant are completely mixed, the pipette is placed in the rack.

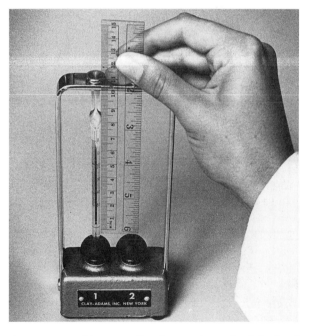

Figure 57.4 After standing for one hour, the fall of red blood cells is measured.

58

Blood Typing

The surfaces of red blood cells possess various antigenic molecules that have been designated by letters such as A, B, C, D, E, c, d, e, M, N, etc. Although the exact chemical composition of all these substances is undetermined, some of them appear to be carbohydrate residues (oligosaccharides).

The presence or absence of these various substances determines the type of blood possessed by an individual. Since the chemical makeup of cells is genetic, an individual's blood type is the same in old age as it is at birth. It never changes.

The only factors that we are concerned with here in this exercise are A, B, and D (Rh) antigens since they are most commonly involved in transfusion reactions.

To determine an individual's blood type, drops of blood-typing sera are added to suspensions of red blood cells to detect the presence of **agglutination** (clumping) of the cells. ABO typing may be performed at room temperature with saline suspensions of red blood cells as shown in figure 58.1. Rh typing for the D factor, on the other hand, requires higher temperatures (around 50° C) and whole blood instead of diluted blood. For ABO typing the diluted blood procedure is preferable. For convenience, however, the warming box method may be used for combined ABO and Rh typing. Two procedures are provided here. Your instructor will indicate which method will be used.

ABO Blood Typing

(Saline Dilution Method)

Materials:
    small vial (10 mm dia × 50 mm long)
    disposable lancets (*B-D Microlance,
        Sera-sharp,* etc.)
    70% alcohol and cotton
    wax pencil and microscope slides
    typing sera (anti-A and anti-B)
    applicators or toothpicks
    saline solution (0.85%)
    1 ml pipettes

1. Mark a slide down the middle with a marking pencil, dividing the slide into two halves (see figure 58.1). Write ANTI-A on the left side and ANTI-B on the right side.
2. Pipette approximately 1 ml of saline solution into a small vial or test tube.
3. Scrub the middle finger with a piece of cotton saturated with 70% alcohol and pierce it with a sterile disposable lancet.
4. Allow two or three drops of blood to mix with the saline by holding the finger over the end of the vial and washing it with the saline by inverting the vial several times.
5. Place a drop of this red cell suspension on each side of the slide.
6. Add a drop of anti-A serum to the left side of the slide and a drop of anti-B serum to the right side. *Do not contaminate the tips of the serum pipettes with the material on the slide.*
7. After mixing each side of the slide with separate applicators or toothpicks, look for agglutination. The slide should be held about 6 inches above an illuminated white background and rocked gently for 2 or 3 minutes.
8. Record your results in section C of the Laboratory Report as of 3 minutes.

Combined ABO and Rh Typing

(Warming Box Method)

As stated above, Rh typing must be performed with heat on blood that has not been diluted with saline. A warming box such as the one in figure 58.2 is essential in this procedure. In performing this test, two factors are of considerable importance: first, only a small amount of blood must be used (a drop of about 3 mm diameter on the slide), and second, proper agitation must be executed. The agglutination that occurs in this antibody-antigen reaction results in finer clumps; therefore, closer examination is also essential. If the agitation is not properly performed, agglutination may not be as apparent as it should be.

In this combined method, we will use whole blood for the ABO typing also. Although this method works out satisfactorily as a classroom demonstration for the ABO groups, it is not as

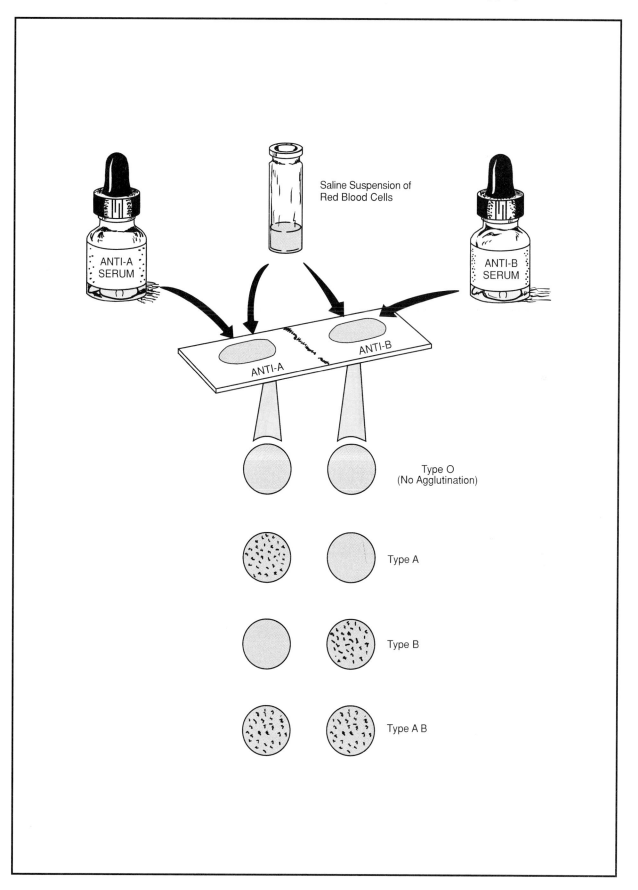

Saline Suspension of
Red Blood Cells

ANTI-A
SERUM

ANTI-B
SERUM

ANTI-B

ANTI-A

Type O
(No Agglutination)

Type A

Type B

Type A B

Figure 58.1 Blood typing ABO groups.

reliable as the other method in which saline and room temperature are used. *For clinical applications, whole undiluted blood, with heat, is not recommended.*

Materials:
slide warming box with a special marked slide
anti-A, anti-B, and anti-D typing sera
applicators or toothpicks
70% alcohol
cotton
disposable sterile lancets *(B-D Microlance, Sera-sharp, etc.)*

1. Scrub the middle finger with a piece of cotton saturated with 70% alcohol and pierce it with a sterile disposable lancet. Place a small drop in each of these squares on the marked slide on the warming box. To get the proper proportion of serum to blood, do not use a drop that is larger than 3 mm diameter on the slide.

2. Add a drop of anti-D serum to the blood in the anti-D square, mix with a toothpick, and note the time. **Only 2 minutes should be allowed for agglutination.**

3. Add a drop of anti-B serum to the anti-B square and a drop of anti-A to the anti-A square. Mix the sera and blood in both squares with *separate fresh* toothpicks.

4. Agitate the mixtures on the slide by slowly rocking the box back and forth on its pivot. At the end of two minutes, examine the anti-D square carefully for agglutination. If no agglutination is apparent, consider the blood to be Rh negative. By this time, the ABO type can also be determined.

Laboratory Report

Record your results in section C of the Laboratory Report and answer all the questions.

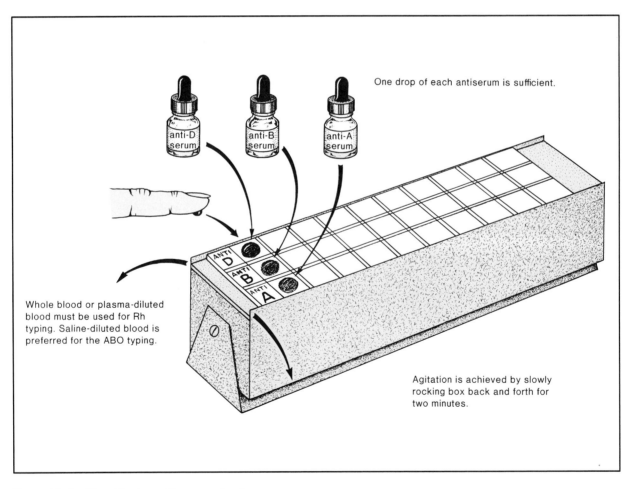

One drop of each antiserum is sufficient.

anti-D serum

anti-B serum

anti-A serum

ANTI D

ANTI B

ANTI A

Whole blood or plasma-diluted blood must be used for Rh typing. Saline-diluted blood is preferred for the ABO typing.

Agitation is achieved by slowly rocking box back and forth for two minutes.

Figure 58.2 Blood typing with a warming box.

PART 11

The Cardiovascular System

This unit consists of twelve exercises. The first eight of them are related directly to the anatomy and physiology of the heart. The remainder are concerned with general circulation. The extensive use of electronic equipment in many of the experiments may necessitate review of certain exercises in Part 5. A cat that has been latex injected will be used for studying arteries and veins in Exercise 68. The role of the thymus gland in cellular and humoral immunity will be studied along with the lymphatic system in Exercise 70.

Histology Self-Quiz No. 2

If at this point you have completed all the histology assignments in Exercises 18, 48, and 50 that pertain to nerve tissue, the eye, and ear, you may wish to test your knowledge of the histology of these areas by taking self-quiz no. 2 on page 488. It consists of twelve photomicrographs and fifty questions. After recording your answers on a sheet of paper, check your answers against the key, which you will find on page 414.

59

Anatomy of the Heart

In this study of the anatomy of the human heart we will use a sheep heart for dissection and comparison. Before beginning the dissection, however, study figures 59.1 and 59.2.

Internal Anatomy

Figure 59.1 reveals the internal anatomy of the human heart. Red-colored vessels carry oxygenated blood and blue ones carry deoxygenated blood.

Chambers of the Heart

Note that the heart has four chambers: two small upper atria and two larger ventricles. The atria receive all the blood that enters the heart and the ventricles pump it out. Note that the **right atrium** in the frontal section is on the left side of the illustration and the **left atrium** is on the right. Observe, also, that although the **right ventricle** is somewhat larger than the left ventricle, the **left ventricle** has a thicker wall. Separating the two ventricles is a partition, the **ventricular septum.**

Wall of the Heart

The wall of the heart consists of three layers: the myocardium, the endocardium, and the epicardium. The **myocardium** is the muscular portion of the wall that is composed of cardiac muscle tissue. Note in the enlarged section of the wall that the myocardium makes up the bulk of the wall thickness.

Lining the inner surface of the heart is the **endocardium.** It is a thin serous membrane that is continuous with the endothelial lining of the arteries and veins. An infection of this membrane is called *endocarditis.*

Attached to the outer surface of the heart is another serous membrane, the **epicardium,** or *visceral pericardium.* The production of serous fluid by this membrane on the outer surface of the heart enables the heart to move freely within the pericardial sac. Note that the **pericardial sac,** or *parietal pericardium,* consists of two layers: an inner serous layer and an outer fibrous layer. Between the heart

and the parietal pericardium is the **pericardial cavity** (label 19). Its dimension has been exaggerated here.

Vessels of the Heart

The major vessels of the heart are the two venae cavae, the pulmonary trunk, the four pulmonary veins, and the aorta. The **superior vena cava** is the upper blue vessel on the right atrium; it conveys deoxygenated blood to the right atrium from the head and arms. The **inferior vena cava** is the lower blue vessel that empties deoxygenated blood from the trunk and legs into the right atrium. From the right atrium blood passes to the right ventricle, where it leaves the heart through the **pulmonary trunk.** This large vessel branches into the **right** and **left pulmonary arteries,** which carry blood to the lungs.

Blood is drained from the lungs by means of the four **pulmonary veins** (four small red vessels), which carry it to the left atrium. From the left atrium the blood passes into the left ventricle, where it exits the heart through the *aorta.* Although the initial emergence of the aorta is obscured, it can be seen as a large curved red vessel, the **aortic arch,** that has three smaller arteries emerging from its superior surface. The aorta carries oxygenated blood to the entire systemic circulatory system.

Note what appears to be a short vessel between the pulmonary trunk and the aortic arch. It is called the **ligamentum arteriosum.** During prenatal life this ligament is a functional blood vessel, the *ductus arteriosus,* that allows blood to pass from the pulmonary trunk to the aorta.

Valves of the Heart

The heart has two atrioventricular and two semilunar valves. Between the right atrium and the right ventricle is the **tricuspid valve.** Between the left atrium and the left ventricle is the **bicuspid,** or *mitral,* **valve.** The differences between these two valves are seen in the superior sectional view of figure 59.1. Note that the tricuspid valve has three flaps or cusps, and the bicuspid valve has only two

_____ Anterior Interventricular Artery

_____ Aorta

_____ Aortic Semilunar Valve

_____ Apex

_____ Bicuspid Valve

_____ Chordae tendineae

_____ Circumflex Artery

_____ Coronary Sinus

_____ Endocardium

_____ Inferior Vena Cava

_____ Left Atrium

_____ Left Coronary Artery

_____ Left Pulmonary Artery

_____ Left Ventricle

_____ Ligamentum arteriosum

_____ Myocardium

_____ Opening to Coronary Arteries

_____ Opening to Coronary Sinus

_____ Papillary Muscle

_____ Parietal Pericardium

_____ Pericardial Cavity

_____ Pulmonary Trunk

_____ Pulmonary Veins

_____ Pulmonic Semilunar Valve

_____ Right Atrium

_____ Right Coronary Artery

_____ Right Pulmonary Artery

_____ Right Ventricle

_____ Superior Vena Cava

_____ Tricuspid Valve

_____ Ventricular Septum

_____ Visceral Pericardium

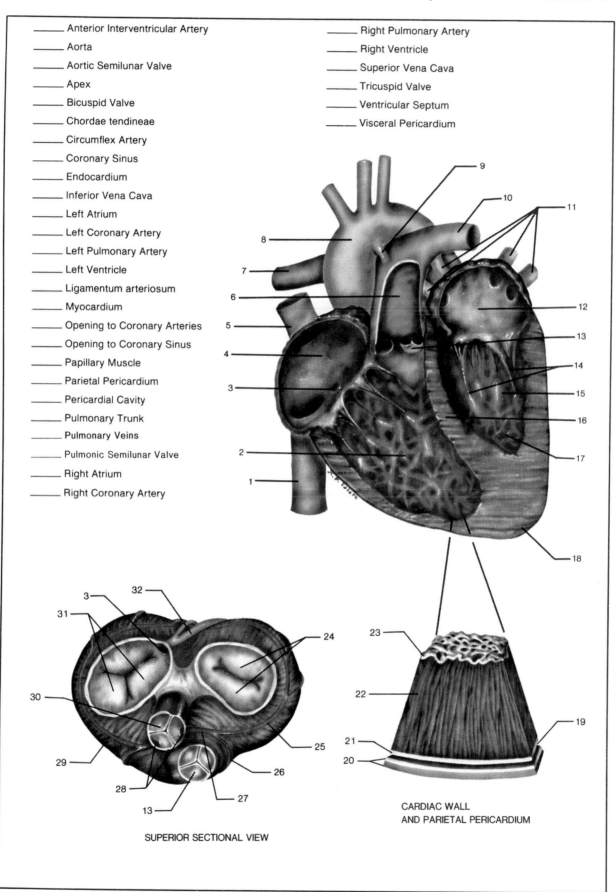

SUPERIOR SECTIONAL VIEW

CARDIAC WALL
AND PARIETAL PERICARDIUM

Figure 59.1 Internal anatomy of the heart.

cusps. Observe, also, that the edges of the cusps have fine cords, the **chordae tendineae,** which are anchored to **papillary muscles** on the wall of the heart (see frontal section). These cords and muscles prevent the cusps from being forced up into the atria during systole (ventricular contraction).

The other two valves are the **pulmonic semilunar** and **aortic semilunar valves.** They are located at the bases of the pulmonary trunk and the aorta. They prevent blood in those vessels from flowing back into the heart during diastole (relaxation phase). Note in the superior sectional view that each of these valves has three small cusps.

Coronary Circulation

The vessels that supply the heart muscle with blood make up the *coronary circulatory system.* Oxygenated blood in the aorta passes into right and left coronary arteries through two small openings at the point just superior to the aortic semilunar valve. These **openings to the coronary arteries** are seen in the wall of the aorta in the superior sectional view in figure 59.1. One opening leads into a **left coronary artery** (label 27) and the other opening leads into the **right coronary artery.**

. The left coronary artery has two branches: a **circumflex artery** that passes around the heart in the left atrioventricular sulcus and an **anterior interventricular artery** (label 26) that passes down the interventricular sulcus on the anterior surface of the heart. The right coronary artery lies in the right atrioventricular sulcus and has branches that supply the posterior and anterior surfaces of the ventricular muscle.

Once the coronary blood has been relieved of its oxygen and nutrients in the heart muscle, it is picked up by the various veins that parallel the arteries and is emptied into a large vein, the **coronary sinus** (label 32). Blood in the coronary sinus is emptied into the right atrium through a portal, the **opening to the coronary sinus** (label 3).

To identify the remainder of the vessels of the coronary system, proceed to study the external features of the heart in figure 59.2.

Assignment:
Label figure 59.1

External Anatomy

The study of the external anatomy of the heart is relatively simple once the internal anatomy is understood. Figure 59.2 reveals two external views of the heart.

Anterior Aspect

Note that the heart lies within the **mediastinum** at a slight tilt to the left so that the *apex,* or tip of the heart, is somewhat on the left side. The obscured appearance of the coronary vessels in this view is due to the fact that the **parietal pericardium** lies intact over the organ. The clarity of the structures on the posterior view is due to the fact that this pericardium has been removed.

Differentiating the right ventricle from the left ventricle is best done by locating the *interventricular sulcus* first. This sulcus is the slight depression over the ventricular septum that contains the **anterior interventricular artery** and the **great cardiac vein.** The yellow material seen in the sulcus consists of fat deposits. The left ventricle is on the left side of the interventricular sulcus, and the right ventricle is to the right of it.

Three portions of the aorta are labeled on the anterior aspect: a short **ascending aorta** that emerges from the left ventricle, a curving **aortic arch,** and a portion of the **descending aorta** that passes down through the mediastinum behind the heart.

Locate the **right** and **left pulmonary arteries** that emerge from behind the aortic arch. The pulmonary trunk that gives rise to these arteries is obscured by the aorta. Also, identify the **ligamentum arteriosum.** The coronary vessels in the atrioventricular sulcus are the **right coronary artery** and the **small cardiac vein.**

Posterior Aspect

The posterior view of the heart in figure 59.2 reveals more clearly the position of the four pulmonary veins and additional coronary vessels. The two **right pulmonary veins** are seen near the superior vena cava. The **two left pulmonary veins** are situated to the left of the right pulmonary veins. The appearance of a single blue vessel lying below the aorta is actually the site of the division of the pulmonary trunk into the **right** and **left pulmonary arteries.**

The following coronary arteries are seen on this side of the heart: the **right coronary artery** (label 4), which lies in the right atrioventricular sulcus, the **posterior descending right coronary artery,** which is a downward extension of the right coronary artery, the **posterior interventricular artery** (label 20), and the **circumflex artery** (label 23).

The major coronary veins on this side are the **middle cardiac vein,** which parallels the posterior descending right coronary artery, the **left posterior**

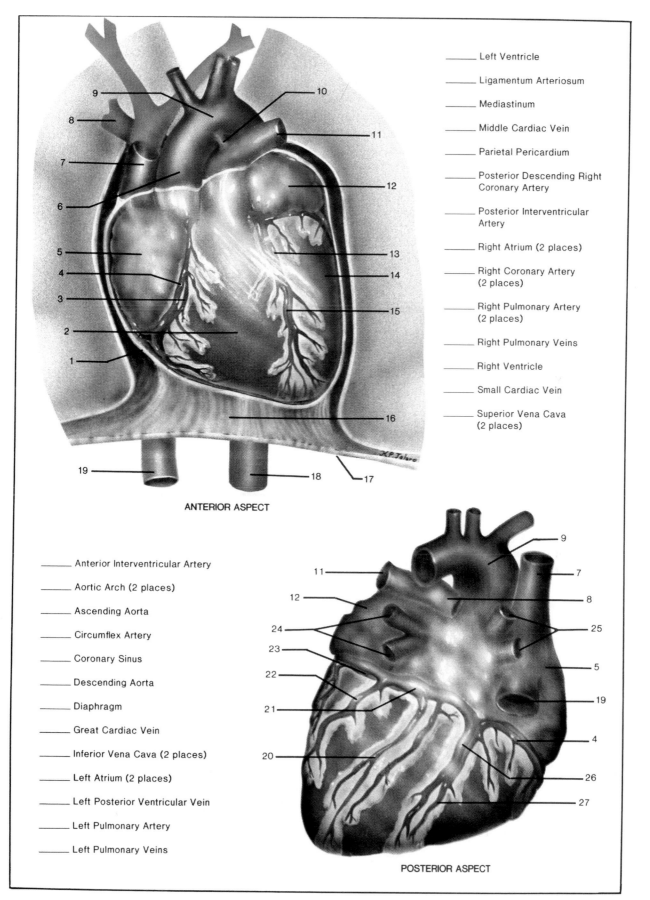

Left Ventricle

Ligamentum Arteriosum

Mediastinum

Middle Cardiac Vein

Parietal Pericardium

Posterior Descending Right Coronary Artery

Posterior Interventricular Artery

Right Atrium (2 places)

Right Coronary Artery (2 places)

Right Pulmonary Artery (2 places)

Right Pulmonary Veins

Right Ventricle

Small Cardiac Vein

Superior Vena Cava (2 places)

ANTERIOR ASPECT

Anterior Interventricular Artery

Aortic Arch (2 places)

Ascending Aorta

Circumflex Artery

Coronary Sinus

Descending Aorta

Diaphragm

Great Cardiac Vein

Inferior Vena Cava (2 places)

Left Atrium (2 places)

Left Posterior Ventricular Vein

Left Pulmonary Artery

Left Pulmonary Veins

POSTERIOR ASPECT

Figure 59.2   External anatomy of the heart.

ventricular vein (label 22), and the **posterior interventricular vein** (not labeled), which lies in the interventricular sulcus with the similarly named artery. All these coronary veins empty into the large **coronary sinus**, which lies in the atrioventricular sinus.

Assignment:
Label figure 59.2.

Sheep Heart Dissection

While dissecting the sheep heart, attempt to identify as many structures as possible that are shown in figures 59.1 and 59.2.

Materials:
    sheep heart, fresh or preserved
    dissecting instruments and tray

1. Rinse the heart with cold water to remove excess preservative or blood. Allow water to flow through the large vessels to irrigate any blood clots out of its chambers.
2. Look for evidence of the **parietal pericardium.** This fibroserous membrane is usually absent from laboratory specimens, but there may be remnants of it attached to the large blood vessels of the heart.
3. Attempt to isolate the **visceral pericardium** (*epicardium*) from the outer surface of the heart. Since it consists of only one layer of squamous cells, it is very thin. With a sharp scalpel try to peel a small portion of it away from the myocardium.
4. Identify the **right** and **left ventricles** by squeezing the walls of the heart as shown in illustra-

tion 1, figure 59.3. The right ventricle will have the thinner wall.
5. Identify the **anterior interventricular sulcus,** which lies between the two ventricles. It is usually covered with fat tissue. Carefully trim away the fat in this sulcus to expose the **anterior interventricular artery** and the **great cardiac vein.**
6. Locate the **right** and **left atria** of the heart. The right atrium is seen on the left side of illustration 1, figure 59.3.
7. Identify the **aorta,** which is the large vessel just to the right of the right atrium as seen in illustration 1, figure 59.3. Carefully peel away some of the fat around the aorta to expose the **ligamentum arteriosum.**
8. Locate the **pulmonary trunk,** which is the large vessel between the aorta and the left atrium as seen when looking at the anterior side of the heart. If the vessel is of sufficient length, trace it to where it divides into the **right** and **left pulmonary arteries.**
9. Examine the posterior surface of the heart. It should appear as in illustration 2, figure 59.3. Note that only the right atrium and two ventricles can be seen on this side. The left atrium is obscured by fat from this aspect. Look for the four thin-walled **pulmonary veins** that are embedded in the fat. Probe into these vessels and you will see that they lead into the atrium.
10. Locate the **superior vena cava,** which is attached to the upper part of the right atrium. Insert one blade of your dissecting scissors into this vessel (illustration 3, figure 59.3), and cut

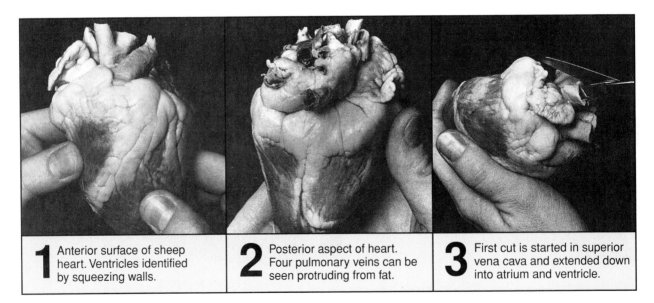

1 Anterior surface of sheep heart. Ventricles identified by squeezing walls.

2 Posterior aspect of heart. Four pulmonary veins can be seen protruding from fat.

3 First cut is started in superior vena cava and extended down into atrium and ventricle.

Figure 59.3   Sheep heart dissection.

through it into the atrium to expose the **tricuspid valve** between the right atrium and right ventricle. Don't cut into the ventricle at this time.

11. Fill the right ventricle with water, pouring it in through the tricuspid valve. Gently squeeze the walls of the ventricle to note the closing action of the valve's cusps.

12. Drain the water from the heart and continue the cut with scissors from the right atrium through the tricuspid valve down to the apex of the heart.

13. Open the heart and flush it again with cold water. Examine the interior. The open heart should look like illustration 1, figure 59.4.

14. Examine the interior wall of the right atrium. This inner surface has ridges, giving it a comb-like appearance; thus, it is called the **pectinate muscle** (*pecten,* comb).

15. Look for the **opening to the coronary sinus.** The white arrow in illustration 1, figure 59.4 shows its location. It is through this opening that blood of the coronary circulation is returned to the venous circulation.

16. Insert a probe under the cusps of the tricuspid valve. Are three flaps evident?

17. Locate the **papillary muscles** and **chordae tendineae** in the right ventricle. How many papillary muscles are there?

18. Look for the **moderator band,** a reinforcement cord that extends from the ventricular septum to the ventricular wall. Its presence prevents excessive stress from occurring in the myocardium of the right ventricle.

19. With scissors, cut the right ventricular wall up along its lower margin parallel to the anterior interventricular sulcus to the pulmonary trunk.

Continue the cut through the exit of the right ventricle into the **pulmonary trunk.** Spread the cut surfaces of this new incision to expose the **pulmonic semilunar valve.** Wash the area with cold water to dispel blood clots.

20. Insert one blade of your scissors into the left atrium, as shown in illustration 2, figure 59.4. Cut through the atrium to the left ventricle. Also, cut from the left ventricle into the aorta, slitting this vessel longitudinally.

21. Examine the **bicuspid** (mitral) **valve,** which lies between the left atrium and left ventricle. With a probe, reveal the two cusps.

22. Examine the pouches of the **aortic semilunar valve.** Compare the number of pouches of this valve with the pulmonic valve.

23. Look for the **openings to the coronary arteries,** which are in the walls of the aorta just above the aortic semilunar valve. Force a blunt probe into each hole. Note how they lead into the two coronary arteries.

Laboratory Report

Complete the Laboratory Report for this exercise.

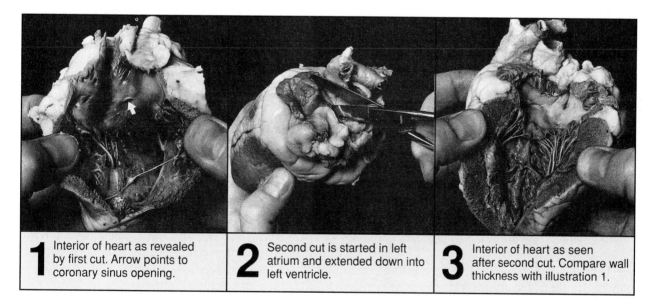

1 Interior of heart as revealed by first cut. Arrow points to coronary sinus opening.

2 Second cut is started in left atrium and extended down into left ventricle.

3 Interior of heart as seen after second cut. Compare wall thickness with illustration 1.

Figure 59.4 Sheep heart dissection.

60

Cardiovascular Sounds

Cardiovascular sounds are created by myocardial contraction and valvular movement. Three distinct sounds are heard.

The **first sound** (lub) occurs simultaneously with the contraction of the ventricular myocardium and the closure of the mitral and tricuspid valves.

The **second sound** (dup) follows the first one after a brief pause. It differs from the first sound in that it has a snapping quality of higher pitch and shorter duration. This sound is caused by the virtually simultaneous closure of the aortic and pulmonic valves at the end of ventricular systole.

The **third sound** is low pitched and is loudest near the apex of the heart. It is difficult to detect, but can be heard most easily when the subject is recumbent. It appears to be caused by the vibration of the ventricular walls and the AV valves during systole.

Abnormal heart sounds are collectively referred to as **murmurs.** They are usually due to damaged valves. Mitral valve damage due to rheumatic fever is a most frequent cause of heart murmur; the aortic valve is next in susceptibility to damage. Damaged valves may result in *stenosis* or *regurgitation* of blood flow.

Examination Procedures

Listening to sounds generated within the body is called *auscultation.* For maximum results in monitoring valvular sounds it is necessary to be familiar with four specific spots in the thorax. Illustration 1, figure 60.1, shows their locations. Note that these auscultatory areas do not coincide with the anatomical locations of the various valves; this is because valvular sounds are projected to different spots on the rib cage.

Three methods can be used for monitoring heart sounds: (1) the conventional stethoscope, (2) electronic recording, and (3) the use of an audio monitor. Only the procedures for the first two methods are presented here. The audio monitor, if available, will be used only as a demonstration. *Since noise is distracting in these experiments, it is imperative that all students keep voices low.*

Audio Monitor Demonstration (By Instructor)

Using a student as a subject in a recumbent position, the instructor will use an audio monitor to demonstrate heart sounds to the class. If a sensitive unit, such as the Grass AM7 is used, it will not be necessary to bare the subject's chest.

When attempting to demonstrate second sound splitting, hand signals should be used instead of verbalization. The subject can be instructed to inspire when the instructor's hand is raised and expire when the hand is lowered.

Stethoscope Auscultation

Work with your laboratory partner to auscultate each other's heart with a stethoscope. Use the following procedure:

Materials:
stethoscope
alcohol and absorbent cotton

1. Clean and disinfect the earpieces of a stethoscope with cotton saturated in alcohol. Fit them to your ears, directing them inward and upward.
2. Try first to maximize the **first sound,** by locating the stethoscope on the tricuspid and mitral ausculatory areas. See illustration 1, figure 60.1.

   Note that the mitral area coincides with the apex of the heart (fifth rib), and the tricuspid area is about 2 to 3 inches medial to this spot.
3. Now move the stethoscope to the aortic and pulmonic areas to hear the **second sound** more clearly.

   Observe that the aortic area is on the right border of the sternum in the space between the second and third ribs, and the pulmonic area is 2 to 3 inches to the left of it.
4. While listening to the second sound, attempt to detect the splitting of this sound while the subject is inhaling. Use hand signals instead of verbalization to communicate with the subject. Have the subject inspire when you raise your hand and expire when you lower your hand.

This splitting characteristic of the second sound during inspiration is normal. It is due to the fact that during inspiration more venous blood is forced into the right side of the heart, causing delayed closure of the pulmonic valve during systole. Result: aortic and pulmonic closure sounds are heard separately.

5. After listening to the sounds with the subject sitting erect, listen to them with the subject lying down, face upward.
6. Can you detect the **third sound?**
7. Is there any evidence of **murmurs?**
8. Compare the sounds before and after exercise, recording your observations on the Laboratory Report. For exercise, leave the room to run up and down stairs or jog for a short distance.
9. Before ending the examination, check all the auscultatory areas in figure 60.1 to make sure that you have identified all valvular sounds.

Electronic Recording (Unigraph Setup)

Produce a phonocardiogram of heart sounds of a subject lying down, face upward:

Materials:
Unigraph
Trans/Med 6605 adapter
Statham P23AA pressure transducer

1. Attach a Trans/Med 6605 adapter to the input end of the Unigraph.

2. Attach a Statham pressure transducer and microphone jack to the 6605 adapter. Refer to illustration 3, figure 60.1.
3. Turn on the Unigraph and set the following controls: c.c. at STBY, heat control at two o'clock position, speed control at slow position, Gain at 1 MV/CM, and Mode on TRANS.
4. Place the microphone against one of the auscultatory areas and start the chart moving by placing the c.c. lever at Chart On. Adjust the temperature control knob to get a good recording. Center the tracing with centering control knob and adjust sensitivity to produce at least 5 mm stylus displacement.
5. Move the speed control lever to the high speed position.
6. Make recordings with microphone over each auscultatory area. Be sure to mark the chart to identify the areas being monitored.
7. Attempt to record splitting of second sound, using the same techniques that were used with a stethoscope.
8. Attach phonocardiograms to the Laboratory Report.
9. Deactivate all controls on the Unigraph at the end of the experiment.

Laboratory Report

Complete the Laboratory Report for this exercise.

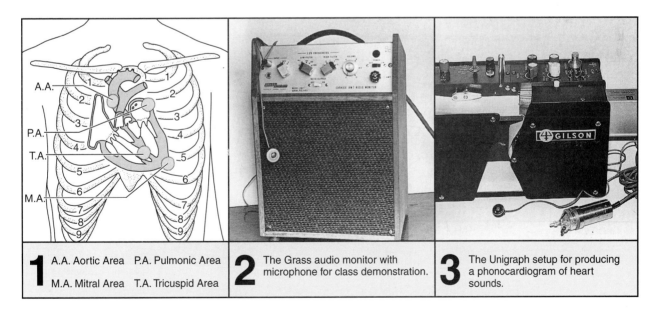

1 A.A. Aortic Area   P.A. Pulmonic Area
  M.A. Mitral Area   T.A. Tricuspid Area

2 The Grass audio monitor with microphone for class demonstration.

3 The Unigraph setup for producing a phonocardiogram of heart sounds.

Figure 60.1 Auscultatory areas and two electronic setups.

61

Heart Rate Control

The rate of heartbeat is governed by internal and external factors. Specialized centers of modified muscle tissue within the heart establish a particular rhythm that may be accelereated or slowed by the influence of extrinsic nerves that supply the heart. These nerves receive their direction from the **cardioregulator center** of the medulla oblongata. Chemical and physical factors, acting on parts of the heart, certain blood vessels, and the cardioregulator center reflexively alter the heart rate.

Intrinsic Tissues

The intrinsic conducting tissues of the heart consist of the sinoatrial (SA) and atrioventricular (AV) nodes. The **SA node** is a small mass of tissue in the wall of the right atrium near the superior vena cava. This center acts as the *pacemaker* of the heart, initiating contractions in the right atrium.

The **AV node** is located in the lower part of the right atrium. It acts as a relay station that picks up impulses of the SA node and transmits them to the ventricles.

Fibers of the **atrioventricular bundle** (bundle of His) in the ventricular septum carry the impulses from the AV node to a network of interlacing fibers, called the **conduction myofibers** (Purkinje fibers) in the walls of the ventricles.

Thus, we see that impulses, originating in the SA node, spread to the ventricles via the AV node, the atrioventricular bundle, and conduction myofibers, causing the ventricles to contract after the two atria have reacted.

Extrinsic Nerve Tissue

The extrinsic nerves consist of the **vagi** of the parasympathetic system and the **accelerators** of the sympathetic system. Fibers of the right vagus nerve innervate the SA node, and those of the left vagus pass to the AV node. The vagus is an inhibitory nerve in that it causes the heart rate to slow down and the

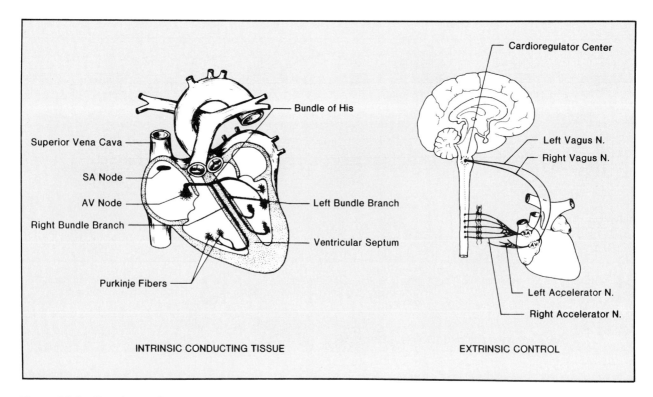

Figure 61.1 Regulatory tissues of the heart.

systole to be weakened. Extremely strong stimulation of the vagi may bring the heart to a temporary standstill. The accelerator nerve fibers arise from cervical and thoracic sympathetic ganglia. The accelerator nerves of the right side innervate the atrial region of the SA node, while those of the left side terminate in the atrial region of the AV node. Stimulation of these sympathetic fibers increases the heartbeat by shortening the systolic period and increasing the force of the atrial and ventricular contractions.

Hormonal Control

The speeding up and slowing down of the heart by these nerves is accomplished by the production of norepinephrine at the ends of the accelerator nerves and acetylcholine at the ends of the vagi. The action of norepinephrine and epinephrine is essentially the same.

The Experiment

In this experiment we will administer epinephrine and acetylcholine to a frog via the liver to note the effects on the heart rate. Figure 61.2 illustrates the equipment setup.

This experiment consists of three steps: (1) animal preparation, (2) equipment setup, and (3) monitoring. Except for the absence of an electronic stimulator, this experiment has essentially the same equipment setup as Exercise 29, in which we studied frog skeletal muscle contraction.

If, due to limited time, it is necessary to work in groups of four or more, divide your group up into teams and assign each one of them to one of the operational phases.

Materials:
for dissection:
    small frog, one per team
    dissection pan, wax bottom
    decapitation scissors
    scalpel, small scissors, forceps
    squeeze bottle of frog Ringer's solution
    spool of cotton thread
    small fishhook
    dissecting pins

for Unigraph setup:
    Unigraph, T/M Biocom #1030 transducer
    ring stand and 2 double clamps
    isometric clamp (Harvard Apparatus, Inc.)

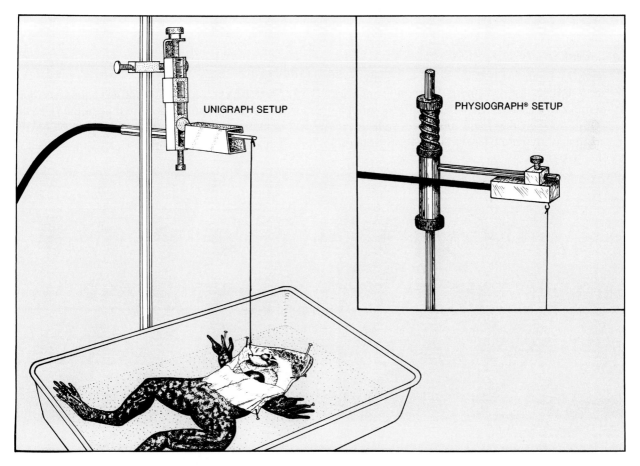

Figure 61.2 Frog heart monitoring.

for Physiograph® setup:
  Physiograph® IIIs with transducer coupler
  transducer stand
  myograph
  myograph tension adjuster

reagents:
  epinephrine 1:10,000
  acetylcholine, 1:10,000

Dissection

1. Decapitate and spinal pith a frog in the manner described in Exercise 29 (page 127).
2. With a pair of sharp scissors, remove the skin from the thorax region, exposing the underlying muscle layer.
3. Remove the muscle layer with scissors, exposing the underlying organs without injuring them. Also, cut away the sternum to expose the beating heart. Note how it beats within the thin pericardium.
4. Lift the pericardium with forceps and slit it open with scissors to expose the heart. The heart is now ready to accept the hook.
5. Moisten the heart with Ringer's solution. From now on it is essential that the heart be kept moist.

Equipment Setup

1. Note in figure 61.2 that the Gilson Unigraph setup utilizes a ring stand to hold the isometric clamp and transducer. The Narco Physiograph® setup, on the other hand, utilizes a transducer stand and myograph tension adjustor to support the myograph.

Hook up the various components as shown and plug the cord for the transducer (myograph) into the appropriate socket of the recorder. Turn on the power switch of the recorder.

2. **Unigraph Settings:** Mode at TRANS, Gain at 1 MV/CM, sensitivity knob completely counterclockwise, speed control lever at slow position (2.5 mm/sec), and c.c. switch at STBY.
3. **Physiograph® Settings:** Amplifier control at 1 MV/CM with inner knob turned completely clockwise; chart speed at 0.25 cm/sec.
4. Tie an 18″ length of thread to a small fishhook or bent pin. Connect the free end of thread to the transducer (myograph). For the Gilson setup, use only the single stationary leaf on the transducer.
5. As shown in figure 61.3, force the fishhook through the tip of the ventricle of the frog's heart. Use forceps to hold the hook. Unless you use a finger to support the back of the heart, it will be hard to force the hook through the ventricular muscle.
6. Adjust the tension of the thread with the isometric clamp or the myograph tension adjustor. The thread tension should be just enough to take up the slack, without pulling the heart straight up.
7. Continue to moisten the heart with Ringer's solution.

Monitoring

1. Start the chart moving to observe the trace, and center the tracing to a desired position. (Use the centering control or Trans-Bal control on the Unigraph for centering.)

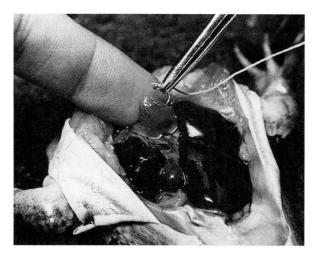

Figure 61.3 Support the heart with the index finger of the left hand when forcing the hook into the ventricles.

Figure 61.4 While injecting drugs into the liver, firm up the liver with the fingers of the left hand.

2. Increase the Gain control settings to 2, 5, or more MV/CM to get 1 cm stylus deflection with each heartbeat. For in-between sensitivity adjustments use the sensitivity control on the Unigraph, or the inner control knob on the amplifier knob on the Physiograph®.

3. Observe the contractions for 3–5 minutes, keeping the heart moist with Ringer's solution.

4. Administer 0.05 ml of 1:10,000 epinephrine by injecting this amount *very slowly* into the liver with a hypodermic needle. Refer to figure 61.4. *Mark the chart* to indicate this injection. Observe the trace for 3–5 minutes. Is there a change in the number of beats per minute? Does the strength of contraction change?

5. With a different syringe, inject 0.05 ml of 1:10,000 acetylcholine into the liver, marking the chart to indicate the time of this injection. How long does it take for the heart to respond to this injection? Observe the trace for another 3–5 minutes.

6. Now, administer another 0.05 ml of epinephrine in a similar manner. Be sure to mark the chart again. Observe the trace for 3 minutes. Does the heart seem to recover completely?

7. Inject 0.25 ml of epinephrine into the liver, marking the chart again at the time of injection. Observe the trace for 3–5 minutes. How does the heart respond to this massive dose?

8. Finally, inject 0.25 ml of acetylcholine and observe the trace for another 3–5 minutes.

Laboratory Report

Complete the Laboratory Report for this exercise.

62 Electrocardiogram Monitoring: Using Chart Recorders

In this laboratory period we will learn the procedures that are involved in making an electrocardiogram (ECG), utilizing a chart recorder such as the Gilson Unigraph or the Narco Physiograph®. Separate instructions will be provided for both types of equipment. If the Cardiocomp™ (Exercise 63) is to be used instead of one of these chart recorders, it will be necessary for you to read over the applications provided here in this exercise since this background information is lacking in the Cardiocomp™ exercise.

The ECG Wave Form

Myocardial contractions of the heart originate with depolarization of the sinoatrial (SA) node of the right atrium. As the myocardium of the right atrium is depolarized and contracts in response to depolarization of the SA node, the atrioventricular (AV) node is activated, causing it to send a depolarization wave via the atrioventricular bundle and conduction myofibers to the ventricular myocardium. The electrical potential changes that result from this depolarization and repolarization can be monitored to produce a record called an **electrocardiogram,** or **ECG.**

To produce such a record it is necessary to attach a minimum of two and as many as ten electrodes to different portions of the body to record the differences in electrical potential that occur. If one electrode is placed slightly above the heart and to the right, and another is placed slightly below the heart to the left, one can record the wave form at its maximum potential.

Figure 62.1 reveals a typical ECG as recorded on a Unigraph, and figure 62.2 depicts the characteristics of an individual normal ECG wave form. The initial depolarization of the SA node, which causes atrial contraction, manifests itself as the **P wave.** This depolarization is immediately followed by repolarization of the atria. Atrial repolarization, not usually seen with surface electrodes, can be demonstrated with implanted electrodes; such a wave is designated as the *TA wave.*

Depolarization of the ventricles via the atrioventricular bundle and conduction myofibers results in the production of the **QRS complex.** This depolarization causes ventricular systole.

As soon as depolarization of the ventricular muscle is completed, repolarization takes place, producing the **T wave.** Some ECG wave forms show an additional wave occurring after the T wave. It is designated as the **U wave.** This small wave is attributed to the gradual repolarization of the papillary muscles.

The final result of this depolarization wave is that the myocardium of the ventricles contracts as a

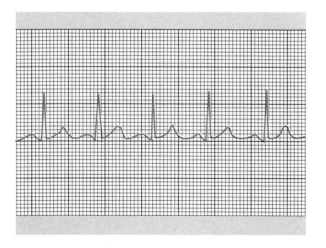

Figure 62.1 An electrocardiogram (Unigraph tracing).

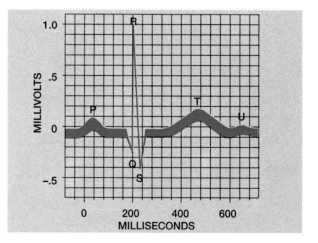

Figure 62.2 The ECG wave form.

unit, forcing blood out through the pulmonary trunk and aorta. Repolarization of the affected tissues quickly follows in preparation for the next contraction cycle.

The value of the ECG to the cardiologist is immeasurable. Almost all serious abnormalities of the heart muscle can be detected by analyzing the contours of the different waves. Interpretation of the ECG wave forms to determine the nature of heart damage is a highly developed technical skill involving vector analysis and goes considerably beyond the scope of this course; however, figure 62.4 illustrates three examples of abnormal ECGs in which heart damage has occurred.

Electrodes and Leads

As indicated earlier, as few as two and as many as ten electrodes can be used to produce an ECG. An understanding of Einthoven's triangle (figure 62.3) will clarify the significance of the different electrode positions and the meaning of leads I, II, and III. Einthoven was the first person to produce an electrocardiogram around the turn of the nineteenth century.

The number of **electrodes** used for the Unigraph is three: the two wrists and the left ankle as shown in figure 62.7. For the Physiograph® setup, five electrodes will be used (figure 62.6): the four appendages and one on the chest over the heart. The right leg acts as a ground in the five-electrode setup.

Three leads, designated as Leads I, II, and III, are derived from these electrode hookups. A **lead,** as defined by Einthoven, is *the electrical potential in the fluids around the heart between two electrodes during depolarization.* On the Physiograph® and Cardiocomp™ setups you will have an opportunity to make lead selections.

Note in figure 62.3 that the apexes of the upper part of the triangle represent the points at which the two arms connect electrically with the fluids around the heart. The lower apex is the point at which the left leg (+) connects electrically with the pericardial fluids. **Lead I** records the electrical potential between the right (−) and left (+) arm electrodes. **Lead II** records the potential between the right arm (−) and the left leg (+). **Lead III** records the potential between the left arm (−) and the left leg (+). These potentials are recorded graphically on the ECG in millivolts.

When we put Leads I, II, and III together on one diagram, as in figure 62.3, we see that the three leads form a triangle around the heart, known as **Einthoven's triangle.** The significance of this trian-

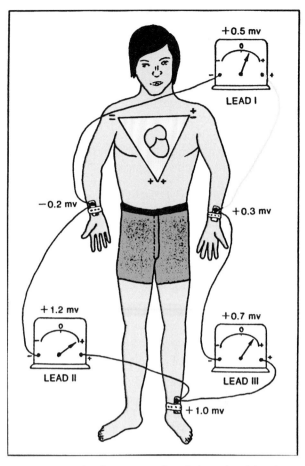

Figure 62.3 In the conventional three-lead hookup, the two arms and left leg form apexes of a triangle (Einthoven's) surrounding the heart. Note that the sum of the voltages in Leads I and III equals the voltage in Lead II.

gle is that if the potentials of two of the leads are recorded, the third one can be determined mathematically. Or, stated another way: the sum of the voltages in Leads I and III equals the voltage of Lead II. This principle is known as **Einthoven's law:**

$$\text{Lead II} = \text{Lead I} + \text{Lead III}$$

Procedure

Students will work in teams of three or four, with one individual being the subject and the others making the electrode hookups and operating the controls. It will be necessary to calibrate the recorder before recording the ECG; separate calibration instructions must be followed for the type of recorder being used.

Failure in this experiment usually results from (1) poor skin contact, (2) damaged electrode cables, or (3) incorrect hookup. Proceed as follows:

Materials:

for Unigraph setup:
Unigraph
3-lead patient cable
ECG plate electrodes (3) and straps

for Physiograph® setup:
Physiograph® with cardiac coupler
5-lead patient cable
4 ECG plate electrodes and straps
1 suction-type chest electrode

for both setups:
Scotchbrite pad
70% alcohol
electrode paste

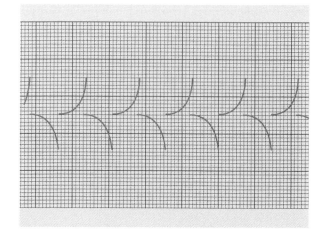

Figure 62.5 Calibration tracing for Unigraph.

Unigraph Calibration

While one member of your team attaches the electrodes to the subject, calibrate the Unigraph so that one millivolt produces one centimeter of stylus deflection (1 mV/1 cm). Follow these steps:

1. With the main power switch still in the OFF position, set the controls as follows: c.c. switch at STBY, speed control lever at the slow position, Gain at 1 MV/CM, sensitivity knob completely counterclockwise, Mode at CC-Cal, and the Hi Filter and Mean switches on NORM.
2. Turn on the power switch and rotate the stylus heat control to the two o'clock position.
3. Place the c.c. switch on Chart On, observe the trace and center the stylus, using the centering control.
4. Increase the sensitivity by rotating the sensitivity control clockwise one-half turn.

5. Depress the 1 MV button, hold it down for about 2 seconds, and then release it. The upward deflection should measure 1 cm. The downward deflection should equal the upward deflection. Refer to figure 62.5. If the travel is not 1 cm, adjust the sensitivity control and depress the 1 MV button until exactly 1 cm is achieved.
6. Place the c.c. switch on STBY and label the chart with the date and subject's name.
7. Place the Mode control on ECG and move the speed selector level to the high speed (25 mm per sec) position. The Unigraph is now ready to accept the patient cable.

Physiograph® Calibration

While one member of your team attaches the electrodes to the subject, calibrate the channel sensitivity

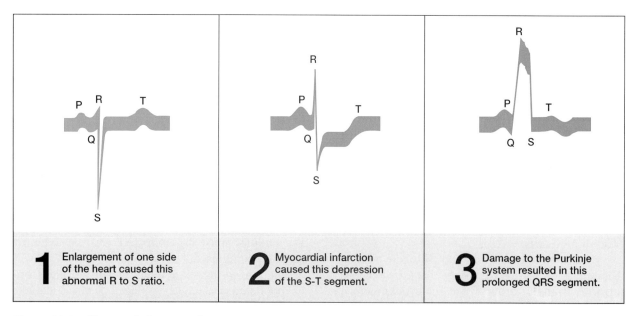

Figure 62.4 Abnormal electrocardiograms.

1 Enlargement of one side of the heart caused this abnormal R to S ratio.

2 Myocardial infarction caused this depression of the S-T segment.

3 Damage to the Purkinje system resulted in this prolonged QRS segment.

of the Physiograph® so that one millivolt produces two centimeters of pen deflection (1 mV/ 2 cm). Follow these steps:

1. Place the Time Constant switch of the cardiac coupler in the 3.2 position.
2. Place the Gain switch on the coupler to the X2 position.
3. Place the Lead Selector Control in the calibrate (Cal) position.
4. Set the outer knob of the channel Amplifier Sensitivity Control on the 10 MV/CM position, and rotate the inner sensitivity knob completely clockwise until it clicks.
5. Start the chart moving at 0.25 cm/sec and lower the pens onto the paper. Adjust the pen to the center of its arc with the Position Control.
6. Place the channel amplifier record button into the ON position.
7. Place the CAL toggle switch on the cardiac coupler into the 1 MV position. Observe that the channel recording pen will deflect *upward* 2 centimeters.
8. Hold the CAL switch in the 1 MV position until the recording pen returns to the center line, then release the switch. Observe that upon release, the channel recording pen will now deflect *downward* 2 cm.
9. If the upward and downward deflections are not exactly 2 centimeters, make adjustments with the amplifier sensitivity controls.
10. Stop recording. The channel is now calibrated.

Preparation of the Subject

While the recorder is being calibrated, the electrodes should be attached to the subject by other members of the team. Although it is not absolutely essential, it is desirable to have the individual lying on a comfortable cot.

Note in figure 62.7 that the two wrists and left ankle will be used. As shown in figure 62.6, both wrists, both ankles, and the heart region will have electrodes attached if the Physiograph® is to be used. Proceed as follows to attach the electrodes to the subject:

1. To achieve good electrode contact on the wrists and ankles, rub the contact areas with a Scotchbrite pad, disinfect the skin with 70% alcohol, and add a little electrode paste to the contact surface of each electrode.
2. Attach each of the plate electrodes with a rubber strap, as shown in figure 62.8.
3. If you are using the five-electrode setup, affix the suction-type electrode over the heart region, using an ample amount of electrode paste on the contact surface.
4. Attach the leads of the patient cable to the electrodes as follows:
   Unigraph Setup: Attach the black unnumbered lead to the left ankle, one of the numbered leads to the right wrist, and the other numbered lead to the left wrist.
   Physiograph® Setup: Attach each lead according to its label designation: LL—left leg,

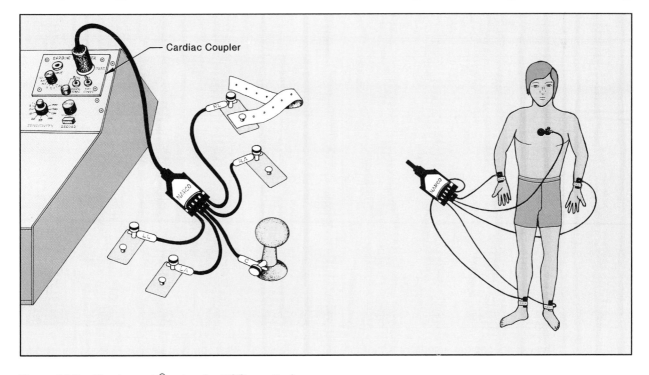

Figure 62.6 Physiograph® setup for ECG monitoring.

RL—right leg, LA—left arm, RA—right arm, and C—cardiac.

5. Make certain that the leads are securely connected to the electrode binding posts by firmly tightening the knobs on the binding posts. If the patient cable has alligator clamps instead of straight jacks, take care to see that the clamps can't slip off easily.

6. If the subject is sitting in a chair instead of lying down, *make sure that the chair is not metallic.*

7. Connect the free end of the patient cable to the appropriate socket on the Unigraph or Physiograph® cardiac coupler.

8. The subject is now ready for monitoring.

Monitoring

Tell the subject to relax and be still. Turn on the recorder and observe the trace. Follow the steps outlined below for the type of recorder you are using.

Unigraph Adjust the stylus heat control so that optimum trace is achieved.

1. Compare the ECG with figures 62.1 and 62.2 to see if the QRS spike is up or down (positive or negative). If it is negative (downward), reverse the leads to the two wrists.

2. After recording for about 30 seconds, stop the paper and study the ECG. Does the wave form appear normal? Determine the pulse rate by counting the spikes between two margin lines and multiplying by 20.

3. Disconnect the leads to the electrodes and have the subject exercise for 2–5 minutes by leaving the room to run up and down stairs or jog outside the building.

4. After reconnecting the electrodes, record for another 30 seconds and compare this ECG with the previous one. Save the record for attachment to the Laboratory Report.

Physiograph® Place the lead selector control at the LEAD I position. With the paper moving, lower the pens onto the recording paper. Make certain that the timer is activated.

1. Place the RECORD button in the ON position and record one sheet of ECG activity.

2. If the pen goes off scale, simply press the TRACE RESET switch on the cardiac coupler and the pen will return to the center line.

3. Place the RECORD button in the OFF position, turn the lead selector control to the LEAD II position.

4. Place the RECORD button back to the ON position and record another sheet of ECG activity.

5. Repeat step 4 above for the LEAD III position on the lead selector control.

6. Stop the recording and compare your results with the sample tracing in figure 62.9.

Laboratory Report

Complete the Laboratory Report for this exercise.

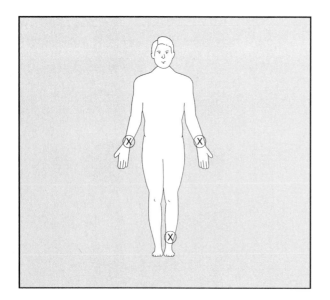

Figure 62.7 Three-lead hookup used for Unigraph setup.

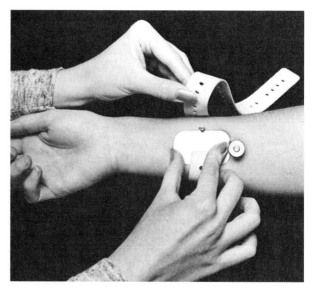

Figure 62.8 After applying electrode paste to electrode, it is strapped in place.

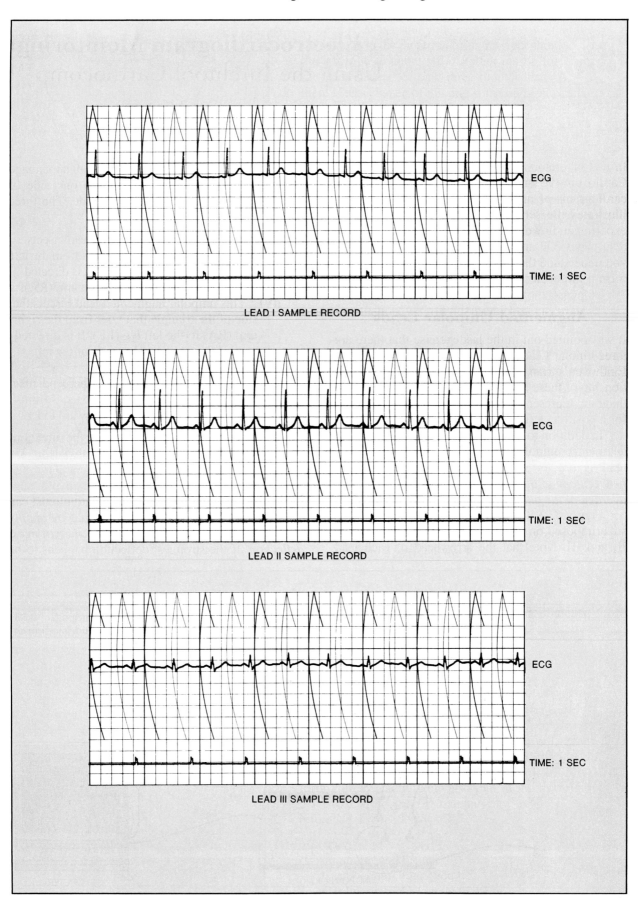

Figure 62.9   Physiograph® electrocardiogram sample recordings at different lead settings.

1. **Balance the ESG coupler** as follows:
   a. Start the chart moving at 0.25 cm/sec and lower the pens onto the paper.
   b. Keep the amplifier RECORD button in the OFF position.
   c. Orient the recording pen to the centerline by using the POSITION control knob.
   d. Start recording by pressing the RECORD button to the ON position (down).
   e. Using the balance control knob on the ESG coupler, return the recording pen to the centerline established in step *c* above.
   f. Check the balance by placing the amplifier RECORD button in the OFF position. If the system is balanced, the pen will remain on the centerline.
   g. If the pen does not remain on the centerline, repeat steps *e* and *f* until balancing is achieved.

2. **Calibrate the coupler** to produce 2.5 cm of pen deflection, which is equivalent to 100 mm Hg. Figure 65.5 is a sample recording of the procedure. Proceed as follows:
   a. Start recording by pressing the RECORD button to the ON position.
   b. With the POSITION control knob, set the recording pen on a baseline that is exactly 2.5 cm below the centerline; i.e., 2.5 cm above the "0" pressure line.
   c. While depressing the "100 mm Hg CAL" button on the coupler, rotate the inner knob of the amplifier sensitivity control until the pen is moved back to the centerline (i.e., 2.5 cm above the 0 pressure line).
   d. Release the CAL button and see if it returns exactly to the 0 pressure line.
   e. Repeat steps *c* and *d* several times to make sure that exact calibration exists.
   Note: The ESG coupler is now calibrated so that each block on the recording paper represents 20 mm Hg pressure and each centimeter of pen deflection is equivalent to 40 mm Hg.

3. *From this point on do not touch* the balance control knob, the POSITION control knob, or the inner knob of the amplifier sensitivity control. Any readjustments of these controls will produce inaccurate results.

4. Place the pressure cuff on the subject's right arm in such a way that the microphone is positioned over the brachial artery. Secure the cuff tightly around the upper arm with the Velcro surfaces holding it closed.

5. **Calibrate the cuff microphone sensitivity** as follows:

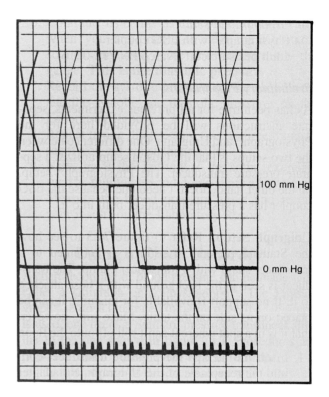

Figure 65.5   Pressure Calibration: Zero pressure line "0" is calibrated so that it is 2.5 cm below the centerline. Each block on the recording paper represents 20 mm Hg; thus, the centerline represents 100 mm Hg.

a. Start the chart moving at 0.5 cm/sec and lower the pens to the paper.
b. Place the RECORD button in the ON position.
c. Close the valve on the hand bulb by rotating it clockwise.
d. With the subject's forearm resting on the table in a relaxed position, pump up the cuff until the recording pen indicates approximately 160 mm Hg (4 cm of pen deflection).
e. Release the bulb valve slightly so that the recorded pressure begins to fall at a rate of approximately 10 mm Hg/sec. Note the Korotkoff sounds on the recording.
f. Taking care not to touch the inner knob of the amplifier sensitivity control, adjust the outer knob of this control so that you get an amplitude of 1–1.5 cm on the recorded sound.
   Usually, a setting of 10 or 20 on the outer knob produces good sound recordings. On some individuals, the setting may be as low as 5 or 2; on others it may be as high as 100.

6. You are now ready to perform your investigation.

Investigative Procedure

Four recordings will be made of the subject's blood pressure under the following conditions: (1) while seated in a chair, (2) after standing for 4 or 5 minutes, (3) while lying down, and (4) after exercising for 5 minutes. Use the following procedure:

With Subject in Chair

1. Start the paper advance at 0.5 cm/sec, and lower the pens onto the chart. Activate the timer.
2. Press the amplifier RECORD button to the ON position.
3. Tighten the pressure bulb valve (turn fully clockwise) and inflate the cuff until you are recording a pressure of approximately 40 mm Hg above the expected systolic pressure. Usually, 160–180 mm Hg is sufficient.
4. Release the hand bulb valve slightly so that the recorded pressure begins to fall at a rate of approximately 10 mm Hg/sec.
5. Watch for the appearance of the **first** and **last Korotkoff sounds** on the recording. These pressures can be considered to be the "normal" systolic and diastolic pressures of your subject.
6. Terminate the recording and refer to Appendix C, illustration A, figure SR-4, for Physiograph® sample record.

With Subject Standing Up

Allow the subject to stand up for 4 or 5 minutes and repeat steps 1 through 6 above while the individual is still standing. Refer to illustration B, figure SR-4, for sample record.

With Subject Reclining

Allow the subject to lie on a laboratory table or cot for 4 or 5 minutes and then follow steps 1 through 6 again to record the blood pressure in this position. Refer to illustration C, figure SR-4 for sample record.

After Exercising

Remove the cuff from the subject and allow him or her to perform 5 minutes of strenuous exercise, such as running, jumping, or climbing stairs.

After exercising, seat the subject in a chair and *immediately* attach the cuff. Record the systolic and diastolic pressures, using the same procedures as above. Remember, it will be necessary to inflate the cuff to a higher pressure (240 mm Hg) to occlude the brachial artery. Refer to illustration A, figure SR-5, for Physiograph® sample record.

Make a series of recordings one minute apart until the pressure has returned to what it was in the original (seated) test. Refer to illustrations B and C, figure SR-5 for sample records.

Laboratory Report

Complete the Laboratory Report for this exercise.

66 A Polygraph Study of the Cardiac Cycle

In Exercises 60, 62, and 64 we performed individual studies of heart sounds, pulse, and the ECG. In this exercise these three cardiac manifestations will be studied, simultaneously, by utilizing three channels on a polygraph to see the relationship of the ECG to the other two phenomena.

What is accomplished by the heart during a lifetime is, too frequently, taken for granted. Consider for example, that at intervals of less than a second, the heart is automatically triggered, pressures are generated, valves are opened and closed, blood flows, and the heart is refilled. This sequence is repeated 100,000 times a day to pump 2000 gallons of blood. This action continues, faithfully and consistently, without stopping, into old age. It is this amazing automatic sequence of events that will be studied here in detail.

The **cardiac cycle** is divided into two major events, systole and diastole; both phases, in turn, are further subdivided into five phases. In showing the relationship between the ECG wave form, heart sounds, and pulse, a review of some of the statements made in Exercise 62 will be necessary.

Five Phases of the Cardiac Cycle

An arbitrary starting point of the cardiac cycle is at the very end of diastole, when the ventricles are nearly filled with blood, the AV valves are still open, and the aortic and pulmonic valves are still closed. Figure 66.1 reveals the relationship of the three polygraph recordings to the cardiac cycle. Note at the top of the diagram are headings identifying the five phases of the cardiac cycle.

Before seeing a demonstration or doing this experiment, read over the following explanation.

Atrial Systole

We learned in Exercise 62 that atrial systole is initiated by a depolarization wave that starts in the SA node. This wave moves over both atria, affecting the right atrium first. Its electrical potential is seen as the **P wave** on the ECG.

Note in figure 66.1 that *no heart sounds are heard during this phase.* Since the ventricles are relaxed here, and the intraventricular pressure is less than the pressure in the aorta or pulmonary trunk, the *pulse wave is not augmented.*

Ventricular Systole (Isometric Phase)

When the depolarization wave reaches the AV node, it moves down the atrioventricular bundle into the conduction myofibers to initiate depolarization of the ventricular myocardium. This depolarization manifests itself as the **QRS complex,** which effectively masks the weaker repolarization wave that takes place in the atria.

Immediately after its depolarization, ventricular contraction takes place, compressing the blood in these chambers. This causes closure of the AV valves, with no significant reduction of ventricular size. This phase is also known as *isometric ventricular contraction.* The closure of AV valves produces a vibration that is conducted through the chest wall and is heard as the **first heart sound.** This event marks the beginning of ventricular systole.

Ventricular Systole (Ejection Phase)

This third phase in the cardiac cycle starts after the conducting system and ventricles have repolarized. The process of ventricular repolarization produces the **T wave.**

Ventricular ejection of blood through the aorta and pulmonary trunk occurs as the ventricular pressures exceed the diastolic pressures in the aorta and pulmonary trunk. It should be noted in figure 66.1 that the pulse pressure is at its lowest level just as the aortic and pulmonary valves open. The up-slope of the pulse pressure corresponds to the elevation of the systolic pressure in the left ventricle as blood is ejected from the heart.

During ventricular ejection the AV valves remain closed while the aortic and pulmonic valves are open. **No heart sounds** are heard during this interval.

Early Diastole (Isometric Ventricular Relaxation)

In this fourth phase, repolarization of the conducting system and myocardium is nearly complete and

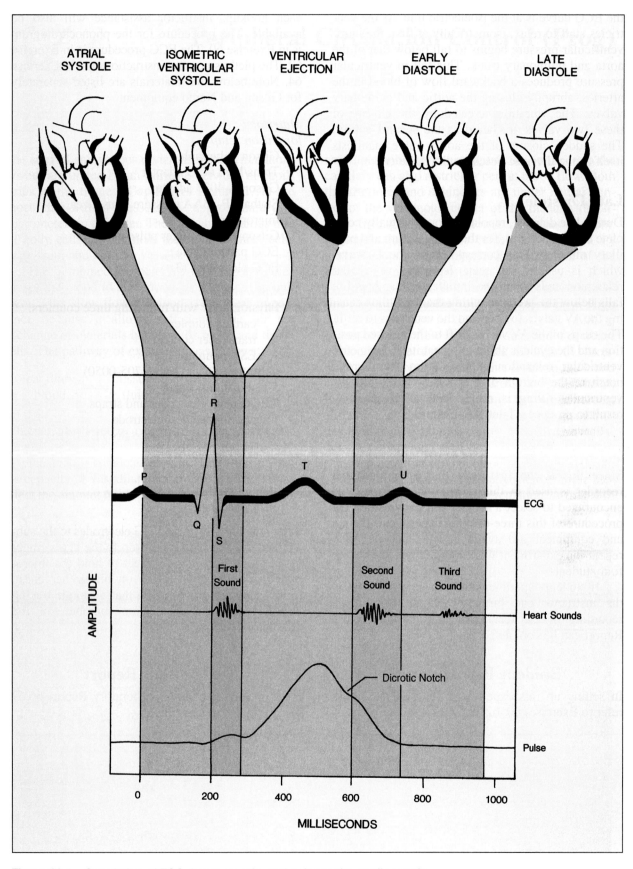

Figure 66.1   Correlation of ECG, heart sounds, and pulse to the cardiac cycle.

the **internal iliac** *(hypogastric)* artery and a larger branch, the **external iliac** artery, that continues on down into the leg.

The external iliac becomes the **femoral** artery in the upper three-fourths of the thigh. Near the origin of the femoral artery, the **deep femoral** branches off and passes backward and downward along the medial surface of the femur. In the knee region the femoral becomes the **popliteal** artery. Just below the knee the popliteal divides to form the **posterior tibial** and **anterior tibial** arteries.

Assignment:
Label the arteries in figure 68.2.

The Venous Division

Figure 68.3 illustrates most of the larger veins of the body. Starting at the heart in figure 68.3, we see the **superior vena cava** emptying into the upper part of the right atrium and the **inferior vena cava** leading into the lower part of the right atrium.

In the neck region are seen four veins: two medial **internal jugular** veins and two lateral **external jugulars.** The internal jugulars empty into the brachiocephalic veins, and the smaller externals empty into the **subclavian** veins, which pass behind the clavicles. The **brachiocephalic** *(innominate)* veins are the short veins that empty directly into the superior vena cava.

Three veins, the basilic, brachial, and cephalic, collect blood in the upper arm. The **basilic** is on the medial side of the arm, and the **brachial** vein lies along the posterior surface of the humerus. The basilic and brachial veins unite to form the **axillary** vein in the armpit region. The **cephalic** vein courses along the lateral aspect of the arm and enters the axillary vein at its proximal end near the subclavian vein.

Between the basilic and cephalic veins in the elbow region is seen the **median cubital** vein. An **accessory cephalic** vein lies on the lateral portion of the forearm and empties into the cephalic in the elbow region.

Blood in the legs is returned to the heart by superficial and deep sets of veins. The superficial veins are just beneath the skin. The deep veins accompany the arteries. Both sets are provided with valves that are more numerous in the deep ones.

The presence of valves in veins can be vividly demonstrated on the back of your own hand. If you apply pressure on a prominent vein on the back of your left hand near the knuckles with the middle finger of your right hand and then strip the blood

in the vein away from the point of pressure with your index finger, you will note that the blood does not flow back toward the point of pressure. This indicates the presence of valves that prevent backward flow.

The **posterior tibial** vein (label 24) is one of the deep veins that lies behind the tibia. It collects blood from the calf and foot. In the knee region the posterior tibial becomes the **popliteal** vein, and above the knee this vessel becomes the **femoral** vein, which in turn, empties into the **external iliac** vein. The external and **internal** iliac (label 10) empty into the **common iliac** vein.

A large superficial vein of the leg, the **great saphenous,** originates from the **dorsal venous arch** on the superior surface of the foot; it enters the femoral vein at the top of the thigh.

Additional veins shown in figure 68.3 that empty into the inferior vena cava are the right gonadal, renal, and hepatic veins. The **hepatic** vein is the short one just under the heart that carries blood from the liver to the inferior vena cava. The two **renals** that drain the kidneys are inferior to the hepatic. Note that the left renal vein has a downward-extending branch, the **left gonadal** *(spermatic* or *ovarian),* which drains blood from the left testis. Note, also, that the **right gonadal** empties directly into the inferior vena cava.

Assignment:
Label the veins in figure 68.3.

Portal Circulation

Figure 68.4 reveals the venous system that constitutes the human *portal circulation.* All the veins shown here that lead from the stomach, spleen, pancreas, and intestine drain into the **portal vein** (label 2), which, in turn, empties into the liver. As a result of this drainage system, all blood from the digestive organs passes through the liver before entering the general circulation. This arrangement enables the liver, with its multiplicity of metabolic functions, to balance out blood composition before allowing the blood to be transported throughout the body.

The large vessel that collects blood from the ascending colon (label 4) and the ileum (label 5) is the **superior mesenteric** vein. It empties directly into the portal vein. Blood from the descending colon (label 7) and rectum is collected by the **inferior mesenteric,** which empties into the **splenic** *(lienal)* vein. The latter blood vessel parallels the length of the pancreas, receiving blood from the pancreas via several short **pancreatic** veins. Near

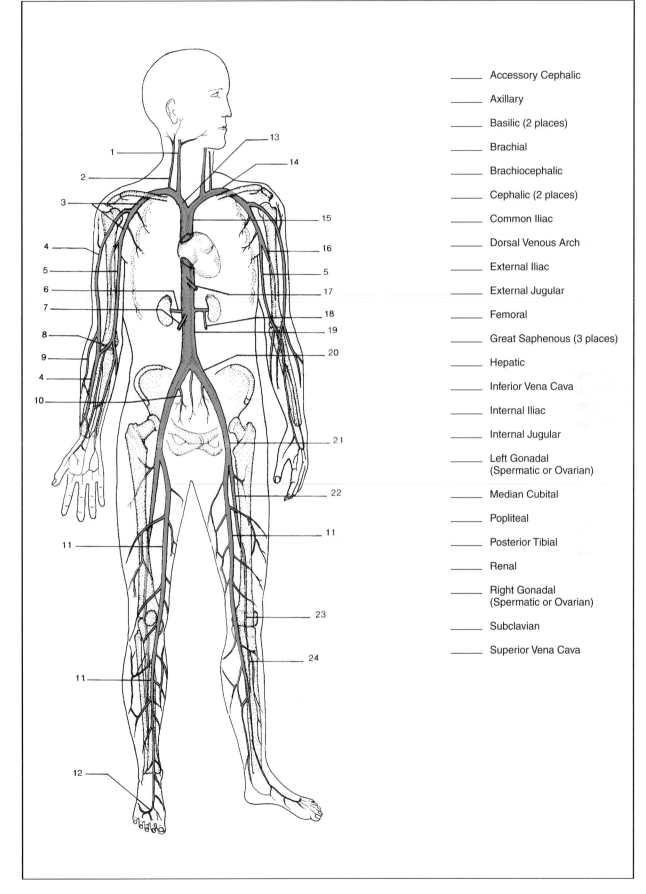

Accessory Cephalic

Axillary

Basilic (2 places)

Brachial

Brachiocephalic

Cephalic (2 places)

Common Iliac

Dorsal Venous Arch

External Iliac

External Jugular

Femoral

Great Saphenous (3 places)

Hepatic

Inferior Vena Cava

Internal Iliac

Internal Jugular

Left Gonadal
(Spermatic or Ovarian)

Median Cubital

Popliteal

Posterior Tibial

Renal

Right Gonadal
(Spermatic or Ovarian)

Subclavian

Superior Vena Cava

Figure 68.3 **Major veins of the body.**

70

The Lymphatic System and the Immune Response

The lymphatic system consists of a drainage system of lymphatic vessels that returns tissue fluid from the interstitial cell spaces to the blood. On its journey back to the blood, the lymph in these vessels is cleansed of bacteria and other foreign material by macrophages in the lymph nodes. In addition to macrophages, the lymph nodes are packed with lymphocytes that play a leading role in cellular and humoral immunity. In this exercise we will explore the genesis and physiology of lymphocytes as well as the overall anatomy of the lymphatic system.

Lymph Pathway

The smallest vessels of the lymphatic system are called *lymph capillaries.* These tiny vessels have closed ends, are microscopic, and are situated among the cells of the various tissues of the body. The lymph capillaries unite to form the **lymphatic vessels** *(lymphatics),* which are the visible vessels shown in the arms and legs in figure 70.1. The irregular appearance of the walls of these vessels is due to the fact that they contain many valves to restrict backward flow of lymph.

As lymph moves toward the center of the body through the lymphatics, it eventually passes through one or more of the small oval **lymph nodes.** Note that these nodes are clustered in the neck, groin, axillae, and abdominal cavity; they are also found in smaller clusters in other parts of the body. They may be as small as a pinhead or as large as an almond.

Two collecting vessels, the thoracic and right lymphatic ducts, collect the lymph from different regions of the body. The largest one is the **thoracic duct,** which collects all the lymph from the legs, abdomen, left half of the thorax, left side of the head, and left arm. Its lower extremity consists of a saclike enlargement, the **cisterna chyli.** Lymph from the intestines passes through lymphatics in the mesentery to the cisterna chyli. The lymph from this region contains a great deal of fat and is usually referred to as *chyle.* The thoracic duct empties into the left subclavian vein near the left internal jugular vein. The **right lymphatic duct** is a short vessel that drains lymph from the right arm and right side

of the head into the right subclavian vein near the right internal jugular.

Lymph Node Structure

The enlarged node shown in figure 70.1 reveals the structure of a typical lymph node. Note that it has a slight depression at its upper end that is called the **hilum.** It is through this depression that blood vessels and an **efferent lymphatic vessel** emerge. Although each node has only one hilum and one efferent lymphatic vessel, there are usually two or more **afferent lymphatic vessels** that carry lymph into the node (two are shown in figure 70.1).

Surrounding the entire node is a **capsule** of fibrous connective tissue, and just inside the capsule is a **cortex** that consists of sinuses reinforced with reticular tissue, and many **germinal centers** *(nodules).* Thirteen ovoid germinal centers are shown in figure 70.1. These centers are the basic structural units of a lymph node. They arise from small nests of lymphocytes or lymphoblasts and reach about one millimeter in diameter. The central portion of a lymph node is called the **medulla.**

Assignment:
Label figure 70.1.

Role of Lymphocytes in Immunity

The immune response to foreign invaders involves the integrated action of an army of different cell types, including monocytes, macrophages, eosinophils, basophils, and lymphocytes. Although each of these cells plays a different role, they interact with each other, even to the extent of regulating each other's activities. The "commander" and predominant "foot soldier" of this defensive force is the lymphocyte.

Although all lymphocytes are morphologically similar, there are considerable differences between individual cells. In addition to being present in blood, lymph, and lymph nodes, they are also seen in the bone marrow, thymus gland, spleen, and lymphoid masses associated with the digestive, respiratory, and urinary passages.

Lymphocytes are subdivided into two subclasses: B-cells and T-cells. The B-cells (*B* for bursa) govern what is called *humoral,* or *antibody-mediated,* immunity. The T-cells (*T* for thymus) are responsible for *cell-mediated immunity.*

Each B-cell is able to recognize only a single foreign antigen, which may be on a bacterium, virus, or some other invader. In other words, all the antibodies on a B-cell are of a single type. Once a B-cell receptor recognizes the foreign antigen, it does two things: it begins to secrete antibodies into the blood, and it begins to multiply very rapidly, increasing the number of identical B-cells. Once these *circulating antibodies* bind to antigens or antigen-bearing targets, they mark them for destruction by other components of the immune system. Since there is a preponderance of this type of lymphocytes in the spleen, it is believed by some that one of the functions of this organ is to generate new B-cells.

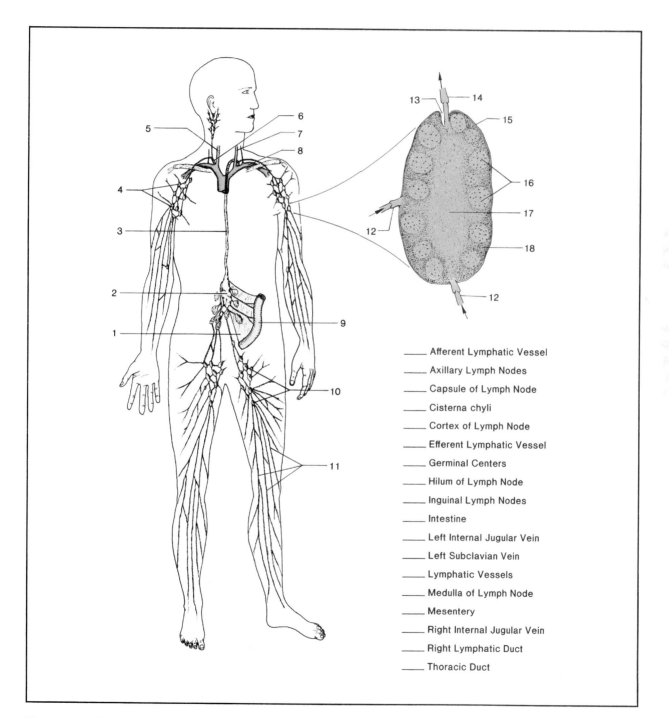

_____ Afferent Lymphatic Vessel

_____ Axillary Lymph Nodes

_____ Capsule of Lymph Node

_____ Cisterna chyli

_____ Cortex of Lymph Node

_____ Efferent Lymphatic Vessel

_____ Germinal Centers

_____ Hilum of Lymph Node

_____ Inguinal Lymph Nodes

_____ Intestine

_____ Left Internal Jugular Vein

_____ Left Subclavian Vein

_____ Lymphatic Vessels

_____ Medulla of Lymph Node

_____ Mesentery

_____ Right Internal Jugular Vein

_____ Right Lymphatic Duct

_____ Thoracic Duct

Figure 70.1 **The lymphatic system.**

373

There are three principal types of T-cells: *helper, suppressor,* and *cytotoxic,* or *"killer," cells.* A T-cell that recognizes and binds to an antigen on the surface of another cell becomes *activated.* An activated T-cell can multiply, and if it is a cytotoxic cell, it can kill the bound cell. For example, cancer cells that display antigens not found on healthy cells can potentially activate T-cells, resulting in their destruction.

The way that cells of the immune system regulate each other is by secreting potent hormones called **cytokines.** If they are produced by lymphocytes they are called *lymphokines.* Hormones produced by monocytes and macrophages are called *monokines.* These hormones differ from hormones of the endocrine gland system in that they act *locally* and do not circulate in the blood.

Lymphocytes, as well as *all* other blood cells, develop from stem cells in the bone marrow before birth and shortly after birth. Stem cells are released into the blood and follow one of two pathways for "processing," as illustrated in figure 70.2.

Cells that settle in the thymus for modification become T-cells. Those that become B-cells are processed in the spleen or some other unknown area. It may well be that they form spontaneously, at random, as suggested by some workers. The reason they were originally named B-cells is that their place of processing in birds was found to be in the *bursa of Fabricius,* a lymphoidal structure on the hindgut. At the present time an equivalent structure in humans may be the spleen.

Once the T-cells have matured in the thymus, they leave the thymus via the blood and colonize in specific loci of the lymph nodes and spleen. T-cells have a long life, possibly years or a lifetime. Once activated by antigens, they are transformed to **lymphoblasts** to perform their specific functions.

Like the T-cells, the B-cells also colonize in specific regions of the lymph nodes and spleen. When B-cells come in contact with the appropriate antigens, they are transformed into **plasma cells** that produce the antibodies specific for the foreign protein.

Laboratory Assignment

Materials:
    prepared slide of a lymph node (H13.11)
    microscope

1. Examine a prepared slide of a lymph node and compare it with the photomicrographs in figure HA-29 of the Histology Atlas. Locate a **germinal center.**
2. Examine the **lymphocytes** under high-dry or oil immersion to see if you can identify cells that are dividing.
3. Make drawings, if required.
4. Answer all questions on Laboratory Report 69,70 that pertain to this exercise.

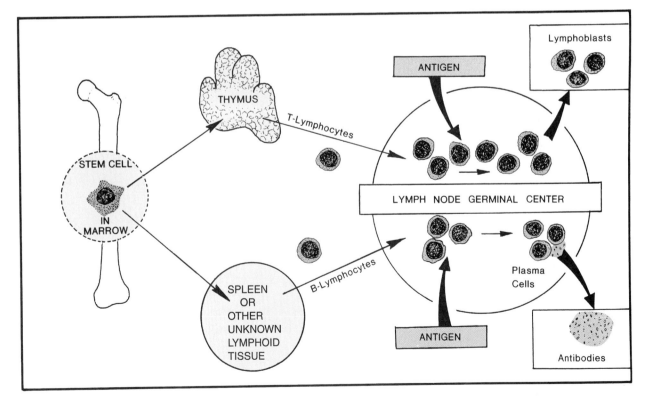

Figure 70.2   The genesis of lymphoblasts and plasma cells.

PART 12

The Respiratory System

This unit comprises seven exercises that cover the anatomy of the respiratory system, hyperventilation, the diving reflex, the Valsalva maneuver, and spirometry. Four of the experiments will involve the use of chart recorders. One experiment utilizes computerized hardware to monitor spirometry experiments. If all the recording experiments are performed in this unit, the protocol for performing the multichannel experiments in Part 13 should be relatively easy to perform.

71

The Respiratory Organs

Both gross and microscopic anatomy of the respiratory system will be studied in this exercise. Although the cat will be the primary dissection specimen here, sheep and frog materials are also available for certain observations. Before any laboratory dissections are performed, however, it is desirable that figures 71.1, 71.2, and 71.3 be labeled.

Materials:
  models of median section of head and larynx
  frog, pithed
  sheep pluck
  dissecting instruments and trays
  Ringer's solution
  bone or autopsy saw

The Upper Respiratory Tract

In addition to breathing, the upper respiratory tract also functions in eating and speech; thus, a study of the respiratory organs in this region must include the organs concerned with those other activities. Figure 71.1 is of this area.

The two principal cavities of the head are the nasal and oral cavities. The upper **nasal cavity,** which serves as the passageway for air, is separated from the lower **oral cavity** by the *palate.* The anterior portion of the palate is reinforced with bone and is called the **hard palate.** The posterior part, which terminates in a fingerlike projection, the **uvula,** is the **soft palate.**

The oral cavity consists of two parts: the **oral cavity proper,** which contains the tongue, and the **oral vestibule,** which is between the lips and teeth.

During breathing, air enters the nasal cavity through the nostrils, or **nares** (*naris,* singular). The lateral walls of this cavity have three pairs of fleshy lobes, the superior, middle, and inferior **nasal conchae.** These lobes serve to warm the air as it enters the body.

Above and behind the soft palate is the **pharynx.** That part of the pharynx posterior to the tongue, where swallowing is initiated, is called the **oropharynx.** Above it is the **nasopharynx.** Two openings to the **auditory** *(Eustachian)* **tubes** are seen on the walls of the nasopharynx.

After air passes through the nasal cavity, nasopharynx, and oropharynx, it passes through the larynx and trachea to the lungs. The cartilages of the larynx (labels 8, 10, and 12) vary considerably in size and histology. The **epiglottis,** which is the uppermost cartilage, is composed of elastic cartilage. Its function is to prevent food from entering the respiratory passages during swallowing. Inferior to the epiglottis is the **thyroid cartilage,** which forms the side walls of the larynx and protrudes externally as the "Adam's apple" in the throat region. Just below the thyroid cartilage is the **cricoid cartilage,** a signet ring-shaped structure. The thyroid and cricoid cartilages are composed of hyaline cartilage.

Note that on the inner wall of the larynx is depicted a **vocal fold,** which is, essentially, the true vocal cord of the larynx. The paired vocal cords are actuated by muscles through the **arytenoid cartilages** (label 5, figure 71.3) to produce sounds. One of these vocal folds exists on each side of the larynx. The vocal folds should not be confused with the **ventricular folds** (label 9, figure 71.1), which lie superior to the vocal folds. The ventricular folds are also known as "false vocal cords." Posterior to the larynx and trachea lies the **esophagus,** the tube that carries food from the pharynx to the stomach.

In several areas of the oral cavity and nasopharynx are seen islands of lymphoidal tissue called *tonsils.* The **pharyngeal tonsils,** or *adenoids,* are situated on the roof of the nasopharynx; the **lingual tonsils** are on the posterior inferior portion of the tongue; and the **palatine tonsils** are on each side of the tongue on the lateral walls of the nasopharynx. These masses of tissue are as important as the lymph nodes in protection against infection. The palatine tonsils are the ones most frequently removed by surgical methods.

Assignment:
Label figure 71.1.
  Study models of the head and larynx. Be able to identify all structures.

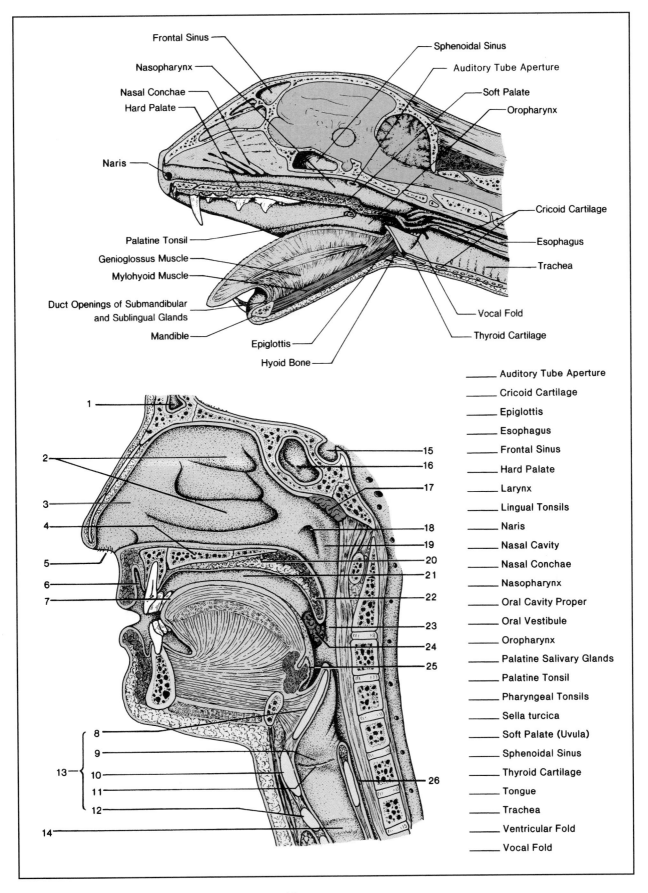

Frontal Sinus
Nasopharynx
Nasal Conchae
Hard Palate
Naris
Palatine Tonsil
Genioglossus Muscle
Mylohyoid Muscle
Duct Openings of Submandibular and Sublingual Glands
Mandible
Epiglottis
Hyoid Bone

Sphenoidal Sinus
Auditory Tube Aperture
Soft Palate
Oropharynx
Cricoid Cartilage
Esophagus
Trachea
Vocal Fold
Thyroid Cartilage

_____ Auditory Tube Aperture
_____ Cricoid Cartilage
_____ Epiglottis
_____ Esophagus
_____ Frontal Sinus
_____ Hard Palate
_____ Larynx
_____ Lingual Tonsils
_____ Naris
_____ Nasal Cavity
_____ Nasal Conchae
_____ Nasopharynx
_____ Oral Cavity Proper
_____ Oral Vestibule
_____ Oropharynx
_____ Palatine Salivary Glands
_____ Palatine Tonsil
_____ Pharyngeal Tonsils
_____ Sella turcica
_____ Soft Palate (Uvula)
_____ Sphenoidal Sinus
_____ Thyroid Cartilage
_____ Tongue
_____ Trachea
_____ Ventricular Fold
_____ Vocal Fold

Figure 71.1   Upper respiratory passages, cat and human.

The Lungs, Trachea, and Larynx

Figure 71.2 illustrates the respiratory passages from the larynx to the lungs. Note that the larynx consists of three cartilages: an upper **epiglottis**, a large middle **thyroid cartilage,** and a smaller **cricoid cartilage.** Figure 71.3 reveals these cartilages in greater detail.

Below the larynx extends the **trachea** to a point in the center of the thorax where it divides to form two short **bronchi.** Note that both the trachea and bronchi are reinforced with rings of cartilage of the hyaline type. Each bronchus divides further into many smaller tubes called **bronchioles.** At the terminus of each bronchiole is a cluster of tiny sacs, the **alveoli,** where gas exchange with the blood takes place. Each lung is made up of thousands of these sacs.

Free movement of the lungs in the thoracic cavity is facilitated by the pleural membranes. Covering each lung is a **pulmonary pleura,** and attached to the thoracic wall is a **parietal pleura.** Between these two pleurae is a potential cavity, the **pleural** *(intrapleural)* **cavity.** Normally, the lungs are firmly pressed against the body wall with little or no space between the two pleurae. The bottom of the lungs rests against the muscular **diaphragm,** the

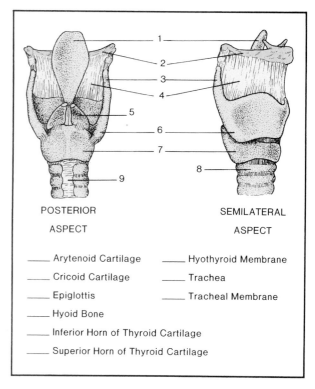

POSTERIOR
ASPECT

SEMILATERAL
ASPECT

_____ Arytenoid Cartilage   _____ Hyothyroid Membrane

_____ Cricoid Cartilage   _____ Trachea

_____ Epiglottis   _____ Tracheal Membrane

_____ Hyoid Bone

_____ Inferior Horn of Thyroid Cartilage

_____ Superior Horn of Thyroid Cartilage

Figure 71.3   Human larynx.

principal muscle of respiration. The parietal pleura is attached to the diaphragm also.

Assignment:
Label figures 71.2 and 71.3.

Cat Dissection

Examination of the upper and lower respiratory passages of the cat will require some dissection. The amount of dissection to be performed will depend on what systems have been previously studied on your specimen. Figures 71.1, 71.4, and 71.5 will be used for reference.

Materials:
    autopsy saw
    embalmed cat
    dissecting tray and dissecting instruments

Upper Respiratory Tract

To identify all the cavities and structures of the cat's head that are shown in figure 71.1, it is necessary to use a mechanical bone saw or an electric autopsy saw to cut the head down the median line. The instructor may designate certain students to make sagittal sections that can be studied by all members of the class. Most specimens will be left intact to facilitate the study of other systems.

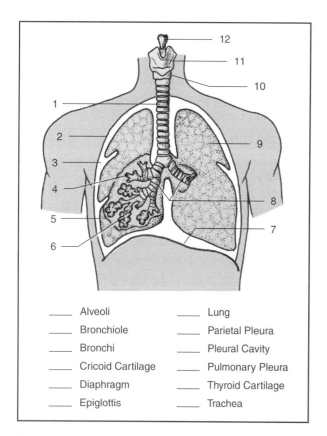

_____ Alveoli   _____ Lung

_____ Bronchiole   _____ Parietal Pleura

_____ Bronchi   _____ Pleural Cavity

_____ Cricoid Cartilage   _____ Pulmonary Pleura

_____ Diaphragm   _____ Thyroid Cartilage

_____ Epiglottis   _____ Trachea

Figure 71.2   The respiratory tract.

If an autopsy saw is used, it is essential that the head of the specimen be held by one student while another student does the cutting. The instructor will demonstrate the procedure. Although an electric autopsy saw is relatively safe because of its reciprocating action, there are certain precautions that must be observed.

The principal rules of safety are: (1) cut only away from the assistant's hands; (2) while cutting, brace your hands against the table for steadiness: don't try to "free-hand" it; and (3) the cutter's hands should never be allowed to tire: frequent rests are essential.

Use the large cutting edge for straight cuts and the small cutting edge for sharp curves. Cut only deep enough to get through the bone; use the scalpel for soft tissues. Once the head has been cut through, it should be washed free of all loose debris. After completing the cutting, identify the structures using figure 71.1 for reference.

1. First, identify the **nares** and **nasal cavity.** As in humans, the nasal cavity lies superior to the palate.

2. Locate the **nasal conchae,** which are shaped somewhat differently in cats than humans.
3. Identify the region designated as the **nasopharynx.** Note that a small **auditory tube aperture** is located in this region. Insert a probe into it.
4. Press against the palate with a blunt probe to note where the **hard palate** ends and the **soft palate** begins.
5. Locate the **oropharynx,** which is located at the back of the mouth near the base of the tongue.
6. Observe that the cat has a very small **palatine tonsil** on each side of the oropharynx.Does it appear to lie in a recess, as is true of the same tonsil in humans?

Lower Respiratory Passages

After identifying the aforementioned structures on the sagittal section, open the thoracic cavity if it has not already been done. If the circulatory system has been studied, it will already be open. The exposed organs should appear as in figure 71.4.

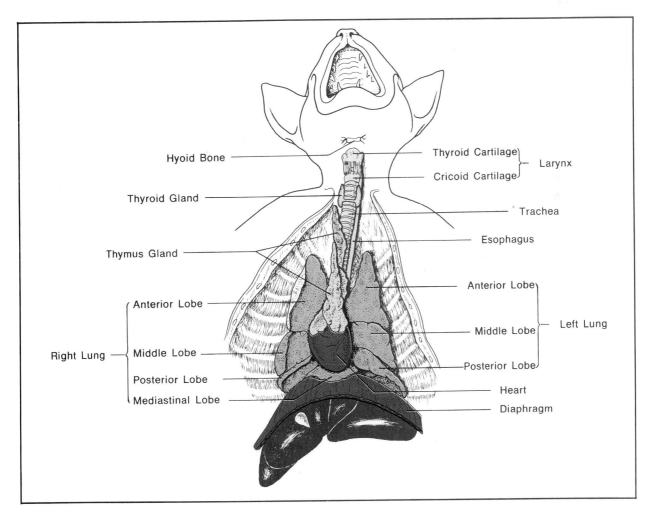

Figure 71.4 Thoracic organs of the cat.

To open the chest cavity, make a longitudinal incision about one centimeter to the right or left of the midline. Such a cut will be primarily through muscle and rib cartilage rather than bone. The incision may be cut partially through with a scalpel and completed with scissors. Avoid damaging the internal organs. Extend the cut up the throat to the mandible.

Continue the incision down the abdominal wall past the liver. Make two cuts laterally from the midline in the region of the liver. Spread apart the thoracic walls and sever the diaphragm from the wall with scissors. To enable the thoracic walls to remain open, make a shallow longitudinal cut with a scalpel along the inside surface of each side that is sufficiently deep to weaken the ribs. When the walls are folded back, the thoracic wall on each side should break at the cut.

Proceed to identify the organs of the lower respiratory passages according to the following procedure. Use figure 71.4 and 71.5 for reference.

1. Examine the larynx to identify the large flap-like **epiglottis**, the **thyroid cartilage**, and the **cricoid cartilage.** Consult figure 71.5.
2. Locate the **arytenoid cartilages** and the paired **vocal cords** on the larynx.
3. Trace the **trachea** down into the **lungs.** In humans the right lung has three lobes and the left lung has only two. How does this compare with the cat?
4. Cut out a short section of the trachea and examine a cross section through the **cartilaginous rings.** Are they continuous all the way around the trachea?
5. Can you see where the trachea branches to form the **bronchi?** Make a frontal section through one lung and look for further branching of the bronchus.

Sheep Pluck Dissection

A "sheep pluck" consists of the trachea, lungs, larynx, and heart of a sheep as removed during routine slaughter. Since it is fresh material rather than formalin-preserved, it is more lifelike than the organs of an embalmed cat. If plucks are available, proceed as follows to dissect one.

Materials:
    sheep plucks, less the liver
    dissecting tray and instruments
    drinking straws or sterile serological pipettes

1. Lay out a fresh sheep pluck on a tray. Identify the major organs, such as the **lungs, larynx, trachea, diaphragm** remnants, and **heart.**
2. Examine the larynx more closely. Can you identify the **epiglottis, thyroid,** and **cricoid cartilages?** Look into the larynx. Can you see the **vocal folds** (cords)?
3. Cut a cross section through the upper portion of the trachea and examine a sectioned cartilaginous ring. Is it similar to the cat in structure?
4. Force your index finger down into the trachea, noting the smooth and slimy nature of the inner lining. What kind of tissue lines the trachea that makes it so smooth and slippery?
5. Note that each lung is divided into lobes. How many lobes make up the right lung? The left lung? Compare with the human lung (figure 71.2).

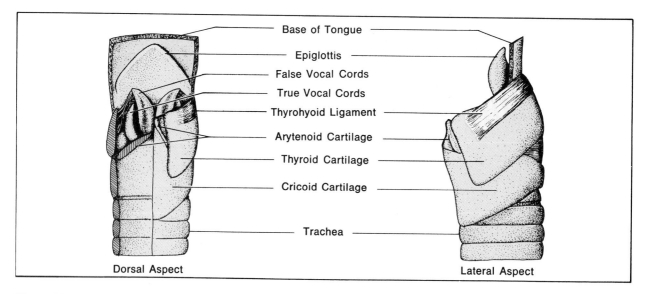

Base of Tongue

Epiglottis

False Vocal Cords

True Vocal Cords

Thyrohyoid Ligament

Arytenoid Cartilage

Thyroid Cartilage

Cricoid Cartilage

Trachea

Dorsal Aspect

Lateral Aspect

Figure 71.5 Larynx of the cat.

6. Rub your fingers over the surface of the lung. What membrane on the surface of the lung makes it so smooth?

7. Free the connective tissue around the **pulmonary trunk** and expose its branches leading into the lung. Also, locate the **pulmonary veins** that empty into the left atrium. Can you find the **ligamentum arteriosum,** which is between the pulmonary trunk and aorta?

8. With a sharp scalpel, free the trachea from the lung tissue and trace it down to where it divides into the **bronchi.**

9. With a pair of scissors, cut off the trachea at the level of the top of the heart.

10. Now, cut the trachea down its posterior surface on the median line with a pair of scissors. The posterior surface of the trachea is opposite the heart.

    Extend this cut down to where it branches into the two primary bronchi. (Observe that the upper right lobe has a separate bronchus leading into it. This bronchus branches off some distance above where the primary bronchi divide from the trachea.)

11. Continue opening up the respiratory tree until you get down deep into the center of the lung. Note the extensive branching.

12. Insert a plastic drinking straw or a serological pipette into one of the bronchioles and blow into the lung.

    If a pipette is used, put the mouthpiece of the pipette into the bronchiole and blow on the small end.

    Note the great expansive capability of the lung.

13. Cut off a lobe of the lung and examine the cut surface. Note the sponginess of the tissue.

14. Record all your observations on the Laboratory Report.

Important: Wash your hands with soap and water at the end of the period.

Frog Lung Observation

To study the nature of living lung tissue in an animal, we will do a microscopic study of the lung of a frog shortly after pithing and dissection. A frog lung is less complex than the human lung, but it is made of essentially the same kinds of tissue. Proceed as follows:

Materials:
    frog, recently double pithed
    dissecting instruments and dissecting trays
    medicine droppers (optional)
    dissecting microscope
    Ringer's solution in wash bottle

1. Pin the frog, ventral side up, in a wax-bottomed dissecting pan.

2. With your scissors, make an incision through the skin along the midline of the abdomen.

3. Make transverse cuts in both directions at each end of the incision and lay back the flaps of skin.

4. Carefully, make an incision through the right or left side of the abdomen over the lung, parallel to the midline. Take care not to cut too deeply. The inflated lung should now be visible.

5. With a probe, gently lift the lung out through the incision. *Do not perforate the lung!* From this point on keep the lung moist with Ringer's solution.

6. If the lungs of your frog are not inflated, insert a medicine dropper into the slitlike glottis on the floor of the oral cavity and pump air into them by squeezing the bulb. Deflated lungs may be the result of excessive squeezing during pithing.

7. Observe the shape and general appearance of the frog lung. Note that it is basically saclike.

8. Place the frog under a dissecting microscope and examine the lung surface.

9. Locate the network of ridges on the inner walls. The thin-walled regions between the ridges represent the **alveoli.**

10. Look carefully to see if you can detect blood cells moving slowly across the alveolar surface. Careful examination of the lungs may reveal the presence of parasitic worms. They are quite common in frogs.

11. When you have completed this study, dispose of the frog, as directed, and clean the pan and instruments. Wash your hands with soap and warm water.

Histological Study

Prepared slides of the nasal cavity, trachea, and lung tissue will be available for study. If Turtox slides are used, the tissues will probably be from monkey or human organs. Tissues from other mammals, such as the mouse or rabbit, may also be available.

Use figures HA-30 and HA-31 in the Histology Atlas for reference, and follow these suggestions in doing your study:

Materials:
    prepared slides of nasal septum (H6.1), trachea (H6.41), and lung tissue (H6.52)

Nasal Septum   Scan this slide first with low power to locate the **nasal septum** and **nasal concha.** Refer to illustration A, figure HA-30 for reference. Identify the bony tissue and nasal epithelium.

Study the ciliated epithelium and the tissues beneath it with high-dry magnification. Look for the **olfactory** (Bowman's) **glands** (illustration B), which are located in the lamina propria. The secretion produced by these glands helps to keep the epithelial surface moist and facilitates the solution of substances that stimulate the olfactory receptors.

Observe that a large number of unmyelinated **olfactory fibers** lie between the glands and the ciliated epithelium. These fibers carry nerve impulses from the olfactory receptors that are dispersed among the cells of the epithelium.

Trachea Identify the hyaline cartilage, ciliated epithelium, lamina propria, and muscularis mucosae. Look for goblet cells on the epithelial layer. Refer to illustrations C and D, figure HA-30.

Lung Tissue Examine with the high-dry objective. Look for the thin-walled **alveoli,** a **bronchiole,** and blood vessels. Consult illustrations A and B of figure HA-30 for normal lung tissue.

Laboratory Report

Complete the Laboratory Report for this exercise.

72 Hyperventilation and Rebreathing

Breathing is a vital activity in which the respiratory muscles are stimulated to increase and decrease the volume of the thoracic cavity to ventilate the alveoli sufficiently to satisfy the exact oxygen requirements of all cells in the body. The center of control for this activity is the respiratory center, which is located in the medulla oblongata. The amounts of CO_2, H^+, and O_2 in the blood and cerebrospinal fluid are the chemical stimuli that act directly, or indirectly, on the respiratory center to regulate the muscles of respiration.

Of the three, the most important factor influencing the respiratory rate is the hydrogen ion (H^+) concentration. The least important factor is the amount of oxygen in the blood. Carbon dioxide facilitates the hydrogen ion concentration by forming carbonic acid in the blood and cerebrospinal fluid.

The pH of blood and tissue fluid during normal breathing is around 7.4. If forced deep breathing (hyperventilation) takes place, the pH of these fluids may be elevated to 7.5 or 7.6 as carbon dioxide is blown off. The reduced hydrogen ion concentration depresses the respiratory center, lessening the desire for increased alveolar ventilation.

Hyperventilation in individuals caused by unconscious deep breathing or sighing, can cause a drop in blood pressure, extreme discomfort, dizziness, and even unconsciousness. All the symptoms are due to the washing out of CO_2 from the blood, causing *alkalosis.* The carbon dioxide depletion can be quickly restored to the blood by **rebreathing** into a paper bag for several minutes. It is these two phenomena that will be studied here.

The symptoms of hyperventilation, and the means of compensating for it, are studied here by two methods: by recording on a Physiograph® or Duograph or by not recording. The nonrecording procedures are presented first. Separate procedures are provided for the Physiograph® and Duograph. If recording equipment is to be used, it is recommended that students do the nonrecording portion as well. Both experiments have unique characteristics.

Nonrecording Experiment

Proceed as follows to observe the characteristics of hyperventilation and to compensate for it with proper rebreathing techniques.

Materials:
    paper bag

1. Breathe very deeply at a rate of about 15 inspirations per minute for 1 or 2 minutes. Do not hurry the rate; concentrate on breathing deeply. Observe that the following symptoms occur:
   • It will become increasingly difficult to breathe deeply.
   • A feeling of dizziness will develop.
   Two events probably account for this: first, the drop in blood pressure due to dilation of the splanchnic vessels lessens blood to the brain; and second, the reduced CO_2 content of the blood causes vasoconstriction of the cerebral blood vessels, further reducing the blood supply.
2. Now, place a paper bag over your nose and mouth, holding it tightly to your face and breathing deeply into it for about **3 minutes.** Note how much easier it is to breathe into the bag than into free air. Why is this?
3. Allow your breathing rate to return to normal and breathe, normally, for 3 or 4 minutes. At the end of a normal inspiration, *without deep inhaling,* pinch your nose shut with the fingers of one hand, and hold your breath as long as you can. Do this three times, timing each duration carefully. Record these times and calculate the average on the Laboratory Report.
4. Now, breathe deeply, as in step 1, for **2 minutes,** and then hold your breath as long as you can. Record your results on the Laboratory Report.
5. Exercise by running in place for **2 minutes,** and then hold your breath as long as you can. Record your results on the Laboratory Report.

Recording Experiment

To produce a record of ventilation during hyperventilation and rebreathing, it will be necessary to use a bellows transducer as shown in figures 72.1 and 72.2. Separate instructions will be provided for the Physiograph® and Duograph. The Duograph is preferable to the Unigraph here for Gilson equipment users since the experiment can be expanded to two-channel recording experiments in the next two exercises. It is conceivable that, where time permits, more than one of these experiments may be performed in a single laboratory period.

Materials:

for Physiograph® setup:
> Physiograph® with transducer coupler
> bellows pneumograph transducer
>   (Narco 705-0190)
> cable for bellows pneumograph transducer

for Duograph setup:
> Duograph
> Statham T–P231D pressure transducer
> Gilson T–4030 chest bellows
> stand and clamp for transducer

for both setups:
> nose clamp
> paper bag

Physiograph® Setup

Set up one channel, using the transducer coupler and bellows, as shown in figure 72.1.

1. Connect the bellows pneumograph transducer to the transducer coupler with a transducer cable. Be sure to match up the yellow dots on the connectors.
2. Balance the recording channel according to the instructions in Appendix B.
3. Attach the bellows pneumograph to the subject as follows:
   a. With the channel amplifier RECORD button in the OFF position, and the valve at the end of the rubber bellows open (turned counterclockwise), place the transducer on the subject's chest and fasten it with the attached leather strap.
   b. Locate the bellows as high as possible where it will get the greatest chest movement during breathing. Note that the transducer cable should be positioned over the shoulder to help support the weight of the transducer.
4. Start the paper advance at 0.25 cm/sec and lower the pens to the recording paper. Activate the timer.
5. Position the recording pen with the POSITION control to a point that is 2 cm below the center line.

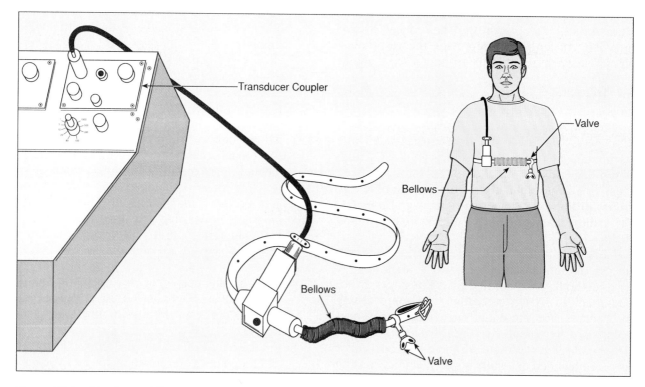

Figure 72.1 Physiograph® setup for monitoring respiration.

6. Close the valve on the bellows, while pinching the bellows slightly. The valve is closed by rotating it completely clockwise.
7. Place the channel amplifier RECORD button in the ON (down) position.
8. Starting at the 1000 position, rotate the outer channel amplifier sensitivity control knob to higher sensitivities (lower numbers) until normal shallow ventilation by the subject gives pen deflections of approximately 2 centimeters. Proceed to perform the following investigations.

Normal Respiration   With the subject seated and relaxed, start the paper moving at 0.25 cm/sec, lower the pens, and activate the timer.

1. Place the amplifier RECORD button in the ON position.
2. Record one page of normal respiratory activity.
3. Increase the paper speed to 5 cm/sec and record several pages of normal respiratory activity.
4. Compare your tracings with sample records A and B in figure SR-6 of Appendix C.

Rebreathing Experiment   Place a nose clamp on the subject and proceed as follows to note the effects of rebreathing into a paper bag.

1. Set the paper speed at 0.25 cm/sec and continue recording.
2. Record about a half page of normal rebreathing, and then place a paper bag tightly over the subject's mouth.
3. Allow the subject to breathe into and out of the bag until visible changes occur in the inspiratory rate and depth.
4. Remove the bag and allow the respiratory rate to return to normal.
5. Compare your tracings with sample records C and D of figure SR-6, Appendix C.

Hyperventilation   While recording at 0.25 cm/sec, have the subject hyperventilate by inhaling and exhaling as deeply as possible for 1½ to 2 minutes.

1. **When the subject begins to feel severe discomfort,** tell the subject to stop the deep breathing and allow the respiratory movements to normalize.
2. During normalization, look for changes that occur in the rate and amplitude of respiratory movements as compared with prehyperventilation recordings.
3. Terminate the recordings.
4. Compare your tracings with the Physiograph® sample records in figure SR-7, Appendix C.

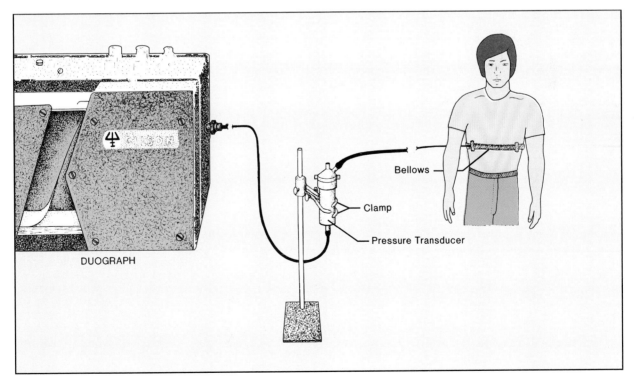

DUOGRAPH

Bellows

Clamp

Pressure Transducer

Figure 72.2   Duograph setup for monitoring respiration.

Duograph Setup

Utilize one channel of the Duograph to hook up a subject according to the following instructions. Figure 72.2 illustrates the setup.

1. Mount the Statham pressure transducer on the transducer stand with a clamp and plug the jack from the transducer into the receptacle of channel 1 on the Duograph.
2. Attach the chest bellows to the subject at the highest point where the greatest respiratory movements occur.
3. Attach the tubing from the bellows to the Statham transducer.
4. Activate the Duograph at slow speed, noting the trace.
5. Adjust the excursion for normal breathing at about 1 centimeter. Proceed as follows to record normal, rebreathing, and hyperventilation activities.

Normal Respiration   With the subject seated and relaxed, make the following recordings of normal respiratory activity.

1. For about 1½ to 2 minutes record normal breathing at the slow speed (2.5 mm/sec).
2. Increase the speed to fast (25 mm/sec) and record for another 1½ to 2 minutes.

Rebreathing Experiment   Place a nose clamp on the subject and proceed as follows to note the effects of rebreathing into a paper bag.

1. Return the paper speed to slow and continue recording.
2. For about 30 seconds, record normal ventilation and then place a paper bag tightly over the subject's mouth.
3. Allow the subject to breathe into and out of the bag until visible changes occur in the respiratory rate and depth.
4. Remove the bag and allow the respiratory rate to return to normal.

Hyperventilation   Continue recording at slow speed and have the subject hyperventilate by inhaling and exhaling, as deeply as possible, at a rate of one respiratory movement per second.

1. **When the subject begins to feel severe discomfort,** tell the subject to stop the deep breathing and continue the recording as the respiratory movements normalize.
2. During normalization, look for changes that occur in the rate and amplitude of respiratory movements as compared with prehyperventilation recordings.
3. Terminate the recording.

Laboratory Report

Complete the Laboratory Report for this exercise.

73

The Diving Reflex

Many aquatic air-breathing animals have evolutionary adaptations that enable them to remain submerged under water for quite long periods of time. The "diving reflex" (or dive reflex) is one such adaptation. In these animals there are prompt cardiovascular and respiratory adjustments that are triggered by partial entry of water in to the air passages. These adjustments include a change in the heart rate, the shunting of blood from less essential tissues, and a cessation of breathing. In humans, these changes are less pronounced, but some adjustments are made. It is this series of changes that has accounted for the survival of some individuals, usually young children, who have fallen through the ice and remained under water for unusually long periods of time.

In an attempt to observe the existence of this reflex in humans, we will monitor cardiovascular changes with a pulse transducer while a subject's face is immersed in cold water. Needless to say, the subject for this experiment should be one who has no aversion to facial submersion.

The equipment setup for this experiment requires two channels on a recorder. To record the respiratory rate on one channel, a chest bellows will be used as in the last experiment (hyperventilation). To detect changes in the heart rate, a pulse transducer will be used on the second channel. Figure 73.1 and 73.2 illustrate the setups for a Physiograph® and a Duograph. Separate instructions are provided for setting up each system.

The immersion maneuver may take place at your laboratory station if large pans are available, or it may be performed in a sink at a perimeter counter. In either case the recording equipment will have to be set up wherever the immersion is to take place.

In this experiment the subject's face will be submerged under water at two different temperatures: 70° F and 60° F (cooled with ice). Before submersion, however, recordings will be made to establish pulse rates during (1) normal breathing and (2) while holding one's breath.

After norms are established, the subject will submerge his or her face in water, first at 70° F and then at 60° F, to note changes in the heart rate.

Hyperventilation will also be performed. The purpose of these tests is to determine if any cardiovascular changes are detectable during submersion, and, to what extent, if any, does temperature affect heart rate.

Equipment Setup

If this experiment is being performed along with Exercise 72, the only additional equipment needed is a pulse transducer.

Physiograph® Setup

By referring to figure 73.1 follow these instructions to hook up the equipment to a subject.

Materials:
Physiograph® with two transducer couplers
bellows pneumograph transducer (Narco PN 705–0190)
cable for bellows pneumograph transducer
pulse transducer (Narco 705–0050)
large pan of water (or sink basin)
ice, towel, thermometer

1. Attach the finger pulse transducer and chest bellows to the subject. As in the previous experiment, locate the bellows high enough to get maximum excursion during breathing.
2. Make all connections in such a way that immersion movement does not tug on the cables.
3. While the subject is being prepared, fill the pan or sink with tap water, adjusting the temperature to 70° F.
4. Turn on the power switch and set the paper speed at 0.25 cm/sec.
5. Start the chart paper moving and lower the pens to the paper. Make certain that the timer is activated.
6. Use the POSITION controls on both channels to position the pens so that they are recording about one centimeter below the centerline.
7. Place the RECORD button in the ON position.
8. Adjust the amplifier sensitivity controls so that the pulse wave amplitude is 2 cm and the tidal ventilation amplitude is 1 cm. You are now

74

The Valsalva Maneuver

Antonio Valsalva, an Italian anatomist, discovered in 1723 that the middle ear could be filled with air if one expires forcibly against closed mouth and nostrils. This action became known as the *Valsalva maneuver*. The term also signifies attempting to expire air from the lungs against a closed glottis. In this maneuver, the abdominal muscles and internal intercostal muscles greatly increase intraabdominal and intrathoracic pressures. This momentary increase in intrathoracic pressure affects both the pulse and blood pressure.

Figure 74.1 illustrates what happens to the blood pressure and pulse in this maneuver. Note that during the maneuver, the blood pressure goes up at the start of the straining, and then it falls a short time later. After the maneuver, the blood pressure increases considerably, and the heart rate slows somewhat.

The initial rise in blood pressure is caused by the application of intrathoracic pressure on the thoracic aorta. The subsequent drop in blood pressure is caused by decreased cardiac output resulting from reduced venous return. The reduced venous return is caused by the restricted flow of blood into the right side of the heart through the compressed venae cavae. The final surge in increased cardiac output is a compensation for the reduced prior output.

The Valsalva maneuver is more than just a stunt to observe in the laboratory. Its real merit is in the diagnosis of two pathological conditions: autonomic insufficiency and hyperaldosteronism.

For example, patients who suffer from *autonomic insufficiency* will exhibit no pulse changes during this maneuver. The cause of this condition is not well understood.

Patients who have *hyperaldosteronism* (excess production of aldosterone by the adrenal cortex) fail to show pulse rate changes and blood pressure rise after this maneuver. This test is often used, diagnostically, on patients suspected of having adrenal cortex tumors that are producing excessive amounts of aldosterone.

Students will work in teams of four or five to monitor blood pressure and pulse of one of the team members during this maneuver. Note that the equipment setups for this experiment are similar to those used in the last experiment, except that the bellows pneumograph transducer will be replaced by a pulse transducer. Figure 74.2 and 74.3 illustrate the setups for the Physiograph® and Duograph.

Caution: Although the Valsalva maneuver is a perfectly safe, normal physiological phenomenon for

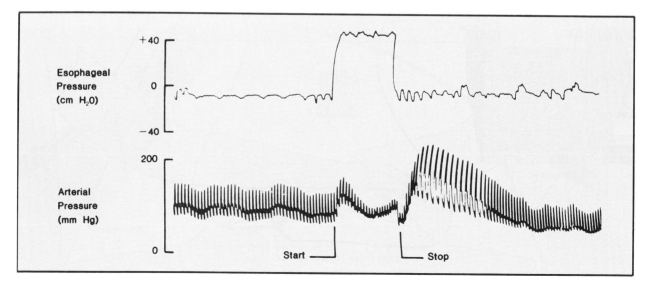

Figure 74.1 Blood pressure and pulse response in the Valsalva maneuver.

healthy individuals, any team member who has a history of heart disease should not be used as a subject.

Equipment Setup

In the last experiment a chest bellows and pulse transducer were used. In this experiment the chest bellows will be replaced with an arm cuff. Make the following hookups and adjustments.

Physiograph® Setup

By referring to figure 74.2, follow these instructions to hook up the equipment to a subject. Note that an ESG coupler replaces one of the transducer couplers that was used in the last experiment.

Materials:

Physiograph® with transducer and ESG couplers
adult pressure cuff (Narco PN 712-0016)
pulse transducer (Narco 705-0050)

1. Hook up the pressure cuff to the ESG coupler according to figure 74.2.
2. Turn on the power switch, and set the paper speed at 0.25 cm/sec.
3. Balance and calibrate the ESG coupler according to the instructions on page 350.

4. From this point on do not touch the balance and position control knobs.
5. Place the cuff on the upper right arm, making certain that the microphone is positioned over the brachial artery.
6. Calibrate the cuff microphone according to the instructions on page 350.
7. Insert the subject's left index finger into the finger pulse pickup, and plug the jack of the pulse transducer into the transducer channel.
8. Start the chart paper moving, again, lower the pens to the paper, and make certain the timer is activated.
9. Adjust the amplifier sensitivity controls on the transducer channel so that the pulse wave amplitude is 1.5 to 2 cm.
10. You are now ready to start monitoring. Proceed to Test Procedure.

Duograph Setup

Set up the equipment according to the following instructions, using figure 74.3 for reference. Note that the same Statham pressure transducer that was used in the last experiment with a chest bellows will be used here with an arm cuff.

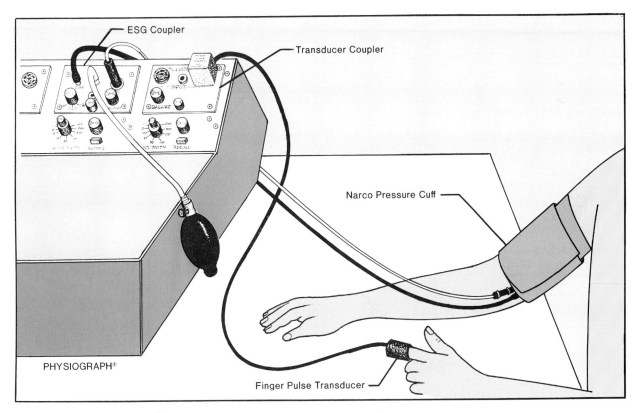

Figure 74.2 Physiograph® setup for monitoring effects of the Valsalva maneuver.

Materials:
>   Duograph
>   Statham T-P231D pressure transducer
>   adult arm cuff
>   stand and clamp for T-P231D transducer
>   Gilson T-4020 finger pulse pickup
>   Gilson A-4023 adapter for pulse pickup

1. Place the cuff on the upper right arm, securing it tightly in place. Attach the tubing from the cuff to the Statham transducer.
2. Test the cuff by pumping some air into it and noting if the pressure holds when the metering valve is closed on the bulb. Release the pressure for the comfort of the subject, while other adjustments are made.
3. Insert the subject's left index finger into the finger pulse pickup.
4. Plug in the Duograph and turn on the power switch.
5. Set the speed control lever in the slow position and the stylus heat control knobs of both channels in the two o'clock position.
6. Set the Gain for channel 1 at 2 MV/CM and for channel 2 at 1 MV/CM.
7. Turn the sensitivity control knobs of both channels completely counterclockwise (lowest values).
8. Set the Mode for channel 1 on DC and for channel 2 on TRANS.
9. Calibrate the Statham transducer on channel 2 by depressing the TRANS button and, simulta-

neously, adjusting the sensitivity knobs to get 2 cm deflection. The calibration line represents 100 mm Hg pressure.
10. Adjust the sensitivity on channel 1 so that you get 1.5 to 2 cm deflection on the pulse.
11. You are now ready to start monitoring. Proceed to Test Procedure.

Test Procedure

Now that the equipment is properly adjusted and activated, proceed as follows:

1. Pump air into the cuff and establish a baseline for one minute.
2. Have the subject take a deep breath and perform a Valsalva maneuver, exerting as much internal pressure as possible.
3. Mark the chart at the instant the maneuver is begun. On the Duograph a mark can be made by pressing the event marker button and holding it down until the maneuver ends.
4. Repeat the maneuver several times to provide enough chart material for each team member.
5. Repeat the experiment, using other team members as subjects.

Laboratory Report

Complete the last portion of combined Laboratory Report 73, 74.

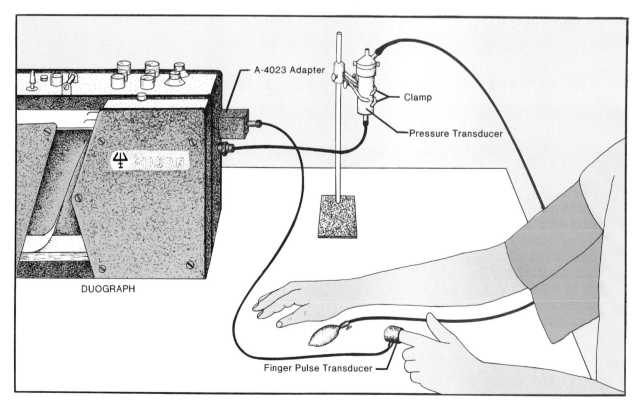

Figure 74.3   Duograph setup for monitoring the effects of the Valsalva maneuver.

75

Spirometry:
Lung Capacities

The volume of air that moves into and out of the lungs during breathing is measured with an apparatus called a **spirometer.** Two types are available: the hand-held (Propper) type shown in figure 75.2 and the tank-type recording spirometer (figure 76.1). When a recording spirometer is used, a **spirogram** similar to figure 75.1 is made. The Propper spirometer is designed, primarily, for screening measurements of vital capacity.

In this exercise we will use the Propper spirometer to determine individual respiratory volumes. Although this convenient device cannot measure inhalation volumes, it is possible to determine most of the essential lung capacities. While most members of the class are working with Propper spirometers, some students can be working with a recording spirometer as directed in Exercise 76.

Proceed as follows to measure or calculate the various respiratory volumes using a Propper spirometer.

Materials:
    Propper spirometer
    disposable mouthpieces (Ward's Nat'l Science
      Est., Catalog No. 14W5070)
    70% alcohol

Tidal Volume (TV)

The amount of air that moves into and out of the lungs during a normal respiratory cycle is called the *tidal volume.* Although sex, age, and weight determine this and other capacities, the average normal tidal volume is around 500 ml. Proceed as follows to determine your tidal volume.

1. Swab the stem of the spirometer with 70% alcohol and place a disposable mouthpiece over the stem.
2. Rotate the dial of the spirometer to zero as illustrated in figure 75.2.
3. After three normal breaths, expire three times into the spirometer while inhaling through your nose. Do not exhale forcibly. **Always hold the spirometer with the dial upward.**
4. Divide the total volume of the three breaths by 3. This is your tidal volume. Record this value on the Laboratory Report.

Minute Respiratory Volume (MRV)

Your *minute respiratory volume* is the amount of Ptidal air that passes into and out of your lungs in one minute. To determine this value, count your

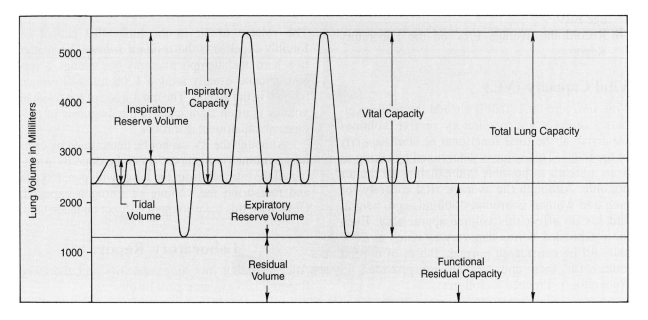

Figure 75.1   Spirogram of lung capacities.

78 Response to Physical Stress:
Monitoring Physical Fitness Training

An increasing awareness of the importance of physical fitness has been apparent for some time. Sedentary life-styles, overeating, and the lack of exertion in our daily activities have caused more and more individuals to take up weight lifting, jogging, bicycling, tennis, and other activities to pursue a program of physical fitness building. Although much can be said for the benefits of these activities to the individual, there is, generally, a lack of scientific evaluation of the accompanying improvement in physical well-being.

In many instances, the only barometer for one's improvement in a physical fitness program is a sense of well-being. The individual feels good, and, therefore, has an incentive to continue with the program. What is needed is a convenient, inexpensive means of scientifically monitoring physiological changes that occur in such a program. It is the goal of this exercise to provide a relatively simple means of evaluating an exercise program to see if beneficial physiological changes are actually taking place.

In general, exercise testing is divided into two categories: *recovery tests* and *effort tests*. In the former case, data are acquired immediately after exercise; in the latter, the data are compiled during exercise from a number of different ergometers. The testing program in this exercise is based on recovery tests, operating on the assumption that *quicker recovery signifies improvement of physical fitness.*

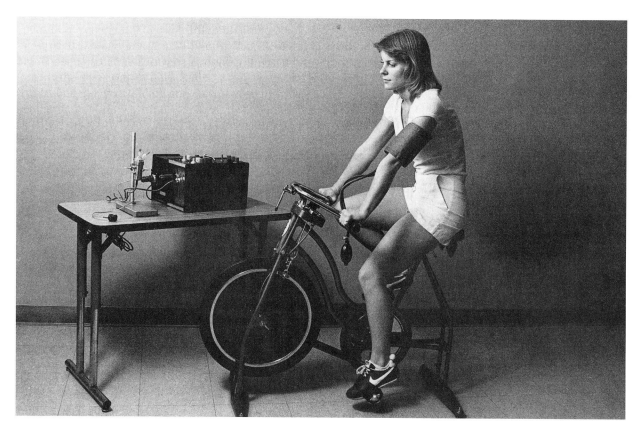

Figure 78.1 A subject who has been involved in a physical fitness program can periodically monitor her recovery rates of blood pressure and heart rate on a Duograph (or Physiograph®) after exercising on an ergometer such as a stationary bicycle.

The program outlined here is intended to last over a period of ten to sixteen weeks—normally, the time of an entire quarter or semester. The procedure will be as follows: first, certain tests and a questionnaire will establish the subject's qualifications for the experiment; second, anthropomorphic data such as height, weight, limb dimensions, etc., will be recorded; and finally, blood pressure and heart rate measurements will be made of the subject while at rest and after a controlled exercise session in the laboratory.

These preliminaries will be followed by weekly monitoring the pulse rate and blood pressure as the exercise program proceeds. Careful records will be kept during the entire period of time, and all data will be plotted on graphs to note the progress.

The results of this experiment will depend on the length of the training period, the extent of the training, the level of fitness that exists at the start, whether or not weight reduction is included, the presence or absence of smoking, and many other factors. Since physical training tends to *reduce the heart rate* and to *promote faster recovery to resting levels after exercise,* it will be evidence of these two changes that will be looked for.

Test Parameters

The assessment of one's level of physical fitness is probably best determined by taking measurements of the metabolic rate under a variety of environmental conditions. Unless a well-equipped research facility is available, this type of testing cannot be done. To be able to measure oxygen consumption, blood gases, blood lactate, work levels performed, etc., one must have some rather sophisticated and expensive equipment, to say nothing of the technical skill to operate it.

Fortunately, numerous studies reveal that **heart rate** bears a linear relationship to oxygen consumption. Since the correlation between heart rate and aerobic capacity is in the range of 50%–90%, we have in the heart rate a simple test parameter that the average person can use to monitor the degree of physical fitness.

Closely associated with the heart rate is **blood pressure,** another easily monitored parameter. For our tests here, we will use a finger pulse pickup and Statham pressure transducer with a Duograph as shown in figure 78.1. If a Physiograph® is used one ESG coupler with arm cuff and one transducer coupler with pulse pickup will be used. These setups were used in Exercise 74 when doing the Valsalva maneuver experiment.

To get meaningful measurements with this setup, it will be necessary to carefully control the following six variables.

1. Time of Day. Humans exhibit circadian rhythms to varying degrees that can influence pulse rate and body temperature. To minimize this variable, tests should be made at about the same time of day, preferably before meals.

2. Ambient Temperature. Increased temperature imposes an additional demand upon cardiac output needed to dissipate the heat load. Thus, higher readings could be expected. Conducting the test in an air-conditioned building can simplify matters. Light clothing and comfortable shoes are essential.

3. Diet and Mealtimes. Increased metabolism after a heavy meal tends to elevate heart rate and ventilation for over an hour. Conversely, fasting diminishes blood sugar levels so that performance may be reduced. Testing one hour to one and one-half hours after breakfast may be a suitable compromise. Stimulants such as tea, coffee, and chocolate should be avoided.

4. Medications. Drugs such as antihistamines or decongestants tend to influence cardiovascular function and should be avoided for at least four hours before testing. If medications must be taken, the fact should be entered into the records.

5. Rest. Unusually strenuous workouts should be avoided on the day before testing. The reasons for this precaution are related to peripheral sequestration leading to reduced central blood volume and stiffness of the affected musculature. Furthermore, depletion of muscle glycogen stores reduces the capacity for sustained muscular effort.

6. Anxiety. Anxiety is a natural subtle complication that frequently accompanies testing because the subject may be self-conscious about poor results. Apprehension is often responsible for seriously high readings, even in individuals who are outwardly calm.

To minimize the effects of anxiety, the participants should become accustomed to the procedure by undertaking several practice trials. Single readings should be interpreted cautiously because even in normal, healthy individuals, the ECG, heart rate, and blood pressure values are sometimes erratic.

Ergometric Standardization

The type of exercise employed in this physical fitness program will vary considerably from student to student. Although no rules will be established here as to how to proceed, the activity should be one that is enjoyable. People who jog, lift weights, or play tennis persist in these activities because they derive great personal satisfaction from the sport.

The use of ergometers in the laboratory for making recovery measurements will be used in a standardized way by all participants. Whether the ergometer is a stationary bicycle or a specific flight of stairs, the standardization is based on the premise that the participant will achieve heart rates of 120 ± 5 beats per minute following a 2–5 minute exertion. An exception to this ruling would be the exceptionally physically fit student whose heart rate is very low to begin with. In such a case, pulse rates of 100 ± 5 beats per minute will suffice.

Experimental Procedures

This lengthy experiment is divided into four phases: (1) screening, (2) anthropomorphic measurements, (3) recovery tests, and (4) final evaluation. Your instructor may decide to modify several aspects of each portion of the experiment. Time and the availability of equipment will determine exactly how the experiment proceeds.

Materials:
> tape measure
> scale for body weight
> for ECG see Ex. 62
> for heart rate and blood pressure see Ex. 74
> for urine analysis see Ex. 86

Screening (First Laboratory Period)

The performance of this experiment is voluntary and will be performed on the student's own time. Since the details of the exercise program are completely student-oriented without instructor control or responsibility, all training will be performed outside of the jurisdiction of this course. However, if the student has any physiological problem that could be life-threatening, the experiment must not be performed. It is the student's responsibility to have medical clearance to enter any type of physical fitness training that will be monitored in this course. As a screening tool, the following tests will be performed in the first laboratory period.

Since many of these tests are difficult to perform without help, work with a laboratory partner when necessary.

Electrocardiogram Work with your laboratory partner to record each other's ECG, following the instructions in Exercise 62. Make two recordings, one before exercise and one right after exercise. Attach both recordings to Screening Report 78S in the back of this manual.

Heart Rate and Blood Pressure With a finger pulse pickup and arm cuff make recordings of the heart rate and blood pressure before and after exercise. See Exercise 74.

Urine Analysis Do a urine analysis according to instructions in Exercise 86, testing for bilirubin, urobilinogen, blood, ketones, glucose, and protein. Check the pH also. Record the tests on table II of Screening Report 78S.

After completing the above tests, answer all questions on Screening Report 78S. Read the "Release" paragraph and sign the report. Turn it in to your instructor for approval.

Anthropomorphic Measurements

The recording of anthropomorphic dimensions before and after the experimental period may reveal physical change improvements in addition to recovery improvement. The following suggestions concerning each measurement will provide for standardization.

Height Subject stands barefoot with back to wall looking straight ahead. Measurement is made to the closest one-half centimeter.

Weight Subject is weighed without shoes, wearing light clothing. Weigh to nearest 0.1 kg.

Chest Measure chest circumference at end of normal expiration. Tape is applied to bare chest at nipple level in males; highest point under breasts in females.

Abdomen Measure maximum circumference of bare abdomen at end of normal expiration.

Thigh Measure the *maximum* circumference of the bare thigh while standing.

Calf Measure the *maximum* circumference of the bare right calf while in the standing position.

Arm Measure the *maximum* circumference of the upper arm while extended and relaxed.

Vital Capacity Determine with Propper or wet spirometers your vital capacity. See Exercise 75 or 77.

Recovery Tests

Whether recovery tests will be performed once a week, every two weeks, or once a month depends on how the trainee views the overall experiment. An advantage of frequent testing is that recovery progress can be used to increase the levels of exertion. For example, if the trainee feels that recovery is slower than desired, it is possible to experiment with a variety of stress-increasing activities to see if recovery rates can be improved.

The type of ergometer used will depend on the kind of equipment that is available. Steps, treadmills, arm cranks, and stationary bicycles are most frequently used. Each type of ergometer has its advantages and disadvantages. An elaborate setup is not required here since all we are concerned with is exerting enough effort to get the pulse rate up to 120 ± 5 beats per minute (100 ± 5 for conditioned athletes). Experimentation will reveal how much exercise (usually 3–5 minutes) will be enough to produce this heart rate.

Before exercising, record your blood pressure and pulse rate on the chart recorder (slow speed), taking into consideration the six controlling factors mentioned previously (time of day, temperature, relation to mealtime, medications, rest, and anxiety). Record your results in the table on Laboratory Report 78.

Before exercising, disconnect the tubing to the arm cuff and remove the finger pulse pickup. Leave the cuff on the arm, however. Exercise for 3–5 minutes with a stationary bicycle, flight of stairs, or whatever is available to get the desired pulse rate. Hook up to the Duograph or Physiograph® again as soon as possible, and record for 2–3 minutes. Record the results on the table on Laboratory Report 78.

Final Evaluation

After performing the last recovery tests and recording the data, proceed as follows:

1. **Anthropomorphic Measurements** Repeat all the anthropomorphic measurements that were made during the first week of the experiment, recording them in the table on the Laboratory Report in Part B. Compare the data and record any significant changes.
2. **Graphs** Plot all the data from the tables onto the appropriate graphs.
3. **Final Statement** Write a few paragraphs on the Laboratory Report, summarizing the results of the experiment. If results seem negative, attempt to interpret what went wrong.

Laboratory Report

Complete Laboratory Report 78.

79

Response to Psychological Stress:
Lie Detection

All of us are familiar with the physiological effects of attempting to conceal the truth. Surely, everyone has experienced sensations of increased heart rate, rush of blood to the face, uncontrollable impulse to swallow, and other symptoms that occur during attempted deception. When we feel that the lie has been successfully accomplished, isn't there often a sigh of relief and heavier breathing?

It is a fact, of course, that some individuals are better at deception than others. They are more successful either because they are not very disturbed by lying or because they have better cortical control over the responses of the autonomic nervous system. Nevertheless, even among the best deceivers, there exist autonomic responses that cannot be masked. It is these physiological aberrations that

form the basis for the success of the lie detector (polygraph) in criminal investigation.

To demonstrate the mechanics of lie detection we will select a subject from the class to monitor five physiological parameters, as illustrated in figure 79.1. When the experiment has been completed, students may be allowed to work in small groups to repeat the process, trying various techniques.

Considerable cooperation will be required by all observing class members or the experiment will not succeed. Ideally, the experiment should be performed in a quiet room, completely lacking in distraction. For obvious reasons, the presence of observers does not add to test reliability. Extraneous noises, such as subdued conversation or telephone ringing, can induce distorted recordings. Even seemingly insignificant objects, such as pictures on the wall or

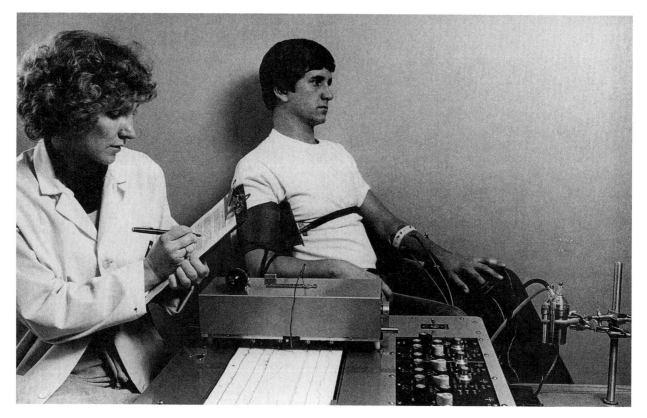

Figure 79.1 The five physiological responses monitored in lie detection are (1) blood pressure (cuff on right arm) (2) respiration (chest bellows), (3) heart rate (finger pulse pickup on index finger of left hand), (4) GSR (fourth finger of left hand), and (5) EMG (left forearm).

room ornaments, can either distract the attention of the subject or permit him or her to use them as a focus for practicing evasion. So it is obvious that the setting for this experiment must be very carefully controlled.

In addition to minimizing the amount of distraction caused by the audience, success of this experiment will depend on (1) skilled interrogation, (2) proper subject preparation, and (3) correct equipment hookup. The following points on each of these topics should be observed.

Role of the Interrogator

The interrogator's demeanor plays a very significant role. The interrogator should maintain a noncommittal expression and ask questions in an even, well-modulated tone free of insinuation, incredulity, and provocation. Distracting activity, such as making notations, organizing materials, and operating the polygraph should be kept to a minimum.

During the test the interrogator should sit to one side of the subject—not in front of him or her. If the subject is not a class member and is not acquainted with the polygraph and the procedure, the interrogator should answer any questions by the subject to satisfy curiosity or allay anxiety.

The Subject

Position, comfort, and lack of movement are the prime considerations for the subject. The subject should be seated in a comfortable armchair, facing neither the audience nor the interrogator. Ideally, the subject should face a blank wall of neutral color and moderate lighting. He or she should not face a window.

Binding straps and wire leads should be judiciously placed to permit maximum comfort. Since the pressure cuff on the arm will cause the greatest discomfort, it should be inflated only while recording and always deflated during resting periods between tests.

The subject should refrain from coughing, sniffing, or clearing the throat to prevent distortion of breathing traces. He or she should look straight ahead, never turning the head, moving the arms, or shifting body position. He or she should remain still, but not rigid; any movement will cause artifacts while recording.

Performing the Test

Since this lie detection demonstration will not pertain to the actual commission of a crime, we are denied the benefit of heightened anxiety as might be expected with an actual criminal suspect. Nevertheless, the subject in this demonstration is inclined to *beat the machine,* and the challenge for the observers is to detect lies by skillfully formulating questions and interpreting the results.

The protocol that follows will utilize regular playing cards. The subject will select a card from a group of eight cards and show the card to two witnesses, but not to the examiner. The examiner will mix the cards and show them one at a time to the subject, asking the subject each time a card is shown to deny that the card is the one originally picked. As the denials are made, the five parameters are monitored. The examiner may then tell the subject and the class which card the subject had lied about. Class members and subject will study the tracings to see what polygraph evidence supported the interrogator's decision.

The five-channel polygraph will be used to record respiration, blood pressure, heart rate, galvanic skin response (GSR), and muscle tension. Except for the GSR, all of these measurements have been made in previous experiments. Note that separate lists of materials are provided for the Gilson and Narco polygraphs.

Three steps in the procedure involve (1) prequestioning, (2) subject preparation, and (3) monitoring. Proceed as follows:

Materials:
    armchair
    electrode paste
    Scotchbrite pad
    self-adhering wrap
    8 nonface cards from deck of playing cards

for Gilson Polygraph (5-channel) setup:
    2 Statham T-P231D pressure transducers
    pneumograph
    blood pressure cuff
    finger pulse pickup
    Gilson A-4023 adapter for pulse pickup
    Gilson E-4041 GSR electrodes
    Gilson A-4042 adapter for GSR electrodes
    3 ECG plate electrodes
    cable for muscle (EMG) electrodes
    2 clamps for Statham transducers
    1 stand for mounting Statham transducers

for Narco Physiograph® setup:
    2 transducer couplers on Physiograph®
    1 EEG/EMG coupler on Physiograph®
    1 GSR coupler on Physiograph®
    1 cardiac coupler on Physiograph® (optional)
    pneumograph transducer (Narco 705–0190)
    cable for pneumograph transducer
    finger pulse pickup (Narco 705–0050)
    cable for EEG/EMG coupler
    GSR electrode kit (Narco PN 710–0008)
    3 ECG plate electrodes and straps

Prequestioning the Subject

Before the subject is connected to the polygraph, the interrogator should talk privately with the subject to get enough personal information about the subject to enable the interrogator to prepare ten simple questions for establishing a control baseline. The questions will be phrased in such a way that the answer is either "yes" or "no." A typical list of questions is as follows:

1. Is your full name *Mary Ann Jones?**
2. Do your friends call you *Peggy?*
3. Did you have a cup of coffee this morning?
4. Do you drive a *Toyota* to school?
5. Your real name is *Hillary Rodham Clinton,* isn't it?
6. Do you smoke?
7. Did you *play tennis* yesterday?
8. Isn't your favorite dessert a *hot fudge sundae?*
9. You're wearing a *red pair of shoes,* aren't you?
10. Today is *Thursday,* isn't it?

*Italics represent words or phrases that can be changed to match the subject.

Subject Preparation

Seat the subject in the armchair and attach the pneumograph, finger pulse pickup, EMG plate electrodes, and blood pressure cuff to the appropriate parts of the body. To refresh your memory on details of the various hookups, consult the following exercises for information:

Pneumograph—Exercise 72.
Finger pulse pickup—Exercise 64.
EMG hookup—Exercise 32.
ECG hookup—Exercise 62.

Observe the following modifications to the setups as described in the various exercises.

Pneumograph   Note in figure 79.1 that two Statham transducers are mounted on a single stand. One of these transducers is for the pneumograph and the other is for the arm cuff on Gilson setups.

Blood Pressure   Cuff pressure should be adjusted to maintain the dicrotic notch in the midposition on the downslope of the pulse, and to achieve this, about 90 mm Hg will be needed. If a lower pressure still produces a high-amplitude tracing, reduce the cuff pressure even further. **Keep the cuff deflated until just before recording and release the cuff pressure as soon as recording is completed.** Keep in mind that it is this parameter that causes most subject discomfort.

Heart Rate   The finger pulse pickup is used for monitoring the heart rate. If the Gilson polygraph has a cardiotachometer module (IC-CT) it will not be necessary to use the A-4023 adapter. The adapter is used only with IC-MP modules.

Muscle Tension   Note in figure 79.1 that muscle tension is monitored by using three plate electrodes on the left forearm. They can also be attached to the calf of either leg. On the Gilson polygraph one can use either the IC-EMG or the IC-MP module for this parameter. On the Physiograph® you should use the EEG/EMG coupler.

GSR Electrodes   Attach the GSR electrode to the fourth finger of the left hand as shown in figure 79.2. Before attaching it, scrub the skin with a Scotchbrite pad and apply a light film of electrode paste to the surface. The electrode can be secured to the finger with self-adhering wrap or Scotch tape.

Figure 79.2   GSR electrodes and finger pulse pickup on the left hand.

Since the GSR is a comparatively slow response, it should be connected directly to the Servo channel. On a Gilson polygraph, a 4042 adapter is needed.

The galvanic skin response measures changes in conductivity of a low DC voltage on the skin. Skin conductivity is affected by blood flow and perspiration, which are regulated by the autonomic nervous system. Detection of emotional stress shows up clearly with this parameter.

Establishing a Control Baseline

1. Pump air into the arm cuff and turn on the polygraph. The speed should be set at 2.5 mm per second.
2. Check the trace on all channels to make certain all are adequate.
3. To establish the appearance of artifacts, ask the subject to do the following as you record on the chart each activity:
   - Cough.
   - Move the arm with the pressure cuff.
   - Move the other arm.
   - Shift position in his or her chair.
4. To establish a baseline for truthful answers, ask the subject the ten questions that must be answered *truthfully*. Record on the chart not only the number of the control question, but + for "yes" and − for "no."
5. Deflate the cuff and allow the subject to rest for 5 minutes.

Deception Test No. 1

We are now ready to test for deception. Proceed as follows with 8 nonface cards of different suits from a deck of playing cards. Two witnesses will be used. One will handle the cards and reveal them to the subject and other witnesses.

1. *Examiner to subject:* "A witness will select one of the playing cards and show it to you and the other witness. Don't show it to me or the audience."
2. Cards are shuffled by witness A and fanned out in front of the subject, *face down*. Witness B picks one card. All three look at it. Witness A returns it to the other 7 cards, mixes them up, and places all 8 cards in a pile in front of subject.
3. Witness A turns top card face up for all four to see.
4. *Examiner to subject:* "Did you select the ____?" (Card is identified by number and suit, such as *three of diamonds*.) Subject will answer with a "no" as the card is shown.
5. Examiner marks card and suit on the chart.
6. Twenty seconds elapse and then the second card is turned over.
7. Subject is asked the same question about the second card. Subject answers "no."
8. Every 20 seconds a card is turned over, question is asked, and the chart is marked until all 8 cards have been exposed.

9. Now the cards are shuffled by witness A and exposed one at a time again, only this time the subject answers "yes" to each question. Chart is marked accordingly.
10. Finally, cards are shuffled and exposed a third time in the same manner, except that the *subject remains silent*.
11. Polygraph chart is stopped, arm cuff pressure is released, and the chart is studied. On the basis of tracing analysis, the examiner tells the subject which card had been picked. The entire class examines the chart.

Deception Test No. 2

Another way in which the same playing cards may be used to detect lying is to employ a *peak of tension test*. The rationale behind this stratagem is to lead the subject progressively to a climax of anxiety in which the greatest autonomic response presumably betrays the truth. In this case the subject answers "no" to all questions asked. The procedure is as follows:

1. At 20-second intervals, ask the subject:
   "Did you select a card of spades?"
   "Did you select a card of diamonds?"
   "Did you select a card of hearts?"
   "Did you select a card of clubs?"
2. While step 1 is being conducted, one of the assistants arranges the 8 cards in numerical order, with the lowest number on top. The examiner then looks at each card and asks the subject questions like:
   "Was your card a 2?"
   "Was your card a 3?"
3. At the end of the test release the pressure cuff, stop the paper movement, and evaluate the chart.

Student Group Projects

The card test protocol can be adapted to finding out concealed information, such as one's weight or IQ. If time is available, the class can be divided into several groups of four or five students to design a test and formulate questions to disclose information. Subjects for tests should be volunteers from another team. The object is to devise a test that creates emotional arousal without inciting anger or embarrassment to the subject. In addition to weight or IQ, one can use the following subject unknowns: birthday, age, GPA, etc.

80 A Biofeedback Demonstration

It is obvious from the results of the experiments in Exercise 79 that autonomic reflexes are greatly influenced by cerebral activity. We have seen that deliberate efforts to conceal the truth will result in changes in heart rate, blood pressure, respiration and galvanic skin response. The occurrence of these physiological changes is solid evidence that visceral reflexes are *unconsciously* influenced by higher brain centers. The whole concept of psychosomatic pathologies is based on this premise. There is no denying that ailments such as peptic ulcers, heart palpitation, and even heart attacks have been induced psychosomatically.

There is also much evidence to support the concept that cerebral activity can *consciously* regulate certain autonomic reflexes. The ability of the mind to deliberately affect autonomic reflexes to improve one's well-being is called **biofeedback.** In recent years there has been much research in the development of techniques along these lines. Considerable

success has been reported by individuals using these methods to gain relief from certain types of headache, muscle tension, tachycardia, and hypertension. Although it is doubtful that biofeedback can completely eliminate pain or greatly intensify pleasure, we must admit that considerable benefit has been derived from these techniques by some people.

In this exercise we will attempt to demonstrate the existence of biofeedback, using four channels on a polygraph. Since biofeedback is essentially a learned skill, the best success will be achieved if the subject has had some training in the technique.

To observe cerebral activity, we will use one channel to monitor brain waves with EEG electrodes as in Exercise 47. The other three parameters will be the GSR, heart rate, and skin temperature. Since two of these parameters were used in the last experiment, this exercise is a logical extension of Exercise 79.

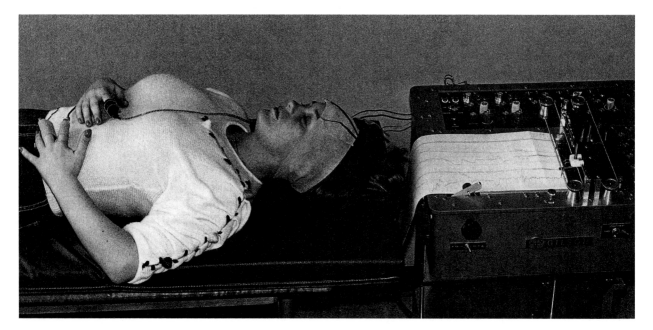

Figure 80.1 In this four-channel biofeedback demonstration, EEG electrodes are attached to the scalp, a pulse pickup is on the right index finger, a GSR electrode is on the left middle finger, and a thermistor is on the left index finger.

The same problems that plague the demonstration of lie detection exist in this experiment. Distraction of the subject by observers is a paramount problem. Blindfolding the subject and using earplugs may be necessary, if desired by the subject. Other problems pertain primarily to maintaining good electrode contact on the scalp and skin. A review of the procedures in Exercise 47 that are related to EEG monitoring will be necessary. The thermistor probe, which will be used for picking up skin temperature, must be kept in close contact with the skin all the time; momentary separation from the skin will cause air cooling and artifacts on the chart.

Although this experiment is set up primarily for the Gilson polygraph, it can be easily adapted to the Physiograph®. Proceed as follows:

Materials:

Gilson polygraph
Gilson T-4060 thermistor probe
Gilson A-4062 thermistor adapter
Gilson T-4020 finger pulse pickup
Gilson A-4023 adapter for pulse pickup
Gilson E-4041 GSR electrodes
electrode paste
Scotchbrite pad
Scotch tape
felt pen for marking chart
patient cable for EEG (3 leads)
three EEG disk-type electrodes
three circular Band-Aids
alcohol swabs
elastic wrapping for EEG
earplugs or cotton
blindfold
elastic banding for thermistor probe

Preparation of Subject

Attach three disk electrodes to the scalp, according to the procedures outlined in Exercise 47. As illustrated in figure 80.1, the finger pulse pickup is attached to the index finger of the right hand.

To the middle and index fingers of the left hand attach the GSR electrodes and thermistor. Figure 80.2 illustrates the method of slipping the thermistor under the wrapping to maintain good skin contact.

After all attachments are secured, allow the subject to assume a comfortable horizontal position. If a blindfold and earplugs are to be used, they should be applied before the subject assumes a horizontal position.

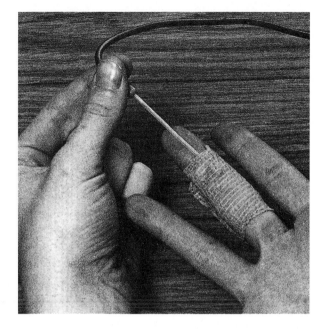

Figure 80.2 To get good thermistor-to-skin contact, wrap the finger with an adhesive wrapping and then slip the thermistor between it and the finger.

Equipment Setup

Connect all cables to the appropriate adapters and modules. The finger pulse pickup and GSR should be connected to the same modules as in Exercise 79. An IC-MP or IC-UM module may be used for the three-lead cable from the EEG electrodes. The thermistor probe will require the Servo, IC-MP, or IC-UM module.

Monitoring

With the chart paper moving at a speed of 25 mm/sec, establish a baseline for about 3 minutes. Adjust sensitivities when necessary to get the desired stylus deflection.

Now, ask the subject to attempt the type of biofeedback he or she has been trained to perform. Mark the chart at the point of beginning. Watch the brain wave pattern to note any changes that occur. If the subject is blindfolded, the operator may wish to inform the subject in a subdued voice of any changes that occur. Look for alpha waves and changes in the other three parameters. Record continuously as long as necessary to get results.

Laboratory Report

There is no Laboratory Report for this exercise.

Answers to Histology Self-Quiz No. 1

1. Fibrous connective tissue
2. Fibroblasts
3. Collagenous (white) fibers
4. Squamous
5. Lining of mouth
6. Plain columnar
7. Basement lamina
8. Adipose
9. Food storage, insulation, protection
10. Stratified squamous
11. Stratum basale
12. Dermis
13. Loose fibrous (areolar) connective
14. Collagenous (white)
15. Elastic (yellow)
16. Epidermis
17. Hair shaft
18. Hair follicle
19. Internal root sheath
20. External root sheath
21. Hypodermis (subcutaneous)
22. Loose fibrous (areolar) connective
23. Mast cell
24. Fibroblasts
25. Transitional epithelial
26. Basement lamina
27. Lamina propria
28. Cuboidal (glandular)
29. Glands
30. Elastic cartilage
31. Reticular connective
32. Liver, lymph glands, etc.
33. Binucleate
34. Transitional
35. Plain columnar
36. Goblet
37. Brush border
38. Fibrocartilage
39. Chondrocytes
40. Collagenous (white)
41. Compact bone
42. Perforating canal
43. Smooth muscle tissue
44. Nuclei
45. Pseudostratified ciliated
46. Cilia
47. Goblet cell
48. Smooth muscle
49. Canaliculi
50. Osteocytes
51. Central canal
52. Cardiac muscle tissue
53. Intercalated disks
54. Heart muscle, walls of vena cava
55. Striated muscle tissue
56. Endomysium
57. Cuboidal epithelial tissue
58. Stratum spinosum
59. Dermis
60. Keratinization
61. Striated muscle
62. Sarcolemma
63. Sarcomere
64. Pacinian corpuscle
65. Sensitive to pressure
66. Bone tissue (membranous)
67. Osteoclasts
68. Bone reabsorption
69. Sweat gland
70. Ducts of sweat gland
71. Dermis
72. External root sheath
73. Hair shaft
74. Bulb
75. Follicular papilla

Answers to Histology Self-Quiz No. 2

1. Cerebellum
2. Molecular layer
3. Granule cell layer
4. Purkinje cells
5. Pyramidal cell
6. Cerebrum
7. Axon
8. Apical dendrite
9. Cochlea of the ear
10. Vestibular membrane
11. Scala vestibuli
12. Spiral ligament
13. Scala tympani
14. Endolymph
15. Tectorial membrane
16. Rod of Corti
17. Outer hair cells
18. Basilar membrane
19. Motor neuron
20. Gray matter of spinal cord
21. Motor neuron
22. Nissl bodies
23. Dendrite
24. Axon
25. Crista ampullaris
26. Ampulla of semicircular duct
27. Hair cells
28. Endolymph
29. Sensory neuron
30. Sensory ganglion
31. Satellite cells
32. Lens of the eye
33. Lens capsule
34. Suspensory ligament
35. Ciliary processes
36. Nerve fiber layer
37. Inner nuclear layer
38. Outer synaptic layer
39. Rods and cones of the eye
40. Inner nuclear layer
41. Melanocytes
42. Rods and cones
43. Fovea centralis
44. Critical vision
45. Iris of the eye
46. Cornea
47. Corneal epithelium
48. Corneal stroma
49. Aqueous humor in anterior chamber
50. Vitreous body in posterior chamber

PART 14

The Digestive System

Although this unit contains only four exercises, one of them, Exercise 81, is rather lengthy and covers a considerable amount of material. It is anticipated that the student will label most of the illustrations of human anatomy before coming to the laboratory to dissect the cat and study the teeth of skulls.

Exercise 82 pertains to the factors that influence intestinal motility. Since this physiological activity is due to smooth muscle fibers and the autonomic nervous system, it may well be that this exercise will be studied with skeletal muscle physiology or with the nervous system. It is equally applicable in all three systems.

Exercise 83 is an in-depth study of the three basic hydrolytic reactions that occur in digestion: i.e., carbohydrate, protein, and fat hydrolysis. After removing the pancreas from a rat, the enzymes will be extracted and analyzed. In Exercise 84 the effects of environmental factors on digestive enzymes will be studied.

81 Anatomy of the Digestive System

The study of the alimentary tract, liver, and pancreas will be pursued in this exercise. Cat dissection will provide an in-depth study of the various components.

The Alimentary Canal

Figure 81.1 is a simplified illustration of the digestive system. The alimentary canal, which is about 30 feet long, has been foreshortened in the intestinal area for clarity. The following text, which pertains to this illustration, is related to the passage of food through its entire length.

When food is taken into the oral cavity, it is chewed and mixed with saliva that is secreted by many glands of the mouth. The most prominent of these glands are the parotid, submandibular, and sublingual glands. The **parotid glands** are located in the cheeks in front of the ears, one on each side of the head. The **sublingual glands** are located under the tongue and are the most anterior glands of the three pairs. The **submandibular glands** are situated posterior to the sublinguals, just inside the body of the mandible. (Figure 81.4 illustrates the position of these glands more precisely.)

After the food has been completely mixed with saliva it passes to the **stomach** by way of a long tube, the **esophagus.** The food is moved along the esophagus by wavelike constrictions called *peristaltic waves*. These constrictions originate in the **oropharynx,** which is the cavity at the top of the esophagus, posterior to the tongue. The upper opening of the stomach through which the food enters is the **cardiac sphincter.** The upper rounded portion, or **fundus,** of the stomach holds the bulk of the food to be digested. The lower portion, or **pyloric region,** is smaller in diameter, more active, and accomplishes most of the digestion that occurs in the stomach. That region between the fundus and the pyloric portion is called the **body.** After the food has been acted upon by the various enzymes of the gastric fluid, it is forced into the small intestine through the **pyloric sphincter** of the stomach.

The *small intestine* is approximately 23 feet long and consists of three parts: the duodenum, jejunum, and ileum. The first 10 to 12 inches make up the **duodenum.** The **jejunum** comprises the next 7 or 8 feet, and the last coiled portion is the **ileum.** Complete digestion and absorption of food take place in the small intestine.

Indigestible food and water pass from the ileum into the large intestine, or **colon,** through the **ileocecal valve.** This valve is shown in a cutaway section. The large intestine has four sections: the ascending, transverse, descending, and sigmoid colons. The **ascending colon** is the portion of the large intestine that the ileum empties into. At its lower end the ascending colon has an enlarged compartment, or pouch, called the **cecum.** Note that the cecum has a narrow tube, the **appendix,** extending downward from it.

The ascending colon ascends on the right side of the abdomen until it reaches the undersurface of the liver, where it bends abruptly to the left, becoming the **transverse colon.** The **descending colon** passes down the left side of the abdomen, where it changes direction again, becoming the **sigmoid colon.** The last portion of the alimentary canal is the **rectum,** which is about 5 inches long and terminates with an opening, the **anus.**

Leading downward from the inferior surface of the liver is the **hepatic duct.** This duct joins the **cystic duct,** which connects with the round saclike **gallbladder.** Bile, a secretion of the liver, passes down the hepatic duct and up the cystic duct to the gallbladder, where it is stored until needed. When the gallbladder contracts, bile is forced down the cystic duct into the **common bile duct,** which extends from the juncture of the cystic and hepatic ducts to the intestine.

Between the duodenum and the stomach lies another gland, the **pancreas.** Its duct, the **pancreatic duct,** joins the common bile duct and empties into the duodenum.

Assignment:
Label figure 81.1.

Disassemble a laboratory manikin and identify all of these structures.

Answer the questions on the Laboratory Report pertaining to this part of the exercise.

Oral Anatomy

A typical normal mouth is illustrated in figure 81.2. To provide maximum exposure of the oral structures, the lips *(labia)* have been retracted away from the teeth and the cheeks *(buccae)* have been cut.

The lips are flexible folds that meet laterally at the angle of the mouth where they are continuous with the cheeks. The lining of the lips, cheeks, and other oral surfaces consists of mucous membrane, the *mucosa.* Near the median line of the mouth on the inner surface of the lips, the mucosa is thickened to form folds, the **labial frenula,** or *frena.* Of the two frenula, the upper one is usually stronger. The delicate mucosa that covers the neck of each tooth is the **gingiva.**

The hard and soft palates are distinguishable as differently shaded areas on the roof of the mouth. The lighter shaded area adjacent to the teeth is the **hard palate.** Posterior to the hard palate is a darker area, the **soft palate.** The **uvula,** a fingerlike projection seen above the back of the tongue, is a part of the soft palate. It varies considerably in size and shape among individuals.

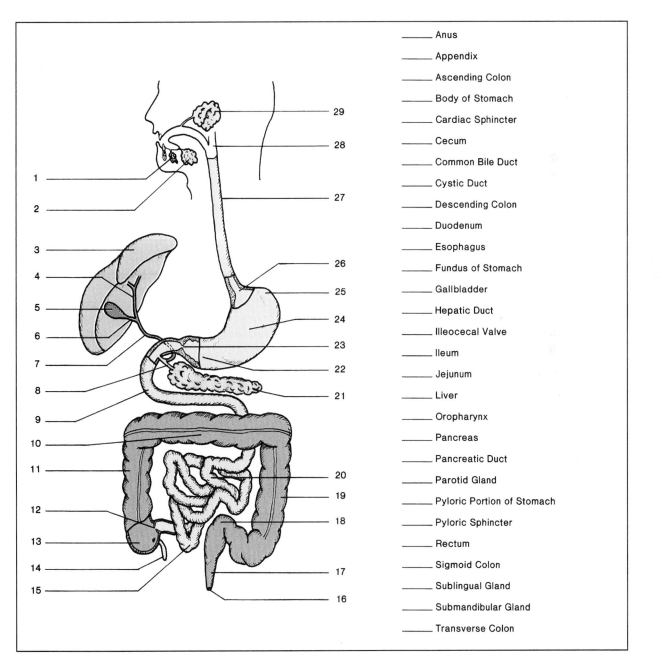

_____ Anus

_____ Appendix

_____ Ascending Colon

_____ Body of Stomach

_____ Cardiac Sphincter

_____ Cecum

_____ Common Bile Duct

_____ Cystic Duct

_____ Descending Colon

_____ Duodenum

_____ Esophagus

_____ Fundus of Stomach

_____ Gallbladder

_____ Hepatic Duct

_____ Illeocecal Valve

_____ Ileum

_____ Jejunum

_____ Liver

_____ Oropharynx

_____ Pancreas

_____ Pancreatic Duct

_____ Parotid Gland

_____ Pyloric Portion of Stomach

_____ Pyloric Sphincter

_____ Rectum

_____ Sigmoid Colon

_____ Sublingual Gland

_____ Submandibular Gland

_____ Transverse Colon

Figure 81.1 The alimentary canal.

On each side of the tongue at the back of the mouth are the **palatine tonsils.** Each tonsil lies in a recess bounded anteriorly by a membrane, the **glossopalatine arch,** and, posteriorly, by a membrane, the **pharyngopalatine arch.**

The tonsils consist of lymphoid tissue covered by epithelium. The epithelial covering of these structures dips inward into the lymphoid tissue forming glandlike pits called **tonsillar crypts** (label 11, figure 81.3). These crypts connect with channels that course through the lymphoid tissue of the tonsil. If they are infected, however, their protective function is impaired and they may actually serve as foci of infection. Inflammation of the palatine tonsils is called *tonsillitis.* Enlargement of the tonsils tends to obstruct the throat cavity and interfere with the passage of air to the lungs.

Assignment:
Label figure 81.2.

The Tongue

Figure 81.3 shows the tongue and adjacent structures. It is a mobile mass of striated muscle completely covered with mucous membrane. It is subdivided into three parts: the apex, body, and root. The **apex** of the tongue is the most anterior tip that rests against the inside surfaces of the anterior teeth. The **body** *(corpus)* is the bulk of the tongue that extends posteriorly from the apex to the root. The body is that part of the tongue that is visible by simple inspection: i.e., without the aid of a mirror. The posterior border of the body is arbitrarily located somewhere anterior to the tonsillar material of the organ. Extending down the median line of the body is a groove, the **central** *(median)* **sulcus.** The **root** of the tongue is the most posterior portion. Its surface is primarily covered by the **lingual tonsil.**

Extending upward from the posterior margin of the root of the tongue is the **epiglottis.** On each side of the tongue are seen the oval **palatine tonsils.**

The dorsum of the tongue is covered with several kinds of projections called *papillae* (see figure HA-20). The cutout section is an enlarged portion of its surface to reveal the anatomical differences between these papillae. The majority of the projections have tapered points and are called **filiform papillae.** These projections are sensitive to touch and give the dorsum a rough texture. This roughness provides friction for the handling of food. Scattered among the filiform papillae are the larger rounded **fungiform papillae.** Two fungiform papillae are

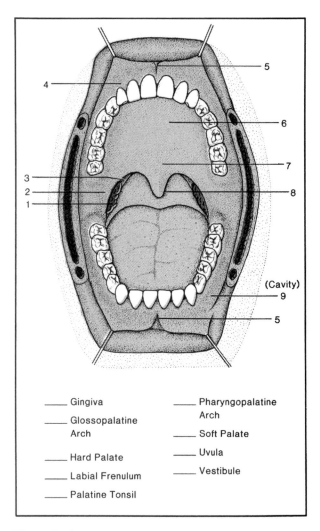

_____ Gingiva	_____ Pharyngopalatine Arch
_____ Glossopalatine Arch	_____ Soft Palate
_____ Hard Palate	_____ Uvula
_____ Labial Frenulum	_____ Vestibule
_____ Palatine Tonsil	

Figure 81.2   The oral cavity.

shown in the sectioned portion. A third type of papilla is the large donut-shaped **vallate** *(circumvallate)* **papilla.** These papillae are arranged in a "V" near the posterior margin of the dorsum. Although the exact number may vary in individuals, eight vallate papillae are shown.

The fungiform and vallate papillae contain **taste buds,** the receptors for taste. Note on the walls of the circular furrow surrounding each vallate papilla are clusters of these round receptors. Comparatively speaking, the vallate papillae contain many more taste buds than the fungiform papillae.

A fourth type of papilla is the **foliate papilla.** These projections exist as vertical rows of folds of mucosa on each side of the tongue, posteriorly. A few taste buds are also scattered among these papillae.

Assignment:
Label figure 81.3.

The Salivary Glands

The salivary glands are grouped according to size. The largest glands, of which there are three pairs, are referred to as the *major salivary glands.* They are the parotids, submandibulars, and sublinguals. The smaller glands, which average only 2–5 mm in diameter, are the *minor salivary glands.*

Major Salivary Glands

Figure 81.4 illustrates the relative positions of the three major salivary glands as seen on the left side of the face. Since all of these glands are paired, it should be kept in mind that the other side of the face has another set of these glands.

The largest gland, which lies under the skin of the cheek in front of the ear is the **parotid gland.** Note that it lies between the skin layer and the *masseter muscle.* Leading from this gland is the **parotid (Stensen's) duct,** which passes over the masseter and through the *buccinator muscle* into the oral cavity. The drainage of this duct usually exits near the upper second molar.

The secretion of the parotid glands is a clear watery fluid that has a cleansing action in the mouth. It contains the digestive enzyme *salivary amylase,* which splits starch molecules into disaccharides (double sugars). The presence of sour (acid) substances in the mouth causes this gland to increase its secretion.

Inside the arch of the mandible lies the **submandibular** *(submaxillary)* **gland.** In Figure 81.4 the mandible has been cut away to reveal two cut surfaces. Note that the lower margin of the submandibular gland extends down somewhat below the inferior border of the mandible. Also, note that a flat muscle, the *mylohyoid,* extends somewhat into the gland. The secretion of this gland empties into the oral cavity through the **submandibular** *(Wharton's)* **duct.** The **opening of the submandibular duct** is located under the tongue near the lingual frenulum. The **lingual frenulum** is a mucosal fold on the median line between the tongue and the floor of the mouth. It is shown in figure 81.4 as a triangular membrane.

The secretion of the submandibulars is quite similar in consistency to that of the parotid glands,

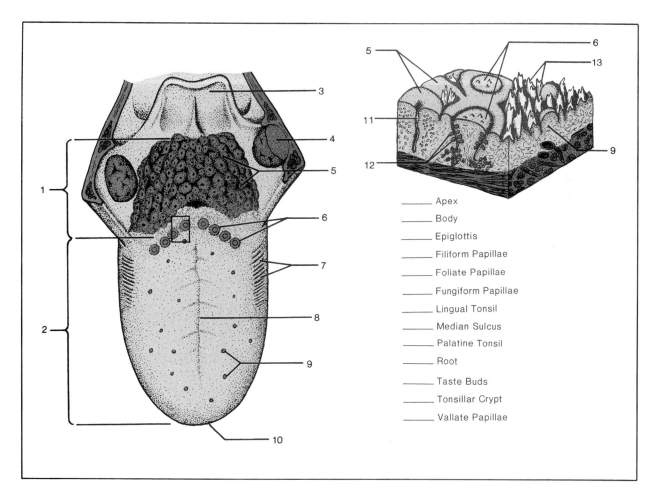

_____ Apex

_____ Body

_____ Epiglottis

_____ Filiform Papillae

_____ Foliate Papillae

_____ Fungiform Papillae

_____ Lingual Tonsil

_____ Median Sulcus

_____ Palatine Tonsil

_____ Root

_____ Taste Buds

_____ Tonsillar Crypt

_____ Vallate Papillae

Figure 81.3 Tongue anatomy.

except that it is slightly more viscous due to the presence of some *mucin*. This ingredient of saliva aids in holding the food together in a bolus. Bland substances such as bread and milk stimulate this gland.

The **sublingual gland** is the smallest of the three major salivary glands. As its name would imply, it is located under the tongue in the floor of the mouth. It is encased in a fold of mucosa, the **sublingual fold,** under the tongue. This fold is shown in figure 81.5 (label 5). Drainage of this gland is through several **lesser sublingual ducts** *(ducts of Rivinus).* The number of openings is

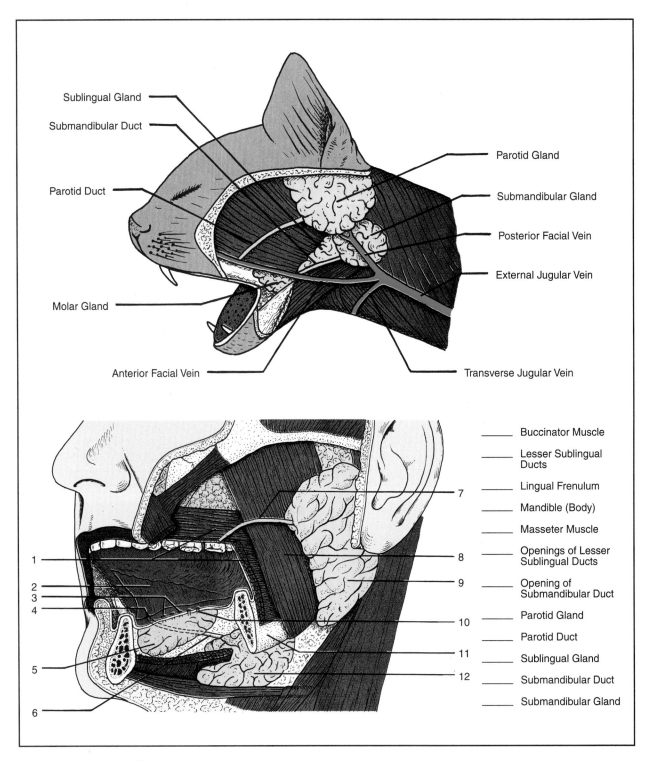

Sublingual Gland
Submandibular Duct
Parotid Duct
Molar Gland
Anterior Facial Vein

Parotid Gland
Submandibular Gland
Posterior Facial Vein
External Jugular Vein
Transverse Jugular Vein

_____ Buccinator Muscle
_____ Lesser Sublingual Ducts
_____ Lingual Frenulum
_____ Mandible (Body)
_____ Masseter Muscle
_____ Openings of Lesser Sublingual Ducts
_____ Opening of Submandibular Duct
_____ Parotid Gland
_____ Parotid Duct
_____ Sublingual Gland
_____ Submandibular Duct
_____ Submandibular Gland

Figure 81.4 Major salivary glands, cat and human.

variable among individuals. The gland differs from the other two in that it is primarily a mucous gland. The high mucin content of its secretion lends a certain degree of ropiness to it.

Minor Salivary Glands

There are four principal groups of these smaller salivary glands in the mouth: the palatine, lingual, buccal, and labial glands. In illustration A, figure 81.5, the palatine mucosa has been partially removed to reveal the closely packed nature of the **palatine glands.** These glands, which are about 2–4 mm in diameter, almost completely cover the roof of the oral cavity. Where the mucosa has been removed near the lower third molar we see that the glands occur even next to the teeth. The glands in this particular area are called **retromolar glands.**

Illustration B in figure 81.5 shows the oral cavity with the tongue pointed upward and the inferior mucosal surface of it partially removed. The gland exposed in this area of the tongue is the **anterior lingual gland.**

The **buccal glands** cover the majority of the inner surface of the cheeks, and the **labial glands** are found under the inner surface of the lips. All of the minor glands except for the palatine glands produce salivary amylase.

Assignment:
Label figures 81.4 and 81.5.

The Teeth

Every individual develops two sets of teeth during the first twenty-one years of life. The first set, which begins to appear at approximately six months of age, are the *deciduous teeth.* Various other names such as *primary, baby,* or *milk teeth* are also used in reference to these teeth.

The second set of teeth, known as *permanent* or *secondary teeth,* begins to appear when the child is about six years of age. As a result of normal growth from the sixth year on, the jaws enlarge, the secondary teeth begin to exert pressure on the primary teeth, and *exfoliation,* or shedding, of the deciduous dentition occurs.

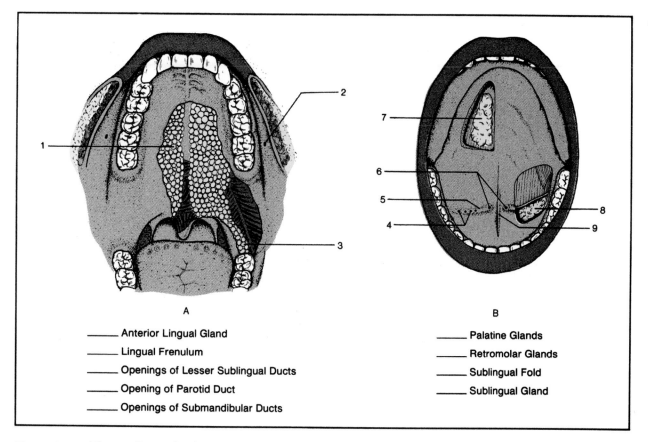

A

_____ Anterior Lingual Gland
_____ Lingual Frenulum
_____ Openings of Lesser Sublingual Ducts
_____ Opening of Parotid Duct
_____ Openings of Submandibular Ducts

B

_____ Palatine Glands
_____ Retromolar Glands
_____ Sublingual Fold
_____ Sublingual Gland

Figure 81.5 Minor salivary glands.

The Deciduous Teeth

The deciduous teeth number twenty in all—five in each quadrant of the jaws. Normally, all twenty have erupted by the time the child is two years old. Starting with the first tooth at the median line, they are named as follows: **central incisor, lateral incisor, cuspid** *(canine),* **first molar,** and **second molar.** The naming of the lower teeth follows the same sequence.

Once the two-year-old child has all of the deciduous teeth, no visible change in the teeth occurs until around the sixth year. At this time the first permanent molars begin to erupt.

The Permanent Teeth

For approximately five years (seventh to twelfth year) the child will have a *mixed dentition,* consisting of both deciduous and permanent teeth. As the submerged permanent teeth enlarge in the tissues, the roots of the deciduous teeth undergo *resorption.* This removal of the underpinnings of the deciduous teeth results eventually in exfoliation.

Figure 81.6 illustrates the permanent dentition. Note that there are sixteen teeth in one half of the mouth—thus, a total of thirty-two teeth. Naming them in sequence from the median line of the mouth

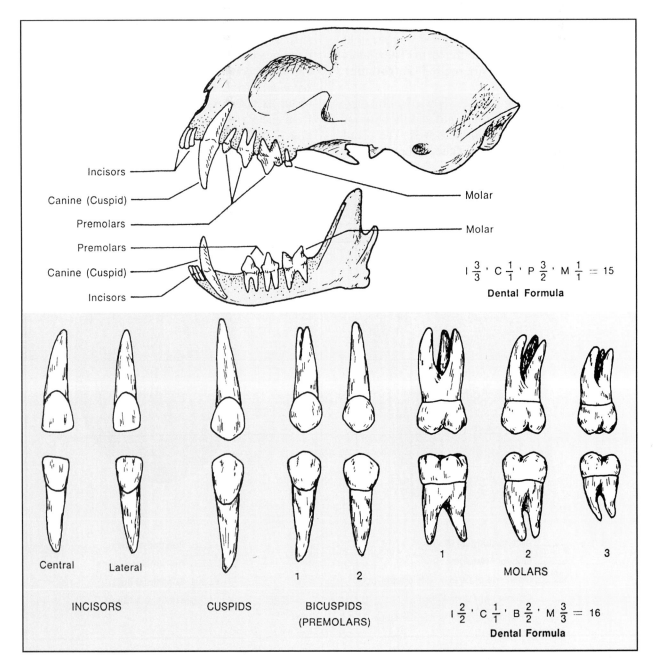

$$I\,\frac{3}{3}\cdot C\,\frac{1}{1}\cdot P\,\frac{3}{2}\cdot M\,\frac{1}{1}=15$$
Dental Formula

$$I\,\frac{2}{2}\cdot C\,\frac{1}{1}\cdot B\,\frac{2}{2}\cdot M\,\frac{3}{3}=16$$
Dental Formula

Figure 81.6 The permanent teeth, cat and human.

they are **central incisor, lateral incisor, cuspid** (*canine*), **first bicuspid, second bicuspid, first molar, second molar,** and **third molar.** The third molar is also called the *wisdom tooth.*

Comparisons of the teeth in figure 81.6 reveal that the anterior teeth have single roots and the posterior ones have several roots. Of the bicuspids, the only ones that have two roots *(bifurcated)* are the maxillary first bicuspids. Although all of the maxillary molars are shown with three roots *(trifurcated),* there may be considerable variability, particularly with respect to the third molars. The roots of the mandibular molars are generally bifurcated. The longest roots are seen on the maxillary cuspids.

Assignment:
Identify the teeth of skulls that are available in the laboratory.

Tooth Anatomy

The anatomy of an individual tooth is shown in figure 81.7. Longitudinally, it is divided into two portions, the crown and root. The line where these two parts meet is the **cervical line,** or *cemento-enamel juncture.*

The dentist sees the crown from two aspects: the anatomical and clinical crowns. The **anatomical crown** is the portion of the tooth that is covered with enamel. The **clinical crown,** on the other hand, is that portion of the crown that is exposed in the mouth. The structure and physical condition of the soft tissues around the neck of the tooth will determine the size of the clinical crown.

The tooth is composed of four tissues: the enamel, dentin, pulp, and cementum. **Enamel** is the most densely mineralized and hardest material in the body. Ninety-six percent of enamel is mineral. The remaining 4% is a carbohydrate-protein complex. Calcium and phosphorus make up over 50% of the chemical structure of enamel. Microscopically, this tissue is made up of very fine rods or prisms that lie approximately perpendicular to the outer surface of the crown. It has been estimated that the upper molars may have as many as 12 million of these small prisms in each tooth. The hardness of enamel enables the tooth to withstand the abrasive action of one tooth against the other.

Dentin is the material that makes up the bulk of the tooth and lies beneath the enamel. It is not as

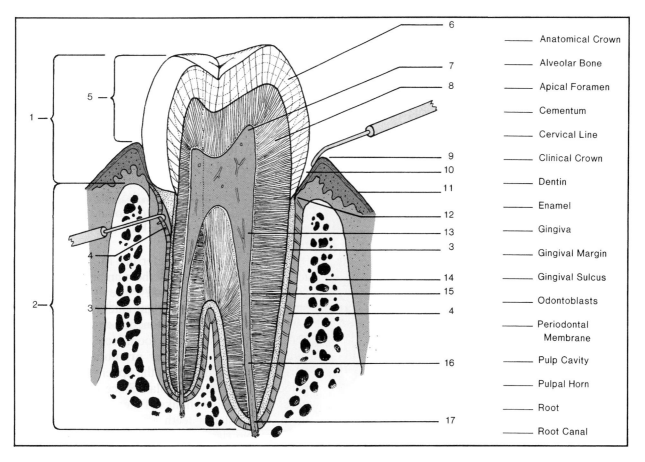

	Anatomical Crown
	Alveolar Bone
	Apical Foramen
	Cementum
	Cervical Line
	Clinical Crown
	Dentin
	Enamel
	Gingiva
	Gingival Margin
	Gingival Sulcus
	Odontoblasts
	Periodontal Membrane
	Pulp Cavity
	Pulpal Horn
	Root
	Root Canal

Figure 81.7 Tooth anatomy.

82

Intestinal Motility

Digestion and absorption of food in the intestine are greatly facilitated by a variety of movements caused by the smooth muscle fibers of the intestinal walls. The principal movements are described as being *peristaltic, segmenting,* and *pendular.* In addition, *rippling movements* of the muscularis mucosa and *pumping action* of the microscopic villi are also seen.

Segmenting contractions occur rhythmically at approximately 17 per minute in the duodenal area and 10 per minute in the ileal region. The degree of motility appears to be related to the metabolic rate since experimentally induced temperature variations greatly affect the rate. Most regulation, however, is accomplished by neural elements in the wall of the digestive tract.

In this experiment, a small segment of rat intestine will be mounted on a Combelectrode. The intestinal movements will be monitored as the ambient temperature is changed and as the hormones epinephrine and acetylcholine are administered. Figure 82.1 illustrates the setup.

As in previous experiments of this type, students will work in teams of four members. Two members of each team will prepare the tissue, while the other two balance the transducer and set up the equipment. Procedures are provided for both Gilson and Narco equipment.

Equipment Setup

While other members of your team are dissecting the rat and preparing the tissue, set up the equipment as follows:

Materials:
for Unigraph setup:
    Unigraph
    electronic stimulator (Grass SD-9)
    force transducer (Trans-Med Biocom #1030)
    event synchronization cable

for Physiograph® setup:
    Physiograph® with transducer coupler
    stimulator (Narco SM-1)
    event marker cable
    myograph (Narco)
    steel rod for myograph

for both setups:
    ring stand
    ring and wire gauze
    alcohol lamp, or small Bunsen burner
    Harvard isometric clamp
    2 double clamps
    oxygen supply or air pump

1. Set up a ring stand, isometric clamp, and transducer in the arrangement shown in figure 82.1. To mount a Narco myograph to the Harvard clamp refer to figure 31.2 on page 146.
2. Place a wire gauze on the ring and adjust the ring to the height that will accommodate an alcohol burner or small Bunsen burner.
3. Plug the transducer jack into the recorder.
4. Plug in the power cords of the stimulator and recorder.
5. Connect the event marker (synchronization) cable between the stimulator and recorder.
6. Balance the transducer (myograph) according to instructions in Appendix B.
7. Calibrate the recorder according to instructions in Appendix B.

You are now ready to accept the Combelectrode with mounted rat intestine tissue prepared by the other team members.

Anesthetization of Rat

One rat will provide enough tissue for six or seven setups. To anesthetize the rat, it will be necessary for one person to hold the animal while another individual injects Nembutal into the abdominal cavity. Work with a team member to anesthetize an animal in the following manner:

Materials:
    rat, 200 grams weight
    Nembutal
    1 cc syringe and needle

1. Draw 0.25 ml of Nembutal into the syringe.
2. Remove the rat from the cage by pulling out by the tail with the right hand. When you have it near the door of the cage, grasp the animal firmly by the skin of the back of the neck with the fingers of the left hand.

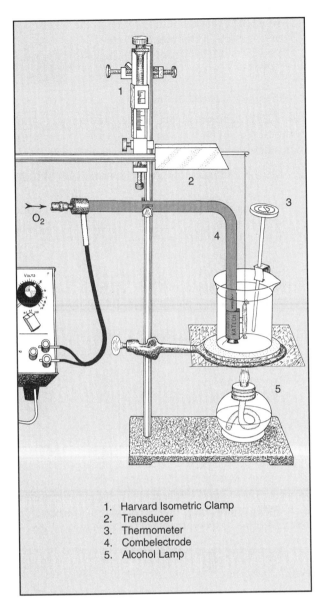

Figure 82.1 Instrumentation setup.

1. Harvard Isometric Clamp
2. Transducer
3. Thermometer
4. Combelectrode
5. Alcohol Lamp

3. Have your assistant inject the Nembutal into the peritoneal cavity about ½ cm below the sternum. The animal will become unconscious within 30 seconds.

Dissection and Tissue Preparation

As soon as the animal is unconscious, the abdomen must be opened to expose the intestines. Before sections of the intestines are excised, motility will be observed. Sections of the intestine are then removed, placed in individual dishes of Tyrode's solution, and dispensed to individual teams for attachment to Combelectrodes. Proceed as follows:

Materials:
dissecting pan
dissecting instruments
petri dishes
Tyrode's solution

1. Fill a beaker with 200 ml of Tyrode's solution. In addition, pour about ¼″ of Tyrodes solution into as many petri dishes as needed to supply tissue to each group.
2. Open up the abdominal wall of the rat with a pair of scissors, exposing the intestines. Examine the intestines before disturbing them to observe the degree of motility.
3. Before cutting off lengths of the intestine, cut away the mesentery from the portion that is to be used.
4. Cut off as many lengths (each about 2 cm long) as needed for the whole class. Place each section into a separate petri dish of Tyrode's solution.
5. Present the tissue to team members who will attach it to the Combelectrode.

Final Hookup

Attach the tissue to a Combelectrode and complete the hookup as shown in figure 82.1, using the following steps:

Materials:
thermometer
Combelectrode
medicine droppers
250 ml graduated beaker
thread

1. While the tissue is still in the petri dish, flush out the lumen of the intestine with a medicine dropper and Tyrode's solution.
2. Tie an 18″ length of thread to each end of the section of tissue.
3. Attach one thread to the lower end of the Combelectrode, securing it in place with the stopper. Be sure the tissue is located so that it completely covers the electrical contacts. Cut off excess length of the thread.
4. Insert the end of the Combelectrode into a beaker of Tyrode's solution on the ring stand, allowing the free thread to hang over the edge of the beaker.
5. Secure the Combelectrode to the clamp on the ring stand.
6. Attach the free thread to the first two leaves of the transducer. Cut off the excess thread. Adjust the tension to the thread with the isometric clamp, taking up any slack.

7. Plug the Combelectrode jacks into the stimulator posts.
8. Connect the oxygen supply hose to the Combelectrode and adjust the oxygen metering valve so that 5–10 bubbles per second are produced.
9. Attach the thermometer to the side of the beaker. You are now ready to begin the investigation.

Investigative Procedure

The contractions of the intestine will be monitored under three conditions. First the temperature will be increased, gradually, from 22° C (73° F); next the tissue will be electrically stimulated; and, finally, it will be subjected to epinephrine and acetylcholine.

Materials:

    epinephrine (1:20,000)
    acetylcholine (1:20,000)
    1 cc syringe and needle

Effect of Temperature

Adjust the temperature of the Tyrode's solution in the beaker to 22° C (73° F). Turn on the recorder and proceed as follows:

1. Set the speed at 2.5 mm/sec (0.25 cm/sec), and start the chart moving. Observe the trace.
2. If no contractions show up, increase the gain by one step (2 MV/CM to 1 MV/CM) without altering the sensitivity knob on the Unigraph or the inner knob on the Physiograph®. (*Calibration will be altered by changing the fine adjustment.*)

    If no contractions show up at this setting, increase the gain to 0.5 MV/CM. The contractions at this setting will be weak and quite slow. Record for about 2 minutes at this temperature. Mark the temperature on the paper.
3. Place an alcohol burner or very small Bunsen burner under the beaker to warm up the Tyrode's solution. An alcohol burner can usually be left under the beaker without overheating; Bunsen burners require more attention.
4. Control the heat so that the temperature increases **at a rate of about 2° C per minute.** Record the temperature every minute on the paper.
5. Continue heating until a temperature of 41° C is reached. Normal body temperature of the rat is 38.1° C (100.5° F). A temperature of 42.5° C is lethal to the animal.
6. Note how the regularity of contractions increases as the temperature increases. Also, observe that the frequency of contraction increases with temperature.

    Do contractions maintain constancy of contraction, or do they cyclically strengthen and weaken? At what temperature is the frequency of contraction the greatest?
7. Record your results on the Laboratory Report.

Effects of Electrical Stimulation

Allow the solution in the beaker to cool to a temperature somewhere around 32° C (90° F) and stimulate the tissue with single pulses as follows:

1. Set the stimulator controls: Duration (or Width) at 15 msec on the Grass SD-9, or 2 msec on the Narco SM-1; Voltage at 10 volts; and the Mode switch at OFF.
2. Turn on the power switch of the stimulator.
3. Depress the Mode switch to SINGLE, and record "10 volts" on the chart. Does the muscle respond? Repeat several times to see if it responds at any particular point on the contraction cycle.
4. If you get no response at 10 volts, increase the voltage by 10-volt increments to see if a response occurs.
5. Record your results on the Laboratory Report.

Effects of Epinephrine and Acetylcholine

Nerve endings of the autonomic nervous system produce hormones that stimulate and inhibit contractions of smooth muscle in the intestinal wall. To simulate these conditions, we will subject the tissue to acetylcholine and epinephrine. We have seen that acetylcholine is produced by the nerve endings of parasympathetic fibers. Epinephrine is similar, if not identical, to the action of norepinephrine of the sympathetic nervous system. Proceed as follows:

1. With the temperature in the beaker adjusted to 38° C, observe the contractions for about one minute and then add 0.5 cc of epinephrine to the solution with a hypodermic syringe. Record on paper the temperature and "epinephrine." Watch for recovery, noting how long it takes. Is the muscle stimulated or inhibited?
2. Remove the Tyrode's solution from the beaker, replacing it with fresh solution at 38° C. This may be accomplished by lowering the ring on the ring stand so that the beaker can be lifted off for emptying. Under no circumstances should the Combelectrode be disturbed.
3. After contractions have stabilized, add 0.25 cc of acetylcholine to the beaker and observe the effects.

Laboratory Report

Complete the Laboratory Report for this exercise.

83

The Chemistry of Hydrolysis

The basic nutrients of the body consist of small quantities of vitamins and minerals plus large amounts of carbohydrates, fats, and proteins. Although vitamins and minerals play a vital role in all physiological activities, it is the carbohydrates, fats, and proteins that provide the energy and raw materials for growth and tissue repair.

Food, as it is normally ingested, consists of large complex molecules of carbohydrates, fats, and proteins that cannot pass through the intestinal wall unless they are converted to smaller molecules by digestion in the stomach and intestines. The conversion of these large molecules to smaller ones is accomplished by digestive enzymes. All digestive enzymes split molecules by hydrolysis.

Hydrolysis is a process whereby the larger molecules are split into smaller units by combining with water. Although hydrolysis without enzymes will occur automatically at body temperature, the process is extremely slow; the catalytic action of digestive enzymes, on the other hand, greatly hastens the process.

The end result of this action is to reduce carbohydrates to monosaccharides, fats to fatty acids and glycerol, and proteins to amino acids. The large size of some food molecules requires that different enzymes work at different levels to produce intermediate-size molecules before the end products are produced.

Digestive enzymes are produced by the salivary glands, the stomach wall, the intestinal wall, and the pancreas. Since the pancreas produces all three of the basic types of enzymes (proteases, lipases, and carbohydrases), we will focus our attention here on this gland.

An enzyme extract from the pancreas of a freshly killed rat will be analyzed for the presence of the three types. Figure 83.5 reveals the steps that will be employed for extracting the pancreatic juice. Since enzymes are relatively unstable and easily denatured by certain adverse environmental conditions, it will be essential that chemical purity, cleanliness, and temperature control be maintained.

Once the pancreatic juice has been extracted, tests will be made for evidence of carbohydrase,

lipase, and protease activity. Small amounts of the pancreatic juice will be added to substrates of starch, fat, and protein to observe the hydrolytic action of the various enzymes. Color changes that occur in each test will indicate the breaking of bonds within the large molecules to produce smaller molecules. Positive and negative test controls will be used for comparisons with the actual tests.

The class will be divided into groups of four students. While two of the students are dissecting the rat, the other two will set up the necessary supplies and equipment for the extraction and assay procedures.

Extraction Procedure

Remove the pancreas from a freshly killed rat and produce the pancreatic extract using the following procedures.

Materials:
　freshly killed rat
　refrigerated centrifuge
　centrifuge tubes
　balance
　Sorval blender
　beaker, 30 ml size
　Erlenmeyer flask, 125 ml size
　graduated cylinder, 100 ml size
　crushed ice
　cold mammalian Ringer's solution
　dissecting pan (wax bottom)
　dissecting instruments
　dissecting pins
　Parafilm®

1. Follow the steps outlined in figures 83.1 through 83.4 to open up the abdominal cavity of a freshly killed rat and remove the pancreas.
2. Place the gland into a beaker containing a small amount (about 20 ml) of cold mammalian Ringer's solution.
3. Snip off a small piece of the intestine and add it to the beaker.
4. After washing the tissues in the Ringer's solution, macerate and homogenize them in a

blender. The tissues of several groups should be blended together to get ample bulk.

5. Distribute the blended mixture into evenly balanced centrifuge tubes and spin the tubes in a refrigerated centrifuge (Beckman or other) for 7 minutes at 10,000 rpm. Tubes must be balanced by weighing carefully and adjusting the contents until they are equal.

6. Decant supernatant from the tubes into Erlenmeyer flasks for each group. The supernatant will be a milky pink mixture of digestive enzymes.

7. Place the flask of pancreatic extract into a container of ice to prevent enzyme deterioration.

Tube Preparation

Figure 83.6 illustrates the number of test tubes that will be set up to assay the pancreatic extract for carbohydrase, protease, and lipase. The detection of the presence of each enzyme will depend on a specific color test after the tubes have been incubated in a 38° C water bath for one hour.

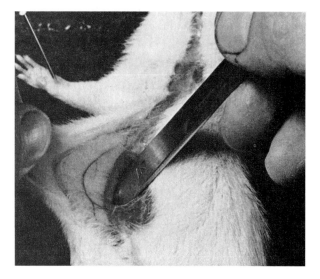

Figure 83.1 After pinning down the feet and cutting through the skin, separate the skin from the musculature with the scalpel handle.

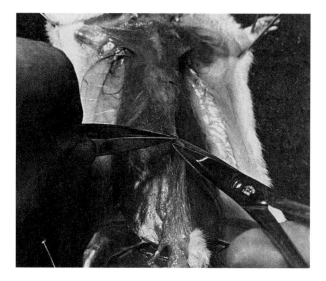

Figure 83.2 Hold the musculature up with forceps while cutting through the wall. Be careful not to damage any organs.

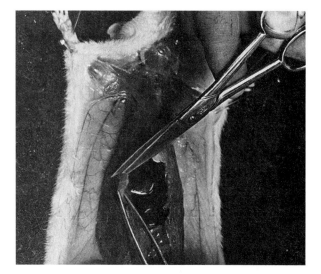

Figure 83.3 As the muscular body wall is opened, pull flaps of muscle wall laterally and pin them down to keep the cavity open.

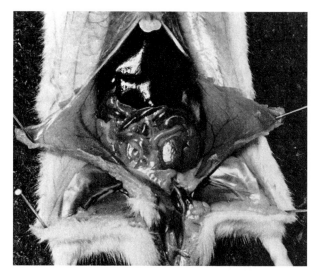

Figure 83.4 Once the abdominal organs are exposed as shown here, lift the liver with forceps to expose the pancreas for removal.

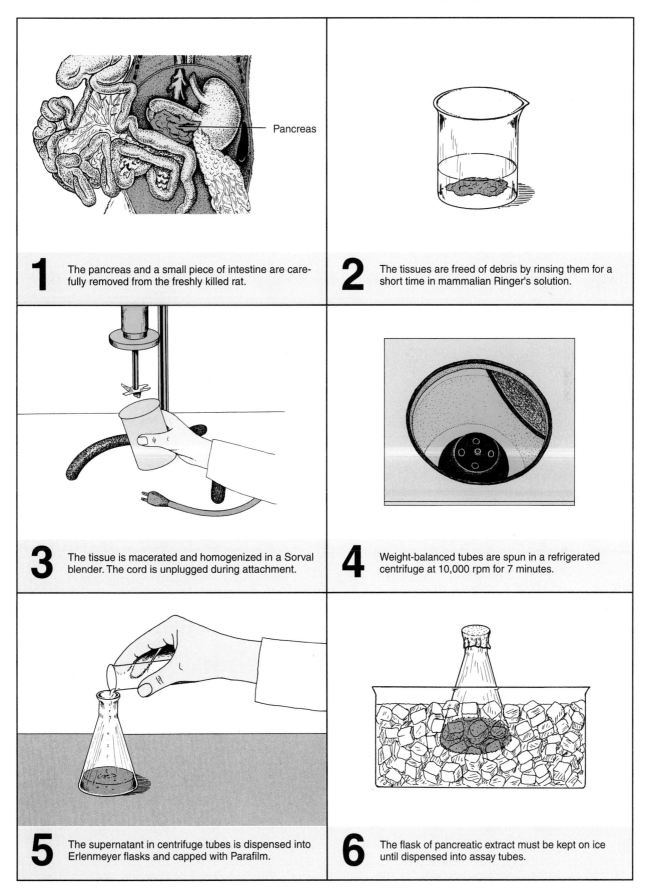

1 The pancreas and a small piece of intestine are carefully removed from the freshly killed rat.

2 The tissues are freed of debris by rinsing them for a short time in mammalian Ringer's solution.

3 The tissue is macerated and homogenized in a Sorval blender. The cord is unplugged during attachment.

4 Weight-balanced tubes are spun in a refrigerated centrifuge at 10,000 rpm for 7 minutes.

5 The supernatant in centrifuge tubes is dispensed into Erlenmeyer flasks and capped with Parafilm.

6 The flask of pancreatic extract must be kept on ice until dispensed into assay tubes.

Figure 83.5   Procedure for extracting pancreatic enzymes.

Tube 1 in each series contains pancreatic extract, a substrate, and necessary buffering agents. The last tube in each series is a **positive test control** that contains the enzyme we are testing for in tube 1.

All other tubes in each series should give negative results because of certain deficiencies. Note that tube 2 in each series contains the same ingredients as tube 1 except for two significant variables: (1) the tube is placed on ice while the others are incubated; and (2) the pancreatic extract is not added until after one hour on ice. This tube is designated as our **zero time control.**

All four members of each team will work cooperatively to set up these 16 tubes. After the tubes have been in the water baths for one hour, 1 ml of pancreatic juice will be added to the three zero time tubes. Various tests will then be performed to detect the presence of hydrolysis.

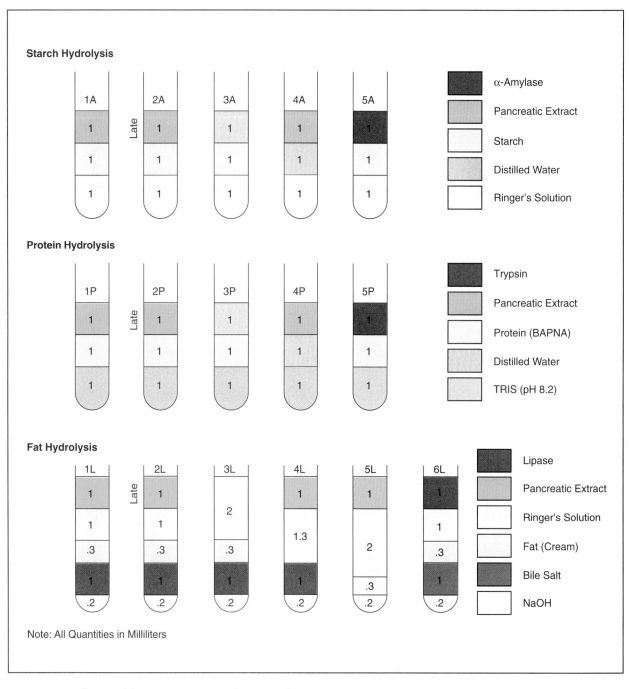

Figure 83.6   Protocol for enzyme assay of pancreatic extract.

Materials:

water baths (38° C)
Vortex mixer
pipetting devices
test-tube rack (wire type)
test tubes (16 mm × 150 mm), 16 per group
hot plates
pipettes, 1 ml, 5 ml, and 10 ml sizes
graduated cylinder (10 ml size)
bromthymol blue pH color standards
china marking pencil
spectrophotometer (Spectronic 20)
spot plates

solutions in dropping bottles

IKI solution
Benedict's solution
Barfoed's solution
bromthymol blue
amylase, 1%
trypsin, 1%
lipase, 1%
TRIS (pH 8.2)

2% sodium taurocholate
  (a bile salt)
soluble starch, 0.1%
cream, unhomogenized
albumin, 0.1%
NaCl, 0.2%
NaOH, 0.1N
BAPNA solution

Carbohydrase

Carbohydrate in plants is stored, primarily, in the form of starch. Starch is composed of 98% amylose and 2% amylopectin. Amylose is a straight-chain polysaccharide of glucose units attached by α-**1,4-glucosidic bonds.** Amylopectin is a branching polysaccharide with side chains attached to the main chain by α-**1,6-glucosidic bonds.** These linkages are illustrated in figure 83.7.

The digestion of starch is catalyzed by α-*amylase,* an enzyme present in saliva, pancreatic juice, and intestinal juice. This enzyme randomly attacks the α-1,4 bonds. An α-*1,6-glucosidase* is necessary to break the α-1,6 bond. The action of α-amylase on starch results in the formation of molecules of maltose and isomaltose. These two molecules are converted to glucose by the enzymes maltase and isomaltose in the epithelial cells of the small intestine.

Set up the five tube test series for α-amylase according to the following protocol:

1. With a china marking pencil, label the test tubes 1A through 5A ("A" for amylase).
2. With a 5 ml or 10 ml pipette, deliver 1 ml of **mammalian Ringer's solution** to each of the five tubes. Use a mechanical pipetting device such as the one illustrated in figure 83.10.
3. Flush out the pipette with water and deliver 1 ml of **distilled water** to tubes 3 and 4.
4. Using the same pipette, deliver 1 ml of **starch solution** to tubes 1, 2, 3, and 5. Note that tube 4 does not get any starch.
5. With a fresh 5 ml pipette, deliver 1 ml of **pancreatic juice** to tubes 1 and 4.
6. With a fresh 1 ml pipette, deliver 1 ml of **α-amylase to tube 5.**
7. Mix each tube on a Vortex mixer (figure 83.11) for a few seconds.
8. Remove tube 2 from the series, cover it with Parafilm, and place it in a rack that is immersed in ice water.

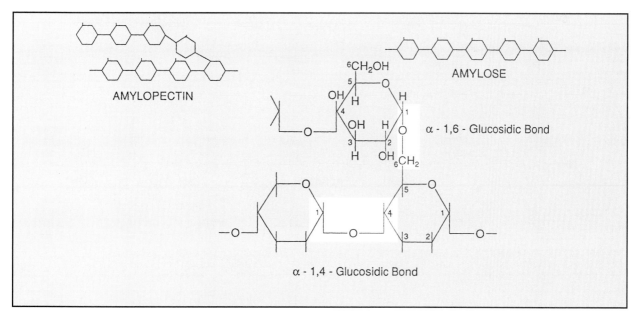

Figure 83.7 Carbohydrate linkage where hydrolysis occurs.

9. Cover the other four tubes with Parafilm and place the test tube rack with these tubes in a 38° C water bath for **one hour.**

Protease

The dietary proteins are derived, primarily, from meats and vegetables. These proteins consist of long chains of amino acids that are connected to each other by **peptide linkages.** Note in figure 83.8 that when a peptide linkage is hydrolyzed, the carbon of one amino acid takes on OH⁻ to form a carboxyl group (COOH), and H⁺ is added to the other amino acid to form an amine (NH₂) group.

Protein digestion begins in the stomach with the action of *pepsin* and is completed in the small intestine. Pancreatic juice contains *trypsin, chymotrypsin,* and *carboxypolypeptidase,* which convert proteoses, peptones, and polypeptides to polypeptides and amino acids. *Amino polypeptidase* and *dipeptidase* within the epithelial cells of the small intestine hydrolyze the final peptide linkages as small polypeptides and dipeptides pass through the intestinal mucosa into the portal blood.

Since there are considerable physical differences between the linkages of different amino acids, there need be, and are, a multiplicity of proteolytic enzymes; no single enzyme can possibly digest protein all the way to its amino acids.

In our assay of pancreatic juice, we will use a synthetic peptide substrate that is designated as BAPNA (benzoyl D-L p-arginine nitroaniline). If our pancreatic extract contains trypsin, this protein will be hydrolyzed to release a yellow aniline dye, giving visual evidence of hydrolysis. Trypsin will be used in our positive test control.

1. With a china marking pencil, label the test tubes 1P through 5P ("P" for protease).
2. With a 5 ml or 10 ml pipette, deliver 1 ml of **pH 8.2 buffered solution** to each of the five tubes. Use a mechanical pipetting device such as the one illustrated in figure 83.10.
3. Flush out the pipette with water and deliver 1 ml of **distilled water** to tubes 3 and 4.
4. Using the same pipette, deliver 1 ml of **BAPNA solution** to tubes 1, 2, 3, and 5. Note that tube 4 does not get any BAPNA; it is a negative test control.
5. With a fresh 5 ml pipette, deliver 1 ml of **pancreatic juice** to tubes 1 and 4.
6. With a fresh 1 ml pipette, deliver 1 ml of **trypsin** to tube 5.
7. Mix each tube on a Vortex mixer (figure 83.11) for a few seconds.
8. Remove tube 2 from the series, cover it with Parafilm, and place it in a rack that is immersed in ice water.
9. Cover the other four tubes with Parafilm and place the test tube rack with these tubes in a 38° C water bath for **one hour.**

Lipase

The hydrolysis of fats to glycerol and fatty acids is catalyzed by *lipase.* Although this enzyme is present

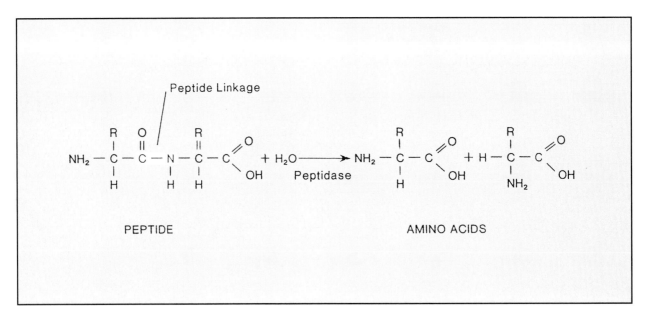

Figure 83.8 Protein hydrolysis.

in both gastric and pancreatic secretions, nearly all fat digestion occurs in the small intestine.

Lipase hydrolysis differs from carbohydrase and protease hydrolysis in that it requires an emulsifying agent to hasten the process. *Bile salts,* produced by the liver, act like detergents to break up large fat globules into smaller globules, greatly increasing the surface of fat particles. Since lipases, like all enzymes, are only able to work on the surfaces of their substrate molecules, the increased surface area greatly enhances the speed of fat hydrolysis.

The most common fats in food are neutral fats, or **triglycerides.** As illustrated in figure 83.9, each triglyceride molecule is composed of a glycerol nucleus and three fatty acids. Although the final end products of triglyceride hydrolysis are fatty acids, and glycerol, there are intermediate products of monoglycerides and diglycerides. Diagrammatically, the process looks like this:

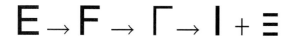

TRIGLYCERIDE DIGLYCERIDE MONOGLYCERIDE GLYCEROL FATTY ACIDS

Set up six test tubes to assay pancreatic extract for the presence of lipase:

1. With a china marking pencil, label the test tubes 1L through 6L ("L" for lipase).
2. Using a 1 ml pipette, deliver 0.2 ml of **0.1N NaOH** to each of the tubes. Use a mechanical pipetting device such as the one in figure 83.10.

3. With a 5 ml pipette, deliver 0.2 ml of **mammalian Ringer's solution** to tubes 1, 2, and 6; 2 ml to tubes 3 and 5; and 1.3 ml to tube 4.
4. Flush out the pipette with water and deliver 1 ml of **bile salt solution** to each of the tubes except tube 5.
5. Flush out the pipette with water again and deliver 0.3 ml of **cream** to all tubes except tube 4.
6. With a fresh 5 ml pipette, deliver 1 ml of **pancreatic juice** to tubes 1, 4, and 5.
7. Deliver 1 ml of **lipase** to tube 6, using a fresh 1 ml pipette.
8. Mix each tube on a Vortex mixer (figure 83.11) for a few seconds.
9. Add 2 drops of bromthymol blue to each tube. Check the color against a bromthymol blue standard of pH 7.2.

   If the correct color does not develop, add NaOH, drop by drop, slowly, until the tube reaches the correct color of pH 7.2.
10. Remove tube 2 from the series, cover it with Parafilm, and place it in a rack that is immersed in ice water.
11. Cover the other five tubes with Parafilm and place the test tube rack with these tubes in a 38° C water bath for **one hour.**

Evaluation of Tests

Remove the tubes from both water baths after one hour and proceed as follows to complete the experiment:

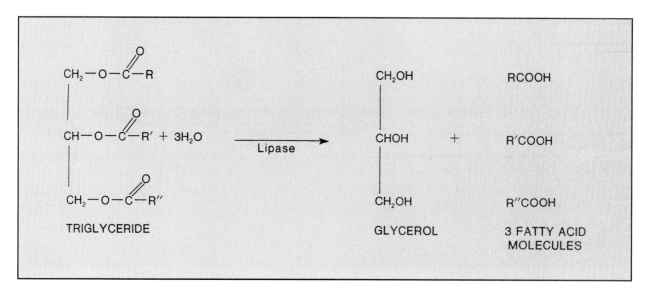

Figure 83.9 Fat hydrolysis.

1. To tube 2 of each series add 1 ml of pancreatic juice and mix for a few seconds on a Vortex mixer. Place each no. 2 tube in place within its series.

2. **Carbohydrase Test:** Remove a sample of test solution from each of the five tubes (1A through 5A) and place in separate depressions of a spot plate.

   Add 2 drops of IKI solution to each of the solutions on the plate. If starch is present due to lack of hydrolysis, the spot solution will turn blue.

   If hydrolysis has occurred, test the remaining solution in the tube with Barfoed's and Benedict's tests to determine degree of digestion.

   See Appendix B for these tests. Record your results on the Laboratory Report.

3. **Protease Test:** Since hydrolysis produces a yellow color in the test tube, record the color of each tube on the Laboratory Report.

   If a spectrophotometer is set up and calibrated at 410 nanometers, take a reading for each tube and record the optical densities (OD) in the table on the Laboratory Report. Be sure to use appropriate cuvettes for the contents of each tube.

   Note: Before taking any photocolorimeter readings, the photocolorimeter should be standardized with a blank made up of 3 ml of BAPNA and 0.1 ml of 0.001 M HCl.

4. **Lipase Test:** Since fat hydrolysis results in a lowering of pH due to the formation of fatty acids, the bromthymol blue in the tubes will change from blue to yellow.

   Compare the six tubes with a bromthymol blue color standard to determine the pH changes that have occurred.

Laboratory Report

Complete the first portion of combined Laboratory Report 83,84.

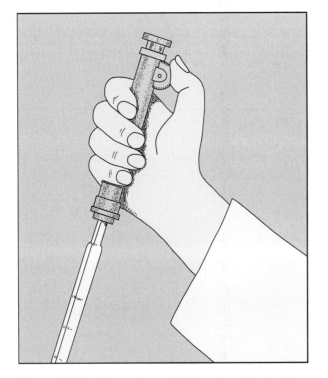

Figure 83.10 The volume of deliveries with pipetting device is controlled with the thumb.

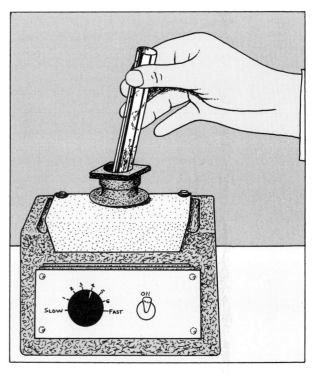

Figure 83.11 Each tube in the assay procedure must be mixed on the Vortex mixer.

84 Factors Affecting Hydrolysis

Enzymes are highly complex protein molecules of somewhat unstable nature due to the presence of weak hydrogen bonds. Many factors influence the rates at which hydrolysis occurs. Temperature and hydrogen ion concentration are probably the most important factors in this respect. It is the influence of these two conditions that will be studied in this exercise.

Salivary amylase (ptyalin) has been selected for this study. This enzyme hydrolyzes the starch amylose to produce soluble starch, maltose, dextrins, achrodextrins, and erythrodextrin. When iodine is used as an indicator of amylase activity, starch and soluble starch produce a blue color and erythrodextrin yields a red color. To detect the presence of maltose, Benedict's solution is used. With heat, maltose is a reducing sugar causing the reduction of soluble cupric to insoluble red cuprous oxide in Benedict's solution.

In this exercise, the saliva of one individual will be used. The saliva will be diluted to 10% with distilled water. To determine the effect of temperature on the rate of hydrolysis, iodine solution (IKI) will be added to a series of ten tubes (figure 84.1) containing starch and saliva. By adding the iodine at one-minute intervals to successive tubes, it is possible to determine, by color, when the hydrolysis of starch has been completed. For temperature comparisons, two water baths will be used. One will be 20° C and the other will be at 37° C. To determine the effects of hydrogen ion concentration on the action of amylase, we will use buffered solutions of pH 5, 7, and 9.

The dispensing of solutions in this exercise may be performed with medicine droppers, graduates, or serological pipettes.

If laboratory time is limited, it might be necessary for the instructor to divide the class into thirds that will perform only a portion of the entire experiment. After each group completes its portion, the results can be tabulated on the chalkboard so that all students can record results for the complete experiment. The best arrangement is for students to work in pairs. The following assignments will work well for a two-hour laboratory period:

GROUP	ASSIGNMENT
A (⅓ of Class)	Controls, 20° C, Boiling
B (⅓ of Class)	Controls, 37° C, Boiling
C (⅓ of Class)	Controls, pH Effects

Saliva Preparation

Before this experiment, the instructor will prepare a 10% saliva solution for the entire class as follows:

Materials:
  50 ml graduate
  250 ml graduated beaker
  paraffin
  12 dropping bottles with labels

Place a small piece (¼″ cube) of paraffin under your tongue and allow it to soften for a few minutes before starting to chew it. As it is chewed, expectorate all saliva into a 50 ml graduate.

When you have collected around 15 to 20 ml of saliva, measure out enough distilled water with a graduate and add it to the saliva, making up a 10% solution.

Dispense the diluted saliva into labeled ("10% saliva") dropping bottles. For a class of 24 students, you will need 12 bottles.

Controls

A set of five control tubes is needed for color comparisons with test results. Tubes 1 and 2 will be used for detection of starch. Tubes 3, 4, and 5 will be used for detection of the presence of maltose. Table 84.1 shows the ingredient in each tube and the significance of test results. Proceed as follows to prepare your set of controls.

Materials:

> 10% saliva solution (in dropping bottle)
> 0.1% starch solution
> 1% maltose solution
> Benedict's solution (in dropping bottle)
> IKI solution (in dropping bottle)
> test tubes (13 mm dia × 100 mm long)
> test-tube rack (Wassermann type)
> medicine droppers
> 250 ml beaker
> electric hot plate
> 1 ml sterile pipettes
> pipetting device

1. Label five clean test tubes: *1, 2, 3, 4,* and *5.*
2. Fill the five tubes with the following reagents, *using a different pipette for each reagent.*

> **Tube 1:** 1 ml starch solution and two drops of IKI solution.
> **Tube 2:** 1 ml distilled water and two drops of IKI solution.
> **Tube 3:** 1 ml starch solution, two drops of saliva solution, and five drops Benedict's solution.
> **Tube 4:** 1 ml distilled water, two drops saliva solution, and five drops Benedict's solution.
> **Tube 5:** 1 ml maltose solution and five drops Benedict's solution.

3. Place tubes 3, 4, and 5 into a beaker of warm water and boil on an electric hot plate for **5 minutes** to bring about the desired color changes in tubes 3 and 5.
4. Set up the five tubes in a test-tube rack to be used for color comparisons. Tube 1 will be your positive starch control. Tubes 3 and 5 will be your positive sugar controls. Tubes 2 and 4 will be your negative controls for starch and sugar.

Effect of Temperature (20° C and 37° C)

In this portion of the experiment we will determine the effect of room temperature (20° C) and body temperature (37° C) on the rate at which amylase acts. Note in figure 84.1 that tubes of starch and saliva are given two drops of IKI solution at one-minute intervals to observe how long it takes for hydrolysis to occur at a given temperature.

When IKI is added to the first tube, the solution will be blue, indicating no hydrolysis. At some point along the way, however, the blue color will disappear, indicating complete hydrolysis. Note that the materials list is for performing the test at only one temperature.

Materials:

> 10 serological test tubes
>   (13 mm dia × 100 mm)
> test-tube rack (Wassermann type)
> water bath (20° C or 37° C)
> 10% saliva solution
> 0.1% starch solution
> china marking pencil
> thermometer (Centigrade scale)
> 1 ml serological pipettes or 10 ml graduate
> mechanical pipetting device

1. Label ten test tubes 1 through 10 with a china marking pencil and arrange them sequentially in the front row of the test-tube rack.
2. With a 1 ml pipette dispense 1 ml of starch solution to each tube.
3. Place the rack of tubes and bottle of saliva in the water bath (20° C or 37° C).
4. Insert a clean thermometer into the bottle of saliva. When the temperature of the saliva has reached the temperature of the water bath (3–5 minutes), proceed to the next step.

Table 84.1 Control tube contents and functions.

TUBE NUMBER	CARBOHYDRATE	SALIVA	REAGENT	TEST RESULTS
1	Starch	—	IKI	Positive for Starch
2	—	—	IKI	Negative for Starch
3	Starch	2 Drops	Benedict's	Positive for Sugar Positive for Hydrolysis
4	—	2 Drops	Benedict's	Negative for Sugar
5	Maltose	—	Benedict's	Positive for Sugar

5. Lift the bottle of saliva from the water bath, **record the time,** and put **one drop** of saliva into each test tube, starting with tube 1. Leave the rack of tubes in the water bath and be sure that the drop falls directly into the starch solution without touching the sides of the tube.
6. Shake the rack of tubes to mix their contents.
7. **Exactly one minute** after tube 1 has received saliva, add 2 drops of IKI solution to it and mix again. Note the color.
8. **One minute later** add two drops of IKI to tube 2 and repeat this process every minute until all ten tubes have received IKI solution.
9. Compare the colors of the tubes with the control tubes and determine the time required to digest the starch.
10. Record your results on the Laboratory Report.
11. Wash the insides of all test tubes with soap and water and rinse thoroughly.

Effect of Boiling

In this portion of this experiment, we will compare the action of amylase that has been boiled for a few seconds with unheated amylase. Figure 84.2 illustrates the procedure.

Materials:

    3 test tubes
    test-tube holder
    Bunsen burner
    10% saliva solution
    0.1 starch solution
    IKI solution

1. Label three test tubes *1, 2,* and *3.*
2. Dispense 1 ml of saliva solution to tubes 1 and 2, and 1 ml of distilled water to tube 3.
3. Heat tube 1 over a Bunsen burner flame so that it comes to a boil for a few seconds. Use a test-tube holder.
4. Add 1 ml of starch solution to each of the three tubes and place them in a 37° C water bath for 5 minutes.
5. Remove the tubes from the water bath and test for the presence of starch by adding 2 drops of IKI to each tube. Record the results on the Laboratory Report.
6. Wash and rinse the test tubes.

Effect of pH on Enzyme Activity

To determine the effect of different hydrogen ion concentrations on amylase activity, the enzyme, starch, and buffered solutions will be incubated at 37° C. The overall procedure is illustrated in figure 84.3.

Materials:

    test-tube rack (Wassermann type)
    12 test tubes (13 mm dia × 100 mm)
    1 ml pipettes and pipetting device
    10 ml graduate
    water bath (37° C)
    thermometer
    10% saliva solution
    0.1% starch solution
    IKI solution
    pH 5 buffer solution
    pH 7 buffer solution
    pH 9 buffer solution

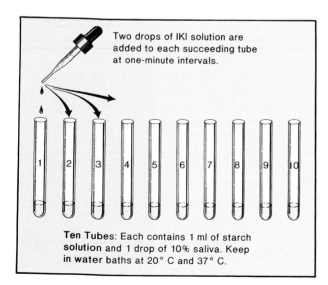

Ten Tubes: Each contains 1 ml of starch solution and 1 drop of 10% saliva. Keep in water baths at 20° C and 37° C.

Two drops of IKI solution are added to each succeeding tube at one-minute intervals.

Figure 84.1 Amylase activity at 20° C and 37° C.

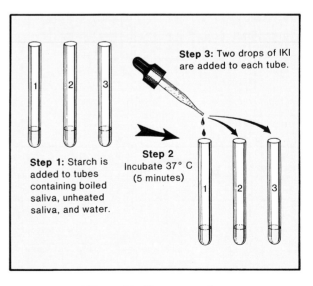

Step 1: Starch is added to tubes containing boiled saliva, unheated saliva, and water.

Step 2 Incubate 37° C (5 minutes)

Step 3: Two drops of IKI are added to each tube.

Figure 84.2 Effect of boiling on amylase.

441

Hormones having some relationship to the reproductive organs might, thus, be expected from the cortex.

The Adrenal Cortex

Figure 87.4 and 87.5 reveal the three layers of the cortex. Note that immediately under the capsule lies a layer called the **zona glomerulosa.** The innermost layer of cells, that interfaces with the medulla, is the **zona reticularis.** The middle layer, which is the thickest, is the **zona fasciculata.**

In all layers are seen vascular channels, called **sinusoids,** that are larger in diameter than capillaries but resemble capillaries in that their walls are one cell thick. It is in these channels that the hormones collect before passing directly into the general circulation.

The adrenal cortex secretes many different hormones, all of which belong to a group of substances called *steroids.* More specifically, they are referred to as **corticosteroids.** All of them are synthesized from cholesterol and have similar molecular structure. Whereas destruction of the adrenal medulla is more or less inconsequential to survival, obliteration of the adrenal cortex results in death.

The corticosteroids fall into three groups: the mineralcorticoids, the glucocorticoids, and the androgenic hormones. The *mineralocorticoids* are hormones that control the excretion of Na^+ and K^+ and therefore affect the electrolyte balance. The *glucocorticoids* affect primarily the metabolism of glucose and protein. The *androgenic hormones* produce some of the same effects on the body as the male sex hormone (testosterone). Of the 30 or more corticosteroids that are produced by the adrenal cortex, the most important two are aldosterone and cortisol. Descriptions of these two hormones follow.

Aldosterone This mineralocorticoid is produced by cells of the zona glomerulosa. The most important function of this hormone is to increase the rate of renal tubular absorption of sodium. Any condition that impairs the production of aldosterone will result in death within two weeks if salt replacement or mineralocorticoid therapy is not provided.

In the absence of aldosterone, the K^+ concentration in extracellular fluids rises, Na^+ and Cl^- concentrations decrease, and the total volume of body fluids becomes greatly reduced. These conditions cause reduced cardiac output, shock and death.

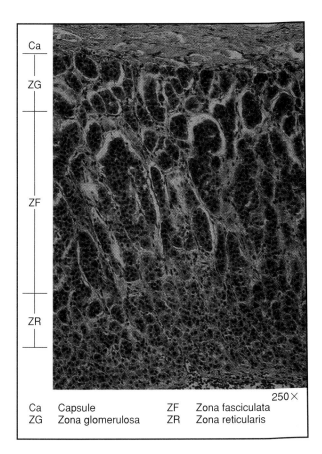

Ca	Capsule
ZG	Zona glomerulosa
ZF	Zona fasciculata
ZR	Zona reticularis

250×

Figure 87.4 The adrenal cortex.

Factors that affect aldosterone production are ACTH, K^+ concentration, Na^+ concentration, and the renin-angiotensin system.

Cortisol (Hydrocortisone, Compound F) Approximately 95% of the glucocorticoid activity results from the action of cortisol. This hormone is produced primarily by cells in the zona fasciculata; some of it is also produced by the zona reticularis.

When an excess of cortisol exists, the following physiological activities take place in the body: (1) the rate of glucogenesis is stepped up; (2) glucose utilization by cells is decreased; (3) protein catabolism in all cells except the liver is increased; (4) amino acid transport to muscle cells is decreased; and (5) fatty acids are mobilized from adipose tissue.

Cushing's syndrome is a condition in which many of the above reactions occur. It is caused by an *excess of glucocorticoids,* induced by a pituitary tumor. Protein depletion in these patients causes them to have poorly developed muscles, weak bones (due to bone dissolution), slow healing of wounds, hyperglycemia, and hair that is thin and scraggly.

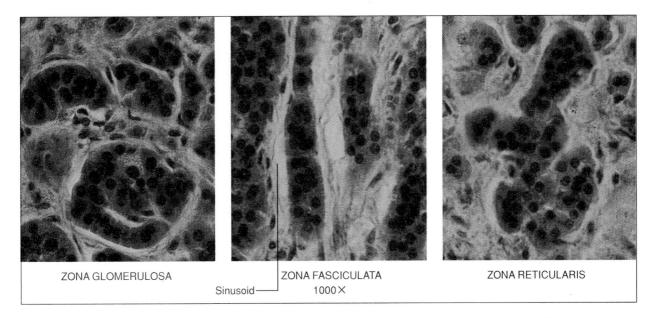

ZONA GLOMERULOSA   ZONA FASCICULATA   ZONA RETICULARIS

Sinusoid⎯   1000×

Figure 87.5   Histology of the three layers of the adrenal cortex.

Addison's disease is a condition in which total glucocorticoid insufficiency exists. It can be caused by destruction of the adrenal cortex by cancer, tuberculosis, or some other infectious agent.

Although several substances such as vasopressin, serotonin, and angiotensin II stimulate the adrenal cortex, it is primarily the action of ACTH that induces the cells of the zona fasciculata and zona reticularis to produce cortisol.

ACTH production by the anterior pituitary is partially initiated by stress through the hypothalamus. Stress induces the hypothalamus to produce CRF (corticotropin-releasing factor), which passes to the anterior pituitary, causing the latter to produce ACTH.

The Adrenal Medulla

The medulla of the adrenal gland produces the two *catecholamines,* norepinephrine and epinephrine. Although different cells in the medulla produce each of these hormones, cell differences are difficult to detect microscopically.

Approximately 80% of the medullary secretion is epinephrine. Norepinephrine makes up the remaining 20%. Norepinephrine is also produced by certain nerve endings of the autonomic nervous system.

Action   The effects of norepinephrine and epinephrine in different tissues depend on the types of receptors that exist in the tissues. There are two classes of adrenergic receptors: alpha and beta. In addition, there are two types of alpha receptors (α_1 and α_2) and two types of beta receptors (β_1 and β_2).

Both norepinephrine and epinephrine increase the force and rate of contraction of the isolated heart. These reactions are mediated by the β_1 receptors. Norepinephrine causes vasoconstriction in all organs through α_1 receptors.

Epinephrine causes vasoconstriction everywhere except in the muscles and liver, where beta receptors bring about vasodilation. Increased mental alertness is also induced by both catecholamines, which may be induced by the increased blood pressure.

Blood glucose levels are increased by both hormones. This is accomplished by glycogenolysis (glycogen → glucose) in the liver. β_2 receptors account for this reaction.

Norepinephrine and epinephrine are equally potent in mobilizing free fatty acids through beta receptors. The metabolic rate is also increased by these hormones; how this takes place is not precisely understood at this time.

Regulation   Production of both hormones is initiated when the adrenal medulla is stimulated by sympathetic nerve fibers.

The Thymus Gland

The thymus consists of two long lobes joined by connective tissue. It lies in the upper chest region above the heart. It is most highly developed before birth and during the growing years. After puberty, it begins to regress and continues to do so throughout life.

Histologically, this gland consists of a cortex and medulla. Figure 87.6 reveals its microscopic appearance. The **cortex** consists of lymphoidlike tissue filled with a large number of lymphocytes. The **medulla** contains a smaller number of lymphocytes and structures called **thymic corpuscles** (Hassall's bodies). The function of the latter structures is unknown.

Function It appears that the principal role of the thymus is to process lymphocytes into T-lymphocytes, as described on page 373. It is speculated that this gland produces one or more hormones that function in T-lymphocyte formation. Thymosin and several other extracts from this gland have been under study.

The Pineal Gland

The pineal gland is situated on the roof of the third ventricle under the posterior end of the corpus callosum. It is supported by a stalk that contains postganglionic sympathetic nerve fibers that do not seem to extend into the gland. The gland consists of neuroglial and parenchymal cells that suggest a secretory function. Figure 87.6, right-hand illustration, reveals its histological appearance.

In young animals and infants the gland is large and more glandlike; cells tend to be arranged in alveoli. Just before puberty, however, the gland begins to regress, and small concretions of calcium carbonate form that are called **pineal sand.** There is

some evidence, though not conclusive, that the gland contains some gonadotropin peptides. Most attention, however, is on the production of melatonin by the gland. **Melatonin** is an indole that is synthesized from serotonin. Its production appears to be regulated by daylight (circadian rhythm).

During daylight the production of melatonin is suppressed. At night the gland becomes active and produces considerable quantities of melatonin. Regulation occurs through the eyes. Light striking the retina sends messages to the pineal gland through a retina-hypothalmic pathway. Norepinephrine reaches the cells from postganglionic sympathetic nerve endings in the pineal stalk. Beta adrenergic receptors to the cells are mediators in the inhibition of melatonin production.

Although considerable speculation exists that melatonin inhibits the estrus (menstrual) cycle in humans, as it probably does in lower animals, there is no definitive proof at this time that this is the case.

The Testes

A section through the testis (figure 87.7) reveals that it consists of coils of **seminiferous tubules,** where spermatozoa are produced, and **interstitial cells,** where the male sex hormone, **testosterone,** is secreted. Note that the interstitial cells lie in the spaces between the seminiferous tubules.

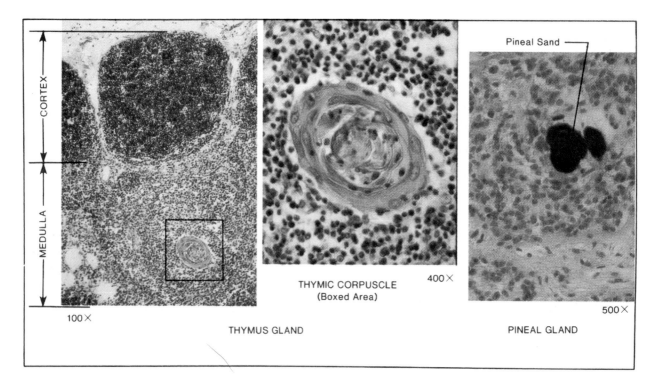

Figure 87.6   The thymus and pineal glands.

Testosterone is produced in large amounts during embryological development, but from early childhood to puberty very little of the hormone is produced. With the onset of puberty at the age of 10 or 11 years the testes begin to produce large quantities of testosterone as the development of sexual maturity takes place.

During embryological development, testosterone is responsible for the development of the male sex organs. If the testes are removed from a fetus at an early stage, the fetus will develop a clitoris and vagina instead of a penis and scrotum, even though the fetus is a male. The hormone also controls the development of the prostate gland, seminal vesicles, and male ducts while suppressing the formation of female genitals.

Embryological development of the testes occurs within the body cavity. During the last two months of development the testes descend into the scrotum through the inguinal canals. The descent of the testes is controlled by testosterone.

During puberty the testes become larger and produce a great deal of testosterone, which causes considerable enlargement of the penis and scrotum. At the same time, the male secondary sexual characteristics develop. These characteristics are (1) the appearance of hair on the face, axillae, chest, and pubic region; (2) the enlargement of the larynx accompanied by a voice change; (3) increased skin thickness; (4) increased muscular development; (5) increased bone thickness and roughness; and (6) baldness.

During embryological development the production of testosterone is regulated by **chorionic gonadotropin,** a hormone produced by the placenta. At the onset of puberty the testes are stimulated to produce testosterone by the luteinizing hormone (LH), which is produced by the anterior lobe of the pituitary gland. This hormone stimulates the interstitial cells to produce testosterone. Maturation of spermatozoa in the testes is controlled by the follicle-stimulating hormone (FSH), which is also produced by the anterior pituitary gland. Testosterone assists FSH in this process.

The Ovaries

During fetal development small groups of cells move inward from the germinal epithelia of each ovary and develop into primordial follicles. As seen in figure 87.8, the **germinal epithelium** is a layer of cuboidal cells near the surface of the ovary, and the **primordial follicles** are the small round bodies that contain **ova.**

The ovaries of a young female at the onset of puberty are believed to contain between 100,000 and 400,000 immature follicles. During all the reproductive years of a woman, only about 400 of these follicles will reach maturity and expel their ova.

During puberty, selected primordial follicles enlarge to form **Graafian follicles,** and every 28 days a maturing Graafian follicle expels an ovum in the ovulation process. The development of these follicles and the subsequent corpus luteum results in

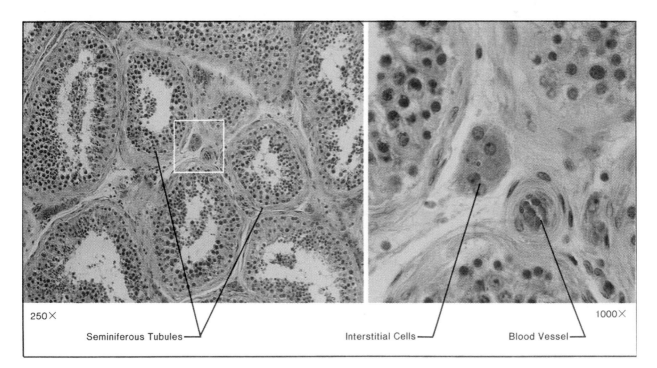

250×

Seminiferous Tubules

Interstitial Cells

Blood Vessel

1000×

Figure 87.7   Testicular tissue.

the production of **estrogens** and **progesterone,** the principal female sex hormones. A discussion of each hormone follows.

Estrogens

The principal functions of estrogens are to promote the growth of specific cells in the body and to control the development of female secondary sex characteristics. In addition to being produced in Graafian follicles, estrogens are also produced by the corpus luteum, placenta, adrenal cortex, and testes. The production of estrogens at puberty is initiated by the secretion of FSH by the anterior lobe of the pituitary gland.

Of the six or seven estrogens that have been isolated from the plasma of women, **β-estradiol, estrone,** and **estriol** are the most abundant ones produced. The most potent of all the estrogens is β-estradiol.

During puberty when the estrogens are produced in large quantities the following changes occur: (1) the female sex organs become fully developed; (2) the vaginal epithelium changes from cuboidal to stratified epithelium; (3) the uterine lining (endometrium) becomes more glandular in preparation for implantation of the fertilized ovum; (4) the breasts form; (5) osteoblastic activity increases with more rapid bone growth; (6) calcification of the epiphyses in long bones is hastened; (7) pelvic bones enlarge and change shape, increasing the size of the pelvic outlet; (8) fat deposition under the skin, in the hips, buttocks, and thighs is increased; and (9) skin vascularization is increased.

Progesterone

Once ovulation takes place in the ovary the Graafian follicle is replaced by a **corpus luteum.** Progesterone is the principal hormone produced by this body. The ovary in figure 87.13 shows the development of Graafian follicles and the corpus luteum.

The most important function of this hormone is to promote secretory changes in the endometrium in preparation for implantation of the fertilized ovum. In addition, the hormone causes mucosal changes in the uterine tubes and promotes the proliferation of alveolar cells in the breasts in preparation for milk production.

When the plasma level of progesterone falls, due to regression of the corpus luteum, deterioration of the endometrium takes place and menstruation occurs. If pregnancy occurs the corpus luteum enlarges and produces additional quantities of progesterone. The conversion of a Graafian follicle into a corpus luteum is completely dependent on LH production.

The Pituitary Gland

The pituitary gland, or hypophysis, consists of two lobes: anterior and posterior. The anterior lobe, or **adenohypophysis,** develops from the roof of the oral

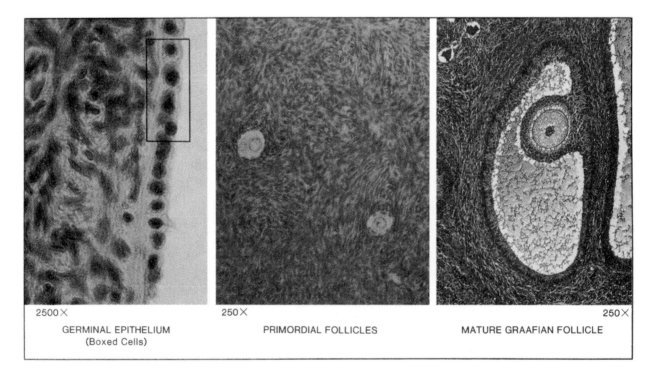

2500×	250×	250×
GERMINAL EPITHELIUM (Boxed Cells)	PRIMORDIAL FOLLICLES	MATURE GRAAFIAN FOLLICLE

Figure 87.8 Ovarian tissue.

cavity of the embryo. This part of the hypophysis is made up of glandular cells. The posterior lobe, or **neurohypophysis,** develops as an outgrowth of the floor of the brain during embryological development. Unlike the adenohypophysis, the cells of this lobe are nonsecretory and resemble neuroglial tissue. The entire gland is invested by an extension of the dura mater.

Figure 87.9 reveals portions of the two lobes. Note that the neurohypophysis in this region consists of two parts: the **pars intermedia** and the **pars nervosa.**

The Adenohypophysis

Two basic types of glandular cells are seen in this portion of the hypophysis: chromophobes and chromophils. Both types of cells can be seen in figure 87.10.

Chromophobes are cells that lack an affinity for routine dyes used in staining tissues. **Chromophils** stain readily and are of two types: acidophils and basophils. **Acidophils** take on the pink stain of eosin. **Basophils,** on the other hand, stain readily with basic stains, such as methylene blue and crystal violet.

Six hormones are produced by these various cells in the anterior lobe. Except for somatotropin, all these hormones are targeted specifically for glands. Production of the hormones is regulated by the hypothalamus. Between the hypothalamus and adenohypophsis is a vascular connection, the *hypophyseal portal system,* that transports releasing factors from the hypothalamus to the secretory cells. The releasing factors are secreted by cells in the hypothalamus. The six hormones of the adenohypophysis are as follows:

Somatotropin (SH) This hormone, also called the **growth hormone (GH)**, is produced by **somatotrophs,** a type of acidophil. In males and nonpregnant females most of the acidophils are of this type.

Somatotropin increases the growth rate of all cells in the body by enhancing amino acid uptake and protein synthesis. Excess production of the hormone during the growing years causes **gigantism** due to the stimulation of growth in the epiphyses of the long bones. An adult with excess SH production develops **acromegaly,** which is characterized by enlargement of the mandible and forehead as well as of the small bones of the hands and feet. Deficiency of the hormone can cause **dwarfism.**

Prolactin (Luteotropic Hormone, LTH) This hormone is produced by another type of acidophil, called a **mammotroph.** With suitable staining methods, it is possible to differentiate these cells from the somatotrophs.

Prolactin promotes the production of milk in the breasts after childbirth. The release of this hormone prior to childbirth is inhibited by *PIF (prolactin-inhibiting factor),* which is produced in the hypothalamus. The presence of large amounts of estrogens and progesterone prior to childbirth causes the hypothalamus to produce the inhibiting factor.

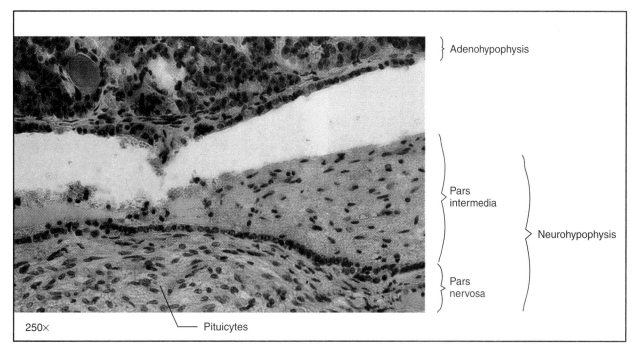

Figure 87.9 **The pituitary gland.**

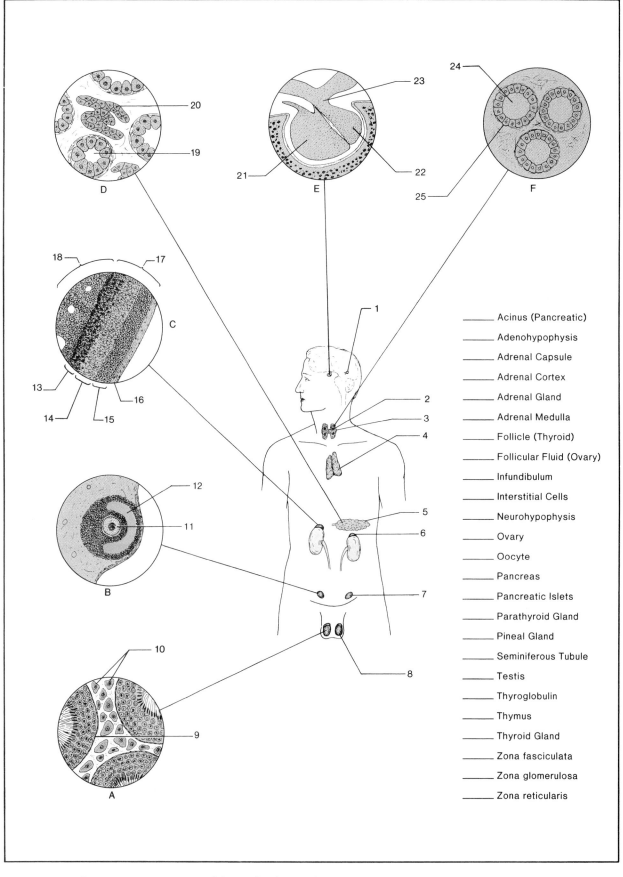

Figure 87.12 Secretory components of the endocrine system.

Acinus (Pancreatic)
Adenohypophysis
Adrenal Capsule
Adrenal Cortex
Adrenal Gland
Adrenal Medulla
Follicle (Thyroid)
Follicular Fluid (Ovary)
Infundibulum
Interstitial Cells
Neurohypophysis
Ovary
Oocyte
Pancreas
Pancreatic Islets
Parathyroid Gland
Pineal Gland
Seminiferous Tubule
Testis
Thyroglobulin
Thymus
Thyroid Gland
Zona fasciculata
Zona glomerulosa
Zona reticularis

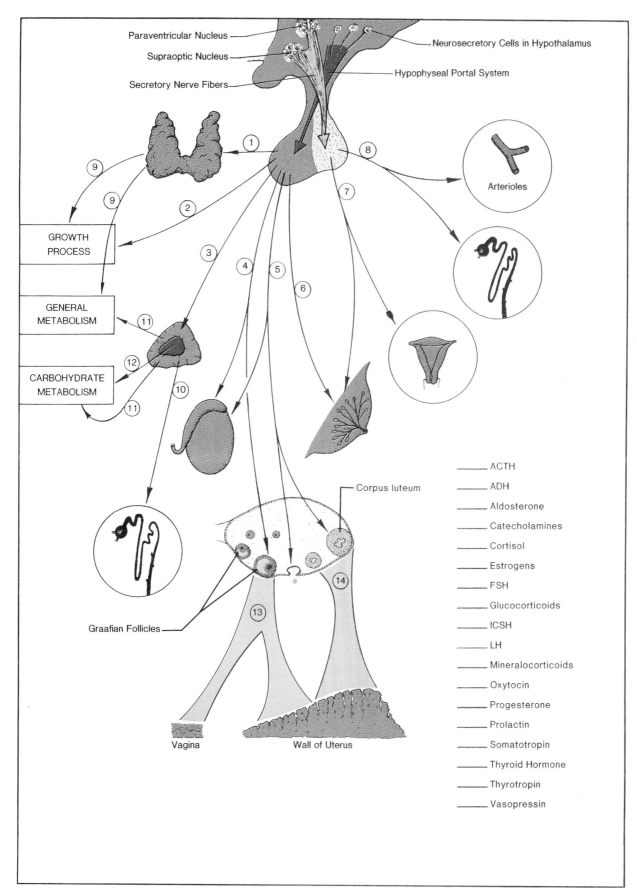

Paraventricular Nucleus

Supraoptic Nucleus

Secretory Nerve Fibers

Neurosecretory Cells in Hypothalamus

Hypophyseal Portal System

Arterioles

GROWTH PROCESS

GENERAL METABOLISM

CARBOHYDRATE METABOLISM

Corpus luteum

Graafian Follicles

Vagina

Wall of Uterus

_____ ACTH

_____ ADH

_____ Aldosterone

_____ Catecholamines

_____ Cortisol

_____ Estrogens

_____ FSH

_____ Glucocorticoids

_____ ICSH

_____ LH

_____ Mineralocorticoids

_____ Oxytocin

_____ Progesterone

_____ Prolactin

_____ Somatotropin

_____ Thyroid Hormone

_____ Thyrotropin

_____ Vasopressin

Figure 87.13   The pituitary regulatory mechanism.

88

The Reproductive System

This exercise pertains to the human reproductive system, using cat dissection for anatomical studies. The histology of the various organs, as well as microscopic studies of spermatogenesis and oogenesis, are also included.

Materials:

models of human reproductive organs
prepared slides of human testis
ocular micrometer
cat and dissection instruments

The Male Organs

Figure 88.1 is a sagittal section of the male reproductive organs. The primary sex organs are the paired oval **testes** *(testicles)*, which lie enclosed in a sac, the **scrotum.** Partially surrounding each testis is a double-layered serous membrane, the **tunica vaginalis** (label 1), which is a downward extension of the peritoneum into the scrotal sac. It is formed during the descent of the testes from the body cavity prior to birth.

Between the tunica vaginalis and the outer skin of the scrotum is a superficial fascia (not shown) and the cremaster muscle. Only a small portion of the **cremaster muscle** (label 6) is shown on the spermatic cord in figure 88.1. Since this muscle is temperature sensitive, it is capable of raising or lowering the testis to maintain an internal testicular temperature of 94°–95° F. Higher temperatures tend to inhibit spermatozoa maturation.

Note in the enlarged sectional view of the testis that it is divided into several chambers by **testicular septa.** Within each chamber lie several coiled-up **seminiferous tubules** where the spermatozoa are produced. Note also that the seminiferous tubules unite to form a plexus of canals called the **rete testis,** and leading from this plexus are several ducts called **vasa efferentia.**

Mature spermatozoa pass from the seminiferous tubules through the rete testis and vasa efferentia to storage ducts within the epididymis. The **epididymis** (label 26) consists of a mass of coiled-up tubules on the posterior, exterior surface of the testis.

In the act of ejaculation the spermatozoa leave the epididymis by way of a tube, the **vas deferens** *(ductus deferens),* which is the central canal of the **spermatic cord** (label 5). In addition to the vas deferens, the spermatic cord also contains nerves and blood vessels that supply the testis.

Tracing the vas deferens upward, we see that it passes through the inguinal canal, over the pubic bone and bladder into the pelvic cavity. The terminus of the vas deferens is enlarged to form the **ampulla of the vas deferens.**

Ducts from a pair of glands, the **seminal vesicles** (color code, blue) and the ampulla join to form a **common ejaculatory duct** that passes through the prostate gland to join the **prostatic urethra.** From here the urethra passes through the **urogenital diaphragm** (label 27) and into the penis, where it becomes the **penile urethra.**

The prostate gland, seminal vesicles, and bulbourethral glands contribute alkaline secretions to the seminal fluid. This alkalinity stimulates sperm motility. The **prostate gland,** which is positioned inferior to the urinary bladder, is the largest of these secondary sex glands.

The **bulbourethral** *(Cowper's)* **glands** (color code, blue) are the smallest of the three glands. These pea-sized glands have ducts about one inch long that empty into the urethra at the base of the penis. The secretion of these latter glands is a clear mucoid fluid that lubricates the end of the penis and prepares the urethra for seminal fluid.

The **penis** consists of three cylinders of erectile tissue: one corpus spongiosum and two corpora cavernosa. The **corpus spongiosum** is a cylinder of tissue that surrounds the penile urethra. Both ends of this body are enlarged: its proximal end forms the **bulb of the penis,** and its distal enlarged end is the **glans penis.** The enlarged portion of the urethra within the glans is the **navicular fossa.**

The two **corpora cavernosa** are located in the dorsal part of the organ and are separated on the midline by a **septum penis.** The cross section of the penis in figure 88.1 shows best the relationship of the three corpora to the penile urethra and septum. This view also reveals that a **tunica albuginea** ensheathes the corpora cavernosa and a **tunica**

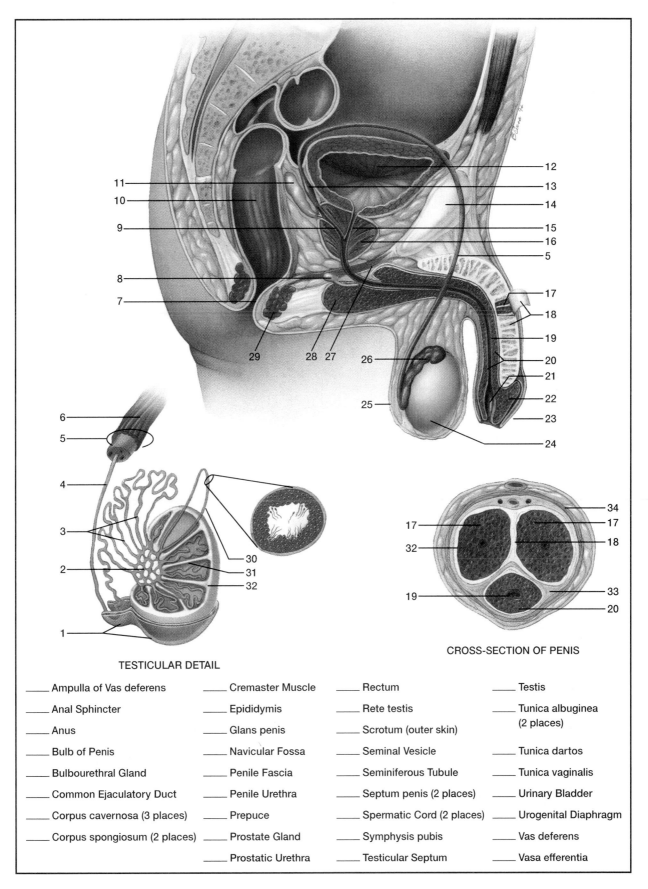

Figure 88.1 Male reproductive organs.

TESTICULAR DETAIL

CROSS-SECTION OF PENIS

_____ Ampulla of Vas deferens

_____ Anal Sphincter

_____ Anus

_____ Bulb of Penis

_____ Bulbourethral Gland

_____ Common Ejaculatory Duct

_____ Corpus cavernosa (3 places)

_____ Corpus spongiosum (2 places)

_____ Cremaster Muscle

_____ Epididymis

_____ Glans penis

_____ Navicular Fossa

_____ Penile Fascia

_____ Penile Urethra

_____ Prepuce

_____ Prostate Gland

_____ Prostatic Urethra

_____ Rectum

_____ Rete testis

_____ Scrotum (outer skin)

_____ Seminal Vesicle

_____ Seminiferous Tubule

_____ Septum penis (2 places)

_____ Spermatic Cord (2 places)

_____ Symphysis pubis

_____ Testicular Septum

_____ Testis

_____ Tunica albuginea (2 places)

_____ Tunica dartos

_____ Tunica vaginalis

_____ Urinary Bladder

_____ Urogenital Diaphragm

_____ Vas deferens

_____ Vasa efferentia

dartos surrounds all three of the corpora. Between the tunica dartos and tunica albuginea is a layer of connective tissue called the **penile fascia,** or *Buck's fascia.* It is during sexual arousal that erection of the penis is achieved by filling of these three corpora with blood.

Over the end of the glans penis lies a circular fold of skin, the **prepuce.** Around the neck of the glans, and on the inner surface of the prepuce, are scattered small *preputial glands.* These sebaceous glands produce a secretion that readily undergoes decomposition to form a whitish substance called *smegma. Circumcision* is a surgical procedure that involves the removal of the prepuce to facilitate sanitation.

Assignment:
Label figure 88.1.

The Female Organs

Figures 88.2, 88.3, and 88.4 illustrate the female reproductive organs. Models and wall charts will be helpful in identifying all the structures.

External Genitalia

The **vulva** of the external female reproductive organs includes the mons pubis, labia majora, labia minora, and hymen. All of these structures are shown in figure 88.2. The **mons pubis** *(mons veneris)* is the most anterior portion and consists of a firm cushionlike elevation over the symphysis pubis. It is covered with hair.

Two folds of skin on each side of the **vaginal orifice** (label 11) lie over the opening. The larger exterior folds are the **labia majora.** Their exterior surfaces are covered with hair; their inner surfaces are smooth and moist. The labia majora are homologous (of similar embryological origin) to the scrotum of the male.

Medial to the labia majora are the smaller **labia minora.** These folds meet anteriorly on the median line to form a fold of skin, the **prepuce of the clitoris.** The **clitoris** is a small protuberance of erectile tissue under the prepuce that is homologous to the penis in males. It is highly sensitive to sexual excitation. The fold of skin that extends from the clitoris to each labium minus is called the **frenulum of the clitoris.** Posteriorly, the labia minora join to form a transverse fold of skin, the **posterior commissure,** or *fourchette.* Between the anus and the posterior commissure is the **central tendinous point of the perineum.** The **perineum** corresponds to the outlet of the pelvis.

The *vestibule* is the area between the labia minora that extends from the clitoris to the vaginal orifice, hymen, urethral orifice, and fourchette. Situated within the vestibule are the vaginal orifice,

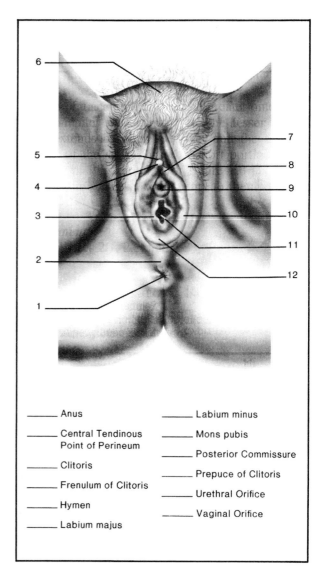

_____ Anus	_____ Labium minus
_____ Central Tendinous Point of Perineum	_____ Mons pubis
_____ Clitoris	_____ Posterior Commissure
_____ Frenulum of Clitoris	_____ Prepuce of Clitoris
_____ Hymen	_____ Urethral Orifice
_____ Labium majus	_____ Vaginal Orifice

Figure 88.2   Female genitalia.

hymen, urethral orifice, and openings of the vestibular glands.

The **urethral orifice** lies about 2–3 centimeters posterior to the clitoris. Many small **paraurethral glands** surround this opening. These glands are homologous to the prostate gland of the male.

On either side of the vaginal orifice are openings from the two **greater vestibular** *(Bartholin's)* **glands.** These glands are homologous to the bulbourethral glands in males. They provide mucous secretions for vaginal lubrication.

The **hymen** is a thin fold of mucous membrane that separates the vagina from the vestibule. It may be completely absent or cover the vaginal orifice partially or completely. Its condition or absence is not a determinant of virginity.

Assignment:
Label figure 88.2.

Internal Organs

Figure 88.3 is a posterior view of the female reproductive organs that shows the relationship of the vagina, uterus, ovaries, and uterine tubes. It also reveals many of the principal supporting ligaments.

The Uterus The uterus is a pear-shaped, thick-walled, hollow organ that consists of a fundus, corpus, and cervix. The **fundus** is the dome-shaped portion at the large end. The body, or **corpus,** extends from the fundus to the cervix. The **cervix,** or neck, of the uterus is the narrowest portion; it is about one inch long. Within the cervix is an **endocervical canal** that has an outer opening, the **external os,** that opens into the **vagina.** An inner opening, the **internal os,** opens into the cavity of the corpus.

The thick muscular portion of the uterine wall is called the **myometrium.** The inner lining, or **endometrium,** consists of mucosal tissue. The ante-rior and posterior surfaces of the corpus, as well as the fundus, are covered with peritoneum.

The Uterine *(Fallopian)* **Tubes** Extending laterally from each side of the uterus is a *uterine tube.* The distal end of each tube is enlarged to form a funnel-shaped fimbriated **infundibulum** that surrounds an **ovary.** Although the infundibulum doesn't usually contact the ovary, one or more of the fingerlike **fimbriae** on the edge of the infundibulum usually do contact it. The infundibula and fimbriae receive oocytes produced by the ovaries.

The uterine tubes are lined with ciliated columnar cells that assist in transporting egg cells from the ovary into the uterus. The tubes also have a muscular wall, which produces peristaltic movements that help propel the egg cells. Fertilization usually occurs somewhere along the uterine tube. Note that each uterine tube narrows down to form an **isthmus** near the uterus.

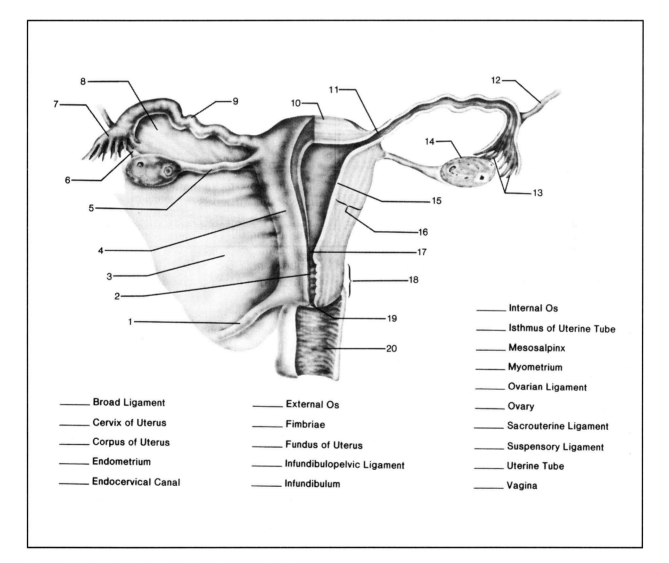

_____ Broad Ligament

_____ Cervix of Uterus

_____ Corpus of Uterus

_____ Endometrium

_____ Endocervical Canal

_____ External Os

_____ Fimbriae

_____ Fundus of Uterus

_____ Infundibulopelvic Ligament

_____ Infundibulum

_____ Internal Os

_____ Isthmus of Uterine Tube

_____ Mesosalpinx

_____ Myometrium

_____ Ovarian Ligament

_____ Ovary

_____ Sacrouterine Ligament

_____ Suspensory Ligament

_____ Uterine Tube

_____ Vagina

Figure 88.3 Posterior view of female reproductive organs.

The Urinary Bladder The sagittal section in figure 88.4 reveals the relationship of the *urinary bladder* to the reproductive organs. Note that it lies between the uterus and the **symphysis pubis** (label 5). Observe, also, that the bladder is in direct contact with the anterior wall of the vagina. The short **urethra** passes from the bladder through the **urogenital diaphragm** (label 25) to the *urethral orifice* within the vulva.

The superior surface of the bladder is covered with peritoneum that is continuous with the peritoneum on the anterior face of the uterus. The small cavity lined with peritoneum between the bladder and uterus is known as the **anterior cul-de-sac.** The space between the uterus and rectum (label 18) is the **rectouterine pouch,** or *pouch of Douglas.*

Ligaments The uterus, ovaries, and uterine tubes are held in place and supported by various ligaments formed from the muscle and connective tissue encased in peritoneum, or simply folds of peritoneum. Figure 88.3 reveals the majority of them. A few are shown in figure 88.4.

The largest supporting structure is the **broad ligament** (label 3, figure 88.3). It is an extension of the peritoneal layers that covers the fundus and corpus of the uterus. Extending up over the uterine tubes, this ligament forms a mesentery, the **mesosalpinx,** on each side of the uterus. This portion of the broad ligament, which is shown in figure 88.3 between the ovary and uterine tube, contains blood vessels that supply nutrients to the uterine tube.

Attached to the cervical region of the uterus are two **sacrouterine ligaments** that anchor the uterus to the sacral wall of the pelvic cavity. A **round ligament** (label 16, figure 88.4) extends from each side of the uterus to the body wall. This ligament is not shown in figure 88.3 because it is anterior in position.

Each ovary is held in place by ovarian and suspensory ligaments. The **ovarian ligament** extends from the medial surface of the ovary to the uterus. The **suspensory ligament** is a peritoneal fold on the other side of the ovary that attaches the ovary to the uterine tube.

Each uterine tube is held in place by an **infundibulopelvic ligament** that is attached to the posterior surface of the infundibulum.

Assignment:
Label figures 88.3 and 88.4.

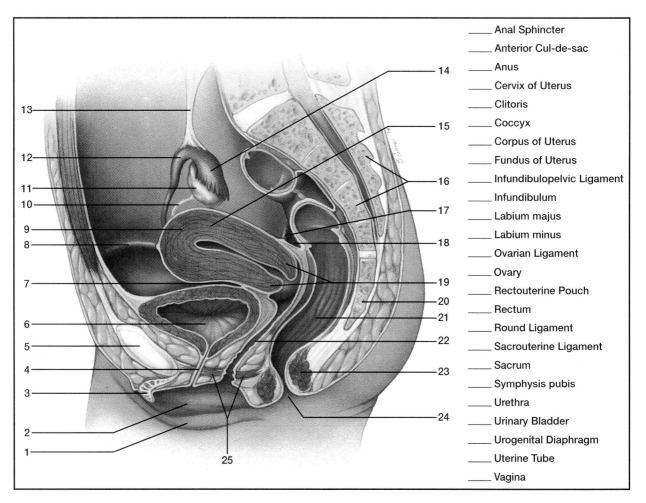

_____ Anal Sphincter
_____ Anterior Cul-de-sac
_____ Anus
_____ Cervix of Uterus
_____ Clitoris
_____ Coccyx
_____ Corpus of Uterus
_____ Fundus of Uterus
_____ Infundibulopelvic Ligament
_____ Infundibulum
_____ Labium majus
_____ Labium minus
_____ Ovarian Ligament
_____ Ovary
_____ Rectouterine Pouch
_____ Rectum
_____ Round Ligament
_____ Sacrouterine Ligament
_____ Sacrum
_____ Symphysis pubis
_____ Urethra
_____ Urinary Bladder
_____ Urogenital Diaphragm
_____ Uterine Tube
_____ Vagina

Figure 88.4 Midsagittal section of female reproductive organs.

474

Gametogenesis

When the ovum is fertilized by the sperm, 23 chromosomes from each gamete unite to form a zygote of 46 chromosomes. When the gametes are produced in the testes and ovaries from primordial germ cells, the reduction of chromosome numbers is achieved by a series of cell divisions. In the testes this process is called **spermatogenesis.** In the ovaries, a similar process is called **oogenesis.** Figures 88.5 and 88.6 illustrate these two processes.

Spermatogenesis

The production of spermatozoa begins at puberty and is continuous without interruption throughout life. Figure 88.5 illustrates a section of a seminiferous tubule and a diagram of the various stages in the development of mature spermatozoa. The formation of these germ cells takes place as the result of both mitosis and meiosis.

In this study of spermatogenesis, slides of human or animal testes will be examined under the microscope. Before examining the slides, however, familiarize yourself with the characteristic differences between meiosis and mitosis.

Materials:
prepared stained slides of:
    human testis (H9.413 or H9.42)
    rat testis (E17.12)
    rabbit testis (E17.22)

Mitosis

All spermatozoa originate from **spermatogonia** of the **primary germinal epithelium.** This layer of cells is located at the periphery of the seminiferous tubule. These cells contain the same number of chromosomes as other body cells (i.e., 23 pairs) and are said to be *diploid.* Mitotic division of these cells occurs constantly, producing other spermatogonia.

As the spermatogonia move toward the center of the tubule, they enlarge to become **primary spermatocytes.** In this stage the homologous chromosomes unite to form chromosomal units called *tetrads.* This union of homologous chromosomes is called *synapsis.*

Meiosis

Each primary spermatocyte divides further to produce two **secondary spermatocytes.** This division occurs by meiosis instead of mitosis. *Meiosis,* or reduction division, results in the distribution of a *haploid* (half) number of chromosomes to each secondary spermatocyte. The result of this process is

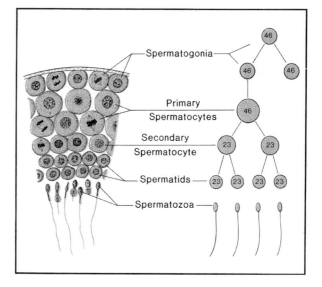

Figure 88.5 Spermatogenesis.

that each secondary spermatocyte has only twenty-three chromosomes instead of forty-six. The individual chromosomes of the secondary spermatocytes are formed by splitting of the tetrads along the line of previous conjugation to form chromosomes called *dyads.*

A second meiotic division occurs when each secondary spermatocyte divides to produce two haploid **spermatids.** In this division the dyads split to form single chromosomes, or *monads.* Each spermatid metamorphoses directly into a mature sperm cell. Note that four **spermatozoa** form from each spermatogonium.

Assignment:
Examine slides of human, rat, or rabbit testicular tissue to identify the various stages of spermatogenesis. Of the three listed in the materials list, above, the rabbit may be best to use.

Refer to figure HA-40 of the Histology Atlas to identify the various types of cells.

Draw a representative section of a tubule, labeling all cell types.

Oogenesis

The production of egg cells by the ovaries also begins at puberty. It was pointed out on page 463 that as many as 400,000 immature follicles develop in the ovaries from the germinal epithelium. At the onset of puberty all the potential ova in these follicles contain 46 chromosomes and are designated as **primary oocytes.** Only about 400 of these oocytes will mature during the reproductive lifetime of a woman.

Note in figure 88.6 that the primary oocyte, which forms from an oogonium of the germinal epithelium during embryological development, has

46 chromosomes. In the prophase stage of a dividing oocyte double-stranded chromosomes *(dyads)* unite to form 23 homologous pairs of chromosomes during synapsis. Each pair is made up of four chromatids; thus, it is called a *tetrad.* When the primary oocyte divides in the first meiotic division each dyad member of each homologous pair enters a different cell. The first meiotic division produces a **secondary oocyte** with all the yolk of the primary oocyte and a **polar body** with no yolk. The secondary oocyte and polar body each contain 23 dyads.

The second meiotic division produces a **mature ovum** with 23 chromosomes that are *monads;* that is, each chromosome consists of a single chromatid. Another polar body is also formed in this division. The first polar body divides also to produce two new polar bodies with monads. The end result is that one primary oocyte produces one mature ovum and three polar bodies. The polar bodies never serve any direct function in the fertilization process.

When ovulation takes place the "ovum" is a secondary oocyte. Usually, it is necessary for the sperm to penetrate the cell to trigger the second meiotic division. When the 23 chromosomes in the sperm unite with the chromosomes of the mature ovum, a zygote of 46 chromosomes is formed.

Assignment:
Study figures 88.6 and 87.8 (ovarian tissue) thoroughly to familiarize yourself with the terminology and kinds of divisions. Several questions on the Laboratory Report pertain to this phenomenon.

Cat Dissection

In Exercise 85 the study of the urinary system necessitated a cursory examination of the reproductive organs in the cat. It will now be necessary to study these organs in greater detail. You may find it necessary to refer to figures 85.4 and 85.5 in some instances in this study as well as to figures 88.7 and 88.8 on these pages when identifying the various organs in this dissection.

The Male Reproductive System

This dissection will begin with an examination of the penis, which will necessitate the eventual removal of the right hind leg to reveal the inner structures. Proceed as follows:

1. Locate the **penis.** The skin that forms the sheath around the organ is called the **prepuce.** Pull back the prepuce to expose the enlarged end, which is the **glans.** The glans in the cat is covered with minute horny papillae.
2. Remove the **scrotum,** which is the skin that encloses the two **testes.** Note that each testis in covered by a *fascial sac.* Do not open this sac at this time.
3. Trace the **spermatic cord** from one of the testes to the body wall. It is composed of the **vas deferens** *(ductus deferens),* blood vessels, nerves and lymphatics that supply and drain the testis. Note that the covering of the spermatic cord is continuous with the fascial sac of the testis.

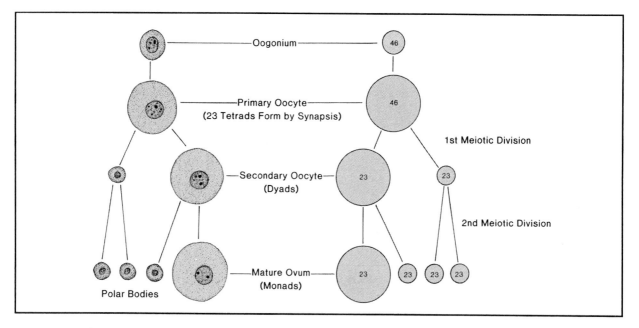

Figure 88.6   Oogenesis.

4. The vas deferens and blood vessels pass through the abdominal wall via the *inguinal canal.* From the inguinal canal the vas deferens passes up over the ureter and enters the penis near the prostate gland.

To be able to study the structures shown in figure 88.7, remove the right leg near the hip joint. To do this it will be necessary to cut through the symphysis pubis.

It will also be necessary to cut through the hipbone above the acetabulum. Clear away the pelvic muscles to expose the structures as shown in figure 88.7.

5. Locate the **prostate gland** where the vas deferens joins the urethra. The tube leading from the prostate through the penis is the **urogenital canal** (penile urethra). At the base of the penis will be seen a pair of **bulbourethral glands.**

6. Cut through the penis to produce a transverse section. Identify the two **corpora cavernosa** that cause erection. Near the symphysis pubis the two corpora diverge forming the **right** and **left crura** (*crus*, singular). Each crus is attached

to a part of the ischium. Refer to figure 85.4 to see both crura.

7. Cut through the fascial sac, exposing the testis. The space between this sac and the testis is called the **vaginal sac.** This cavity is a downward extension of the peritoneal cavity.

8. Trim away the fascial sac and trace the vas deferens toward the testis, noting its convoluted nature. Identify the **epididymis,** which lies on the dorsal side of the testis and is continuous with the vas deferens.

9. With a sharp scalpel or razor blade cut through the testis and the epididymis to produce a section similar to the lower left illustration in figure 88.1. Identify the structures labeled in that diagram.

The Female Reproductive System

Examination of the female reproductive organs will begin with the ovaries. Figure 85.5 will be useful for this portion of the study. Examination of the lower portion of the reproductive tract will necessitate removal of the right leg. Proceed as follows:

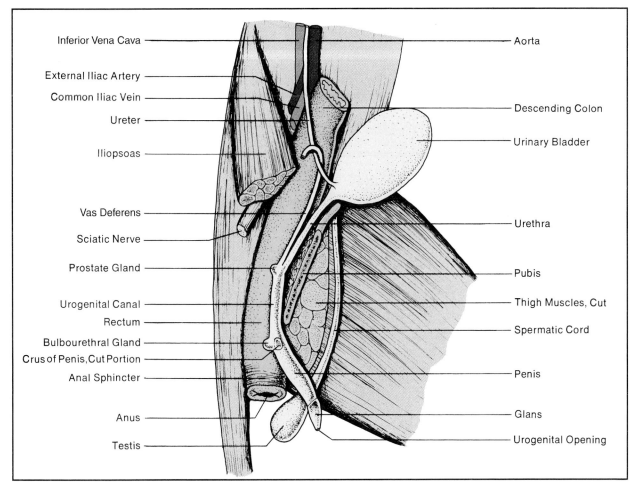

Figure 88.7   Lateral view of the urogenital system of the male cat.

1. Identify the small, light-colored oval ovaries, which lie posterior to the kidneys. Note that they are held by a fold of tissue to the dorsal body wall. This peritoneal fold is called the **mesovarium.** Note, also, that the ovary is attached to the uterine horn by the **ovarian ligament.**

2. Identify the **uterine tube** (oviduct) with its enlarged funnel-like terminis, the **infundibulum** (abdominal ostium). Note that the perimeter of the infundibulum has many irregular projections called **fimbriae.**

   When ova leave the ovary they are drawn into the uterine tube through the infundibulum by ciliated epithelium on the tissues of the area.

3. Observe that the uterus is Y-shaped, having two horns and a body. The **uterine horns** lie along each side of the abdominal cavity. The **uterine body** lies between the urinary bladder and the rectum (see figure 88.8).

4. To be able to observe the relationship of the vagina to other structures, as shown in figure 88.8, remove the right leg near the hip joint. To do this, it will be necessary to cut through the hipbone above the acetabulum. Remove the cut portion of the hipbone and clear away the pelvic muscles to expose the structures, as shown in figure 88.8. Preserve the blood vessels as much as possible.

5. Identify the general areas of the uterine body, vagina, and urogenital sinus.

6. Remove the entire female reproductive tract and open it to reveal the inner structures shown in figure 85.5, which include the **cervix, vagina,** and **urogenital sinus.**

Histological Study

Assuming that spermatogenesis slides have been previously studied, we will focus here, primarily, on a histological study of the various reproductive organs of both sexes. Appropriate pages of the Histology Atlas and some of the illustrations in Exercise 87 (Endocrine Glands) will be used for reference.

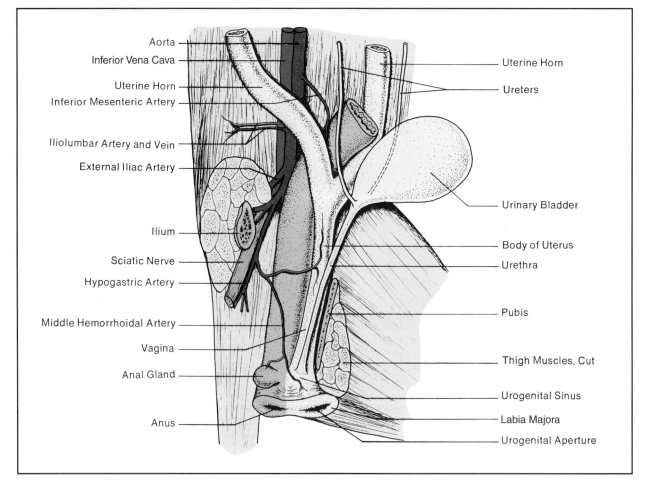

Figure 88.8 Lateral view of the urogenital system of the female cat.

Materials:

prepared slides:
    ovary (H9.310 or H9.3181)
    uterus (H9.331)
    vagina (H9.341)
    uterine tube (H9.321)
    vas deferens (H9.432)
    penis (H9.462)
    prostate gland (H9.443)
    seminal vesicle (H9.442)

Ovary   Study various slides of the ovary to identify **primordial follicles,** the **germinal epithelium,** and **Graafian follicles.** Since the ovaries are not included in the Histology Atlas, use figure 87.8, page 464, for reference.

Note that the developing ova in primordial follicles are surrounded by an incomplete layer of low cuboidal or flattened epithelium.

Mature Graafian follicles are not always seen on some slides. If sectioned properly, you should see a large space surrounding the oocyte that contains large quantities of estrogen produced by the follicular cells.

Uterus   Differentiate the **endometrium** from the **myometrium.** Locate the **uterine glands** and **endometrial epithelium,** which consists of **columnar cells.** Refer to figure HA-34.

Uterine Tube   Scan a slide of the uterine tube with low-power objective to find the lumen of the tube. Refer to figure HA-35. Note the extremely irregular outline of the mucosa, which consists of a mixture of ciliated and nonciliated columnar epithelium. How many layers of smooth muscle tissue do you find in the muscularis?

Vagina   Examine a slide of the mucosa of the vagina and compare it with illustration C, figure HA-36. How does the vaginal mucosa differ from the skin? What function do the lymphocytes perform in the lamina propria?

Uterine Cervix   Study a longitudinal section through the cervical area of the uterus that shows where the vaginal lining meets the cervical region.

Consult illustration A, figure HA-36. Note the large number of spaces, or **cervical crypts,** that lie deep in the cervical tissue. Read the part of the legend in figure HA-36 that pertains to the significance of these crypts.

Vas Deferens   Examine a cross-sectional slide of the vas deferens, using minimum magnification to study the entire structure. Compare your slide with figure HA-38.

Note that between its inner **epithelium** and outer **adventitia** are three layers of smooth muscle tissue: an inner longitudinal layer, an outer longitudinal layer, and a middle layer of circular fibers. Read what the legend states about the **stereocilia** that are seen on the epithelium.

Penis   Examine a cross-sectional preparation of the penis. The photomicrographs in illustrations C and D, figure HA-37, were made from an H9.462 Turtox slide.

Examine the epithelium under high-dry magnification. Does the epithelium on your slide look like illustration D? If not, why not? Read the comments in the legend that pertain to epithelial differences.

Seminal Vesicle   A cross section of a portion of a seminal vesicle is seen in illustrations A and B, figure HA-39. Examine a slide first with low-power magnification and note the striking way in which the mucosa is folded upon itself to produce a maze of pockets.

Examine the epithelium under high-dry or oil immersion. What is the nature and function of the seminal fluid?

Prostate Gland   Examine a slide of this gland under low power first to identify the structures shown in illustration C, figure HA-39. Note the large number of **tubulo-alveolar glands** that secrete the prostatic fluid. Why does the prostate gland tend to become hardened and restrict micturition in older men?

Laboratory Report

Complete the Laboratory Report for this exercise.

Histology Self-Quiz No. 4

If at this point you have completed all the histology assignments in this exercise and Exercises 70, 85, 87, and 88 you might wish to test your knowledge of histology in these areas by taking the last self-quiz on page 495. It consists of nineteen photomicrographs and sixty-six questions. After recording your answers on a sheet of paper, check your answers against the key, which you will find on page 456.

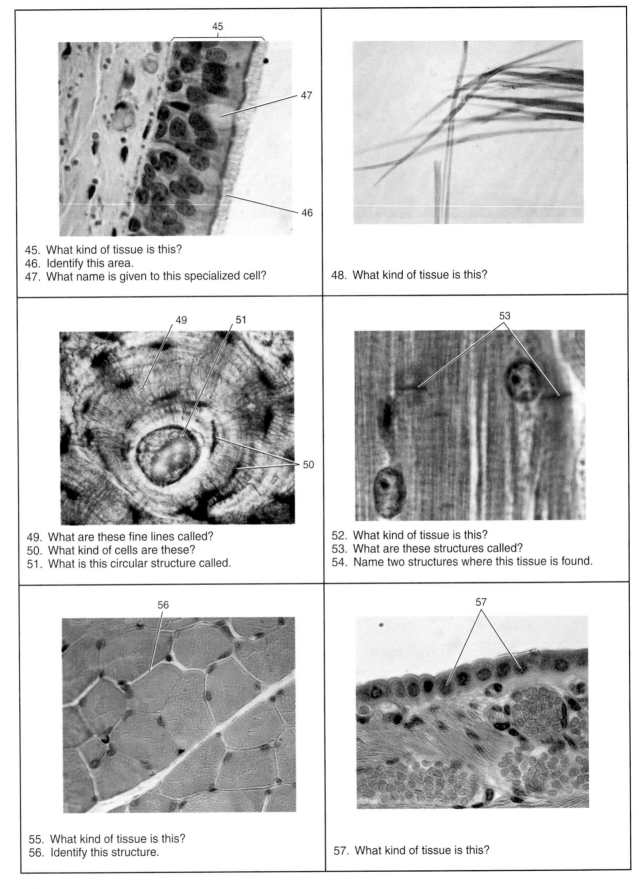

45. What kind of tissue is this?
46. Identify this area.
47. What name is given to this specialized cell?

48. What kind of tissue is this?

49. What are these fine lines called?
50. What kind of cells are these?
51. What is this circular structure called.

52. What kind of tissue is this?
53. What are these structures called?
54. Name two structures where this tissue is found.

55. What kind of tissue is this?
56. Identify this structure.

57. What kind of tissue is this?

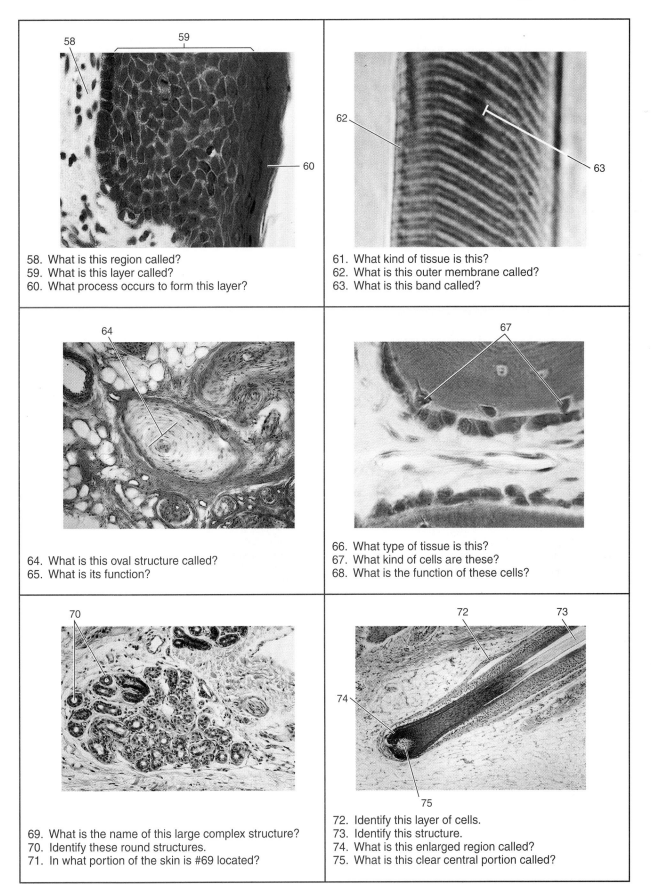

58. What is this region called?
59. What is this layer called?
60. What process occurs to form this layer?

61. What kind of tissue is this?
62. What is this outer membrane called?
63. What is this band called?

64. What is this oval structure called?
65. What is its function?

66. What type of tissue is this?
67. What kind of cells are these?
68. What is the function of these cells?

69. What is the name of this large complex structure?
70. Identify these round structures.
71. In what portion of the skin is #69 located?

72. Identify this layer of cells.
73. Identify this structure.
74. What is this enlarged region called?
75. What is this clear central portion called?

Histology Self-Quiz No. 2

Once you have completed all the exercises in Parts 8 and 9 and have studied the histology slides depicted in HA-13 through HA-18 of the Histology Atlas, test your knowledge of this material by taking this test. The answers to this quiz are on page 414. This will prepare you for a laboratory practical that may be given.

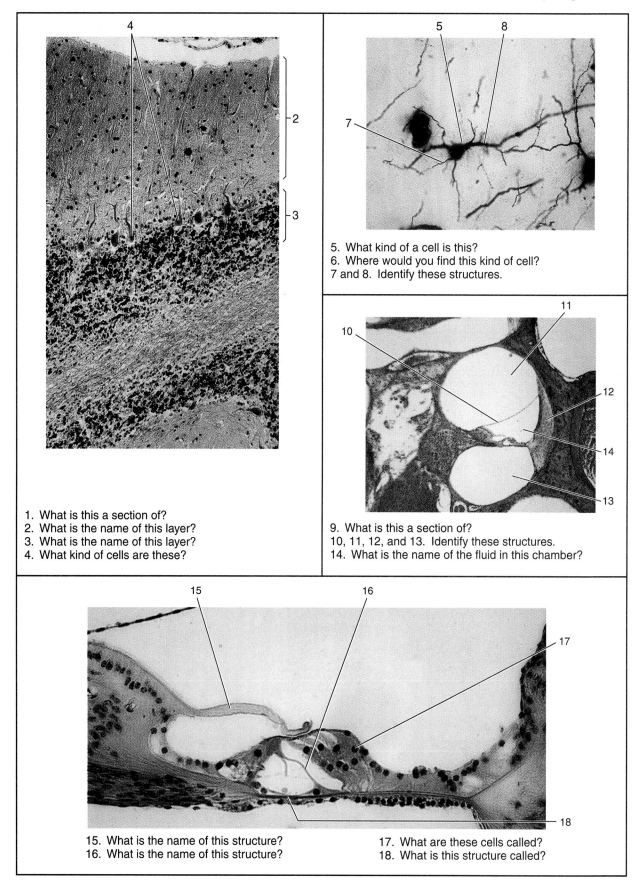

5. What kind of a cell is this?
6. Where would you find this kind of cell?
7 and 8. Identify these structures.

1. What is this a section of?
2. What is the name of this layer?
3. What is the name of this layer?
4. What kind of cells are these?

9. What is this a section of?
10, 11, 12, and 13. Identify these structures.
14. What is the name of the fluid in this chamber?

15. What is the name of this structure?
16. What is the name of this structure?

17. What are these cells called?
18. What is this structure called?

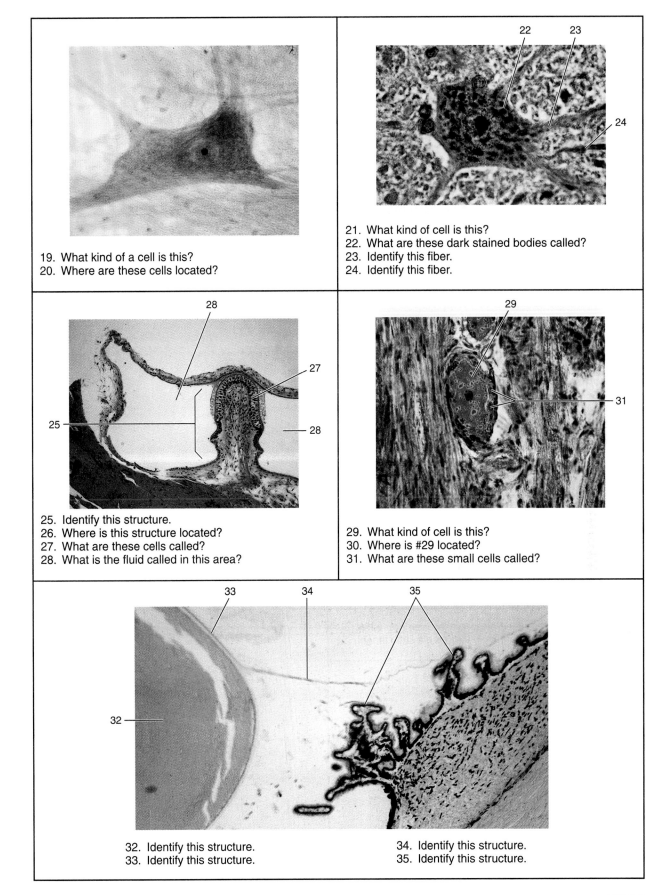

19. What kind of a cell is this?
20. Where are these cells located?

21. What kind of cell is this?
22. What are these dark stained bodies called?
23. Identify this fiber.
24. Identify this fiber.

25. Identify this structure.
26. Where is this structure located?
27. What are these cells called?
28. What is the fluid called in this area?

29. What kind of cell is this?
30. Where is #29 located?
31. What are these small cells called?

32. Identify this structure.
33. Identify this structure.

34. Identify this structure.
35. Identify this structure.

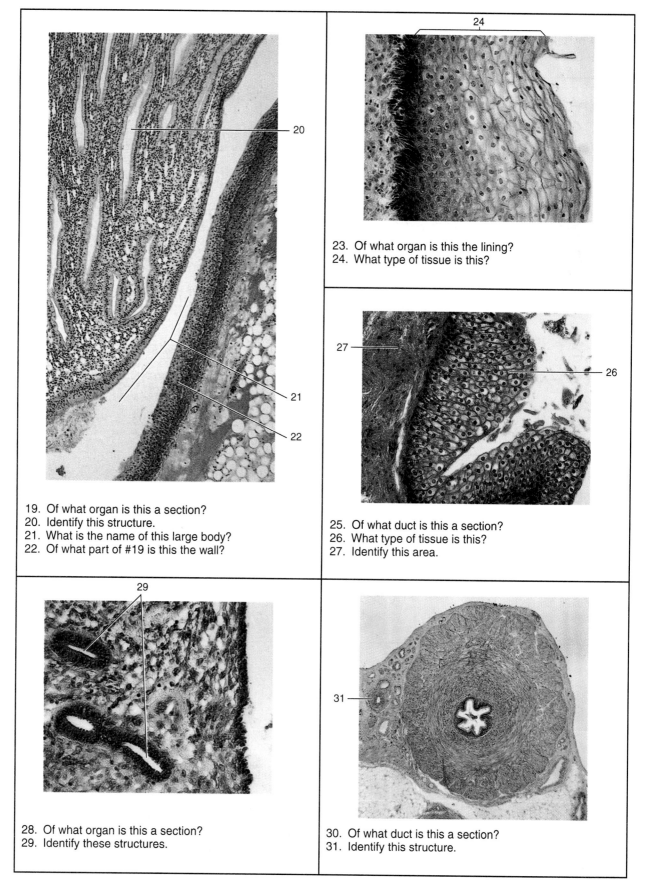

19. Of what organ is this a section?
20. Identify this structure.
21. What is the name of this large body?
22. Of what part of #19 is this the wall?

23. Of what organ is this the lining?
24. What type of tissue is this?

25. Of what duct is this a section?
26. What type of tissue is this?
27. Identify this area.

28. Of what organ is this a section?
29. Identify these structures.

30. Of what duct is this a section?
31. Identify this structure.

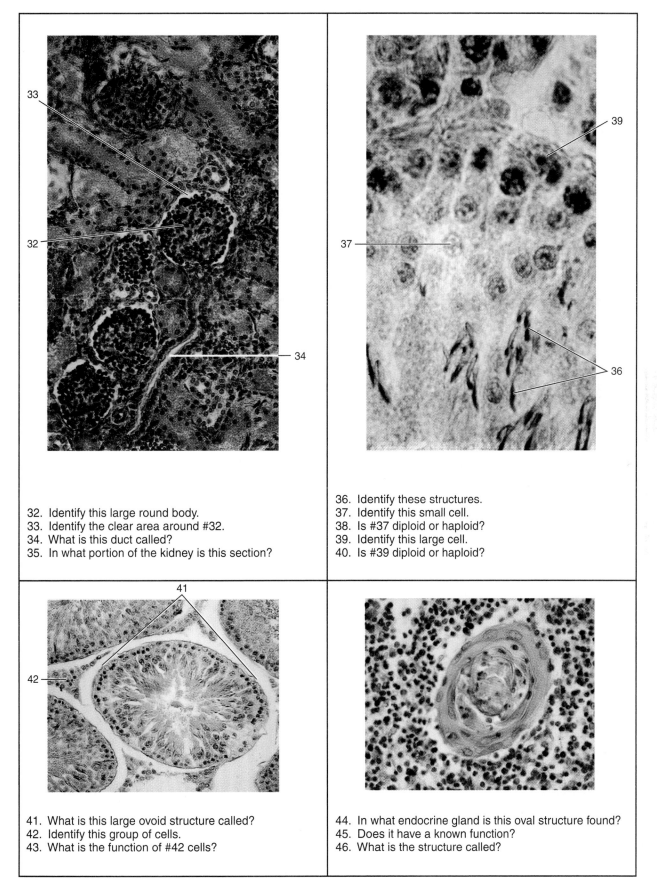

32. Identify this large round body.
33. Identify the clear area around #32.
34. What is this duct called?
35. In what portion of the kidney is this section?

36. Identify these structures.
37. Identify this small cell.
38. Is #37 diploid or haploid?
39. Identify this large cell.
40. Is #39 diploid or haploid?

41. What is this large ovoid structure called?
42. Identify this group of cells.
43. What is the function of #42 cells?

44. In what endocrine gland is this oval structure found?
45. Does it have a known function?
46. What is the structure called?

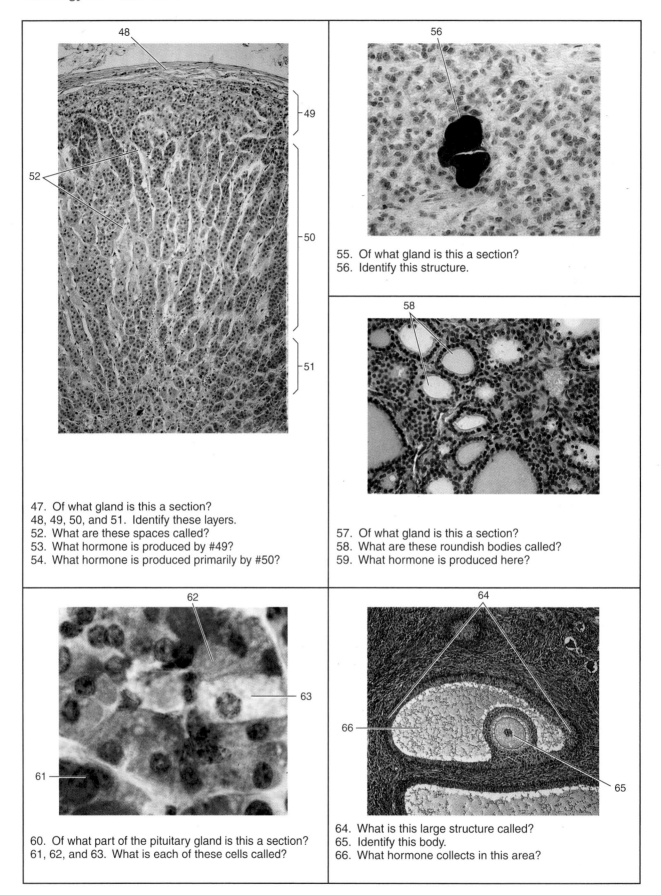

47. Of what gland is this a section?
48, 49, 50, and 51. Identify these layers.
52. What are these spaces called?
53. What hormone is produced by #49?
54. What hormone is produced primarily by #50?

55. Of what gland is this a section?
56. Identify this structure.

57. Of what gland is this a section?
58. What are these roundish bodies called?
59. What hormone is produced here?

60. Of what part of the pituitary gland is this a section?
61, 62, and 63. What is each of these cells called?

64. What is this large structure called?
65. Identify this body.
66. What hormone collects in this area?

LABORATORY REPORT
1,2

Student: _____

Desk No.: _____ Section: _____

Anatomical Terminology, Body Cavities, and Membranes

A. Illustration Labels
Record the label numbers for all the illustrations in the appropriate places in the answer columns.

B. Terminology
From the list of positions, sections, and membranes, select those that are applicable to the following statements. In some cases more than one answer may apply.

Relative Positions		Sections	Membranes
anterior—1	lateral—7	frontal—13	mesentery—17
caudal—2	medial—8	midsagittal—14	pericardium, parietal—18
cranial—3	posterior—9	sagittal—15	pericardium, visceral—19
distal—4	proximal—10	transverse—16	peritoneum, parietal—20
dorsal—5	superior—11		peritoneum, visceral—21
inferior—6	ventral—12		pleura, parietal—22
			pleura, pulmonary—23

Sections:
1. Divides body into front and back portions.
2. Divides body into unequal right and left sides.
3. Three sections that are longitudinal.
4. Divides body into equal right and left sides.
5. Two sections that expose both lungs and heart in each section.

Positions:
6. The tongue is _____ to the palate.
7. The shoulder of a dog is _____ to its hip.
8. A hat is worn on the _____ surface of the head.
9. The cheeks are _____ to the tongue.
10. The fingertips are _____ to all structures of the hand.
11. The human shoulder is _____ to the hip.
12. The shoulder is the _____ (*proximal, distal*) portion of the arm.

Membranes:
13. Serous membrane attached to lung surface.
14. Serous membrane attached to thoracic wall.
15. Double-layered membrane that holds abdominal organs in place.
16. Double-layered membrane that surrounds the heart.

Select the surfaces on which the following are located:
17. Ear
18. Adam's apple
19. Kneecap
20. Palm of hand

Answers

Terms	Fig. 1.1
1. _____	
2. _____	
3. _____	
4. _____	
5. _____	
6. _____	
7. _____	
8. _____	
9. _____	
10. _____	
11. _____	
12. _____	
13. _____	
14. _____	
15. _____	
16. _____	
17. _____	
18. _____	
19. _____	
20. _____	

Fig. 2.1	Fig. 1.3

C. Localized Areas

Identify the specific areas of the body described by the following statements.

antebrachium—1	cubital—8	hypogastric—15
antecubital—2	epigastric—9	iliac—16
axilla—3	flank—10	lumbar—17
brachium—4	gluteal—11	plantar—18
buttocks—5	groin—12	pectoral—19
calf—6	ham—13	popliteal—20
costal—7	hypochondriac—14	umbilical—21

1. The elbow area.
2. The "rump" area.
3. The underarm area.
4. The sole of the foot.
5. The upper arm.
6. The forearm.
7. Upper chest region.
8. Depression on back of leg behind knee.
9. Side of abdomen between lower edge of rib cage and upper edge of hipbone.
10. Back portion of abdominal body wall that extends from lower edge of rib cage to hipbone.
11. Area over the ribs on the dorsum.
12. Posterior portion of lower leg.
13. Anterior surface of the elbow.
14. Abdominal area that surrounds the navel.
15. Abdominal area that is lateral to pubic region.
16. Area of abdominal wall that covers the stomach.
17. Abdominal area that is lateral to epigastric area.
18. Abdominal area that is lateral to umbilical area.
19. Area of arm on opposite side of elbow.
20. Abdominal areas between the transpyloric and transtubercular planes.

D. Body Cavities

From the following list of cavities select those applicable to the following statements.

abdominal—1	dorsal—4	spinal—8
abdominopelvic—2	pelvic—5	thoracic—9
cranial—3	pericardial—6	ventral—10
	pleural—7	

1. Cavity that consists of cranial and spinal cavities.
2. Most inferior portion of the abdominopelvic cavity.
3. Cavity separated into two major divisions by the diaphragm.
4. Cavity inferior to the diaphragm that consists of two parts.

Select the cavity in which the following organs are located:

5. Kidneys	9. Rectum	13. Spleen
6. Duodenum	10. Pancreas	14. Spinal cord
7. Brain	11. Lungs and heart	15. Urinary bladder
8. Heart	12. Liver	16. Stomach

Answers

Areas	Fig. 1.2
1. _____	
2. _____	
3. _____	
4. _____	
5. _____	
6. _____	
7. _____	
8. _____	
9. _____	
10. _____	
11. _____	
12. _____	
13. _____	
14. _____	
15. _____	
16. _____	
17. _____	
18. _____	
19. _____	
20. _____	Fig. 2.2

Cavities	
1. _____	
2. _____	
3. _____	
4. _____	
5. _____	
6. _____	
7. _____	
8. _____	Fig. 2.3
9. _____	
10. _____	
11. _____	
12. _____	
13. _____	
14. _____	
15. _____	
16. _____	

LABORATORY REPORT

3

Student: _____

Desk No.: _____ Section: _____

Organ Systems:
Rat Dissection

A. System Functions

Select the system or systems that perform the following functions in the body. Record the numbers in the answer column. More than one answer may apply.

cardiovascular—1	lymphatic—5	respiratory—9
digestive—2	muscular—6	reticuloendothelial—10
endocrine—3	nervous—7	skeletal—11
integumentary—4	reproductive—8	urinary—12

1. Removes carbon dioxide from the blood.
2. Provides a shield of protection for vital organs such as the brain and heart.
3. Movement of ribs in breathing.
4. Carries heat from the muscles to the surface of the body for dissipation.
5. Provides rigid support for the attachment of muscles.
6. Transportation of food from the intestines to all parts of the body.
7. Disposes of urea, uric acid, and other cellular wastes.
8. Movement of arms and legs.
9. Helps to maintain normal body temperature by dissipating excess heat.
10. Destroys microorganisms once they break through the skin.
11. Phagocytic cells of this system remove bacteria from lymph.
12. Carries messages from receptors to centers of interpretation.
13. Controls the rate of growth.
14. Returns tissue fluid to the blood.
15. Produces various kinds of blood cells.
16. Enables the body to adjust to changing internal and external environmental conditions.
17. Ensures continuity of the species.
18. Eliminates excess water from the body.
19. Carries hormones from glands to all tissues.
20. Acts as a shield against invasion by bacteria.
21. Carries fats from the intestines to the blood.
22. Coordination of all body movements.
23. Breaks food particles down into small molecules.
24. Controls the development of the ovaries and testes.
25. Production of spermatozoa and ova.

Answers
System Functions
1. _____
2. _____
3. _____
4. _____
5. _____
6. _____
7. _____
8. _____
9. _____
10. _____
11. _____
12. _____
13. _____
14. _____
15. _____
16. _____
17. _____
18. _____
19. _____
20. _____
21. _____
22. _____
23. _____
24. _____
25. _____

B. Organ Placement

Select the system at the right that includes the following organs.

1.	Bones	cardiovascular—1
2.	Brain	digestive—2
3.	Bronchi	endocrine—3
4.	Dermis	integumentary—4
5.	Esophagus	lymphatic—5
6.	Hair	muscular—6
7.	Heart	nervous—7
8.	Kidneys	reproductive—8
9.	Lungs	respiratory—9
10.	Pancreas	skeletal—10
11.	Spleen	urinary—11
12.	Stomach	
13.	Teeth	
14.	Trachea	
15.	Toenails	
16.	Tonsils	
17.	Thyroid	
18.	Ureters	
19.	Uterus	
20.	Veins	

Answers

Organ Placement

1. _____
2. _____
3. _____
4. _____
5. _____
6. _____
7. _____
8. _____
9. _____
10. _____
11. _____
12. _____
13. _____
14. _____
15. _____
16. _____
17. _____
18. _____
19. _____
20. _____

LABORATORY REPORT

4

Student: _____

Desk No.: _____ Section: _____

Microscopy

A. Completion Questions

Record the answers to the following questions in the column at the right.

1. List three fluids that may be used for cleaning lenses.
2. How can one greatly increase the bulb life on a microscope lamp if the voltage is variable?
3. What characteristic of a microscope enables one to switch from one objective to another without altering the focus?
4. What effect (*increase* or *decrease*) does closing the diaphragm have on the following?
   a. Image brightness
   b. Image contrast
   c. Resolution
5. In general, at what position should the condenser be kept?
6. Express the maximum resolution of the compound microscope in terms of micrometers (μm).
7. If you are getting 225$\times$ magnification with a 45$\times$ high-dry objective, what is the power of the eyepiece?
8. What is the magnification of objects observed through a 100$\times$ oil immersion objective with a 7.5$\times$ eyepiece?
9. Immersion oil must have the same refractive index as _____ to be of any value.
10. Substage filters should be of a _____ color to get the maximum resolution of the optical system.

B. True–False

Record these statements as True or False in the answer column.

1. Eyepieces are of such simple construction that almost anyone can safely disassemble them for cleaning.
2. Lenses can be safely cleaned with almost any kind of tissue or cloth.
3. When swinging the oil immersion objective into position after using high-dry, one should always increase the distance between the lens and slide to prevent damaging the oil immersion lens.
4. Instead of starting first with the oil immersion lens, it is best to use one of the lower magnifications first, and then swing the oil immersion into position.
5. The 45$\times$ and 100$\times$ objectives have shorter working distances than the 10$\times$ objective.

Answers

Completion

1a. _____

b. _____

c. _____

2. _____

3. _____

4a. _____

b. _____

c. _____

5. _____

6. _____

7. _____

8. _____

9. _____

10. _____

True–False

1. _____

2. _____

3. _____

4. _____

5. _____

C. Multiple Choice

Select the best answer for the following statements.

1. The resolution of a microscope is increased by
    1. using blue light.
    2. stopping down the diaphragm.
    3. lowering the condenser.
    4. raising the condenser to its highest point.
    5. Both 1 and 4 are correct.

2. The magnification of an object seen through the 10× objective with a 10× ocular is
    1. ten times.
    2. twenty times.
    3. 1000 times.
    4. None of these are correct.

3. The most commonly used ocular is
    1. 5×.
    2. 10×.
    3. 15×.
    4. 20×.

4. Microscope lenses may be cleaned with
    1. lens tissue.
    2. a soft linen handkerchief.
    3. an air syringe.
    4. Both 1 and 3 are correct.
    5. 1, 2, and 3 are correct.

5. When changing from low power to high power, it is generally necessary to
    1. lower the condenser.
    2. open the diaphragm.
    3. close the diaphragm.
    4. Both 1 and 2 are correct.
    5. Both 1 and 3 are correct.

LABORATORY REPORT

5,6

Student: _____

Desk No.: _____ Section: _____

Cytology: Basic Cell Structure and Mitosis

A. Figure 5.1

Record the labels for figure 5.1 in the answer column.

B. Cell Drawings

Sketch in the space below one or two epithelial cells as seen under high-dry magnification. Label the **nucleus, cytoplasm,** and **cell membrane.** If mitosis drawings are to be made put them on a separate sheet of paper.

C. Cell Structure

From the list of structures below, select those that are described by the following statements. More than one structure may apply to some of the statements.

centrosome—1
chromatin granules—2
cilia—3
condensing vacuoles—4
cytoplasm—5
flagellum—6
Golgi apparatus—7

lysosome—8
microvilli—9
mitochondria—10
microfilaments—11
microtubules—12
nuclear envelope—13
nucleolus—14

nucleus—15
plasma membrane—16
polyribosomes—17
RER—18
ribosomes—19
secretory granules—20
SER—21

1. Prominent body in nucleus that produces ribosomes.
2. Trilaminar unit membrane structure.
3. Outer membrane of the cell.
4. A nonmembranous organelle consisting of two bundles of microtubules.
5. Bodies within the nucleus that represent chromosomal material.
6. ER that lacks ribosomes.
7. Double-layer unit membrane structure.
8. Small bodies of RNA and protein attached to surfaces of the RER.
9. Extensive system of tubules, vesicles, and sacs within the cytoplasm.
10. All protoplasmic material between the plasma membrane and nucleus.
11. Sacs in the cytoplasm that contain digestive enzymes.
12. A unit membrane organelle in the cytoplasm that contains cristae.
13. Short, hairlike appendages on the surface of some cells.
14. ER that is covered by ribosomes.
15. Chains and rosettes of small bodies scattered throughout the cytoplasm.
16. Unit membrane structure that is continuous with the plasma and nuclear membranes.
17. Extranuclear bodies that possess DNA and are unable to replicate themselves.
18. A layered concave structure in the cytoplasm that is similar to the basic structure of SER.
19. Surface protuberances that form a delicate brush border.
20. Bodies produced by Golgi and exocytosed by the cell.

Answers

Cell Structure	Fig. 5.1
1. _____	_____
2. _____	_____
3. _____	_____
4. _____	_____
5. _____	_____
6. _____	_____
7. _____	_____
8. _____	_____
9. _____	_____
10. _____	_____
11. _____	_____
12. _____	_____
13. _____	_____
14. _____	_____
15. _____	_____
16. _____	_____
17. _____	_____
18. _____	_____
19. _____	_____
20. _____	

D. Organelle Functions

Select the organelles that perform the following functions. More than one structure may apply to some statements.

centrosome—1
flagellum—2
Golgi apparatus—3
lysosome—4

microvilli—5
mitochondrion—6
nucleolus—7
nucleus—8

plasma membrane—9
polyribosomes—10
RER—11
SER—12

1. Forms asters during mitosis.
2. Produces ribonucleoprotein for ribosomes.
3. The microcirculatory system of the cell.
4. Contains genetic code of the cell.
5. Place where oxidative respiration occurs.
6. Synthesize protein for endogenous use.
7. Brings about rapid hydrolysis (digestion) of cell after death.
8. Place where ATP is synthesized and stored.
9. Regulates the flow of materials into and out of cell.
10. Synthesizes protein for extracellular distribution.
11. Disposal organelles in cytoplasm that digest protein.
12. Increases absorptive area of the cell.
13. Synthesis, packaging, and transportation of substances.
14. Plays a role in flagellar development.
15. Accomplishes some motile function for some cells.

E. True–False

Evaluate the validity of the following statements concerning cell structure and function.

1. Secretory cells have well-developed Golgi.
2. Micochondria are able to replicate themselves.
3. Practically all molecules that pass through the plasma membrane are actively assisted by the membrane.
4. Ribosomes are found in the cytoplasm, mitochondria, RER, and SER.
5. The inner and outer layers of the plasma membrane are essentially lipoidal in nature with scattered molecules of protein and glycoprotein.
6. A "brush border" consists of microvilli.
7. Spermatozoa move by means of cilia.
8. Ribosomes are primarily involved in oxidative respiration.
9. Cilia form from basal bodies derived from centrioles.
10. Asters develop from mitochondria.

F. Mitosis (Ex. 6)

Select the phase in which the following events occur.

1. Nuclear membrane disappears.
2. Cleavage furrow forms.
3. Spindle fibers begin to form.
4. New nuclear membrane forms.
5. Asters form from centrioles.
6. Centriole replication occurs.
7. Replication of DNA occurs.
8. Sister chromosomes separate and pass to opposite poles.
9. Formed after telophase is completed.
10. Chromosomes become oriented on equatorial plane.

interphase—1
prophase—2
metaphase—3
anaphase—4
telophase—5
daughter cells—6

Answers

Functions

1. _____
2. _____
3. _____
4. _____
5. _____
6. _____
7. _____
8. _____
9. _____
10. _____
11. _____
12. _____
13. _____
14. _____
15. _____

True–False

1. _____
2. _____
3. _____
4. _____
5. _____
6. _____
7. _____
8. _____
9. _____
10. _____

Mitosis

1. _____
2. _____
3. _____
4. _____
5. _____
6. _____
7. _____
8. _____
9. _____
10. _____

LABORATORY REPORT

7

Student: _____

Desk No.: _____ Section: _____

Osmosis and Cell Membrane Integrity

A. Molecular Movement

1. Brownian Movement

   a. Were you able to see any movement of carbon particles under the microscope? _____

   b. What forces propel the movement of these particles? _____

_____

_____

_____

2. Diffusion Plot the distance of diffusion of molecules for the two crystals on the following graphs.

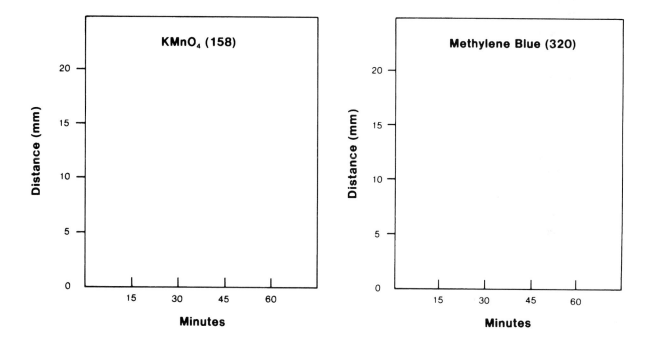

3. Conclusion What is the relationship of molecular weight to the rate of diffusion? _____

_____

_____

_____

_____

B. Osmotic Effects

As you examine each tube macroscopically and microscopically, record your results in the following table:

SOLUTION	MACROSCOPIC APPEARANCE (lysis or no lysis)	MICROSCOPIC APPEARANCE (crenation, lysis, or isotonic)
0.15M Sodium Chloride		
0.30M Sodium Chloride		
0.28M Glucose $C_6H_{12}O_6$		
0.30M Glycerine $C_3H_5(OH)_3$		
0.30 Urea $CO(NH_2)_2$		

1. Are any of the above solutions isotonic for red blood cells? _____

   If so, which ones? _____

2. With the aid of the atomic weight table in Appendix A, calculate the percentages of solutes in those solutions that proved to be isotonic. _____

   _____

   _____

   _____

LABORATORY REPORT
8,9

Student: _____

Desk No.: _____ Section: _____

Hydrogen Ion Concentration (Ex. 8)

A. Comparison Testing

As each solution is tested colorimetrically with pH paper and electronically with a pH meter, record your results in the following table. To promote objectivity, the instructor will not furnish you with the known (actual) pH values until the end of the period. At that time, these values should be added to the chart.

SOLUTION	pH VALUE			SOLUTION	pH VALUE		
	PAPER METHOD	METER METHOD	ACTUAL		PAPER METHOD	METER METHOD	ACTUAL
A				G			
B				H			
C				I			
D				J			
E				K			
F				L			

1. Compare the merits and disadvantages of the two methods: _____

   _____

   _____

   _____

B. Body Fluids

Record the pH of the various body fluids. Obtain several values for each fluid.

Saliva: _____

Urine: _____

Blood: _____

Tears: _____

Perspiration: _____

C. pH Computations

With the aid of the log tables in Appendix A, make the following pH conversions:

pH	gm H$^+$/LITER
5.0	
5.8	
7.9	
	4.8×10^9
	8.3×10^3

Buffering Systems (Ex. 9)

A. Tabulations

Record the number of drops of 0.0001N HCl that were added to each of the solutions:

	TEST SOLUTIONS	DROPS OF 0.0001N HCL
A	Distilled H_2O	
B	Bicarbonate	
C	Phosphate	
D	Albumin	
E	Blood	
F	Plasma	

B. Questions

1. Which mixture exhibits the greatest buffering capacity? _____

   The least? _____

2. Why is the pH meter preferred over colorimetric methods for determining the pH of the blood?

   _____

   _____

   _____

Student: _____

Desk No.: _____ Section: _____

Epithelial and Connective Tissues

A. Microscopic Study

If any drawings of tissue are to be made, make them on separate sheets of paper and label all identifiable structures.

B. Figure 11.5

Record the labels for this illustration in the answer column.

C. Epithelial Tissue Characteristics

From the list of tissues select those that are described by the following statements. More than one tissue may apply.

1. Elongated cells, single layer.
2. Cells with blocklike cross section.
3. Tissue with goblet cells.
4. Some cells are binucleate.
5. Have basal lamina.
6. Lack intercellular matrix.
7. Flattened cells, single layer.
8. Flattened cells, several layers.
9. Much intercellular matrix.
10. Reinforcing fibers between cells.
11. Cells near surface are dome-shaped.
12. Produce mucus on free surface.
13. Elongated cells, with some cells not extending up to free surface.
14. Cells have microvilli on free surface.

columnar
  pseudostratified—1
  simple—2
  stratified—3
cuboidal—4
squamous
  simple—5
  stratified—6
transitional—7
all of above—8
none of above—9

D. Epithelial Tissue Locations

Identify the kind of epithelial tissues found in the following structures.

columnar, ciliated—1
columnar, plain—2
columnar, stratified—3
cubodial—4
pseudostratified, ciliated—5

pseudostratified, plain—6
squamous, simple—7
squamous, stratified—8
transitional—9
none of above—10

1. Epidermis of skin
2. Lining of kidney calyces
3. Peritoneum
4. Auditory tube lining
5. Cornea of the eye
6. Capillary walls
7. Capsule of lens of the eye
8. Lining of the pharynx
9. Lining of the mouth
10. Pleural membranes

11. Intestinal lining
12. Lining of trachea
13. Lining of blood vessels
14. Lining of urinary bladder
15. Lining of stomach
16. Lining of male urethra
17. The conjunctiva
18. Lining of vagina
19. Anal lining
20. Parotid gland

Answers

Characteristics	Fig. 11.5
1. _____	_____
2. _____	_____
3. _____	_____
4. _____	_____
5. _____	_____
6. _____	_____
7. _____	_____
8. _____	_____
9. _____	_____
10. _____	_____
11. _____	_____
12. _____	_____
13. _____	_____
14. _____	_____
	_____

Locations

1. _____	11. _____
2. _____	12. _____
3. _____	13. _____
4. _____	14. _____
5. _____	15. _____
6. _____	16. _____
7. _____	17. _____
8. _____	18. _____
9. _____	19. _____
10. _____	20. _____

E. Connective Tissue Characteristics

From the list of tissues select those that are described by the following statements. More than one tissue may apply.

1. Contains all three types of fibers.
2. Reinforced mostly with collagenous fibers.
3. Cells are contained in lacunae.
4. Cells have large fat vacuoles.
5. Contain mast cells.
6. Nonyielding supporting tissue.
7. Devoid of blood supply.
8. Contain macrophages.
9. Most flexible of supporting tissues.
10. Converted to bone during growing years.
11. Matrix impregnated with calcium and phosphorous salts.
12. Toughest ordinary tissue (for stretching).
13. Cells have signet ringlike appearance.
14. Reinforced mostly with yellow fibers.
15. Has smooth glassy appearance.

adipose—1
areolar—2
bone—3
cartilage,
   elastic—4
   fibro—5
   hyaline—6
dense irregular—7
dense regular—8
reticular—9
none of above—10

F. Connective Tissue Locations

From the above list of tissues select the connective tissue that is found in the following:

1. Deep fasciae
2. Tendon sheaths
3. Intervertebral disks
4. Superficial fasciae
5. Epiglottis
6. Basal lamina
7. Lymph nodes
8. Aponeuroses
9. Nerve sheaths
10. Symphysis pubis
11. Auditory tube
12. Skeletal components
13. Periosteum (outer layer)
14. Sharpey's fiber
15. Ligaments and tendons
16. In hemopoietic tissue
17. In tracheal rings
18. In pinna of outer ear

G. Terminology

Select the terms on the right that are described by the following statements.

1. Protein in white fibers.
2. Cell that secretes mucus.
3. Material between fibroblasts.
4. Cartilage-producing cells.
5. Inner layer of periosteum.
6. Cell that produces fibers.
7. Synonym for Haversian system.
8. Protein in reticular fibers.
9. Bony plates in spongy bones.
10. Anchors periosteum to bone tissue.
11. Layers of bony matrix around central (Haversian) canal.
12. Layer of reticular fibers at base of epithelial tissues.
13. Cell found in loose fibrous tissue.
14. Minute canals radiating out from lacunae in bone tissue.
15. Canal in bone that contains capillaries.
16. Horizontal canals that connect adjacent central canals.

basement lamina—1
canaliculi—2
central canal—3
chondrocytes—4
collagen—5
fibroblast—6
goblet cell—7
lamellae—8
macrophage—9
mast cell—10
matrix—11
osteocyte—12
osteogenic layer—13
osteon—14
perforating canal—15
Sharpey's fibers—16
trabeculae—17

Answers

Characteristics	Terms
1. _____	1. _____
2. _____	2. _____
3. _____	3. _____
4. _____	4. _____
5. _____	5. _____
6. _____	6. _____
7. _____	7. _____
8. _____	8. _____
9. _____	9. _____
10. _____	10. _____
11. _____	11. _____
12. _____	12. _____
13. _____	13. _____
14. _____	14. _____
15. _____	15. _____
	16. _____

Locations

1. _____
2. _____
3. _____
4. _____
5. _____
6. _____
7. _____
8. _____
9. _____
10. _____
11. _____
12. _____
13. _____
14. _____
15. _____
16. _____
17. _____
18. _____

LABORATORY REPORT

12

Student: _____

Desk No.: _____ Section: _____

The Integument

A. Figures 12.1 and 12.2
Record the labels for these illustrations in the answer column.

B. Microscopic Study
If the space provided below is adequate for your histology drawings use it. Use a separate sheet of paper if the space is inadequate.

C. Questions
Select the answer that completes the following statements. Only one answer applies for each question.

1. Granules of the stratum granulosum consist of
   1. melanin.   2. keratin.   3. eleidin.

2. The papillary layer is a part of the
   1. dermis.   2. epidermis.   3. stratum spinosum.

3. The epidermis consists of the following number of distinct layers:
   1. three.   2. four.   3. five.   4. six.

4. Meissner's corpuscles are sensitive to
   1. temperature.   2. pressure.   3. touch.

5. Pacinian corpuscles are sensitive to
   1. temperature.   2. pressure.   3. touch.

6. The outermost layer of the epidermis is the
   1. stratum corneum.   2. stratum basale.
   3. stratum lucidum.   4. stratum spinosum.

7. New cells of the epidermis originate in the
   1. dermis.   2. stratum basale.
   3. stratum corneum.   4. stratum lucidum.

8. Melanin granules are produced by the
   1. stratum granulosum.   2. melanocytes.
   3. stratum lucidum.   4. stratum corneum.

Answers

Multiple Choice	Fig. 12.1
1._____	_____
2._____	_____
3._____	_____
4._____	_____
5._____	_____
6._____	_____
7._____	_____
8._____	_____
	_____
	_____
	_____
	_____
	_____
	_____
	_____
	_____
	_____
	_____
Fig. 12.2	
	_____
	_____
	_____
	_____
	_____

9. The secretions of the apocrine glands usually empty
   1. into a hair follicle.   2. directly out through the skin surface.
   3. Both 1 and 2.   4. None of these is correct.

10. Sebum is secreted by
    1. sweat glands.   2. eccrine glands.
    3. apocrine glands.   4. None of these is correct.

11. Keratin's function is primarily to
    1. destroy bacteria.   2. provide waterproofing.   3. cool the body.

12. Meissner's corpuscles are located in the
    1. papillary layer.   2. reticular layer.   3. subcutaneous layer.

13. Arrector pili assist in
    1. maintaining skin tonus.   2. limiting excessive sweating.
    3. forcing out sebum.   4. None of these is correct.

14. The hair follicle is
    1. a shaft of hair.   2. the root of a hair.
    3. a tube in the skin.   4. a glandular structure.

15. Pacinian corpuscles are located in the
    1. papillary layer.   2. reticular layer.   3. subcutaneous layer.

16. Sweat is produced by
    1. sebaceous glands.   2. eccrine glands.
    3. ceruminous glands.   4. None of these is correct.

17. The coiled-up portion of an apocrine gland is located in the
    1. epidermis.   2. dermis.
    3. subcutaneous layer.   4. hair follicle.

18. Apocrine glands are located
    1. in the axillae.   2. on the scrotum.
    2. in the ear canal.   4. All of these are correct.

Answers
Questions

9. _____

10. _____

11. _____

12. _____

13. _____

14. _____

15. _____

16. _____

17. _____

18. _____

LABORATORY REPORT
13

Student: _____

Desk No.: _____ Section: _____

The Skeletal Plan

A. Illustrations

Record the labels for figures 13.1, 13.2, and 13.3 in the answer columns.

B. Long Bone Structure

Identify the terms described by the following statements.

1. Shaft portion of bone.
2. Hollow chamber in bone shaft.
3. Type of marrow in medullary canal.
4. Enlarged end of a bone.
5. Type of bone in diaphysis.
6. Fibrous covering of bone shaft.
7. Linear growth area of long bone.
8. Smooth gristle covering bone end.
9. Type of bone marrow in bone ends.
10. Lining of medullary canal.
11. Type of bone tissue in bone ends.
12. Lines the central canals.

articular cartilage—1
cancellous bone—2
compact bone—3
diaphysis—4
endosteum—5
epiphyseal disk—6
epiphysis—7
medullary canal—8
periosteum—9
red marrow—10
yellow marrow—11
none of these—12

C. Bone Identification

Select the structures on the right that match the statements on the left.

1. Shoulder blade
2. Collarbone
3. Breastbone
4. Shinbone
5. Kneecap
6. Upper arm bone
7. Bones of spine
8. Thighbone
9. Lateral bone of forearm
10. Joint between ossa coxae
11. Horseshoe-shaped bone
12. Bones of shoulder girdle
13. One half of pelvic girdle
14. Medial bone of forearm
15. Thin bone paralleling tibia (calf bone)

clavicle—1
femur—2
fibula—3
humerus—4
hyoid—5
os coxa—6
patella—7
radius—8
scapula—9
sternum—10
symphysis pubis—11
tibia—12
ulna—13
vertebrae—14

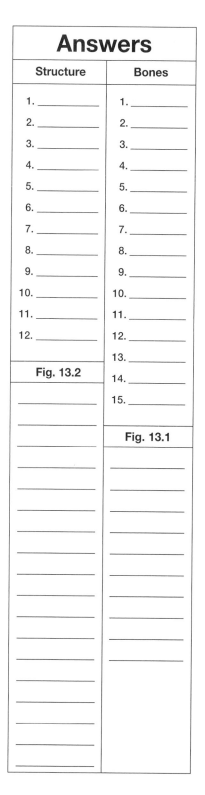

Answers

Structure	Bones
1. _____	1. _____
2. _____	2. _____
3. _____	3. _____
4. _____	4. _____
5. _____	5. _____
6. _____	6. _____
7. _____	7. _____
8. _____	8. _____
9. _____	9. _____
10. _____	10. _____
11. _____	11. _____
12. _____	12. _____
	13. _____
Fig. 13.2	14. _____
	15. _____
_____	**Fig. 13.1**
_____	_____
_____	_____
_____	_____
_____	_____
_____	_____
_____	_____
_____	_____
_____	_____
_____	_____
_____	_____

D. Medical

Select the condition that is described by the following statements. Since not all conditions are described in this manual it will be necessary for you to consult your lecture text or medical dictionary for some of the terminology.

closed reduction—1
Colles' fracture—2
comminuted fracture—3
compacted fracture—4
compound fracture—5

fissured fracture—6
greenstick fracture—7
open reduction—8
osteomalacia—9
osteomyelitis—10
osteoporosis—11

pathological fracture—12
Pott's fracture—13
rickets—14
simple fracture—15
none of these—16

1. Fracture due to weak bone structure, not trauma.
2. Fracture in which the skin is not broken.
3. Fracture caused by torsional forces.
4. Procedure used to set broken bones without using surgery.
5. Fracture caused by severe vertical forces.
6. Skeletal softness in adults.
7. Fracture in which a piece of bone is broken out of shaft.
8. Fracture characterized by two or more fragments.
9. Term applied to bone setting with the aid of surgery.
10. Skeletal softness in children due to vitamin D deficiency.
11. Infection of bone marrow.
12. Bone fracture extends only partially through a bone; incomplete fracture.
13. Fracture that occurs at an angle to the longitudinal axis of a bone.
14. Bone condition in which increased porosity occurs due to widening of central canals.
15. Outward displacement of foot due to fracture of lower part of fibula and malleolus.
16. Incomplete bone fracture in which fracture is apparent only on convex surface.
17. Displacement of hand backward and outward due to fracture of lower end of radius.

E. Terminology

Identify the term at the right that is described by the following statements. More than one term may apply.

1. A narrow slit.
2. A cavity in a bone.
3. Process to which muscle is attached.
4. A small process for muscle attachment.
5. A large process for muscle attachment.
6. A rounded knuckle-like process that articulates with another bone.
7. A depression in a bone.
8. A long tubelike passageway.
9. A hole through which nerves and blood vessels pass.
10. A sharp or long slender process.

condyle—1
fissure—2
foramen—3
fossa—4
meatus—5
sinus—6
spine—7
trochanter—8
tubercle—9
tuberosity—10

Answers

Medical

1. ___
2. ___
3. ___
4. ___
5. ___
6. ___
7. ___
8. ___
9. ___
10. ___
11. ___
12. ___
13. ___
14. ___
15. ___
16. ___
17. ___

Fig. 13.3

Terms

1. ___
2. ___
3. ___
4. ___
5. ___
6. ___
7. ___
8. ___
9. ___
10. ___

LABORATORY REPORT
14

Student: _____

Desk No.: _____ Section: _____

The Skull

A. Illustrations

Record the labels for all the illustrations of this exercise in the answer columns.

B. Bone Names

Select the bones that are described by the following statements. More than one answer may apply.

1. Forehead bone	ethmoid—1
2. Upper jaw	frontal—2
3. Cheekbone	lacrimal—3
4. Lower jaw	mandible—4
5. Sides of skull	maxilla—5
6. On walls of nasal fossa	nasal—6
7. Unpaired bones (usually)	nasal conchae—7
8. Bones of hard palate (2 pairs)	occipital—8
9. Back and bottom of skull	palatine—9
10. Contain sinuses (4 bones)	parietal—10
11. External bridge of nose	sphenoid—11
12. On median line in nasal cavity	temporal—12
	vomer—13
	zygomatic—14

C. Terminology

From the above list of bones select those that have the following structures.

1. Alveolar process	13. Mandibular fossa
2. Angle	14. Mastoid process
3. Antrum of Highmore	15. Mylohyoid line
4. Chiasmatic groove	16. Occipital condyles
5. Coronoid process	17. Ramus
6. Cribriform plate	18. Sella turcica
7. Crista galli	19. Styloid process
8. External acoustic meatus	20. Superior nasal concha
9. Incisive fossa	21. Superior temporal line
10. Inferior temporal line	22. Symphysis
11. Internal acoustic meatus	23. Tear duct
12. Mandibular condyle	24. Zygomatic arch

D. Sutures

Identify the sutures described by the following statements.

1. Between parietal and occipital.	coronal—1
2. Between two maxillae on median line.	lambdoidal—2
3. Between the two parietals on the median line.	median palatine—3
4. Between frontal and parietal.	sagittal—4
5. Between parietal and temporal.	squamosal—5
6. Between maxillary and palatine.	tranverse palatine—6

Answers

Bones	Fig. 14.1	Fig. 14.2
1._____	_____	_____
2._____	_____	_____
3._____	_____	_____
4._____	_____	_____
5._____	_____	_____
6._____	_____	_____
7._____	_____	_____
8._____	_____	_____
9._____	_____	_____
10._____	_____	_____
11._____	_____	_____
12._____	_____	_____

Terms		
1._____	_____	_____
2._____	_____	_____
3._____	_____	_____
4._____	_____	_____
5._____	_____	_____
6._____	_____	_____
7._____	_____	_____
8._____	_____	_____
9._____	_____	_____
10._____	Fig. 14.7	_____
11._____	_____	_____
12._____	_____	_____
13._____	_____	_____
14._____	_____	_____
15._____	_____	_____
16._____	_____	_____
17._____	_____	Sutures
18._____	_____	1._____
19._____	_____	2._____
20._____	_____	3._____
21._____	_____	4._____
22._____		5._____
23._____		6._____
24._____		

E. Foramina Locations

By referring to the illustrations select the bones on which the following foramina or canals are located.

1. Anterior palatine foramen
2. Carotid canal
3. Condyloid canal
4. Foramen rotundum
5. Foramen cecum
6. Foramen ovale
7. Foramen spinosum
8. Infraorbital foramen
9. Foramen magnum
10. Mental foramen
11. Hypoglossal canal
12. Supraorbital foramen
13. Internal acoustic meatus
14. Mandibular foramen
15. Optic canal
16. Zygomaticofacial foramen

frontal—1
mandible—2
maxilla—3
occipital—4
palatine—5
parietal—6
sphenoid—7
temporal—8
zygomatic—9

F. Foramen Function

Select the cranial nerves and other structures that pass through the following foramina, canals, and fissures. Note that the number following each cranial nerve corresponds to the number designation of the particular cranial nerve. (Example: the abducens nerve is also known as the sixth cranial nerve.) Also, note that the fifth cranial nerve (trigeminal) has three divisions.

1. Carotid canal
2. Cribriform plate foramina
3. Foramen magnum
4. Foramen rotundum
5. Foramen ovale
6. Internal acoustic meatus
7. Optic canal
8. Hypoglossal canal
9. Superior orbital fissure

cranial nerves
olfactory—1
optic—2
oculomotor—3
trochlear—4
trigeminal branches
ophthalmic—5.1
maxillary—5.2
mandibular—5.3
abducens—6
facial—7
vestibulocochlear—8
spinal accessory—11
hypoglossal—12
brain stem—13
internal carotid a.—14
internal jugular v.—15

G. Completion

Provide the name of the structures described by the following statements.

1. Lighten the skull.
2. Provides attachment site for falx cerebri.
3. Allow compression of skull bones during birth.
4. Provides anchorage for sternocleidomastoideus muscle.
5. Point of attachment for some tongue muscles.
6. Allow sound waves to enter the skull.
7. Point of articulation between mandible and skull.
8. Points of articulation between skull and first vertebra.

Answers

Locations	Fig. 14.4	Fig. 14.3

Function
Fig. 14.5 / Fig. 14.6
Completion / Fig. 14.8

518

LABORATORY REPORT
15

Student: _____

Desk No.: _____ Section: _____

The Vertebral Column and Thorax

A. Illustrations
Record the labels for the four figures in the answer columns.

B. Bone Names
Select the proper anatomical terminology for the following bones.

1. Tailbone
2. Breastbone
3. False rib
4. Floating rib
5. True rib
6. Second cervical vertebra
7. First cervical vertebra
8. Inferior portion of breastbone
9. Midportion of breastbone
10. Superior portion of breastbone

atlas—1
axis—2
coccyx—3
gladiolus—4
manubrium—5
sacrum—6
sternum—7
vertebral ribs—8
vertebrochondral
   ribs—9
vertebrosternal
   ribs—10
xiphoid—11

C. Structures
Select the structures or foramina that perform the following functions.

body—1
intervertebral disks—2
intervertebral foramina—3
odontoid process—4
rib cage—5

spinal curves—6
spinous process—7
sternum—8
transverse process—9
transverse foramina—10

1. Provides anterior attachment for vertebrosternal ribs.
2. Openings through which spinal nerves exit.
3. Portion of vertebra that supports body weight.
4. Protects vital thoracic organs.
5. Impart springiness to vertebral column (two things).
6. Part of thoracic vertebra that provides attachment for rib.
7. Acts as a pivot for the atlas to move around.
8. Openings through which vertebral artery and vein pass.

Answers	
Bone Names	**Fig. 15.1**
1. _____	
2. _____	
3. _____	
4. _____	
5. _____	
6. _____	
7. _____	
8. _____	
9. _____	
10. _____	
Structures	
1. _____	
2. _____	
3. _____	
4. _____	
5. _____	
6. _____	
7. _____	
8. _____	
Fig. 15.2	
_____	
_____	
_____	
_____	
_____	
_____	
_____	
_____	

C. Evaluation of Recorded Twitches

From the chart record on the previous page, determine the following:

Latent Period (A). . . . . _____ msec

Contraction Phase (B) . . . . _____ msec

Relaxation Phase (C) . . . . _____ msec

Entire Duration (A + B + C) . . . . _____ msec

D. Multiple Motor Unit Summation

Attach a segment of the chart that illustrates multiple motor unit summation. Be sure to indicate the voltages that were used for each stimulation.

E. Wave Summation

Attach a segment of the chart that illustrates wave summation and tetanization.

F. Definitions

Define the following terms:

Motor Unit _____

Threshold Stimulus _____

Recruitment (Motor Unit Summation) _____

Tetany _____

LABORATORY REPORT
31,32

Student: _____

Desk No.: _____ Section: _____

The Effects of Drugs on Muscle Contraction (Ex. 31)

A. Tabulation of Results

Indicate after each drug the effect of that drug on muscle contraction. Use the blank spaces to fill in any optional drugs that were tested.

DRUG	EFFECT
Valium	
Epinephrine	
Acetylcholine	
EDTA	

B. Conclusions

Can any conclusions be drawn from this experiment? Did you encounter any problems that would cause you to do this experiment differently if it were repeated? _____

_____

_____

_____

Electromyography (Ex. 32)

A. Recordings

Attach samples of myograms that were made to the back of this page.

B. Applications

List several applications in which electromyography can be very useful: _____

_____

_____

_____

_____

LABORATORY REPORT
33,34

Student: _____

Desk No.: _____ Section: _____

Muscle Structure (Ex. 33)

A. Figure 33.1

Record the labels for figure 33.1 in the answer column.

B. Muscle Terminology

Match the terms of the right-hand column to the following statements.

1. Bundle of muscle fibers.

2. Sheath of fibrous connective tissue that surrounds each fasciculus.

3. Movable end of a muscle.

4. Immovable end of muscle.

5. Thin layer of connective tissue surrounding each muscle fiber.

6. Layer of connective tissue between the skin and deep fascia.

7. Flat sheet of connective tissue that attaches a muscle to the skeleton or to another muscle.

8. Layer of connective tissue that surrounds a group of fasciculi.

9. Outer connective tissue that covers an entire muscle.

10. Band of fibrous connective tissue that connects muscle to bone or to another muscle.

aponeurosis—1
deep fascia—2
endomysium—3
epimysium—4
fasciculus—5
insertion—6
ligament—7
origin—8
perimysium—9
superficial fascia—10
tendon—11

Body Movements (Ex. 34)

A. Figure 34.1

Record the letters for the various types of movement in this illustration in the answer column.

B. Muscle Types

Determine the types of muscles described by the following statements.

1. Prime movers.
2. Opposing muscles.
3. Muscles that assist prime movers.
4. Muscles that hold structures steady to allow agonists to act smoothly.

agonists—1
antagonists—2
fixation muscles—3
synergists—4

Answers

Terms

1. _____
2. _____
3. _____
4. _____
5. _____
6. _____
7. _____
8. _____
9. _____
10. _____

Types

1. _____
2. _____
3. _____
4. _____

Fig. 33.1

_____
_____
_____
_____
_____
_____
_____
_____
_____
_____
_____

C. Movements

Identify the types of movements described by the following statements.

1. Movement of limb away from median line of body.
2. Movement of limb toward median line of body.
3. Movement around central axis.
4. Turning of sole of foot inward.
5. Turning of sole of foot outward.
6. Flexion of toes.
7. Backward movement of head.
8. Extension of foot at the ankle.
9. Angle between two bones is increased.
10. Angle between two bones is decreased.
11. Movement of hand from palm-down to palm-up position.
12. Conelike rotational movement.
13. Movement of hand from palm-up to palm-down position.

abduction—1
adduction—2
circumduction—3
dorsiflexion—4
eversion—5
extension—6
flexion—7
hyperextension—8
inversion—9
plantar flexion—10
pronation—11
rotation—12
supination—13

Answers

Movements	Fig. 34.1
1. _____	_____
2. _____	_____
3. _____	_____
4. _____	_____
5. _____	_____
6. _____	_____
7. _____	_____
8. _____	_____
9. _____	_____
10. _____	_____
11. _____	
12. _____	
13. _____	

Laboratory Report

36

Student: _____

Desk No.: _____ Section: _____

Head and Neck Muscles

A. Illustrations

Record the labels for the illustrations of this exercise in the answer column.

B. Facial Expressions

Consult figure 36.1 and the text to identify the muscles described in the following statements.

1. Smiling	frontalis—1
2. Horror	orbicularis oculi—2
3. Irony	orbicularis oris—3
4. Sadness	platysma—4
5. Contempt or disdain	quadratus labii
6. Pouting lips	inferioris—5
7. Blinking or squinting	quadratus labii
8. Horizontal wrinkling of forehead	superioris—6
	triangularis—7

Origins   zygomaticus—8
 9. On frontal bone.
10. On zygomatic bone.

Insertions
11. On lips.
12. On eyebrows.
13. On eyelid.

C. Mastication

Consult figure 36.1 and the text to identify the muscles that perform the following movements in mastication of food.

1. Lowers the mandible.	buccinator—1
2. Raises the mandible.	external pterygoid—2
3. Retracts the mandible.	internal pterygoid—3
4. Protrudes the mandible.	masseter—4
5. Holds food in place.	temporalis—5

Origins
 6. On zygomatic arch.
 7. On sphenoid only.
 8. On frontal, parietal, and temporal bones.
 9. On maxilla, mandible, and pteromandibular raphé.
10. On maxilla, sphenoid, and palatine bones.

Insertions
11. On orbicularis oris.
12. On coronoid process of mandible.
13. On medial surface of body of mandible.
14. On lateral surface of ramus and angle of mandible.
15. On neck of mandibular condyle and articular disk.

Answers

Facial	Fig. 36.1
1. _____	_____
2. _____	_____
3. _____	_____
4. _____	_____
5. _____	_____
6. _____	_____
7. _____	_____
8. _____	_____
9. _____	_____
10. _____	_____
11. _____	_____
12. _____	_____
13. _____	_____

Mastication	
1. _____	_____
2. _____	_____
3. _____	_____
4. _____	**Fig. 36.2**
5. _____	_____
6. _____	_____
7. _____	_____
8. _____	_____
9. _____	_____
10. _____	_____
11. _____	_____
12. _____	**Fig. 36.3**
13. _____	_____
14. _____	_____
15. _____	_____
	_____
	_____
	_____

D. Neck Muscles

Consult figures 36.2 and 36.3 and the text to identify the muscles of the neck that are described below.

digastricus—1
longissimus capitis—2
mylohyoideus—3
omohyoideus—4
platysma—5
semispinalis capitis—6

splenius capitis—7
sternocleidomastoideus—8
sternohyoideus—9
sternothyroideus—10
trapezius—11

1. Broad sheetlike muscle on side of neck just under the skin.
2. A broad triangular muscle of the back that functions only partially as a neck muscle.
3. Provides support for the floor of the mouth.
4. A V-shaped muscle attached to hyoid and mandibular bones.
5. Posterior neck muscles that insert on the mastoid process.
6. Anterior neck muscle that is attached to the thyroid cartilage and part of the sternum.
7. Anterior neck muscle that extends from the hyoid bone to the clavicle and manubrium.
8. Long muscle of the neck that extends from the mastoid process to the clavicle and sternum.
9. Muscles that pull the head backward (hyperextension).
10. Muscles that pull the head forward.

Answers
Neck Muscles
1. _____
2. _____
3. _____
4. _____
5. _____
6. _____
7. _____
8. _____
9. _____
10. _____

LABORATORY REPORT

37

Student: _____

Desk No.: _____ Section: _____

Trunk and Shoulder Muscles

A. Illustrations

Record the labels for the illustrations of this exercise in the answer column.

B. Anterior Trunk Muscles

By referring to figure 37.1 and the text, identify the muscles described by the following statements.

Origins
1. On upper eight or nine ribs.
2. On scapular spine and clavicle.
3. On third, fourth, and fifth ribs.
4. On clavicle, sternum, and costal cartilages.

Insertions
5. On coracoid process.
6. On deltoid tuberosity of humerus.
7. On anterior surface of scapula.
8. On upper anterior end of humerus.

Actions
9. Pulls scapula forward, downward, and inward.
10. Rotates arm medially.
11. Raises ribs (inhaling).
12. Lowers ribs (exhaling).
13. Adducts arm; also, rotates it medially.
14. Abducts arm.

deltoideus—1
intercostalis externi—2
intercostalis interni—3
pectoralis major—4
pectoralis minor—5
serratus anterior—6
subscapularis—7

C. Posterior Trunk Muscles

By referring to figure 37.2 and the text, identify the muscles described by the following statements.

infraspinatus—1   rhomboideus major—4   supraspinatus—7
latissimus dorsi—2   rhomboideus minor—5   teres major—8
levator scapulae—3   sacrospinalis—6   teres minor—9
                                                   trapezius—10

Origins
1. Near inferior angle of scapula.
2. On lateral (axillary) margin of scapula.
3. On fossa of scapula above spine.
4. On infraspinous fossa of scapula.
5. On thoracic and lumbar vertebrae, sacrum, lower ribs, and iliac crest.
6. On lower part of ligamentum nuchae and first thoracic vertebra.
7. On transverse processes of first four cervical vertebrae.

Answers	
Anterior	**Fig. 37.1**
1._____	
2._____	
3._____	
4._____	
5._____	
6._____	
7._____	
8._____	
9._____	**Fig. 37.2**
10._____	
11._____	
12._____	
13._____	
14._____	
Posterior	
1._____	
2._____	
3._____	
4._____	
5._____	
6._____	
7._____	

Trunk and Shoulder Muscles

infraspinatus—1	rhomboideus major—4	supraspinatus—7
latissimus dorsi—2	rhomboideus minor—5	teres major—8
levator scapulae—3	sacrospinalis—6	teres minor—9
		trapezius—10

Insertions
8. On lower third vertebral border of scapula.
9. On vertebral border of scapula near origin of scapular spine.
10. On crest of lesser tubercle of humerus.
11. On intertubercular groove of humerus.
12. On middle facet of greater tubercle of humerus.
13. On ribs and vertebrae.

Actions
14. Abducts arm.
15. Raises scapula and draws it medially.
16. Extends spine to maintain erectness.
17. Adducts and rotates arm medially.
18. Adducts and rotates arm laterally.

Answers
Posterior
8. _____
9. _____
10. _____
11. _____
12. _____
13. _____
14. _____
15. _____
16. _____
17. _____
18. _____

Student: _____

Desk No.: _____ Section: _____

Upper Extremity Muscles

A. Illustrations

Record the labels for figures 38.1, 38.2, and 38.3 in the answer column.

B. Arm Movements

Consult figure 38.1 to select the muscles that apply to the following statements. Verify your answers by referring to the text material.

1. Extends the forearm.
2. Adducts the arm.
3. Flexes the forearm.
4. Carries the arm forward in flexion.
5. Flexes forearm and rotates the radius outward to supinate the hand.

coracobrachialis—1
biceps brachii—2
brachialis—3
brachioradialis—4
triceps brachii—5

Origins

6. On lower half of humerus.
7. On coracoid process of scapula.
8. Above the lateral epicondyle of the humerus.
9. Three heads: one on scapula, another on posterior surface of humerus, and the third just below the radial groove of humerus.
10. Two heads: one on coracoid process; other on supraglenoid tubercle of humerus.

Insertions

11. On the olecranon process.
12. On the radial tuberosity.
13. Near the middle, medial surface of humerus.
14. On front surface of coronoid process of ulna.
15. On the lateral surface of the radius just above the styloid process.

C. Hand Movements

Consult figures 38.2 and 38.3 to select the muscles that apply to the following statements. Verify as above.

1. Supinates the hand.
2. Pronates the hand.
3. Flexes and abducts hand.
4. Extends wrist and hand.
5. Flexes the thumb.
6. Extends all fingers except thumb.
7. Abducts the thumb.
8. Flexes all distal phalanges except the thumb.
9. Flexes all fingers except thumb.
10. Extends thumb.

abductor pollicis—1
extensor carpi radialis brevis—2
extensor carpi radialis longus—3
extensor carpi ulnaris—4
extensor digitorum communis—5
extensor pollicis longus—6
flexor carpi radialis—7
flexor carpi ulnaris—8
flexor digitorum profundus—9
flexor digitorum superficialis—10
flexor pollicis longus—11
pronator quadratus—12
pronator teres—13
supinator—14

Answers	
Arm Movements	Fig. 38.1
1. _____	_____
2. _____	
3. _____	
4. _____	
5. _____	
6. _____	
7. _____	Fig. 38.2
8. _____	
9. _____	
10. _____	
11. _____	
12. _____	
13. _____	
14. _____	
15. _____	Fig. 38.3
Hand Movements	_____
1. _____	
2. _____	
3. _____	
4. _____	
5. _____	
6. _____	
7. _____	
8. _____	
9. _____	
10. _____	

C. *Hand Movements*

abductor pollicis—1
extensor carpi radialis brevis—2
extensor carpi radialis longus—3
extensor carpi ulnaris—4
extensor digitorum communis—5
extensor pollicis longus—6
flexor carpi radialis—7
flexor carpi ulnaris—8
flexor digitorum profundus—9
flexor digitorum superficialis—10
flexor pollicis longus—11
pronator quadratus—12
pronator teres—13
supinator—14

Origins

11. On interosseous membrane between radius and ulna.
12. On radius, ulna, and interosseous membrane.
13. On lateral epicondyle of humerus and part of ulna.
14. On humerus, ulna, and radius.
15. On medial epicondyle of humerus.

Insertions

16. On distal phalanges of second, third, fourth, and fifth fingers.
17. On distal phalanx of thumb.
18. On middle phalanges of second, third, fourth, and fifth fingers.
19. On radial tuberosity and oblique line of radius.
20. On upper lateral surface of radius.
21. On first metacarpal and trapezium.
22. On second metacarpal.
23. On proximal portion of second and third metacarpals.
24. On middle metacarpal.
25. On fifth metacarpal.

Answers
Hand Movements
11. _____
12. _____
13. _____
14. _____
15. _____
16. _____
17. _____
18. _____
19. _____
20. _____
21. _____
22. _____
23. _____
24. _____
25. _____

LABORATORY REPORT
39,40

Student: _____

Desk No.: _____ Section: _____

Abdominal and Pelvic Muscles (Ex. 39)

A. Illustrations
Record the labels for figures 39.1 and 39.3 in the answer column.

B. Abdomen
Consult figure 39.1 to identify the muscles that apply to the following statements. Verify your selections from the text.

external oblique—1   rectus abdominis—3
internal oblique—2   transversus abdominis—4

1. Flexion of spine in lumbar region.
2. Maintain intraabdominal pressure.
3. Antagonists of diaphragm.

Origins
4. On pubic bone.
5. On external surface of lower eight ribs.
6. On lateral half of inguinal ligament and anterior two-thirds of iliac crest.
7. On inguinal ligament, iliac crest, and costal cartilages of lower six ribs.

Insertions
8. On linea alba and crest of pubis.
9. On cartilages of fifth, sixth, and seventh ribs.
10. On the linea alba where fibers of the aponeurosis interlace.
11. On costal cartilages of lower three ribs, linea alba, and crest of pubis.

C. Pelvic Region
Consult figure 39.3 to identify the muscles that apply to the following statements. Verify your selections from the text.

iliacus—1
psoas major—2
quadratus lumborum—3

1. Flexes femur on trunk.
2. Flexes lumbar region of vertebral column.
3. Extension of spine at lumbar vertebrae.

Origins
4. On lumbar vertebrae.
5. On iliac crest, iliolumbar ligament, and transverse processes of lower four lumbar vertebrae.
6. On iliac fossa.

Insertions
7. On inferior margin of last rib and transverse processes of upper four lumbar vertebrae.
8. On lesser trochanter of femur.

Answers	
Abdomen	**Fig. 39.1**
1. _____	
2. _____	
3. _____	
4. _____	
5. _____	
6. _____	
7. _____	
8. _____	
9. _____	**Fig. 39.3**
10. _____	
11. _____	
Pelvic	
1. _____	
2. _____	
3. _____	
4. _____	
5. _____	
6. _____	
7. _____	
8. _____	

Lower Extremity Muscles (Ex. 40)

A. Illustrations

Record the labels for this exercise in the answer column.

B. Thigh Movements

Consult figure 40.1 to select the muscles that apply to the following statements. More than one muscle may apply to a statement.

1. Gluteus muscle that extends femur.
2. Gluteus muscles that abduct femur.
3. Three muscles that pull femur toward median line.
4. Gluteus muscle that rotates femur outward.
5. Three muscles attached to linea aspera that flex femur.
6. Small muscle that abducts, extends, and rotates femur.

adductor brevis—1
adductor longus—2
adductor magnus—3
gluteus maximus—4
gluteus medius—5
gluteus minimus—6
piriformis—7

Origins

7. On anterior surface of sacrum.
8. On the pubis.
9. On external surface of ilium.
10. On ilium, sacrum, and coccyx.
11. On inferior surface of ischium and portion of pubis.

Insertions

12. On linea aspera of femur.
13. On anterior border of greater trochanter.
14. On upper border of greater trochanter of femur.
15. On lateral part of greater trochanter of femur.
16. On iliotibial tract and posterior part of femur.

C. Thigh Muscles

Identify the muscles of the thigh that are described by the following statements.

biceps femoris—1
gracilis—2
rectus femoris—3
sartorius—4
semimembranosus—5
semitendinosus—6
tensor fasciae latae—7
vastus intermedius—8
vastus lateralis—9
vastus medialis—10

1. The largest quadriceps muscle that is located on the side of the thigh.
2. That portion of the quadriceps that lies beneath the rectus femoris.
3. Portion of the quadriceps that is on the medial surface of the thigh.
4. Hamstring muscle that is on the lateral surface of the thigh.
5. The longest muscle of the thigh.
6. The most medial component of the hamstring muscles.
7. The smallest hamstring muscle.
8. A superficial muscle on the medial surface of the thigh that inserts with the sartorius on the tibia.
9. Four muscles of a group that extend the leg.
10. Flexes the thigh upon the pelvis.
11. Two muscles other than parts of quadriceps that rotate the thigh medially (inward).
12. Rotates the thigh laterally (outward).
13. Adducts the thigh.

Answers

Thigh Movements	Fig. 40.1	Fig. 40.10 Anterior
1.		
2.		
3.		
4.		
5.		
6.		
7.		
8.		
9.		
10.	Fig. 40.2	
11.		
12.		
13.		
14.		
15.		
16.		

Thigh Muscles		
1.		
2.		Fig. 40.10 Posterior
3.		
4.	Fig. 40.3	
5.		
6.		
7.		
8.		
9.		
10.		
11.	Fig. 40.4	
12.		
13.		

D. Lower Leg and Foot Muscles

Identify the muscles of the lower leg and foot that are described by the following statements. In some instances several answers are applicable.

extensor digitorum longus—1
extensor hallucis longus—2
flexor digitorum longus—3
flexor hallucis longus—4
gastrocnemius—5
peroneus brevis—6
peroneus longus—7
peroneus tertius—8
soleus—9
tibialis anterior—10
tibialis posterior—11

1. Located on the front of the lower leg; causes dorsiflexion and inversion of the foot.
2. Deep portion of triceps surae.
3. Superficial portion of triceps surae.
4. Three muscles that originate on the fibula that cause various movements of the foot.
5. Two muscles that join to form the Achilles tendon.
6. Part of the triceps surae that originates on the femur.
7. Inserts on the proximal superior portion of the fifth metatarsal bone of the foot.
8. Inserts on the proximal inferior portion of the fifth metatarsal bone of the foot.
9. Flexes the great toe.
10. On the posterior surfaces of the tibia and fibula; causes plantar flexion and inversion of foot.
11. Originates on the tibia and fascia of the tibialis posterior; causes flexion of second, third, fourth, and fifth toes.
12. Flexes the calf of the thigh.
13. Originates on the tibia, fibula, and interosseous membrane; extends toes (dorsiflexion) and inverts foot.
14. Two muscles that insert on the calcaneous.
15. Peroneus muscle that causes dorsiflexion and eversion of foot.
16. Two peroneus muscles that cause plantar flexion.
17. Muscle on tibia that provides support for foot arches.

E. General Questions

Record the answers to the following questions in the answer column.

1. List the four muscles that are collectively referred to as the quadriceps femoris.
2. List two muscles associated with the iliotibial tract.
3. List three muscles that constitute the hamstrings.
4. What two muscles are associated with the Achilles tendon?
5. What two muscles constitute the triceps surae?

F. Figure 34.1

Refer back to figure 34.1 and identify the muscles that cause each type of movement. Record answers in column.

Answers

Lower Leg and Foot

1. ___ 10. ___
2. ___ 11. ___
3. ___ 12. ___
4. ___ 13. ___
5. ___ 14. ___
6. ___ 15. ___
7. ___ 16. ___
8. ___ 17. ___
9. ___

General Questions

1a. ___ b. ___ c. ___ d. ___
2a. ___ b. ___
3a. ___ b. ___ c. ___
4a. ___ b. ___
5a. ___ b. ___

Fig. 34.1

A. ___ B. ___ C. ___ D. ___ E. ___ F. ___ G. ___ H. ___ I. ___ J. ___ K. ___ L. ___ M. ___

C. *Reflex Radiation*

Describe the response of the spinal frog to increased electrical stimulation of the foot. _____

_____

_____

_____

_____

D. *Reflex Inhibition*

Was it possible to inhibit right foot withdrawal from the acid by electrically stimulating the left foot?

_____1_____ . If so, how much voltage was required? _____2_____. At what pH? _____. At what

acid concentration was reflex action inhibited? _____ At what voltage? _____

E. *Synaptic Fatigue*

Where do you believe fatigue occurred in this experiment? _____

_____

F. *Diagnostic Reflex Tests*

Record the responses for the first five reflexes with a 0 for *no response,* a + for *diminished response,* and
+ + for *good response.* For Hoffmann's reflex place a check in the proper space. List the nerves in the
last column that are affected in abnormal responses.

REFLEX	RESPONSES		Nerves Affected When Response Is Abnormal
	Right Side	Left Side	
1. Biceps			
2. Triceps			
3. Brachioradialis			
4. Patellar			
5. Achilles			
6. Hoffmann's	Absent (Normal)	Present (Abnormal)	
7. Plantar Flexion	Present (Normal)	Babinski's Sign (Abnormal)	

LABORATORY REPORT

44

Student: _____

Desk No.: _____ Section: _____

Action Potential Velocities

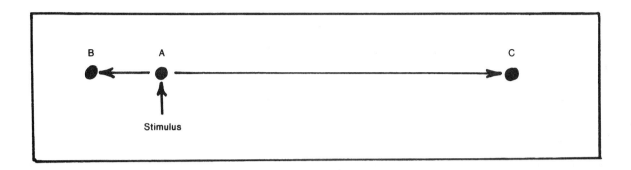

A. Calculations

With a meter stick measure the distances on the arm between AB and AC in millimeters and convert to meters.

DISTANCES
AB = _____ meters
AC = _____ meters

Knowing that each millimeter on the Unigraph chart represents 1/25 second, or 0.04 second, determine how long it took for the action potentials to move from A to B and A to C.

TIMES
A to B = _____ seconds
A to C = _____ seconds

To determine the speed of the action potentials, divide the distances by the times.

VELOCITIES
A to B = _____ m/sec
A to C = _____ m/sec

$$Velocity = \frac{Distance\ (meters)}{Time\ (seconds)}$$

B. Questions

1. What is the threshold for muscle stimulation in this experiment? _____

2. What is the threshold for nerve stimulation in this experiment? _____

3. The motor fibers of the ulnar nerve have a known velocity of 46–73 meters per second. How do your experimental results compare with this? _____

4. Is the velocity of the nerve impulse in motor fibers distinguishable from the velocity of the impulse in sensory fibers? _____ Explain: _____

_____

5. Which would be most susceptible to conduction blockage due to pressure: motor, sensory, or autonomic fibers? _____ Explain: _____

_____

6. Can you identify the reflex potential? _____ What is its latency from the onset of stimulation? _____

7. What factors account for the latency of the reflex potential as the impulse completes the reflex arc?

_____

_____

_____

8. Describe the sensations perceived by the subject in this experiment. _____

_____

_____

To what areas did they seem to be projected? _____

_____

C. Chart
Attach the Duograph chart to the space below:

LABORATORY REPORT

45

Student: _____

Desk No.: _____ Section: _____

Brain Anatomy: External

A. Illustrations

Record the labels for all the illustrations of this exercise in the answer columns.

B. Brain Dissections

Answer the following questions that pertain to your observations of the sheep brain dissections.

1. Does the arachnoid mater appear to be attached to the dura

   mater? _____ to the brain? _____

2. Describe the appearance of the dura mater.

   _____

   _____

   _____

3. Within what structure is the sagittal sinus enclosed?

   _____

4. Differentiate:

   Gyrus: _____

   Sulcus: _____

5. Of what value are the sulci and gyri of the cerebrum?

   _____

   _____

6. List five structures that you were able to identify on the dorsal

   surface of the midbrain: _____

   _____

   _____

7. Is the cerebellum of the sheep brain divided on the median line

   as is the human brain? _____

   _____

8. What significance is there to the size differences of the olfactory
   bulbs on the sheep and human brains?

   _____

   _____

   _____

9. Which cranial nerve is the largest in diameter?

   _____

Answers	
Fig. 45.1	**Fig. 45.9**
_____	_____
_____	_____
_____	_____
_____	_____
_____	_____
_____	_____
_____	_____
_____	_____
_____	_____
_____	_____
_____	_____
_____	_____
Fig. 45.7	_____
_____	_____
_____	_____
_____	_____
_____	_____
_____	_____
_____	_____
_____	**Fig. 45.10**
_____	_____
_____	_____
_____	_____
_____	_____
_____	_____

C. Meninges

Select the meninx or meningeal space described by the following statements.

1. Fibrous outer meninx.
2. The middle meninx.
3. Meninx that surrounds the sagittal sinus.
4. Meninx that forms the falx cerebri.
5. Meninx attached to the brain surface.
6. Space that contains cerebrospinal fluid.

arachnoid mater—1
dura mater—2
epidural space—3
pia mater—4
subarachnoid space—5
subdural space—6

D. Fissures and Sulci

Select the fissure or sulcus that is described by the following statements.

1. Between the frontal and parietal lobes.
2. On the median line between the two cerebral hemispheres.
3. Between the occipital and parietal lobes.
4. Between the temporal and parietal lobes.

anterior central sulcus—1
central sulcus—2
lateral cerebral fissure—3
longitudinal cerebral fissure—4
parieto-occipital fissure—5

E. Brain Functions

Select the part of the brain that is responsible for the following activities.

1. Maintenance of posture.
2. Cardiac control.
3. Respiratory control.
4. Vasomotor control.
5. Consciousness.
6. Voluntary muscular movements.
7. Brightness and sound discrimination (animals only).
8. Contains fibers that connect the two cerebral hemispheres.
9. Contains nuclei of fifth, sixth, seventh, and eighth cranial nerves.
10. Coordination of complex muscular movements.
11. Function in humans is unknown.

cerebellum—1
cerebrum—2
corpora quadrigemina—3
corpus callosum—4
medulla oblongata—5
midbrain—6
pineal body—7
pons Varolii—8

F. Cerebral Functional Localization

Select those areas of the cerebrum that are described by the following statements.

Location
1. On postcentral gyrus.
2. On parietal lobe.
3. On superior temporal gyrus.
4. On temporal lobe.
5. On frontal lobe.
6. On occipital lobe.
7. On precentral gyrus.
8. On angular gyrus.
9. In area anterior to precentral gyrus.

association area—1
auditory area—2
common integrative area—3
motor speech area—4
olfactory area—5
somatomotor area—6
somatosensory area—7
premotor area—8
visual area—9

Answers

Meninges	Fissures and Sulci
1. _____	1. _____
2. _____	2. _____
3. _____	3. _____
4. _____	4. _____
5. _____	
6. _____	

Fig. 45.11

Brain Functions

1. _____
2. _____
3. _____
4. _____
5. _____
6. _____
7. _____
8. _____
9. _____
10. _____
11. _____

Localization

1. _____
2. _____
3. _____
4. _____
5. _____
6. _____
7. _____
8. _____
9. _____

F. Cerebral Functional Localization (continued)

Function

10. Speech production.
11. Speech understanding.
12. Cutaneous sensibility.
13. Sight.
14. Sense of smell.
15. Visual interpretation.
16. Voluntary muscular movement.
17. Integration of sensory association areas.
18. Influences motor area function.

association area—1
auditory area—2
common integrative area—3
motor speech area—4
olfactory area—5
somatomotor area—6
somatosensory area—7
premotor area—8
visual area—9

G. Cranial Nerves

Record the number of the cranial nerve that applies to the following statements. More than one nerve may apply to some statements.

1. Sensory nerves.
2. Mixed nerves.
3. Emerges from midbrain.
4. Emerges from pons Varolii.
5. Emerges from medulla.

Structures Innervated

6. Cochlea of ear.
7. Heart.
8. Salivary glands.
9. Abdominal viscera.
10. Retina of eye.
11. Receptors in nasal membranes.
12. Taste buds at back of tongue.
13. Lateral rectus muscle of eye.
14. Taste buds of anterior two-thirds of tongue.
15. Semicircular canals of ear.
16. Thoracic viscera.
17. Three extrinsic eye muscles (superior rectus, medial rectus, inferior oblique) and levator palpebrae.
18. Superior oblique muscle of eye.
19. Pharynx, upper larynx, uvula, and palate.

1. Olfactory (I)
2. Optic (II)
3. Oculomotor (III)
4. Trochlear (IV)
5. Trigeminal (V)
6. Abducens (VI)
7. Facial (VII)
8. Vestibulocochlear (VIII)
9. Glossopharyngeal (IX)
10. Vagus (X)
11. Accessory (XI)
12. Hypoglossal (XII)

H. Trigeminal Nerve

Select the branch of the trigeminal nerve that innervates the following structures.

1. All lower teeth.
2. All upper teeth.
3. Tongue.
4. Lacrimal gland.
5. Outer surface of nose.
6. Buccal gum tissues of mandible.
7. Lower teeth, tongue, muscles of mastication, and gum surfaces.

incisive—1
infraorbital—2
inferior alveolar—3
lingual—4
long buccal—5
mandibular—6
maxillary—7
ophthalmic—8

Answers

Localization

10. _____
11. _____
12. _____
13. _____
14. _____
15. _____
16. _____
17. _____
18. _____

Cranial Nerves

1. _____
2. _____
3. _____
4. _____
5. _____
6. _____
7. _____
8. _____
9. _____
10. _____
11. _____
12. _____
13. _____
14. _____
15. _____
16. _____
17. _____
18. _____
19. _____

Trigeminal Nerve

1. _____
2. _____
3. _____
4. _____
5. _____
6. _____
7. _____

I. General Questions

Select the correct answers that complete the following statements.

	Answers
	General Questions

1. Shallow furrows on the surface of the cerebrum are called
   (1) sulci.   (2) fissures.   (3) gyri.

2. Deep furrows on the surface of the cerebrum are called
   (1) sulci.   (2) fissures.   (3) gyri.

3. The infundibulum supports the
   (1) mammillary body.   (2) hypophysis.   (3) hypothalamus.

4. The mammillary bodies are a part of the
   (1) medulla.   (2) midbrain.   (3) hypothalamus.

5. The corpora quadrigemina are located on the
   (1) cerebrum.   (2) medulla.   (3) midbrain.

6. The major ganglion of the trigeminal nerve is the
   (1) Gasserian.   (2) sphenopalatine.   (3) ciliary.

7. Convolutions on the surface of the cerebrum are called
   (1) sulci.   (2) fissures.   (3) gyri.

8. The hypophysis is located on the inferior surface of the
   (1) medulla.   (2) midbrain.   (3) hypothalamus.

9. The spinal bulb is the
   (1) medulla oblongata.   (2) pons Varolii.   (3) midbrain.

10. The pineal gland is located on the
    (1) medulla.   (2) pons Varolii.   (3) midbrain.

Answers

General Questions

1. _____
2. _____
3. _____
4. _____
5. _____
6. _____
7. _____
8. _____
9. _____
10. _____

LABORATORY REPORT

46

Student: _____

Desk No.: _____ Section: _____

Brain Anatomy: Internal

A. Illustrations

Record the labels for all the illustrations of this exercise in the answer columns.

B. Location

Identify that part of the brain in which the following structures are located.

cerebrellum—1   diencephalon—3   midbrain—5
cerebrum—2   medulla oblongata—4   pons Varolii—6

1. Caudate nuclei
2. Cerebral aqueduct
3. Cerebral peduncles
4. Corpus callosum
5. Fornix
6. Globus pallidus
7. Hypothalamus
8. Lateral ventricles
9. Mammillary bodies
10. Putamen
11. Rhinencephalon
12. Thalamus
13. Third ventricle

C. Functions

Identify the structures that perform the following functions.

arachnoid granu-
  lations—1
caudate nuclei—2
cerebral aqueduct—3
cerebral peduncles—4
choroid plexus—5
corpus callosum—6
fornix—7
hypothalamus—8
infundibulum—9
intermediate mass—10
interventricular
  foramen—11
lateral aperture—12
lentiform nucleus—13
median aperture—14
rhinencephalon—15
thalamus—16

1. Relay station for all messages to cerebrum.
2. Temperature regulation.
3. Entire olfactory mechanism of cerebrum.
4. Assists in muscular coordination.
5. Fiber tracts of olfactory mechanism.
6. Connects the two halves of the thalamus.
7. Exerts steadying effect on voluntary movements.
8. Allows cerebrospinal fluid to pass from third to fourth ventricle.
9. Consists of ascending and descending tracts.
10. Supporting stalk of hypophysis.
11. Allows cerebrospinal fluid to return to blood.
12. A commissure that unites the cerebral hemispheres.
13. Secretes cerebrospinal fluid into ventricle.
14. Coordinates autonomic nervous system.

D. General Questions

Select the best answer that completes the following statements.

1. The brain stem consists of the
   (1) cerebrum, pons, midbrain, and medulla.
   (2) cerebellum, medulla, and pons.
   (3) pons, medulla, and midbrain.
2. The lateral ventricles are separated by the
   (1) thalamus.   (2) fornix.   (3) septum pellucidum.

Answers		
Location		**Fig. 46.2**
1. _____		_____
2. _____		_____
3. _____		_____
4. _____		_____
5. _____		_____
6. _____		_____
7. _____		_____
8. _____		_____
9. _____		_____
10. _____		_____
11. _____		_____
12. _____		_____
13. _____		_____
Functions		
1. _____		_____
2. _____		_____
3. _____		_____
4. _____		_____
5. _____		_____
6. _____		_____
7. _____		**Fig. 46.3**
8. _____		_____
9. _____		_____
10. _____		_____
11. _____		_____
12. _____		_____
13. _____		_____
14. _____		_____
General		_____
1. _____		_____
2. _____		_____
		_____
		_____
		_____

3. Pain is perceived in the
   (1) somatomotor area.   (2) thalamus.   (3) pons.
4. The hypothalamus regulates
   (1) the hypophysis, appetite, and wakefulness.
   (2) body temperature, hypophysis, and vision.
   (3) reproductive functions, body temperature, and voluntary movements.
5. Feelings of pleasantness and unpleasantness appear to be associated with the
   (1) somatomotor area.   (2) hypothalamus.   (3) thalamus.
   (4) medulla oblongata.
6. The reticular formation consists of
   (1) gray matter.   (2) white matter.
   (3) an interlacement of white and gray matter.
7. Alert consciousness is partially regulated by
   (1) caudate nuclei.   (2) lentiform nuclei.
   (3) nuclei of the reticular formation.
8. The intermediate mass passes through the
   (1) midbrain.   (2) third ventricle.   (3) hypothalamus.
9. The reticular formation is seen in the
   (1) spinal cord.   (2) cerebrum.   (3) brain stem.
   (4) spinal cord, brain stem, and diencephalon.
10. Centers for vomiting, coughing, swallowing, and sneezing are located in the
    (1) pons.   (2) medulla.   (3) midbrain.   (4) thalamus.

E. Cerebrospinal Fluid

You should be able to trace the path of the cerebrospinal fluid from its point of origin to where it is reabsorbed into the blood. If you can complete the following paragraph without referring to the text, you know the sequence fairly well. If you can state the sequence from memory, better yet.

Ventricles
lateral—1
third—2
fourth—3
Passageways
acoustic meatus—4
cerebral aqueduct—5
foramen magnum—6
interventricular foramen—7
lateral aperture—8
median aperture—9

Structures
arachnoid granulations—10
cerebellum—11
cerebrum—12
choroid plexus—13
cisterna cerebellomedullaris—14
cisterna superior—15
dura mater—16
sagittal sinus—17
septum pellucidum—18
subarachnoid space—19

Cerebrospinal fluid is secreted into each ventricle by a __1__ . From the __2__ ventricles, which are located in the cerebral hemispheres, the fluid passes to the __3__ ventricle through an opening called the __4__ . A canal, called the __5__ , allows the fluid to pass from the latter ventricle to the __6__ ventricle. From this last ventricle the cerebrospinal fluid passes into a subarachnoid space called the __7__ through three foramina: one __8__ and two __9__ . From this cavity the fluid passes over the cerebellum into another subarachnoid space called the __10__ . It also passes __11__ *(up, down)* the posterior side of the spinal cord and __12__ *(up, down)* the anterior side of the spinal cord. From the cisterna superior the cerebrospinal fluid passes to the subarachnoid space around the __13__ . This fluid is reabsorbed back into the blood through delicate structures called the __14__ . The blood vessel that receives the cerebrospinal fluid is the __15__ .

Answers	
General	**Fig. 46.4**
3._____	_____
4._____	_____
5._____	_____
6._____	_____
7._____	_____
8._____	_____
9._____	_____
10._____	_____
Cerebrospinal Fluid	_____
1._____	_____
2._____	_____
3._____	_____
4._____	
5._____	
6._____	
7._____	
8._____	
9._____	
10._____	
11._____	
12._____	
13._____	
14._____	
15._____	

Electroencephalography

A. Records

Attach samples of the EEG in spaces provided below. Use tape, glue, or staples for attachment. It will be necessary to trim excess paper from the chart.

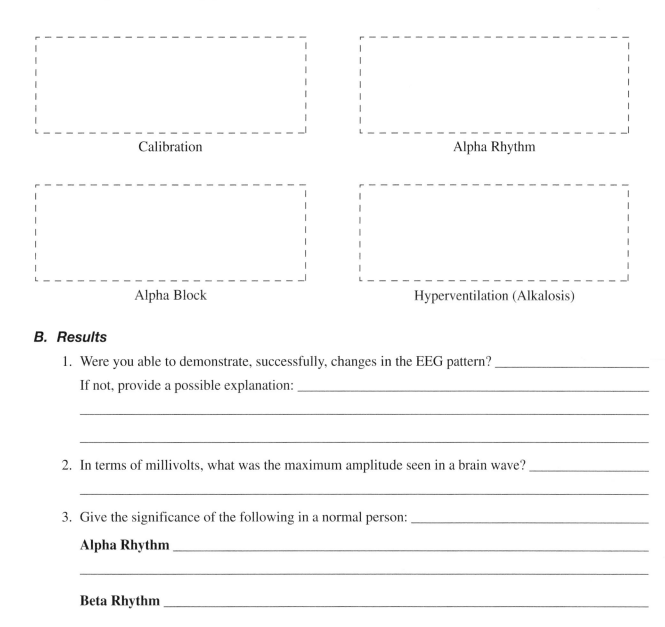

Calibration

Alpha Rhythm

Alpha Block

Hyperventilation (Alkalosis)

B. Results

1. Were you able to demonstrate, successfully, changes in the EEG pattern? _____

   If not, provide a possible explanation: _____

   _____

   _____

2. In terms of millivolts, what was the maximum amplitude seen in a brain wave? _____

   _____

3. Give the significance of the following in a normal person: _____

   Alpha Rhythm _____

   _____

   Beta Rhythm _____

   _____

   Delta Waves _____

   _____

   _____

LABORATORY REPORT
48, 49

Student: _____

Desk No.: _____ Section: _____

Anatomy of the Eye (Ex. 48)

A. Illustrations
Record the labels from figures 48.1, 48.2, and 48.3 in the answer columns.

B. Structures
Identify the structures described by the following statements.

1. Small nonphotosensitive area on retina.
2. Small pit in retina of eye.
3. Outer layer of wall of eye.
4. Fluid between lens and retina.
5. Fluid between lens and cornea.
6. Inner light-sensitive layer.
7. Round yellow spot on retina.
8. Delicate membrane that lines eyelids.
9. Middle vascular layer of wall of eyeball.
10. Drainage tubes for tears in eyelids.
11. Clear transparent portion of front of eyeball.
12. Tube that drains tears from lacrimal sac.
13. Chamber between iris and cornea.
14. Conical body in medial corner of the eye.
15. Chamber between the iris and lens.
16. Circular color band between lens and cornea.
17. Circular band of smooth muscle tissue surrounding lens.
18. Cartilaginous loop through which superior oblique muscle acts.
19. Connective tissue between lens perimeter and surrounding muscle.
20. A semicircular fold of conjunctiva in the medial canthus of the eye.

anterior chamber—1
aqueous humor—2
blind spot—3
caruncula—4
choroid coat—5
ciliary body—6
cornea—7
conjunctiva—8
fovea centralis—9
iris—10
lacrimal ducts—11
lacrimal sac—12
macula lutea—13
medial canthus—14
nasolacrimal duct—15
plica semilunaris—16
posterior chamber—17
pupil—18
retina—19
scleroid coat—20
suspensory ligament—21
trochlea—22
vitreous body—23

C. Extrinsic Muscles
Select the muscles of the eye that are described by the following statements.

1. Inserted on top of eyeball.
2. Inserted on side of eyeball.
3. Inserted on medial surface.
4. Inserted on bottom of eyeball.
5. Inserted on eyelid.
6. Raises the eyelid.
7. Rotates eyeball inward.
8. Rotates eyeball downward.
9. Rotates eyeball outward.
10. Rotates eyeball upward.

inferior oblique—1
inferior rectus—2
lateral rectus—3
medial rectus—4
superior levator palpebrae—5
superior oblique—6
superior rectus—7

Answers

Structures | Fig. 48.1
1. ___
2. ___
3. ___
4. ___
5. ___
6. ___
7. ___
8. ___
9. ___
10. ___
11. ___
12. ___
13. ___ Fig. 48.2
14. ___
15. ___
16. ___
17. ___
18. ___
19. ___
20. ___

Muscles
1. ___
2. ___
3. ___
4. ___
5. ___
6. ___
7. ___
8. ___
9. ___
10. ___

567

D. Functions

Select the part of the eye that performs the following functions. More than one answer may apply in some cases.

aqueous humor—1
blind spot—2
choroid coat—3
ciliary body—4
conjunctiva—5
iris—6
lacrimal ducts—7
lacrimal puncta—8
lacrimal sac—9
macula lutea—10

nasolacrimal duct—11
pupil—12
retina—13
scleral venous sinus—14
scleroid coat—15
suspensory ligament—16
trabeculae—17
trochlea—18
vitreous body—19

Answers	
Functions	Fig. 48.3
1. _____	_____
	_____
2. _____	_____
	_____
3. _____	_____
	_____
4. _____	_____
	_____
5. _____	_____
	_____
6. _____	
7. _____	
8. _____	

1. Place where nerve fibers of retina leave eyeball.
2. Furnishes blood supply to retina and sclera.
3. Large vessel in wall of the eye that collects aqueous humor from trabeculae.
4. Exerts force on the lens, changing its contour.
5. Controls the amount of light that enters the eye.
6. Maintains firmness and roundness of eyeball.
7. Part of retina where critical vision occurs.
8. Provides most of the strength to the wall of the eyeball.

E. Beef Eye Dissection

Answer the following questions that pertain to the dissection of the beef eye.

1. What is the shape of the pupil?_____

2. Why do you suppose it is so difficult to penetrate the sclera with a sharp scalpel? _____

_____

3. What is the function of the black pigment in the eye? _____

_____

_____

4. Compare the consistency of the two fluids in the eye.

   aqueous humor: _____

   vitreous humor: _____

5. When you hold the lens up and look through it what is unusual about the image? _____

_____

6. Does the lens magnify printed matter when placed directly on it? _____

7. Compare the consistencies of the following portions of the lens.

   center: _____

   edge: _____

8. What is the reflective portion of the choroid coat called? _____

9. Is there a macula lutea on the retina of the beef eye? _____

F. Ophthalmoscopy

Record your ophthalmoscope measurements here, and answer the questions.

1. **Diopter Measurements**

    The diopter value of a lens is the reciprocal of its focal length (f) in meters, or $D = \dfrac{1}{f}$.

    A lens of one diopter (1D) has a focal length of 1 meter, $(D = \dfrac{1}{f} 5 \dfrac{1}{1} = 1)$; a 2D lens has a focal length of 0.5 meter, or $\dfrac{1}{f} 5 \dfrac{1}{0.5} = 2$; etc.

    Record here your measurements for: 5D_____; 10D_____; 20D_____; and 40D_____.

    Calculated diopter distances: 5D_____; 10D_____; 20D_____; and 40D_____.

    Do your calculations match the measured distances? _____

2. If you were able to examine your laboratory partner's eye with "0" in the diopter window of the ophthalmoscope, what would this indicate about the curvature of the lens of your eye? _____

    about your laboratory partner's eye? _____

Visual Tests (Ex. 49)

A. The Purkinje Tree

Describe in a few sentences the image you observed. _____

_____

_____

_____

_____

_____

B. Tabulations

Record the results that were obtained in the following five eye tests:

TEST	RIGHT EYE	LEFT EYE
Blind Spot (inches)		
Near Point (inches)		
Scheiner's Experiment (inches)		
Visual Acuity (X/20)		
Astigmatism (present or absent)		

C. Color Blindness Test

Have your laboratory partner record your responses to each test plate. The mark *x* indicates inability to read correctly.

PLATE NUMBER	SUBJECT'S RESPONSE	NORMAL RESPONSE	RESPONSE IF RED-GREEN DEFICIENCY				RESPONSE IF TOTALLY COLOR-BLIND
1		12	12				12
2		8	3				x
3		5	2				x
4		29	70				x
5		74	21				x
6		7	x				x
7		45	x				x
8		2	x				x
9		x	2				x
10		16	x				x
11		traceable	x				x
			Protan		Deuteran		
			Strong	Mild	Strong	Mild	
12		35	5	(3)5	3	3(5)	
13		96	6	(9)6	9	9(6)	
14		can trace two lines	purple	purple (red)	red	red purple	x

x—Indicates inability of subject to respond in any way to test.

Conclusion: _____

D. Pupillary Reflexes

Answer the following questions related to your observations in the two experiments on pupillary reflexes.

1. **Accommodation to Distance**
   a. What pupillary size change occurred in this experiment? _____

   _____

   b. Can you suggest a benefit that might result from this happening? _____

   _____

2. **Light Intensity and Pupil Size**
   a. Did the pupil of the unexposed eye become smaller when the right eye was exposed to light?

   _____

   b. Trace the pathways of the nerve impulses in effecting the response. _____

   _____

   _____

   c. Can you suggest a possible benefit that might result from this phenomenon? _____

   _____

LABORATORY REPORT

50

Student: _____

Desk No.: _____ Section: _____

The Ear:
Its Role in Hearing

A. Figure 50.2

Record the labels for this figure in the answer column.

B. Structure

Identify the structures described by the following statements.

basilar membrane—1	malleus—6	scala tympani—11
endolymph—2	perilymph—7	scala vestibuli—12
hair cells—3	reticular lamina—8	stapes—13
helicotrema—4	rods of Corti—9	tectorial membrane—14
incus—5	scala media—10	tympanic membrane—15
		vestibular membrane—16

1. Ossicle that fits in oval window.
2. Ossicle activated by tympanic membrane.
3. Chamber of cochlea into which round window opens.
4. Chamber of cochlea into which oval window opens.
5. Membrane set in vibration by sound waves in the air.
6. Membrane that contains 20,000 fibers of varying lengths.
7. Membrane in which hair tips of hair cells are embedded.
8. Two fluids found in cochlea.
9. Fluid within scala tympani.
10. Fluid within cochlear duct.
11. Fluid within scala vestibuli.
12. Parts of spiral organ.
13. Transfers vibrations from basilar membrane to reticular lamina.
14. Receptors that initiate action potentials in cochlear nerve.
15. Opening between scala vestibuli and scala tympani.

C. Physiology of Hearing (True–False)

Indicate the validity of each of the following statements with a T or an F.

1. The loudness of sound is directly proportional to the square of the amplitude.
2. The pitch of a sound is determined by the amplitude of the sound wave.
3. Quality and pitch are two different characteristics of sound.
4. Some hearing takes place through the bones of the skull rather than through the auditory ossicles.
5. Whereas there is no cure for nerve deafness, conduction deafness can often be corrected.
6. Endolymph in the scala media flows into the scala vestibuli through the helicotrema.
7. Nerve deafness can be detected by placing a tuning fork on the mastoid process.
8. The frequency range used in speech is from 30 to 20,000 cps.
9. The bending of hairs of hair cells causes an alternating electrical charge known as the endocochlear potential.
10. Before using an audiometer, it is necessary for the operator to calibrate it.

Answers

Structure	Fig. 50.2
1. _____	
2. _____	
3. _____	
4. _____	
5. _____	
6. _____	
7. _____	
8. _____	
9. _____	
10. _____	
11. _____	
12. _____	
13. _____	
14. _____	
15. _____	

Physiology	
1. _____	
2. _____	
3. _____	
4. _____	
5. _____	
6. _____	
7. _____	
8. _____	
9. _____	
10. _____	

D. Screening Test

Record the distances at which the watch tick could be heard with your ears, as well as in three other persons.

Subject	INCHES FROM EAR			
	Right Ear		Left Ear	
	Approaching	Receding	Approaching	Receding

E. Tuning Fork Methods

Record the results of the two tuning fork methods.

Rinne Test

Right Ear _____

Left Ear _____

Weber Test

Right Ear _____

Left Ear _____

F. Audiometry

Record the threshold levels in the right ear with a red "O" and in the left ear with a blue "X." Connect the points with appropriately colored lines.

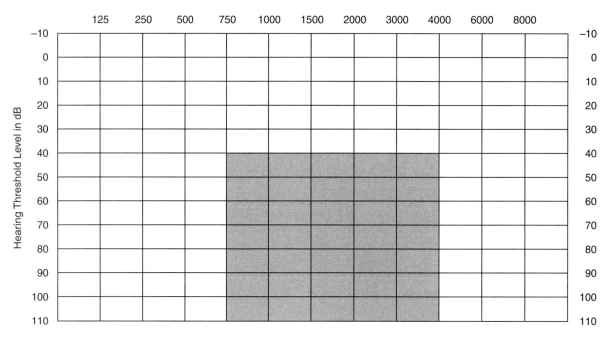

(Shaded area indicates critical area of speech interpretation)

LABORATORY REPORT
51

Student: _____

Desk No.: _____ Section: _____

The Ear:
Its Role in Equilibrium

A. Anatomy

Answer the following questions that pertain to the anatomy of the vestibular apparatus.

1. What are the sensory receptors for static equilibrium called?
2. Where are the static equilibrium sensory receptors located (two places)?
3. What are the small calcium carbonate crystals of the macula called?
4. Where are the sensory receptors of dynamic equilibrium located?
5. What nerve branch of the eighth cranial nerve supplies the vestibular apparatus?
6. What part of the vestibular apparatus connects directly to the cochlear duct?
7. What is the name of the fluid within the vestibular apparatus?
8. What structures in joints prevent a sense of malequilibrium when one tilts the head to one side?
9. What fluid lies between the semicircular ducts and the surrounding bone of the semicircular canals?
10. Within how many planes do the semicircular ducts lie?

Answers
Anatomy
1. _____
2. _____
_____
3. _____
4. _____
5. _____
6. _____
7. _____
8. _____
9. _____
10. _____

B. Nystagmus

After having observed nystagmus, answer these questions:

1. What direction do the eyes move in relation to the direction of rotation? _____

2. What seems to be the function of slow movement of the eye? _____

_____

_____

3. What seems to be the function of fast movement of the eye? _____

_____

_____

4. What part of the vestibular apparatus triggers the reflexes that control the eye muscles? _____

_____

C. Proprioceptive Influences

Record your observations here.

1. At-Rest Reactions

a. Was the subject able to place the heel of his foot on the toes of the other foot with eyes closed?

_____

b. Was the subject able to touch her nose with eyes closed? _____

c. Explain what mechanisms in the body make this possible: _____

_____

d. Describe the subject's ability to touch the pencil eraser with eyes closed after practicing with eyes open. _____

_____

2. Effects of Rotation

a. Was the subject, with eyes open, able to point directly to the pencil eraser with finger after rotation? _____

_____

b. When blindfolded subject tried to locate pencil eraser after rotation, what was the result?

_____

_____

LABORATORY REPORT

52-58

Student: _____

Desk No.: _____ Section: _____

Hematological Tests

Test Results

Except for blood typing, record all test results in table II on page 578. Calculations for blood cell counts should be performed as stated below.

A. Differential White Blood Cell Count (Ex. 52)

As you move the slide in the pattern indicated in figure 52.4, record all the different types of cells in table I. Refer to figures 52.1 and 52.2 for cell identification. Use this method of tabulation: ⊢⊢⊢⊥ ⊢⊢⊢⊥ 1 1. Identify and tabulate 100 leukocytes. Divide the total of each kind of cell by 100 to determine percentages.

Table 1 Leukocyte tabulation.

NEUTROPHILS	LYMPHOCYTES	MONOCYTES	EOSINOPHILS	BASOPHILS
Totals				
Percent				

B. Calculations for Blood Cell Counts (Ex. 53)

1. **Total White Blood Cell Count**
   Total WBCs counted in 4 W areas $\times$ 50 = **WBCs per mm^3**

2. **Red Blood Cell Count**
   Total RBCs counted in 5 R areas $\times$ 10,000 = **RBCs per mm^3**

C. Blood Typing (Ex. 58)

Record your blood type here: _____

1. If you needed a blood transfusion, what types of blood could you be given?

   _____

2. If your blood were to be given to someone in need, what type should that person have?

   _____

Questions

Although most of the answers to the following questions can be derived from this manual, it will be necessary to consult your lecture text or medical dictionary for some of the answers.

A. Blood Components

Identify the blood components described by the following statements. More than one answer may apply.

antibodies—1	fibrin—5	neutrophils—10
basophils—2	fibrinogen—6	platelet factor—11
eosinophils—3	lymphocytes—7	platelets—12
erythrocytes—4	lysozyme—8	prothrombin—13
	monocytes—9	thromboplastin—14

1. Cells that transport O_2 and CO_2.
2. Phagocytic cell concerned, primarily, with local infections.
3. Phagocytic cell concerned, primarily, with generalized infections.
4. Protein substances in blood that inactivate foreign protein.
5. Noncellular formed elements essential for blood clotting.
6. Substances necessary for blood clotting.
7. Gelatinous material formed in blood clotting.
8. Substance released by injured cells that initiates blood clotting.
9. Enzyme in blood plasma that destroys some kinds of bacteria.
10. Leukocyte distinguished by having distinctive red-stained granules.
11. Leukocyte that probably produces heparin.
12. Most numerous type of leukocyte.
13. Contains hemoglobin.
14. Most numerous type of blood cell.

B. Blood Diseases

Identify the pathological conditions characterized by the following statements. More than one condition may apply to some statements.

anemia—1	leukemia—5	neutropenia—9
eosinophilia—2	leukocytosis—6	neutrophilia—10
erythroblastosis fetalis—3	leukopenia—7	pernicious anemia—11
hemophilia—4	lymphocytosis—8	polycythemia—12
		sickle-cell anemia—13

1. Too few red blood cells.
2. Too many red blood cells.
3. Too few leukocytes.
4. Too many neutrophils.
5. Too many lymphocytes.
6. Too little hemoglobin.
7. Defective RBCs due to stomach enzyme deficiency.
8. Rh factor incompatibility present at birth.
9. Hereditary blood disease of blacks; RBCs of abnormal shape.
10. Heritable bleeder's disease.
11. Cancerouslike condition in which there are too many leukocytes.

Answers

Components

1. _____
2. _____
3. _____
4. _____
5. _____
6. _____
7. _____
8. _____
9. _____
10. _____
11. _____
12. _____
13. _____
14. _____

Diseases

1. _____
2. _____
3. _____
4. _____
5. _____
6. _____
7. _____
8. _____
9. _____
10. _____
11. _____

C. Terminology

Differentiate between the following pairs of related terms:

Plasma _____

Lymph _____

Coagulation _____

Agglutination _____

Antibody _____

Antigen _____

D. True–False

Indicate in the answer column whether the following statements are true or false by writing a T or an F.

1. The most common types of blood are types O and A.

2. A universal donor has O-Rh negative blood.

3. Diapedesis is the ability of leukocytes to move between the cells of the capillary walls.

4. Anemia is usually due to an iodine deficiency.

5. Hemoglobin combines with both oxygen and carbon dioxide.

6. The maximum life span of an erythrocyte is about 30 days.

7 White blood cells live much longer than red blood cells.

8. The vitamin that is essential to blood clotting is vitamin D.

9. An infant born with erythroblastosis would best be treated with a transfusion of Rh negative instead of Rh positive blood.

10. Worm infestations may cause eosinophilia.

11. Viral infections may cause an increase in the number of lymphocytes or monocytes.

12. Blood clotting can be enhanced with dicumarol.

13. Megakaryocytes of bone marrow produce blood platelets.

14. Bloods that are matched for ABO and Rh factor can be mixed with complete assurance of no incompatibility.

15. The rarest type of blood is AB negative.

16. Heparin is essential to fibrin formation.

17. Blood typing may be used to prove that an individual could not be the father in a paternity suit.

18. Blood typing may be used to prove that an individual is the father in a paternity suit.

19. Approximately 87% of the population is Rh negative.

20. A deficiency of calcium will result in the failure of blood to clot.

21. Fibrin is formed when prothrombin reacts with fibrinogen.

Answers
True–False
1. _____
2. _____
3. _____
4. _____
5. _____
6. _____
7. _____
8. _____
9. _____
10. _____
11. _____
12. _____
13. _____
14. _____
15. _____
16. _____
17. _____
18. _____
19. _____
20. _____
21. _____

E. Summarization of Results

Record blood cell counts and other test results in the following table:

Table II

TEST	NORMAL VALUES	TEST RESULTS	EVALUATION (over, under, normal)
Differential WBC Count	Neutrophils: 50%–70%		
	Lymphocytes: 20%–30%		
	Monocytes: 2%–6%		
	Eosinophils: 1%–5%		
	Basophils: 0.5%–1%		
Total WBC Count	5000–9000 per mm^3		
RBC Count	Males: 4.8–6.0 million mm^3		
	Females: 4.1–5.1 million mm^3		
Hemoglobin Percentage	Males: 13.4–16.4 gm/100 ml		
	Females: 12.2–15.2 gm/100 ml		
Hematocrit (VPRC)	Males: 40%–54% (Av. 47%)		
	Females: 37%–47% (Av. 42%)		
Coagulation Time	2–6 minutes		
Sedimentation Rate	0–6 mm		

F. Materials

Identify the various blood tests in which the following supplies are used.

1. Capillary tubes
2. Centrifuge
3. Hemacytometer
4. Hemoglobinometer
5. Landau rack
6. Microscope slides
7. Hemolysis applicators
8. Seal-Ease
9. Diluting fluid
10. Sodium citrate
11. Wright's stain

hematocrit (VPRC)—1
clotting time—2
RBC and WBC counts—3
sedimentation rate—4
differential WBC count—5
hemoglobin determination—6

Materials

1. _____
2. _____
3. _____
4. _____
5. _____
6. _____
7. _____
8. _____
9. _____
10. _____
11. _____

Laboratory Report

59

Student: _____

Desk No.: _____ Section: _____

Anatomy of the Heart

A. Illustrations

Record the labels for figures 59.1 and 59.2 in the answer columns.

B. Structures

Identify the structures of the heart described by the following statements.

aorta—1	myocardium—13
aortic semilunar valve—2	papillary muscles—14
bicuspid valve—3	parietal pericardium—15
chordae tendineae—4	pulmonic semilunar valve—16
endocardium—5	pulmonary trunk—17
epicardium—6	right atrium—18
inferior vena cava—7	right pulmonary artery—19
left atrium—8	right ventricle—20
left pulmonary artery—9	superior vena cava—21
left ventricle—10	tricuspid valve—22
ligamentum arteriosum—11	ventricular septum—23
mitral valve—12	visceral pericardium—24

1. Lining of the heart.
2. Partition between right and left ventricles.
3. Fibroserous saclike structure surrounding the heart.
4. Two chambers of the heart that contain deoxygenated blood.
5. Large artery that carries blood from the right ventricle.
6. Blood vessel that returns blood to heart from head and arms.
7. Remnant of a functional prenatal vessel between the pulmonary trunk and the aorta.
8. Structure formed from ductus arteriosus.
9. Valve at the base of the pulmonary trunk.
10. Two arteries that are branches of the pulmonary trunk.
11. Synonym for bicuspid valve.
12. Structures on the cardiac wall to which the chordae tendineae are attached.
13. Muscular portion of cardiac wall.
14. Thin covering on surface of heart (2 names).
15. Large vein that empties blood into top of right atrium.
16. Two chambers of the heart that contain oxygenated blood.
17. Chamber of the heart that receives blood from the lungs.
18. Large artery that carries blood out from left ventricle.
19. Large vein that empties blood into the lower part of the right atrium.
20. Atrioventricular valves.
21. Valve at the base of the aorta.
22. Blood vessel in which openings to the coronary arteries are located.
23. Atrium into which the coronary sinus empties.
24. Valvular restraints that prevent the atrioventricular valve cusps from being forced back into the atria.

Answers

Structures	Fig. 59.1
1. _____	_____
2. _____	_____
3. _____	_____
4. _____	_____
5. _____	_____
6. _____	_____
7. _____	_____
8. _____	_____
9. _____	_____
10. _____	_____
11. _____	_____
12. _____	_____
13. _____	_____
14. _____	_____
15. _____	_____
16. _____	_____
17. _____	_____
18. _____	_____
19. _____	_____
20. _____	_____
21. _____	_____
22. _____	_____
23. _____	_____
24. _____	_____
	_____
	_____
	_____
	_____
	_____
	_____
	_____

C. Coronary Circulation

Identify the arteries and veins of the coronary circulatory system that are described by the following statements.

anterior interventricular artery—1
circumflex artery—2
coronary sinus—3
great cardiac vein—4
left coronary artery—5
middle cardiac vein—6
posterior descending right coronary artery—7
posterior interventricular vein—8
right coronary artery—9
small cardiac vein—10

1. Two principal branches of the left coronary artery.
2. Coronary vein that lies in the anterior interventricular sulcus.
3. Coronary artery that lies in the right atrioventricular sulcus.
4. Vein that receives blood from the great cardiac vein.
5. Vein that lies in the posterior interventricular sulcus.
6. Two major coronary arteries that take their origins in the wall of the aorta.
7. Large coronary vessel that empties into the right atrium.
8. Coronary vein that parallels the right coronary artery in the right atrioventricular sulcus.
9. Branch of the right coronary artery on the posterior surface of the heart.
10. Vein that lies alongside of the posterior descending right coronary artery.

D. Sheep Heart Dissection

After completing the sheep heart dissection answer the following questions. Some of these questions were encountered during the dissection; others pertain to structures not shown in figures 59.1 and 59.2.

1. Identify the following structures:

   Pectinate muscle: _____

   _____

   Moderator band: _____

   _____

   Ligamentum arteriosum: _____

   _____

2. How many papillary muscles did you find in the right ventricle? _____ left ventricle? _____

3. How many pouches are present in each of the following?

   pulmonary semilunar valve: _____

   aortic semilunar valve: _____

4. Where does blood enter the myocardium? _____

   _____

5. Where does blood leave the myocardium and return to the cardiovascular system? _____

Answers	
Coronary Circulation	Fig. 59.2
1. _____	_____
2. _____	_____
3. _____	_____
4. _____	_____
5. _____	_____
6. _____	_____
7. _____	_____
8. _____	_____
9. _____	_____
10. _____	_____
	_____
	_____
	_____
	_____
	_____
	_____
	_____
	_____
	_____
	_____
	_____
	_____
	_____
	_____
	_____

Student: _____

Desk No.: _____ Section: _____

Cardiovascular Sounds

A. *Phonocardiogram*

Attach below a segment of a phonocardiogram that you made on the recorder. Identify on the chart those segments that represent first and second sounds.

B. *Effect of Exercise*

How do heart sounds differ before and after exercise? _____

_____

_____

C. *Causes*

Explain what causes:

1. First Sound _____

_____

_____

2. Second Sound _____

_____

_____

3. Third Sound _____

_____

4. Murmur _____

_____

_____

D. Questions

1. Why does the aortic valve close sooner than the pulmonic valve during inspiration? _____

_____

_____

2. What effect would complete right bundle branch block have on the splitting of the second sound?

_____

_____

_____

3. Explain why pulmonary stenosis is associated with greater splitting of the second heart sound.

_____

_____

_____

Student: _____

Desk No.: _____ Section: _____

Heart Rate Control

A. *Tracings*

Attach six tracings of your experiment in the spaces provided below. It will be necessary to trim excess paper from the tracings to fit them into the allotted spaces. Attach with Scotch tape. Also, record the contractions per minute in the space provided.

NO.	TRACINGS	CONTRACTIONS PER MINUTE
1	Normal Contractions: Ringer's Solution	
2	0.05 ml Epinephrine	
3	0.05 ml Acetylcholine	
4	0.05 ml Epinephrine	
5	0.25 ml Epinephrine	
6	0.25 ml Acetylcholine	

*Since the paper moves at 2.5 mm per second (slow speed), 5 mm on the paper represents 2 seconds. Also, note that the white margin on the Gilson paper has a line every 7.5 centimeters. This distance between these margin lines represents 30 seconds.

B. Results

1. How did the first injection of epinephrine affect the following?

   Heart rate _____

   Strength of contraction _____

   _____

2. How much of a change occurred when the first injection of acetylcholine was administered?

   Previous rate _____ Acetylcholine rate _____

   Previous amplitude _____ (mm) A/C amplitude _____ (mm)

3. Was the heart rate stopped with any injection? _____

   If so, which one? _____

   Did the heart ever recover? _____

C. Questions

1. What happens to acetylcholine in the tissues to prevent it from having a prolonged effect?

   _____

   _____

2. Does acetylcholine affect both the rate and strength of contraction? _____

3. Does epinephrine affect both the rate and strength of contraction? _____

4. What is the pharmacological significance of acetylcholine and epinephrine to the autonomic control

   of the heart? _____

   _____

   _____

LABORATORY REPORT

62

Electrocardiogram Monitoring:
Using Chart Recorders

A. Tracings

Attach three tracings of your experiment in the spaces provided below. It will be necessary to trim excess paper from the tracings to fit them into the allotted spaces. Attach with Scotch tape.

TRACINGS	EVALUATION
Calibration	
	*Heart Rate: _____ per min QR Potential: _____ millivolts **Duration of cycle: _____ msec
Subject at Rest	
	*Heart Rate: _____ per min QR Potential: _____ millivolts **Duration of cycle: _____ msec
After Exercise	

*Space between two margin lines is 3 seconds.
**Time from beginning of P wave to end of T wave.

B. Questions

1. During what part of ECG wave pattern do atrial depolarization and contraction occur? _____

   _____

2. During what part of ECG wave pattern do ventricular depolarization and contraction occur?

   _____

3. What infectious disease causes the highest incidence of heart disease? _____

   _____

4. Would a heart murmur necessarily show up on an ECG? _____

   Explain. _____

   _____

LABORATORY REPORT

64

Student: _____

Desk No.: _____ Section: _____

Pulse Monitoring

A. *Tracings*

Attach samples of tracings made in this experiment in the spaces provided below and on the next page. Use the right-hand column for relevant comments.

TRACINGS	EVALUATION
Transducer at Heart Level	Pulse Rate: _____
Arm Slowly Raised above Heart Level	Pulse Rate: _____
Arm Lowered to Floor	Pulse Rate: _____
Holding Breath for 15 Seconds	Pulse Rate: _____

TRACINGS	EVALUATION
	Pulse Rate: _____
Brachial Artery Occluded for 15–20 Seconds	
	Pulse Rate: _____
Chart Speed Increased to 25 mm per Second	
	Pulse Rate: _____
Subject Startled	

B. Pulse Tabulations

Record your pulse (while at rest) on the chalkboard. Once the pulse rates of all students are recorded, record the highest, lowest, and median pulses:

_____ Highest _____ Lowest _____ Median _____

Significance _____

_____

C. Questions

1. What conceivably might be an explanation for changes occurring in the dicrotic notch with aging?

   _____

   _____

2. What might be an explanation for an alteration in the pulse amplitude during smoking? _____

   _____

3. The finger pulse is a manifestation of blood volume, velocity, direction, and pressure. Which one of these factors is directly responsible for the dicrotic notch and why? _____

   _____

LABORATORY REPORT
65

Student: _____

Desk No.: _____ Section: _____

Blood Pressure Monitoring

A. Stethoscope Method

Record here the blood pressures of the text subject.

	BEFORE EXERCISE	AFTER EXERCISE
Systolic		
Diastolic		
*Pulse Pressure		
**Mean Arterial Pressure		

* Pulse pressure = systolic pressure minus diastolic pressure.
**Mean arterial pressure ≈ diastolic pressure + ⅓ pulse pressure.

B. Recording Method

In the spaces provided below and on the next page tape or staple chart recordings for the various portions of the experiments.

Calibration of Recorder

Subject in Chair

Subject Standing Erect

Subject Reclining

After Exercise

C. Questions

1. Give the systolic and diastolic pressures for normal blood pressure: _____

2. What systolic pressure indicates hypertension? _____

3. Why is a low diastolic pressure considered harmful? _____

4. Why are diuretics of value in treating hypertension? _____
   _____

5. Why is reduced salt intake of value in regulating blood pressure? _____
   _____
   _____

LABORATORY REPORT

66,67

Student: _____

Desk No.: _____ Section: _____

A Polygraph Study of the Cardiac Cycle (Ex. 66)

A. Tracing

Attach a portion of the tracing with Scotch tape or staples in the space provided below.

B. Events

Identify the stages in which each of the following events occur. More than one answer may apply in some cases.

early atrial systole—1	late ventricular ejection—5
late atrial systole—2	early diastole—6
isometric ventricular systole—3	late diastole—7
midventricular ejection—4	

1. First sound.

2. Second sound.

3. Third sound.

4. Dicrotic notch.

5. T wave.

6. P wave.

7. U wave.

8. AV valves open.

9. Aortic and pulmonic valves open.

10. All four valves momentarily closed.

11. Arterial pressure is highest.

12. Arterial pressure is lowest.

13. Ventricular depolarization.

14. Ventricular repolarization.

15. Atrial depolarization.

16. Atrial repolarization.

Answers

Events

1. _____
2. _____
3. _____
4. _____
5. _____
6. _____
7. _____
8. _____
9. _____
10. _____
11. _____
12. _____
13. _____
14. _____
15. _____
16. _____

Peripheral Circulation Control (Ex. 67)

A. Results

1. What observable change occurred in the frog's foot when histamine was applied to the foot?

    _____

    _____

2. What was the effect of histamine on the arterioles to produce this effect (*vasodilation* or *vasocon-striction*)? _____

3. What observable change occurred in the frog's foot when epinephrine was applied to the foot?

    _____

    _____

4. What was the effect of epinephrine on the arterioles to produce this effect (*vasodilation* or *vasocon-striction*)? _____

B. Questions

1. Does epinephrine have the same effect on arterioles of skeletal muscles? _____

    _____

2. Provide an explanation (theoretical) of how epinephrine can cause vasodilation in one part of the body and vasoconstriction in another region. _____

    _____

    _____

    _____

3. What are baroreceptors, and what role do they play in regulating blood flow? _____

    _____

    _____

    _____

4. Why are arteries called "resistance" vessels and veins "capacitance" vessels? _____

    _____

    _____

LABORATORY REPORT

68

Student: _____

Desk No.: _____ Section: _____

The Arteries and Veins

A. Illustrations

Record the labels for figures 68.2, 68.3, and 68.4 in the answer columns.

B. Arteries

Refer to figure 68.2 to identify the arteries described by the following statements.

anterior tibial—1
aorta—2
aortic arch—3
axillary—4
brachial—5
brachiocephalic—6
celiac—7
common iliac—8
deep femoral—9
external iliac—10
femoral—11
inferior mesenteric—12

internal iliac—13
left common carotid—14
left subclavian—15
popliteal—16
posterior tibial—17
radial—18
renal—19
right common carotid—20
subclavian—21
superior mesenteric—22
ulnar—23

1. Artery of the shoulder.
2. Artery of upper arm.
3. Artery of the armpit.
4. Gives rise to right common carotid and right subclavian.
5. Lateral artery of the forearm.
6. Medial artery of the forearm.
7. Three branches of the aortic arch.
8. Gives rise to femoral artery.
9. Major artery of the thigh.
10. Artery of knee region.
11. Artery of calf region.
12. Large branch of common iliac.
13. Small branch of common iliac.
14. Supplies the stomach, spleen, and liver.
15. Major artery of chest and abdomen.
16. Supplies the kidney.
17. Also known as the hypogastric artery.
18. Supplies blood to most of small intestines and part of colon.
19. Artery in interior portion of lower leg.
20. Branch of femoral that parallels medial surface of femur.
21. Curved vessel that receives blood from left ventricle.
22. Supplies the large intestine and rectum.

Answers

Arteries	Fig. 68.2
1. _____	_____
2. _____	_____
3. _____	_____
4. _____	_____
5. _____	_____
6. _____	_____
7. _____	_____
8. _____	_____
9. _____	_____
10. _____	_____
11. _____	_____
12. _____	_____
13. _____	_____
14. _____	_____
15. _____	_____
16. _____	_____
17. _____	_____
18. _____	_____
19. _____	_____
20. _____	_____
21. _____	_____
22. _____	_____

C. Veins

Refer to figure 68.3 and to figure 68.4 to identify the veins described by the following statements.

accessory cephalic—1
axillary—2
basilic—3
brachial—4
brachiocephalic—5
cephalic—6
common iliac—7
coronary—8
dorsal venous arch—9
external iliac—10
external jugular—11
femoral—12
great saphenous—13
hepatic—14
inferior vena cava—15
inferior mesenteric—16
internal iliac—17
internal jugular—18
median cubital—19
popliteal—20
portal—21
posterior tibial—22
pyloric—23
subclavian—24
superior mesenteric—25
superior vena cava—26

1. Largest vein in neck.
2. Small vein in neck.
3. Vein of armpit.
4. Empty into brachiocephalic veins.
5. Collects blood from veins of head and arms.
6. Short vein between basilic and cephalic veins.
7. Collects blood from veins of chest, abdomen, and legs.
8. On posterior surface of humerus.
9. Large veins that unite to form the inferior vena cava.
10. On medial surface of upper arm.
11. On lateral portion of forearm.
12. On lateral portion of upper arm.
13. Vein from liver to inferior vena cava.
14. Receives blood from descending colon and rectum.
15. Empties into popliteal.
16. Collects blood from two brachiocephalic veins.
17. Collects blood from top of foot.
18. Superficial vein on medial surface of leg.
19. Vein of knee region.
20. Collects blood from intestines, stomach, and colon.
21. Vein that receives blood from posterior tibial.
22. Receives blood from ascending colon and part of ileum.
23. Empties into great saphenous.
24. Small vein that empties into common iliac at juncture of external iliac.
25. Vein that great saphenous empties into.
26. Two veins that drain blood from stomach into portal vein.

Answers

Veins	Fig. 68.3
1. ___	___
2. ___	___
3. ___	___
4. ___	___
5. ___	___
6. ___	___
7. ___	___
8. ___	___
9. ___	___
10. ___	___
11. ___	___
12. ___	___
13. ___	___
14. ___	___
15. ___	___
16. ___	___
17. ___	___
18. ___	___
19. ___	___
20. ___	___
21. ___	___
22. ___	___
23. ___	___
24. ___	___
25. ___	
26. ___	Fig. 68.4

LABORATORY REPORT

69,70

Fetal Circulation (Ex. 69)

A. Figure 69.1

Record the labels for this illustration in the answer column.

B. Questions

Record the answers for the following questions in the answer column.

1. What blood vessel in the umbilical cord supplies the fetus with nutrients?
2. What blood vessels in the umbilical cord return blood to the placenta from the fetus?
3. What blood vessel shunts blood from the pulmonary trunk to the aorta?
4. What vessel in the liver carries blood from the umbilical vein to the inferior vena cava?
5. What structure in the liver is formed from the ductus venosus?
6. What structure forms from the ductus arteriousus?
7. What structure forms from the umbilical vein?
8. What structure enables blood to flow from the right atrium to the left atrium before birth?
9. Give two occurrences at childbirth that prevent excessive blood loss by the infant through cut umbilical cord.
10. What structures near the bladder form from the hypogastric arteries?

The Lymphatic System and the Immune Response (Ex. 70)

A. Figure 70.1

Record the labels for this illustration in the answer column.

B. Components

By referring to figures 70.1 and 70.2 identify the following structures in the lymphatic system.

1. Large vessel in thorax and abdomen that collects lymph from lower extremities.
2. Short vessel that collects lymph from right arm and right side of head.
3. Microscopic lymphatic vessels situated among cells of tissues.
4. Saclike structure that receives chyle from intestine.
5. Blood vessel that receives fluid from thoracic duct.
6. Filters of the lymphatic system.
7. Vessels of arms and legs that convey lymph to collecting ducts.
8. Depression in lymph node through which blood vessels enter node.
9. Outer covering of lymph node.
10. Structural units of lymph node that are packed with lymphocytes.
11. Cells involved in organ transplant rejections.
12. Cells that produce antibodies.

capsule—1
cisterna chyli—2
germinal centers—3
hilum—4
left subclavian vein—5
lymphatics—6
lymph capillaries—7
lymphoblasts—8
lymph nodes—9
plasma cells—10
right lymphatic duct—11
thoracic duct—12
none of these—13

Answers

Questions

1. _____
2. _____
3. _____
4. _____
5. _____
6. _____
7. _____
8. _____
9a. _____
b. _____
10. _____

Components	Fig. 69.1
1. _____	_____
2. _____	_____
3. _____	_____
4. _____	_____
5. _____	_____
6. _____	_____
7. _____	_____
8. _____	_____
9. _____	_____
10. _____	_____
11. _____	_____
12. _____	_____
	_____
	_____
	_____
	_____
	_____
	_____

C. Questions

1. List three forces that move lymph through the lymphatic vessels:

   a. _____

   b. _____   c. _____

2. How does lymph in the lymphatics of the legs differ in composition from the lymph in the thoracic duct?

   _____

   _____

3. Where do stem cells for all lymphocytes originate? _____

   _____

4. Give the cell type that accounts for each type of immunity:

   Humoral immunity: _____

   Cellular immunity: _____

5. What tissues program stem cells for the following?

   Humoral immunity: _____

   Cellular immunity: _____

6. What type of tissue is common to the thymus gland and bursal tissues?

   _____

Answers
Fig. 70.1
_____
_____
_____
_____
_____
_____
_____
_____
_____
_____
_____
_____
_____
_____

LABORATORY REPORT

71

The Respiratory Organs

A. Illustrations

Record the labels for the illustrations of this exercise in the answer column.

B. Cat Dissection

After completing the dissection of the cat, answer the following questions. Both of these questions are asked in the course of the dissection.

1. Are the cartilaginous rings of the trachea continuous all the way

   around the organ? _____

2. Which lung has four lobes? _____

   Which has three lobes? _____

C. Sheep Pluck Dissection

After completing the sheep pluck dissection, answer the following questions.

1. Are the cartilaginous rings of the trachea continuous all the way

   around the organ? _____

2. Describe the texture of the surface of the lung. _____

   _____

   _____

3. How many lobes exist on the right lung? _____

   On the left lung? _____

4. Why does lung tissue collapse so readily when you quit blowing

   into it with a straw? _____

   _____

   _____

5. What does the "pulmonary membrane" consist of? _____

   _____

   _____

   _____

D. Histological Study

On a separate sheet of plain paper make drawings, as required by your instructor, of nasal, tracheal, and lung tissues.

Answers	
Fig. 71.1	**Fig. 71.2**
_____	_____
_____	_____
_____	_____
_____	_____
_____	_____
_____	_____
_____	_____
_____	_____
_____	_____
_____	_____
_____	_____
_____	**Fig. 71.3**
_____	_____
_____	_____
_____	_____
_____	_____
_____	_____
_____	_____
_____	_____
_____	_____

E. Organ Identification

Identify the respiratory structures according to the following statements.

alveoli—1
bronchioles—2
bronchi—3
cricoid cartilage—4
epiglottis—5
hard palate—6
larynx—7
lingual tonsils—8
nasal cavity—9
nasal conchae—10

nasopharynx—11
oral cavity proper—12
oral vestibule—13
oropharynx—14
palatine tonsils—15
pharyngeal tonsils—16
pleural cavity—17
soft palate—18
thyroid cartilage—19
trachea—20

1. Partition between nasal and oral cavities.
2. Cartilage of Adam's apple.
3. Small air sacs of lung tissue.
4. Cavity above soft palate.
5. Cavity above hard palate.
6. Cavity that contains the tongue.
7. Cavity near palatine tonsils.
8. Voice box.
9. Small tubes leading into alveoli.
10. Another name for adenoids.
11. Tube between larynx and bronchi.
12. Fleshy lobes in nasal cavity.
13. Cavity between lips and teeth.
14. Tonsils located on sides of pharynx.
15. Tonsils attached to base of tongue.
16. Flexible flaplike cartilage over larynx.
17. Most inferior cartilaginous ring of larynx.
18. Tubes formed by bifurcation of trachea.
19. Potential cavity between lung and thoracic wall.

F. Organ Function

Select the structures in the above list that perform the following functions.

1. Initiates swallowing.
2. Prevents food from entering the larynx when swallowing.
3. Provides supporting walls for vocal folds.
4. Assists in destruction of harmful bacteria in the oral region.
5. Warms the air as it passes through the nasal cavity.
6. Prevents food from entering nasal cavity during chewing and swallowing.
7. Provides surface for gas exchange in the lungs.
8. Essential for speech.

Answers

Organ Identification

1. _____
2. _____
3. _____
4. _____
5. _____
6. _____
7. _____
8. _____
9. _____
10. _____
11. _____
12. _____
13. _____
14. _____
15. _____
16. _____
17. _____
18. _____
19. _____

Organ Function

1. _____
2. _____
3. _____
4. _____
5. _____
6. _____
7. _____
8. _____

LABORATORY REPORT

72

Hyperventilation and Rebreathing

A. Questions for Nonrecording Portion of Experiment

1. Why does breathing into a bag after hyperventilating restore normal breathing more rapidly than breathing into the air? _____

   _____

   _____

   _____

   _____

2. How long were you able to hold your breath with only one deep inspiration? _____

3. How long were you able to hold your breath after hyperventilating? _____

   _____

4. After exercising, how long were you able to hold your breath? _____

   Generalizations on above: _____

   _____

   _____

   _____

B. Tracings

In the spaces provided below and on the next page tape or staple chart recordings for the various portions of this experiment.

Normal Respiration (Slow Speed)	Speed: _____ cm/sec

B. *Tracings* (continued)

Normal Respiration (Fast Speed)   Speed: _____ cm/sec
Rebreathing and Recovery
Hyperventilation

C. *Generalizations*

What generalizations can you make from the above tracings?

_____

_____

_____

_____

_____

_____

_____

Student: _____

Desk No.: _____ Section: _____

The Diving Reflex (Ex. 73)

A. *Tracings*

Tape or staple the chart recordings for the various portions of this experiment in the spaces provided below and on the next page.

Tidal Breathing before Immersion

Holding Breath after Tidal Inspiration

Immersion in Water at 70° Fahrenheit

Immersion in Water at 60° Fahrenheit

A. *Tracings* (continued)

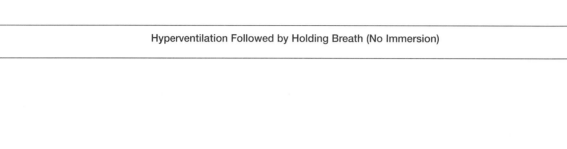

Hyperventilation Followed by Holding Breath (No Immersion)

Hyperventilation Followed by Immersion   Total Time Immersion: _____

B. *Generalizations*

_____

_____

_____

_____

_____

_____

The Valsalva Maneuver (Ex. 74)

A. *Tracing*

In the space below tape or staple a chart recording made during a Valsalva maneuver.

Student: _____

Desk No.: _____ Section: _____

Spirometry:
Lung Capacities (Ex. 75)

A. Tabulation

Record in the following table the results of your spirometer readings and calculations.

LUNG CAPACITIES	NORMAL (ml)	YOUR CAPACITIES
Tidal Volume (TV)	500	
Minute Respiratory Volume (MRV) (MRV = TV × Resp. Rate)	6000	
Expiratory Reserve Volume (ERV)	1100	
Vital Capacity (VC) (VC = TV + ERV + IRV)	See Appendix A (Tables III and IV)	
Inspiratory Capacity (IC) (IC = VC − ERV)	3000	
Inspiratory Reserve Volume (IRV) (IRV = IC − TV)	2500	

B. Evaluation

List below any of your lung capacities that are significantly low (22% or more). _____

_____

_____

_____

Spirometry:
The FEV$_T$ Test (Ex. 76)

A. Expirogram

Trim the tracing of your expirogram to as small a piece as you can without removing volume values and tape it in space shown. Do not tape sides and bottom.

— — — — — — — — — —

Tape Here

B. Calculations

Record your computations for determining the various percentages here.

1. **Vital Capacity** From the expirogram determine the total volume of air expired (Vital Capacity)... _____

2. **Corrected Vital Capacity** By consulting table 76.1 determine the corrected vital capacity (step 4 on page 397).
   VC × Conversion Factor = ... _____

   Math:

3. **One-Second Timed Vital Capacity (FEV$_1$)** After determining the FEV$_1$ (steps 1, 2, and 3 on page 397) record your results here .... _____

4. **Corrected FEV$_1$** Correct the FEV$_1$ for the temperature of the spirometer (step 5 on page 397).

   FEV$_1$ × Conversion Factor = ... _____

5. **FEV$_1$ Percentage** Divide the corrected FEV$_1$ by the corrected vital capacity to determine the percentage expired in the first second (step 6 on page 397).

   $$\frac{\text{Corrected FEV}_1}{\text{Corrected Vital Capacity}} = $$ .... _____

   Math:

6. **Percent of Predicted VC** Determine what percentage your vital capacity is of the predicted vital capacity (step 7 on page 397).

   $$\frac{\text{Your Corrected VC}}{\text{Predicted VC}} = $$ .... _____

   Math:

C. Questions

1. Why isn't a measure of one's vital capacity as significant as the FEV$_T$? _____

   _____

   _____

   _____

2. What respiratory diseases are most readily detected with this test? _____

   _____

   _____

LABORATORY REPORT

78S

Student: _____

Desk No.: _____ Section: _____

Physical Fitness Clearance

A. Cardiovascular (Table I)

Report blood pressure, heart rates, and ECG results here. Attach ECG tracing below the table.

| | ECG | | | OTHER | |
	S-T Segment Depression	PVCs	Other Arrhythmia	Heart Rates	Blood Pressure
Before Exercise					
After Exercise					

ECG Tracing

B. Urine Analysis (Table II)

Record results of urine analysis here.

UROBILINOGEN	BLOOD	BILIRUBIN	KETONES	GLUCOSE	PROTEIN	pH

C. Adverse Postexercise Signs

Record **present** or **absent** for each of the following.

PALLOR	CYANOSIS	COLD, MOIST SKIN	STAGGERING	HEAD NODDING	CONFUSION

D. *Questionnaire*

After moderate to strenuous exercise, do you experience any of the following?

YES	NO	SYMPTOMS
		Severe shortness of breath
		Pains in the chest
		Severe fatigue
		Fainting spells
		Tendency to limp

E. *Release*

I voluntarily participate in this fitness program to monitor my own physical progress while engaged in an exercise regimen of my own choice. Although my instructor has taken all reasonable precautions to prevent injuries, I reserve the right to withdraw from further testing if I feel endangered in any way. Furthermore, I do not hold my instructor or the college liable for any accidents arising from circumstances beyond their control.

Signature _____ Date _____ Age _____

Additional Tracings

LABORATORY REPORT

Student: _____

Desk No.: _____ Section: _____

Monitoring Physical Fitness Training

A. Data

Record the following pertinent information:

SS No: _____ Sex: _____ Age: _____

Nature of Exercise: _____ Starting Date: _____ Completion Date: _____

B. Anthropomorphic Measurements

Record all dimensions here. First line is for beginning of experiment. Second line is for completion date.

DATE	HEIGHT (cm)	WEIGHT (kg)	CHEST (cm)	ABDOMEN (cm)	THIGH (cm)	CALF (cm)	ARM (cm)	VC

C. Cardiovascular Measurements

Record the blood pressures and heart rates before and after exercise in the following tables:

BLOOD PRESSURE		
Date	At Rest	After Exercise
1		
2		
3		
4		
5		

HEART RATE													
Date	At Rest	Minutes after Exercising											
		0	½	1	1½	2	2½	3	3½	4	4½	5	
1													
2													
3													
4													
5													

D. Recordings

Use this page for attaching tracings of the various recovery tests.

E. Graphs

Plot the data from the various tables in the graphs below:

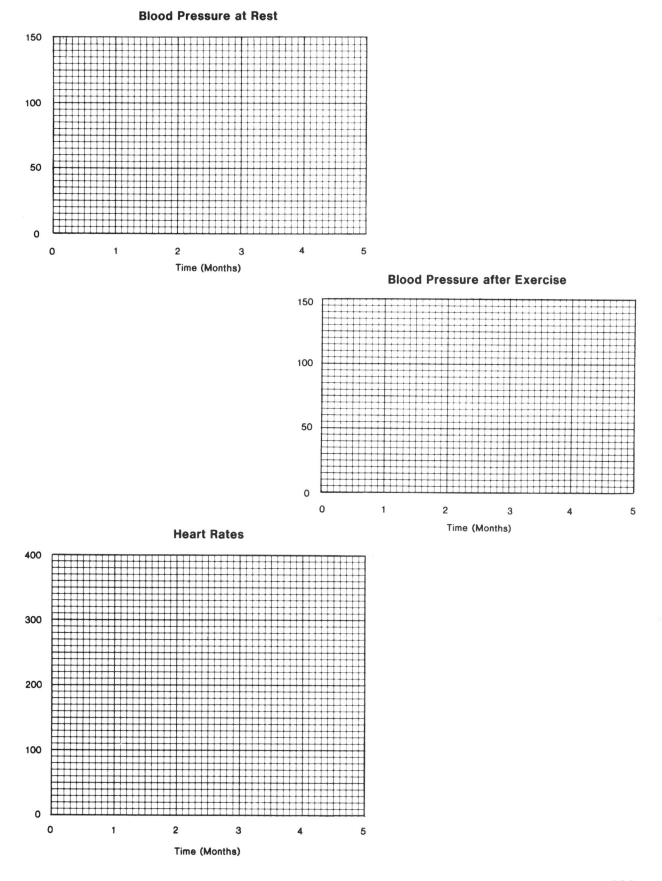

Blood Pressure at Rest

Blood Pressure after Exercise

Heart Rates

F. Statement

Summarize in a few sentences the effectiveness (or ineffectiveness) of the regimen you pursued.

LABORATORY REPORT

81

Anatomy of the Digestive System

A. Illustrations

Record the labels for the illustrations of this exercise in the answer columns.

B. Microscopic Studies

Record here the drawings of the microscopic examinations that are required for this exercise. If there is insufficient space for all required drawings, utilize a separate sheet of paper for all of them.

Answers	
Fig. 81.1	**Fig. 81.2**
	Fig. 81.3

C. Alimentary Canal

Identify the parts of the digestive system described by the following statements.

anus—1
cecum—2
colon—3
colon, sigmoid—4
duct, common bile—5
duct, cystic—6
duct, hepatic—7
duct, pancreatic—8
duodenum—9

esophagus—10
ileum—11
pharynx—12
rectum—13
sphincter, cardiac—14
sphincter, pyloric—15
stomach, fundus—16
stomach, pyloric portion—17
valve, ileocecal—18

1. Place where swallowing (peristalsis) begins.
2. Entrance opening of stomach.
3. Exit opening of stomach.
4. Tube between mouth and stomach.
5. Distal coiled portion of small intestine.
6. First 12 inches of small intestine.
7. Proximal pouch or compartment of large intestine.
8. Section of large intestine between descending colon and rectum.
9. Most active portion of stomach.
10. Duct that drains the liver.
11. Valve between small and large intestines.
12. Duct that drains the gallbladder.
13. Part of small intestine where most digestion occurs.
14. Part of small intestine where most absorption occurs.
15. Duct that joins the common bile duct before entering the intestine.
16. Structure on which appendix is located.
17. Duct that conveys bile to intestine.
18. Exit of alimentary canal.
19. Part of tract where most water absorption (conservation) occurs.
20. Last 6 inches of alimentary canal.

Answers	
Alimentary Canal	**Fig. 81.4**
1. _____	_____
2. _____	_____
3. _____	_____
4. _____	_____
5. _____	_____
6. _____	_____
7. _____	_____
8. _____	_____
9. _____	_____
10. _____	_____
11. _____	_____
12. _____	_____
13. _____	
14. _____	
15. _____	
16. _____	
17. _____	
18. _____	
19. _____	
20. _____	

D. Oral Cavity

Identify the structures of the mouth that are described by the following statements.

arch, glossopalatine—1
arch, pharyngopalatine—2
buccae—3
duct, parotid—4
duct, submandibular—5
ducts, lesser sublingual—6
frenulum, labial—7
frenulum, lingual—8
gingiva—9
glands, parotid—10
glands, sublingual—11
glands, submandibular—12
mucosa—13
papillae, filiform—14
papillae, foliate—15
papillae, fungiform—16
papillae, vallate—17
tonsil, lingual—18
tonsil, palatine—19
tonsil, pharyngeal—20
uvula—21

1. Lining of the mouth.
2. Name for the cheeks.
3. Duct that drains the parotid gland.
4. Fold of skin between the lip and gums.
5. Duct that drains the submandibular gland.
6. Fold of skin between the tongue and floor of mouth.
7. Portion of mucosa around the teeth.
8. Tonsils seen at back of mouth.
9. Tonsils located at root of tongue.
10. Fingerlike projection at end of soft palate.
11. Vertical ridges on side of tongue.
12. Rounded papillae on dorsum of tongue.
13. Small tactile papillae on surface of tongue.
14. Large papillae at back of tongue.
15. Ducts that drain sublingual gland.
16. Place where taste buds are located.
17. Salivary glands located inside and below the mandible.
18. Membrane located in front of palatine tonsil.
19. Membrane located in back of palatine tonsil.
20. Salivary glands located under the tongue.
21. Salivary glands located in cheeks.

E. The Teeth

Select the correct answer that completes each of the following statements.

1. All deciduous teeth usually erupt by the time a child is
   (1) one year.   (2) two years.   (3) four years old.
2. A complete set of primary teeth consists of
   (1) ten teeth.   (2) twenty teeth.   (3) thirty-two teeth.
3. The permanent teeth with the longest roots are the
   (1) incisors.   (2) cuspids.   (3) molars.
4. The smallest permanent molars are the
   (1) first molars.   (2) second molars.   (3) third molars.
5. A complete set of permanent teeth consists of
   (1) twenty-four teeth.   (2) twenty-eight teeth.
   (3) thirty-two teeth.
6. The primary dentition lacks
   (1) molars.   (2) incisors.   (3) bicuspids.

Answers

Oral Cavity	Fig. 81.5
1. _____	
2. _____	
3. _____	
4. _____	
5. _____	
6. _____	
7. _____	
8. _____	
9. _____	
10. _____	
11. _____	Fig. 81.7
12. _____	
13. _____	
14. _____	
15. _____	
16. _____	
17. _____	
18. _____	
19. _____	
20. _____	
21. _____	
Teeth	
1. _____	
2. _____	
3. _____	
4. _____	
5. _____	
6. _____	

7. A child's first permanent tooth usually erupts during the
   (1) second year.   (2) third year.
   (3) fifth year.   (4) sixth year.
8. Trifurcated roots exist on
   (1) upper cuspids.   (2) upper molars.   (3) lower molars.
9. Bifurcated roots exist on
   (1) upper first bicuspids and lower molars.
   (2) cuspids and lower bicuspids.   (3) all molars.
10. A tooth is considered "dead" or devitalized if
    (1) the enamel is destroyed.   (2) the pulp is destroyed.
    (3) the periodontal membrane is infected.
11. Dentin is produced by cells called
    (1) odontoblasts.   (2) ameloblasts.   (3) cementocytes.
12. Cementum is secreted by cells called
    (1) odontoblasts.   (2) ameloblasts.   (3) cementocytes.
13. The wisdom tooth is
    (1) a supernumerary tooth.   (2) a succedaneous tooth.
    (3) the third molar.
14. Caries (cavities) are caused primarily by
    (1) using the wrong toothpaste.   (2) acid production by bacteria.
    (3) fluorides in water.
15. Most teeth that have to be extracted became useless because of
    (1) caries.   (2) gingivitis.   (3) periodontitis (pyorrhea).
16. Enamel covers the
    (1) entire tooth.   (2) clinical crown only.
    (3) anatomical crown only.
17. The tooth receives nourishment through
    (1) the apical foramen.   (2) the periodontal membrane.
    (3) both the apical foramen and the periodontal membrane.
18. The tooth is held in the alveolus by
    (1) bone.   (2) dentin.   (3) periodontal membrane.

Answers
Teeth
7. _____
8. _____
9. _____
10. _____
11. _____
12. _____
13. _____
14. _____
15. _____
16. _____
17. _____
18. _____

LABORATORY REPORT

82

Student: _____

Desk No.: _____ Section: _____

Intestinal Motility

A. Temperature

Record here your observations of intestinal motility changes that occurred as the temperature of Tyrode's solution was gradually increased.

1. Frequency

Plot the frequency of contraction as related to temperature in the graph below.

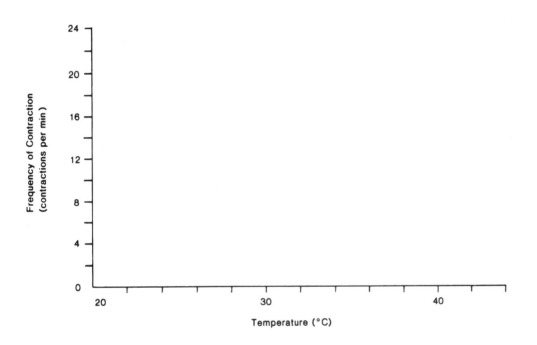

2. Strength of Contraction

a. Did the strength of contraction change with the change in temperature? _____

_____

b. If the answer to the above question is affirmative, what temperature was optimum for strength of

contraction? _____

B. Electrical Stimulation

Describe in detail the results of electrical stimulation. _____

_____

_____

_____

_____

_____

C. Neurohumoral Control

1. What effect does epinephrine have on muscular contraction? _____

   _____

   _____

2. What effect does acetylcholine have on muscular contraction? _____

   _____

   _____

3. Since acetylcholine is produced by the parasympathetic nerves and norepinephrine is produced by the sympathetic nerves, what generalizations can you make concerning nervous control of intestinal motility?

   _____

   _____

   _____

   _____

D. Tracings

Attach samples of tracings with labels in space below.

LABORATORY REPORT

83,84

Student: _____

Desk No.: _____ Section: _____

The Chemistry of Hydrolysis (Ex. 83)

A. Carbohydrate Digestion

    1. IKI Test Record here the presence ($+$) or absence ($-$) of starch as revealed by IKI test on spot plates.

1A	2A	3A	4A	5A

    a. Did the pancreatic extract hydrolyze starch? _____

    b. Which tube is your evidence for this conclusion? _____

    c. How do tubes 1A and 5A compare in color? _____

    d. What do you conclude from question *c?* _____

    e. What is the value of tube 4A? _____

    _____

    f. What is the value of tube 2A? _____

    _____

    2. Barfoed's and Benedict's Tests After performing Barfoed's and Benedict's tests on tubes 1A and 5A, what are your conclusions about the degree of digestion of starch by amylase? _____

    _____

    _____

B. Protein Digestion

Record the color and optical densities (OD) of each tube in the following chart.

	1P	2P	3P	4P	5P
Color					
OD					

    1. Did pancreatic juice hydrolyze the BAPNA? _____

    2. Which tube is your evidence for this conclusion? _____

3. What other tube shows protein hydrolysis? _____

4. What can you conclude from the previous observation in question 3? _____

_____

5. Why is there no hydrolysis in tube 4P? _____

6. Why is there no hydrolysis in tube 3P? _____

7. Why didn't tube 2P show hydrolysis? _____

_____

C. Fat Digestion

Determine the pH of each tube by comparing with a bromthymol blue color standard. Record these values in the table below.

1L	2L	3L	4L	5L	6L

1. Did the pancreatic extract hydrolyze fat in the presence of bile? _____

2. Did the pancreatic extract hydrolyze fat in the absence of bile? _____

3. How do tubes 1L and 6L compare as to degree of hydrolysis? _____

_____

4. What can you conclude from the previous observation in question 3? _____

_____

5. What is the value of tube 2L? _____

_____

D. Summarization

Make an itemized statement of all the facts you learned in this experiment.

Enzyme Review

Consult your text and lecture notes to answer the following questions concerning digestive enzymes.

A. Digestive Juices

Select the enzymes that are present in the following digestive juices. Use the enzymes listed in the right column.

1. Saliva
2. Gastric juice
3. Pancreatic fluid
4. Intestinal juice (succus entericus)

Carbohydrases
amylase, pancreatic—1
amylase, salivary—2
lactase—3
maltase—4
sucrase—5

Proteases
carboxypeptidase—6
chymotrypsin—7
erepsin (peptidase)—8
pepsin—9
rennin—10
trypsin—11

B. Substrates

Select the enzymes that act on the following substrates.

1. Casein
2. Fats
3. Lactose
4. Maltose
5. Peptides
6. Proteins
7. Proteoses and peptones
8. Starch
9. Sucrose

Lipases
lipase, pancreatic—12

C. End Products

Select the enzymes from the above list that produce the following end products in digestion.

1. Amino acids
2. Fatty acids
3. Fructose
4. Galactose
5. Glucose
6. Glycerol
7. Maltose
8. Polypeptides

Answers

Juices

1. _____
2. _____
3. _____
4. _____

Substrates

1. _____
2. _____
3. _____
4. _____
5. _____
6. _____
7. _____
8. _____
9. _____

End Products

1. _____
2. _____
3. _____
4. _____
5. _____
6. _____
7. _____
8. _____

Factors Affecting Hydrolysis (Ex. 84)

A. Tabulations

If the class has been subdivided into separate groups to perform different parts of this experiment, these tabulations should be recorded on the chalkboard so that all students can copy the results onto their own Laboratory Report sheets.

1. **Temperature** After IKI solution has been added to all ten tubes, record the color and degree of amylase activity in each tube in the following tables. Refer to the carbohydrate differentiation separation outline in Appendix B pertaining to the IKI test.

Table I Amylase action at 20° C.

TUBE NO.	1	2	3	4	5	6	7	8	9	10
Color										
Starch Hydrolysis										

Table II Amylase action at 37° C.

TUBE NO.	1	2	3	4	5	6	7	8	9	10
Color										
Starch Hydrolysis										

2. **Hydrogen Ion Concentration** After IKI solution has been added to all nine tubes, record, as above, the color and degree of amylase action for each tube in the following table.

Table III pH and amylase action at 37° C.

TIME	2 MINUTES			4 MINUTES			6 MINUTES		
Tube No.	1	2	3	4	5	6	7	8	9
pH	5	7	9	5	7	9	5	7	9
Color									
Starch Hydrolysis									

B. Conclusions

Answer the following questions that are related to the results of this experiment.

1. How long did it take for complete starch hydrolysis at 20° C? _____ at 37° C? _____

2. Why would you expect the above results to turn out as they have? _____

_____

_____

3. What effect does boiling have on amylase? _____

4. At which pH was starch hydrolysis most rapid? _____

   How does this pH compare with the pH of normal saliva? _____

_____

5. What is the pH of gastric juice? _____

   What do you suppose happens to the action of amylase in the stomach? _____

_____

Student: _____

Desk No.: _____ Section: _____

Anatomy of the Urinary System

A. *Illustrations*

Record the labels for the three illustrations of this exercise in the answer columns.

B. *Microscopy*

If the space provided below is adequate for your histology drawings, use it. Use a separate sheet of paper if several sketches are to be made.

C. *Anatomy*

Identify the structures of the urinary system that are described by the following statements.

calyces—1	medulla—6	renal pelvis—11
collecting tubule—2	nephron—7	renal pyramids—12
cortex—3	renal capsule—8	ureters—13
glomerular capsule—4	renal column—9	urethra—14
glomerulus—5	renal papilla—10	

1. Tubes that drain the kidneys.
2. Tube that drains the urinary bladder.
3. Portion of kidney that contains renal corpuscles.
4. Cone-shaped areas of medulla.
5. Portion of kidney that consists primarily of collecting tubules.
6. Basic functioning unit of the kidney.
7. Two portions of renal corpuscle.
8. Distal tip of renal pyramid.
9. Short tubes that receive urine from renal papillae.
10. Funnel-like structure that collects urine from calyces of each kidney.
11. Structure that receives urine from several nephrons.
12. Cup-shaped membranous structure that surrounds glomerulus.
13. Thin fibrous outer covering of kidney.
14. Cortical tissue between renal pyramids.
15. Tuft of capillaries that produces dilute urine.

Answers

Anatomy	Fig. 85.1
1. _____	_____
2. _____	_____
3. _____	_____
4. _____	_____
5. _____	_____
6. _____	_____
7. _____	_____
8. _____	_____
9. _____	_____
10. _____	_____
11. _____	
12. _____	**Fig. 85.3**
13. _____	_____
14. _____	
15. _____	_____
	_____
	_____
	_____
	_____
	_____
	_____
	_____
	_____

D. Physiology

Select the best answer that completes the following statements concerning the physiology of urine production. For definitions of medical terms consult a medical dictionary or your lecture text.

1. Blood enters the glomerulus through the
   (1) efferent arteriole.   (2) arcuate artery.
   (3) afferent arteriole.   (4) None of these is correct.

2. Surgical removal of a kidney is called
   (1) nephrectomy.   (2) nephrotomy.   (3) nephrolithotomy.

3. Water reabsorption from the glomerular filtrate into the peritubular blood is facilitated by
   (1) antidiuretic hormone   (2) renin.
   (3) aldosterone.   (4) Both 1 and 3 are correct.

4. The following substances are reabsorbed through the walls of the nephron into the peritubular blood:
   (1) glucose and water.   (2) urea and water.
   (3) glucose, amino acids, salts, and water.
   (4) glucose, amino acids, urea, salts, and water.

5. The amount of urine produced is affected by
   (1) blood pressure.   (2) environmental temperature.
   (3) amount of solute in glomerular filtrate.
   (4) 1, 2, 3 and additional factors are correct.

6. The amount of urine normally produced in 24 hours is about
   (1) 100 ml.   (2) 500 ml.   (3) 1.5 liters.   (4) 4.5 liters.

7. The reabsorption of sodium ions from the glomerular filtrate into the peritubular blood draws the following back into the blood:
   (1) potassium ions.   (2) chloride ions.
   (3) water.   (4) chloride ions and water.

8. Cells of the walls of collecting tubules secrete the following substances into urine:
   (1) amino acids.   (2) uric acid.
   (3) ammonia.   (4) Both 2 and 3 are correct.

9. Renal diabetes is due to
   (1) a lack of ADH.   (2) a lack of insulin.
   (3) faulty reabsorption of glucose in the nephron.
   (4) Both 1 and 2 are correct.

10. The presence of glucose in the urine and a low level of insulin in the blood would be diagnosed as
    (1) diabetes insipidus.   (2) diabetes mellitus.
    (3) renal diabetes.   (4) None of these is correct.

11. Approximately 80% of water in the glomerular filtrate is reabsorbed in the
    (1) proximal convoluted tubule. (2) Henle's loop.
    (3) distal convoluted tubule.   (4) None of these is correct.

12. The desire to micturate normally occurs when the following amount of urine is present in the bladder:
    (1) 100 ml.   (2) 200 ml.   (3) 300 ml.   (4) 400 ml.

13. Removal of calculus (kidney stones) is called
    (1) nephrectomy.   (2) nephrotomy.
    (3) nephrolithotomy.   (4) Both 2 and 3 are correct.

Answers
Physiology
1. _____
2. _____
3. _____
4. _____
5. _____
6. _____
7. _____
8. _____
9. _____
10. _____
11. _____
12. _____
13. _____

Fig. 85.2
_____
_____
_____
_____
_____
_____
_____
_____
_____
_____
_____
_____
_____
_____
_____
_____
_____
_____
_____

LABORATORY REPORT

86

Student: _____

Desk No.: _____ Section: _____

Urine: Composition and Tests

A. Test Results

Record on the chart below the results of all urine tests performed. If the tests are done as a demonstration, the results will be tabulated in columns A and B. If students perform the tests on their own urine, the last column will be used for their results.

TEST	NORMAL VALUES	ABNORMAL VALUES	A POSITIVE TEST CONTROL	B UNKNOWN SAMPLE	C STUDENT'S URINE
COLOR	Colorless Pale straw Straw Amber	Milky Reddish amber Brownish yellow Green Smoky brown			
CLOUDINESS	Clear	+ slight ++ moderate +++ cloudy ++++ very cloudy			
SP. GRAVITY	1.001–1.060	Above 1.060			
pH	4.8–7.5	Below 4.8 Above 7.5			
ALBUMIN (Protein)	None	+ barely visible ++ granular +++ flocculent ++++ large flocculent			
MUCIN	None	Visible amounts			
GLUCOSE	None	+ yellow green ++ greenish yellow +++ yellow ++++ orange			
KETONES	None	+ pink-purple ring			
HEMOGLOBIN	None	See color chart			
BILIRUBIN					

B. Microscopy

Record here by sketch and word any structures seen on microscopic examination.

C. Interpretation

Indicate the probable significance of each of the following in urine. More than one condition may apply in some instances.

bladder infection—1   diabetes mellitus—4   glomerulonephritis—7
cirrhosis of liver—2   exercise (extreme)—5   gonorrhea—8
diabetes insipidus—3   kidney tumor—6   hepatitis—9
   normally present—10

1. Albumin (constantly)
2. Albumin (periodic)
3. Bacteria
4. Bile pigments
5. Blood
6. Creatinine
7. Glucose
8. Mucin
9. Porphyrin
10. Pus cells (neutrophils)
11. Sodium chloride
12. Specific gravity (high)
13. Specific gravity (low)
14. Urea
15. Uric acid

D. Terminology

Select the terms described by the following statements.

albuminuria—1   cystitis—4   glycosuria—7
anuria—2   diuretic—5   nephritis—8
catheter—3   enuresis—6   pyelitis—9
   uremia—10

1. Absence of urine production.
2. High urea level in blood.
3. Sugar in urine.
4. Albumin in urine.
5. Inflammation of nephrons.
6. Inflammation of pelvis of kidney.
7. Bladder infection.
8. Device for draining bladder.
9. Involuntary bed-wetting during sleep.
10. Chemical that stimulates urine production.

Answers

Interpretation

1. _____
2. _____
3. _____
4. _____
5. _____
6. _____
7. _____
8. _____
9. _____
10. _____
11. _____
12. _____
13. _____
14. _____
15. _____

Terminology

1. _____
2. _____
3. _____
4. _____
5. _____
6. _____
7. _____
8. _____
9. _____
10. _____

LABORATORY REPORT

87

Student: _____

Desk No.: _____ Section: _____

The Endocrine Glands

A. Illustrations

Record the labels for figures 87.12 and 87.13 in the answer columns.

B. Microscopy

Make drawings of the various microscopic studies on separate drawing paper. Include these drawings with the Laboratory Report.

C. Sources

Select the glandular tissues that produce the following hormones.

1. ACTH
2. ADH
3. Aldosterone
4. Calcitonin
5. Chorionic gonadotropin
6. Cortisol
7. Epinephrine
8. Estrogens
9. FSH
10. Glucagon
11. ICSH
12. Insulin
13. LH
14. Melatonin
15. Norepinephrine
16. Oxytocin
17. Parathyroid hormone
18. PIF
19. Prolactin
20. Progesterone
21. Somatostatin
22. Testosterone
23. Thymosin
24. Thyrotropin
25. Thyroxine
26. Triiodothyronine
27. Vasopressin

adrenal cortex
   zona fasciculata—1
   zona glomerulosa—2
   zona reticularis—3
adrenal medulla—4
hypothalamus—5
ovary
   corpus luteum—6
   Graafian follicle—7
pancreas
   alpha cells—8
   beta cells—9
   delta cells—10
   acini cells—11
parathyroid gland—12
pineal gland—13
pituitary gland
   adenohypophysis—14
   neurohypophysis—15
placenta—16
testis—17
thymus gland—18
thyroid gland
   follicular cells—19
   parafollicular cells—20

D. Physiology

Select the hormones that produce the following physiological effects. Note that a separate list of hormones is provided for each group.

GROUP I

1. Increases metabolism.
2. Increases blood pressure.
3. Increases strength of heartbeat.
4. Promotes sodium absorption in nephron.
5. Promotes gluconeogenesis.
6. Causes vasoconstriction in all organs.
7. Causes vasoconstriction in all organs except muscles and liver.
8. Causes vasodilation in skeletal muscles.
9. Inhibits the estrus cycle in lower animals.

aldosterone—1
epinephrine—2
glucocorticoids—3
melatonin—4
norepinephrine—5
none of these—6

Answers	
Sources	**Fig. 87.12**
1. _____	_____
2. _____	_____
3. _____	_____
4. _____	_____
5. _____	_____
6. _____	_____
7. _____	_____
8. _____	_____
9. _____	_____
10. _____	_____
11. _____	_____
12. _____	_____
13. _____	_____
14. _____	_____
15. _____	_____
16. _____	_____
17. _____	_____
18. _____	_____
19. _____	_____
20. _____	_____
21. _____	_____
22. _____	_____
23. _____	_____
24. _____	_____
25. _____	_____
26. _____	_____
27. _____	_____
Physiology Group 1	
1. _____	_____
2. _____	
3. _____	_____
4. _____	
5. _____	_____
6. _____	_____
7. _____	
8. _____	
9. _____	

D. Physiology (continued)

Select the hormones that produce the following physiological effects.

GROUP II

1. Inhibits insulin production.
2. Inhibits production of SH.
3. Stimulates osteoclasts.
4. Mobilizes fatty acids.
5. Inhibits glucagon production.
6. Raises blood calcium level.
7. Increases cardiac output and respiratory rate.
8. Impedes phosphorus reabsorption in renal tubules.
9. Promotes storage of glucose, fatty acids, and amino acids.
10. Mobilizes glucose, fatty acids, and amino acids.
11. Stimulates metabolism of all cells in body.
12. Lowers blood phosphorus levels.
13. Decreases blood calcium when injected intravenously.

calcitonin—1
glucagon—2
insulin—3
parathyroid hormone—4
somatostatin—5
thyroid hormone—6
none of these—7

GROUP III

1. Promotes breast enlargement.
2. Promotes rapid bone growth.
3. Stimulates testicular descent.
4. Causes milk ejection from breasts.
5. Promotes maturation of spermatozoa.
6. Causes constriction of arterioles.
7. Stimulates production of thyroxine.
8. Stimulates glucocorticoid production.
9. Inhibits prolactin production.
10. Causes thickening and vascularization of endometrium.
11. Causes interstitial cells to produce testosterone.
12. Causes myometrium contraction during childbirth.
13. Promotes milk production after childbirth.
14. Promotes stratification of vaginal epithelium.
15. Suppresses development of female genitalia in male.
16. Prepares the endometrium for egg cell implantation.
17. Promotes growth of all cells in the body.

ACTH—1
ADH—2
chorionic gonadotropin—3
estrogens—4
FSH—5
LH—6
oxytocin—7
PIF—8
progesterone—9
somatotropin—10
testosterone—11
thyrotropin—12
none of these—13

E. Hormonal Imbalance

Select the hormones that are responsible for the following disorders. After each hormone number indicate with a (+) or (−) whether the condition is due to *excess* (+) or *deficiency* (−). More than one hormone may apply in some cases.

1. Acromegaly
2. Addison's disease
3. Cretinism
4. Cushing's syndrome
5. Diabetes insipidus
6. Diabetes mellitus
7. Dwarfism
8. Exophthalmos
9. Gigantism
10. Hyperglycemia
11. Hypothyroidism
12. Myxedema
13. Tetany

ADH—1
glucagon—2
glucocorticoids—3
insulin—4
mineralocorticoids—5
parathyroid hormone—6
somatotropin—7
thyroid hormone—8
none of these—9

Answers

Physiology Group II	Fig. 87.13
1. _____	_____
2. _____	_____
3. _____	_____
4. _____	
5. _____	_____
6. _____	
7. _____	_____
8. _____	_____
9. _____	
10. _____	
11. _____	
12. _____	
13. _____	

Group III	
1. _____	_____
2. _____	_____
3. _____	_____
4. _____	
5. _____	**Imbalance**
6. _____	1. _____
7. _____	2. _____
8. _____	3. _____
9. _____	4. _____
10. _____	5. _____
11. _____	6. _____
12. _____	7. _____
13. _____	8. _____
14. _____	9. _____
15. _____	10. _____
16. _____	11. _____
17. _____	12. _____
	13. _____

LABORATORY REPORT

88

Student: _____

Desk No.: _____ Section: _____

The Reproductive System

A. Labels
Record the labels for the illustrations of this exercise in the answer columns.

B. Microscopy
Draw the various stages of spermatogenesis on a separate piece of paper to be included with this Laboratory Report.

C. Male Organs
Identify the structures described by the following statements.

common ejaculatory
   duct—1
corpora cavernosa—2
corpus spongiosum—3
bulbourethral gland—4

epididymis—5
glans penis—6
penis—7
prepuce—8
prostate gland—9

prostatic urethra—10
seminal vesicle—11
seminiferous tubule—12
testes—13
vas deferens—14

1. Copulatory organ of male.
2. Source of spermatozoa.
3. Erectile tissue of penis (3 answers).
4. Fold of skin over end of penis.
5. Structure that stores spermatozoa.
6. Contribute fluids to semen (3 answers).
7. Duct that passes from urinary bladder through the prostate gland.
8. Coiled-up structures in testes where spermatozoa originate.
9. Tube that carries sperm from epididymis to ejaculatory duct.
10. Distal end of penis.

D. Female Organs
Identify the structures described by the following statements.

cervix—1
clitoris—2
external os—3
fimbriae—4
greater vestibular
   glands—5

hymen—6
infundibulum—7
internal os—8
labia majora—9
labia minora—10

mons pubis—11
myometrium—12
paraurethral glands—13
ovaries—14
uterine tubes—15
vagina—16

1. Copulatory organ of female.
2. Source of ova in female.
3. Muscular wall of uterus.
4. Neck of uterus.
5. Provides vaginal lubrication during coitus.
6. Fingerlike projections around edge of infundibulum.
7. Protuberance of erectile tissue sensitive to sexual excitation.
8. Entrance to uterus at vaginal end.
9. Homologous to bulbourethral glands of male.
10. Homologous to penis of male.
11. Homologous to scrotum of male.
12. Homologous to prostate gland of male.
13. Ducts that convey ovum to uterus.
14. Membranous fold of tissue surrounding entrance to vagina.
15. Open funnel-like end of uterine tube.

Answers

Male Organs	Fig. 88.1
1. _____	_____
2. _____	_____
3. _____	_____
4. _____	_____
5. _____	_____
6. _____	_____
7. _____	_____
8. _____	_____
9. _____	_____
10. _____	_____

Female Organs	
1. _____	_____
2. _____	_____
3. _____	_____
4. _____	_____
5. _____	_____
6. _____	_____
7. _____	_____
8. _____	_____
9. _____	_____
10. _____	_____
11. _____	_____
12. _____	_____
13. _____	_____
14. _____	_____
15. _____	_____

E. Ligaments

Identify the following supporting structures that hold the ovaries, uterus, and uterine tubes in place.

broad ligament—1   mesosalpinx—3   sacrouterine ligament—6
infundibulopelvic   ovarian ligament—4   suspensory ligament—7
   ligament—2   round ligament—5   none of these—8

1. A ligament that extends from the uterus to the ovary.
2. Ligament between the cervix and sacral part of the pelvic wall.
3. A ligament that extends from the infundibulum to the wall of the pelvic cavity.
4. A ligament that extends from the corpus of the uterus to the body wall.
5. A large flat ligament of peritoneal tissue that extends from the uterine tube to the cervical part of the uterus.
6. A mesentery between the ovary and uterine tubes that contains blood vessels leading to the uterine tube.
7. A fold of peritoneum that attaches the ovary to the uterine tube.

F. Germ Cells

Identify the types of cells of oogenesis and spermatogenesis described by the following statements. More than one answer may apply.

Spermatogenesis   *Oogenesis*
primary spermatocyte—1   oogonia—6
secondary spermatocyte—2   polar bodies—7
spermatids—3   primary oocyte—8
spermatogonia—4   secondary oocyte—9
spermatozoa—5

1. Cells that are haploid.
2. Cells that contain tetrads.
3. Cells that contain monads.
4. Cells that contain dyads.
5. Cells that are diploid.
6. Cells at periphery of ovary that produce all ova.
7. Cells that undergo mitosis.
8. Cells in which first meiotic division occurs.
9. Cells at periphery of seminiferous tubule that give rise to all spermatozoa.
10. Cells that develop directly into mature spermatozoa.
11. Cells in which second meiotic division occurs.

G. Physiology of Reproduction and Development

Select the *best answer* that completes the following statements.

1. The ovum gets from the ovary to the uterus by
   (1) amoeboid movement.   (2) ciliary action.
   (3) peristalsis and ciliary action of uterine tube.
2. The vaginal lining is normally
   (1) alkaline.   (2) acid.   (3) neutral.
3. The seminal vesicle secretion is
   (1) alkaline.   (2) acid.   (3) neutral.
4. The birth canal consists of the
   (1) vagina.   (2) uterus.   (3) the vagina and uterus.
5. Spermatozoan viability is enhanced by a temperature that is
   (1) 98.6° F.   (2) above 98.6° F.   (3) below 98.6° F.
6. The prostatic secretion is
   (1) alkaline.   (2) acid.   (3) neutral.

Answers

Ligaments	Fig. 88.2
1. _____	_____
2. _____	_____
3. _____	_____
4. _____	_____
5. _____	_____
6. _____	_____
7. _____	_____
	_____

Germ Cells	
1. _____	_____
2. _____	_____
3. _____	
4. _____	Fig. 88.3
5. _____	_____
6. _____	_____
7. _____	_____
8. _____	_____
9. _____	_____
10. _____	_____
11. _____	_____

Physiology	
1. _____	_____
2. _____	_____
3. _____	_____
4. _____	_____
5. _____	_____
6. _____	_____

7. The fetal stage of the human is
   (1) the first 8 weeks. (2) from the ninth week until birth.
   (3) the entire prenatal term.
8. Circumcision involves excisement of the
   (1) prepuce. (2) perineum. (3) glans penis.
9. Fertilization of the human ovum usually occurs in the
   (1) uterus. (2) vagina. (3) uterine tube.
10. Implantation of the fertilized ovum occurs usually
    (1) within 2 days after fertilization.
    (2) within 6 to 8 days after fertilization.
    (3) after 10 days.
11. The innermost fetal membrane surrounding the embryo is the
    (1) amnion. (2) chorion. (3) decidua.
12. The life span of an ovum without fertilization is
    (1) 6 hours. (2) 24–28 hours. (3) 7 days.
13. Uterine muscle consists of
    (1) smooth muscle. (2) striated muscle.
    (3) both smooth and striated muscle.
14. The outermost fetal membrane surrounding the embryo is the
    (1) amnion. (2) chorion. (3) decidua.
15. The lining of the uterus is the
    (1) myometrium. (2) endometrium. (3) epimetrium.
16. Milk production by the breasts usually occurs
    (1) immediately after birth. (2) on the second day.
    (3) on the third or fourth day.
17. The afterbirth refers to the
    (1) placenta. (2) damaged uterus.
    (3) the placenta, umbilical vessels, and fetal membranes.
18. Parturition is the
    (1) process of giving birth. (2) period of pregnancy.
    (3) first few days of postnatal life.
19. The following substances pass from the mother to the fetus through the placenta:
    (1) nutrients, gases, and blood cells.
    (2) nutrients and all blood cells except red blood cells.
    (3) nutrients, gases, hormones, and antibodies.
20. Abortion is correctly known as
    (1) criminal emptying of the uterus.
    (2) interruption of pregnancy during fetal life.
    (3) interruption of pregnancy during embryonic life.

H. Menstrual Cycle

Select the correct answer to the following statements.

1. The cessation of menstrual flow in a woman in her late forties is known as
   (1) menarche. (2) menopause. (3) amenorrhea.
2. The first menstrual flow is known as
   (1) menarche. (2) climacteric. (3) menopause.
3. A distinct rise in the basal body temperature usually occurs
   (1) at ovulation. (2) during the proliferative phase.
   (3) during the quiescent period.

Answers	
Physiology	Fig. 88.4
7._____	_____
8._____	_____
9._____	_____
10._____	_____
11._____	_____
12._____	_____
13._____	_____
14._____	_____
15._____	_____
16 _____	_____
17._____	_____
18._____	_____
19._____	_____
20._____	_____
Menstrual Cycle	
1._____	_____
2._____	_____
3._____	_____
	_____
	_____
	_____
	_____
	_____
	_____

4. Menstrual flow is the result of
   (1) a deficiency of progesterone and estrogen.
   (2) a deficiency of estrogen.
   (3) an excess of progesterone and estrogen.
5. Ovulation usually (not always) occurs the following number of days before the next menstrual period:
   (1) 3.   (2) 7.   (3) 14.   (4) 18.
6. Pain during ovulation, as experienced by some women, is known as
   (1) dysmenorrhea.   (2) oligomenorrhea.   (3) mittelschmerz.
7. The proliferative phase in the menstrual cycle begins at about the
   (1) second day.   (2) fifth day.   (3) seventh day.
8. Repair of the endometrium after menstruation is due to
   (1) estrogen.   (2) progesterone.   (3) luteinizing hormone.
9. Progesterone secretion ceases entirely within the following number of days after the onset of menses:
   (1) 10.   (2) 14.   (3) 26.   (4) 28.
10. Excessive blood flow during menstruation is known as
    (1) oligomenorrhea.   (2) dysmenorrhea.   (3) menorrhagia.
11. Excessive discomfort and pain during menstruation is known as
    (1) dysmenorrhea.   (2) oligomenorrhea.   (3) dysmenorrhea.
12. Occasional or irregular menses is known as
    (1) amenorrhea.   (2) oligomenorrhea.   (3) dysmenorrhea.
13. Absence of menstruation is called
    (1) dysmenorrhea.   (2) oligomenorrhea.   (3) amenorrhea.

I. Medical

Select the type of surgery or condition that is described by the following statements.

anteflexion—1	hysterectomy—6	retroflexion—11
cryptorchidism—2	mastectomy—7	salpingitis—12
endometritis—3	oophorectomy—8	salpingectomy—13
episiotomy—4	oophorhysterectomy—9	syphilis—14
gonorrhea—5	oophoroma—10	vasectomy—15

1. Surgical removal of breast.
2. Failure of testes to descend into scrotum.
3. Surgical removal of uterus.
4. Spirochaetal sexually transmitted disease.
5. Sterilization procedure in males.
6. Type of incision made in perineum at childbirth to prevent excessive damage to anal sphincter.
7. Surgical removal of one or both ovaries.
8. Ovarian malignancy.
9. Sexually transmitted disease that affects the mucous membranes rather than the blood.
10. Surgical removal or sectioning of uterine tubes.
11. Most common sexually transmitted disease.
12. Inflammation of uterine wall.
13. Malpositioned uterus (2 types).
14. Sexually transmitted disease caused by a coccoidal (spherical) organism.
15. Surgical removal of ovaries and uterus.

Answers

Menstrual Cycle

4. _____
5. _____
6. _____
7. _____
8. _____
9. _____
10. _____
11. _____
12. _____
13. _____

Medical

1. _____
2. _____
3. _____
4. _____
5. _____
6. _____
7. _____
8. _____
9. _____
10. _____
11. _____
12. _____
13. _____
14. _____
15. _____

Student: _____

Desk No.: _____ Section: _____

Fertilization and Early Embryology

A. Drawings

Put all drawings of cleavage stages on a separate sheet of paper. This should include the zygote (1 cell), 2, 4, 8, and 16 cell stages. If prepared slides are available showing blastula and gastrula stages, make drawings of them, also.

B. Questions

1. Differentiate between

   Activation: _____

   _____

   Fertilization: _____

   _____

2. What germ layer forms as a result of gastrula formation? _____

3. Where does mesoderm form? _____

4. Indicate the germ layer that gives rise to the following structures:

   Epidermis: _____

   Nervous system: _____

   Muscles: _____

   Skeleton: _____

   Digestive tract: _____

   Hair: _____

   Kidneys: _____

Table 1 International atomic weights

ELEMENT	SYMBOL	ATOMIC NUMBER	ATOMIC WEIGHT
Aluminum	Al	13	26.97
Antimony	Sb	51	121.76
Arsenic	As	33	74.91
Barium	Ba	56	137.36
Beryllium	Be	4	9.013
Bismuth	Bi	83	209.00
Boron	B	5	10.82
Bromine	Br	35	79.916
Cadmium	Cd	48	112.41
Calcium	Ca	20	40.08
Carbon	C	6	12.010
Chlorine	Cl	17	35.457
Chromium	Cr	24	52.01
Cobalt	Co	27	58.94
Copper	Cu	29	63.54
Fluorine	F	9	19.00
Gold	Au	79	197.2
Hydrogen	H	1	1.0080
Iodine	I	53	126.92
Iron	Fe	26	55.85
Lead	Pb	82	207.21
Magnesium	Mg	12	24.32
Manganese	Mn	25	54.93
Mercury	Hg	80	200.61
Nickel	Ni	28	58.69
Nitrogen	N	7	14.008
Oxygen	O	8	16.0000
Palladium	Pd	46	106.7
Phosphorus	P	15	30.98
Platinum	Pt	78	195.23
Potassium	K	19	39.096
Radium	Ra	88	226.05
Selenium	Se	34	78.96
Silicon	Si	14	28.06
Silver	Ag	47	107.880
Sodium	Na	11	22.997
Strontium	Sr	38	87.63
Sulfur	S	16	32.066
Tin	Sn	50	118.70
Titanium	Ti	22	47.90
Tungsten	W	74	183.92
Uranium	U	92	238.07
Vanadium	V	23	50.95
Zinc	Zn	30	65.38
Zirconium	Zr	40	91.22

Table II Four place logarithms.

N	0	1	2	3	4	5	6	7	8	9
10	0000	0043	0086	0128	0170	0212	0253	0294	0334	0374
11	0414	0453	0492	0531	0569	0607	0645	0682	0719	0755
12	0792	0828	0864	0899	0934	0969	1004	1038	1072	1106
13	1139	1173	1206	1239	1271	1303	1335	1367	1399	1430
14	1461	1492	1523	1553	1584	1614	1644	1673	1703	1732
15	1761	1790	1818	1847	1875	1903	1931	1959	1987	2014
16	2041	2068	2095	2122	2148	2175	2201	2227	2253	2279
17	2304	2330	2355	2380	2405	2430	2455	2480	2504	2529
18	2553	2577	2601	2625	2648	2672	2695	2718	2742	2765
19	2788	2810	2833	2856	2878	2900	2923	2945	2967	2989
20	3010	3032	3054	3075	3096	3118	3139	3160	3181	3201
21	3222	3243	3263	3284	3304	3324	3345	3365	3385	3404
22	3424	3444	3464	3483	3502	3522	3541	3560	3579	3598
23	3617	3636	3655	3674	3692	3711	3729	3747	3766	3784
24	3802	3820	3838	3856	3874	3892	3909	3927	3945	3962
25	3979	3997	4014	4031	4048	4065	4082	4099	4116	4133
26	4150	4166	4183	4200	4216	4232	4249	4265	4281	4298
27	4314	4330	4346	4362	4378	4393	4409	4425	4440	4456
28	4472	4487	4502	4518	4533	4548	4564	4579	4594	4609
29	4624	4639	4654	4669	4683	4698	4713	4728	4742	4757
30	4771	4786	4800	4814	4829	4843	4857	4871	4886	4900
31	4914	4928	4942	4955	4969	4983	4997	5011	5024	5038
32	5051	5065	5079	5092	5105	5119	5132	5145	5159	5172
33	5185	5198	5211	5224	5237	5250	5263	5276	5289	5302
34	5315	5328	5340	5353	5366	5378	5391	5403	5416	5428
35	5441	5453	5465	5478	5490	5502	5514	5527	5539	5551
36	5563	5575	5587	5599	5611	5623	5635	5647	5658	5670
37	5682	5694	5705	5717	5729	5740	5752	5763	5775	5786
38	5798	5809	5821	5832	5843	5855	5866	5877	5888	5899
39	5911	5922	5933	5944	5955	5966	5977	5988	5999	6010
40	6021	6031	6042	6053	6064	6075	6085	6096	6107	6117
41	6128	6138	6149	6160	6170	6180	6191	6201	6212	6222
42	6232	6243	6253	6263	6274	6284	6294	6304	6314	6325
43	6335	6345	6355	6365	6375	6385	6395	6405	6415	6425
44	6435	6444	6454	6464	6474	6484	6493	6503	6513	6522
45	6532	6542	6551	6561	6571	6580	6590	6599	6609	6618
46	6628	6637	6646	6656	6665	6675	6684	6693	6702	6712
47	6721	6730	6739	6749	6758	6767	6776	6785	6794	6803
48	6812	6821	6830	6839	6848	6857	6866	6875	6884	6893
49	6902	6911	6920	6928	6937	6946	6955	6964	6972	6981
50	6990	6998	7007	7016	7024	7033	7042	7050	7059	7067
51	7076	7084	7093	7101	7110	7118	7126	7135	7143	7152
52	7160	7168	7177	7185	7193	7202	7210	7218	7226	7235
53	7243	7251	7259	7267	7275	7284	7292	7300	7308	7316
54	7324	7332	7340	7348	7356	7364	7372	7380	7388	7396
N	0	1	2	3	4	5	6	7	8	9

Table II Four place logarithms—continued.

N	0	1	2	3	4	5	6	7	8	9
55	7404	7412	7419	7427	7435	7443	7451	7459	7466	7474
56	7482	7490	7497	7505	7513	7520	7528	7536	7543	7551
57	7559	7566	7574	7582	7589	7597	7604	7612	7619	7627
58	7634	7642	7649	7657	7664	7672	7679	7686	7694	7701
59	7709	7716	7723	7731	7738	7745	7752	7760	7767	7774
60	7782	7789	7796	7803	7810	7818	7825	7832	7839	7846
61	7853	7860	7868	7875	7882	7889	7896	7903	7910	7917
62	7924	7931	7938	7945	7952	7959	7966	7973	7980	7987
63	7993	8000	8007	8014	8021	8028	8035	8041	8048	8055
64	8062	8069	8075	8082	8089	8096	8102	8109	8116	8122
65	8129	8136	8142	8149	8156	8162	8169	8176	8182	8189
66	8195	8202	8209	8215	8222	8228	8235	8241	8248	8254
67	8261	8267	8274	8280	8287	8293	8299	8306	8312	8319
68	8325	8331	8338	8344	8351	8357	8363	8370	8376	8382
69	8388	8395	8401	8407	8414	8420	8426	8432	8439	8445
70	8451	8457	8463	8470	8476	8482	8488	8494	8500	8506
71	8513	8519	8525	8531	8537	8543	8549	8555	8561	8567
72	8573	8579	8585	8591	8597	8603	8609	8615	8621	8627
73	8633	8639	8645	8651	8657	8663	8669	8675	8681	8686
74	8692	8698	8704	8710	8716	8722	8727	8733	8739	8745
75	8751	8756	8762	8768	8774	8779	8785	8791	8797	8802
76	8808	8814	8820	8825	8831	8837	8842	8848	8854	8859
77	8865	8871	8876	8882	8887	8893	8899	8904	8910	8915
78	8921	8927	8932	8938	8943	8949	8954	8960	8965	8971
79	8976	8982	8987	8993	8998	9004	9009	9015	9020	9025
80	9031	9036	9042	9047	9053	9058	9063	9069	9074	9079
81	9085	9090	9096	9101	9106	9112	9117	9122	9128	9133
82	9138	9143	9149	9154	9159	9165	9170	9175	9180	9186
83	9191	9196	9201	9206	9212	9217	9222	9227	9232	9238
84	9243	9248	9253	9258	9263	9269	9274	9279	9284	9289
85	9294	9299	9304	9309	9315	9320	9325	9330	9335	9340
86	9345	9350	9355	9360	9365	9370	9375	9380	9385	9390
87	9395	9400	9405	9410	9415	9420	9425	9430	9435	9440
88	9445	9450	9455	9460	9465	9469	9474	9479	9484	9489
89	9494	9499	9504	9509	9513	9518	9523	9528	9533	9538
90	9542	9547	9552	9557	9562	9566	9571	9576	9581	9586
91	9590	9595	9600	9605	9609	9614	9619	9624	9628	9633
92	9638	9643	9647	9652	9657	9661	9666	9671	9675	9680
93	9685	9689	9694	9699	9703	9708	9713	9717	9722	9727
94	9731	9736	9741	9745	9750	9754	9759	9763	9768	9773
95	9777	9782	9786	9791	9795	9800	9805	9809	9814	9818
96	9823	9827	9832	9836	9841	9845	9850	9854	9859	9863
97	9868	9872	9877	9881	9886	9890	9894	9899	9903	9908
98	9912	9917	9921	9926	9930	9934	9939	9943	9948	9952
99	9956	9961	9965	9969	9974	9978	9983	9987	9991	9996
100	0000	0004	0009	0013	0017	0022	0026	0030	0035	0039
N	0	1	2	3	4	5	6	7	8	9

Short
Normal
Long

Table III   Predicted vital capacities (in milliliters) for females.

HEIGHT IN CENTIMETERS AND INCHES

AGE	152 / 59.8	154 / 60.6	156 / 61.4	158 / 62.2	160 / 63.0	162 / 63.7	164 / 64.6	166 / 65.4	168 / 66.1	170 / 66.9	172 / 67.7	174 / 68.5	176 / 69.3	178 / 70.1	180 / 70.9	182 / 71.7	184 / 72.4	186 / 73.2	188 / 74.0
16	3,070	3,110	3,150	3,190	3,230	3,270	3,310	3,350	3,390	3,430	3,470	3,510	3,550	3,590	3,630	3,670	3,715	3,755	3,800
17	3,055	3,095	3,135	3,175	3,215	3,255	3,295	3,335	3,375	3,415	3,455	3,495	3,535	3,575	3,615	3,655	3,695	3,740	3,780
18	3,040	3,080	3,120	3,160	3,200	3,240	3,280	3,320	3,360	3,400	3,440	3,480	3,520	3,560	3,600	3,640	3,680	3,720	3,760
20	3,010	3,050	3,090	3,130	3,170	3,210	3,250	3,290	3,330	3,370	3,410	3,450	3,490	3,525	3,565	3,605	3,645	3,695	3,720
22	2,980	3,020	3,060	3,095	3,135	3,175	3,215	3,255	3,290	3,330	3,370	3,410	3,450	3,490	3,530	3,570	3,610	3,650	3,685
24	2,950	2,985	3,025	3,065	3,100	3,140	3,180	3,220	3,260	3,300	3,335	3,375	3,415	3,455	3,490	3,530	3,570	3,610	3,650
26	2,920	2,960	3,000	3,035	3,070	3,110	3,150	3,190	3,230	3,265	3,300	3,340	3,380	3,420	3,455	3,495	3,530	3,570	3,610
28	2,890	2,930	2,965	3,000	3,040	3,070	3,115	3,155	3,190	3,230	3,270	3,305	3,345	3,380	3,420	3,460	3,495	3,535	3,570
30	2,860	2,895	2,935	2,970	3,010	3,045	3,085	3,120	3,160	3,195	3,235	3,270	3,310	3,345	3,385	3,420	3,460	3,495	3,535
32	2,825	2,865	2,900	2,940	2,975	3,015	3,050	3,090	3,125	3,160	3,200	3,235	3,275	3,310	3,350	3,385	3,425	3,460	3,495
34	2,795	2,835	2,870	2,910	2,945	2,980	3,020	3,055	3,090	3,130	3,165	3,200	3,240	3,275	3,310	3,350	3,385	3,425	3,460
36	2,765	2,805	2,840	2,875	2,910	2,950	2,985	3,020	3,060	3,095	3,130	3,165	3,205	3,240	3,275	3,310	3,350	3,385	3,420
38	2,735	2,770	2,810	2,845	2,880	2,915	2,950	2,990	3,025	3,060	3,095	3,130	3,170	3,205	3,240	3,275	3,310	3,350	3,385
40	2,705	2,740	2,775	2,810	2,850	2,885	2,920	2,955	2,990	3,025	3,060	3,095	3,135	3,170	3,205	3,240	3,275	3,310	3,345
42	2,675	2,710	2,745	2,780	2,815	2,850	2,885	2,920	2,955	2,990	3,025	3,060	3,100	3,135	3,170	3,205	3,240	3,275	3,310
44	2,645	2,680	2,715	2,750	2,785	2,820	2,855	2,890	2,925	2,960	2,995	3,030	3,060	3,095	3,130	3,165	3,200	3,235	3,270
46	2,615	2,650	2,685	2,715	2,750	2,785	2,820	2,855	2,890	2,925	2,960	2,995	3,030	3,060	3,095	3,130	3,165	3,200	3,235
48	2,585	2,620	2,650	2,685	2,715	2,750	2,785	2,820	2,855	2,890	2,925	2,960	2,995	3,030	3,060	3,095	3,130	3,160	3,195
50	2,555	2,590	2,625	2,655	2,690	2,720	2,755	2,785	2,820	2,855	2,890	2,925	2,955	2,990	3,025	3,060	3,090	3,125	3,155
52	2,525	2,555	2,590	2,625	2,655	2,690	2,720	2,755	2,790	2,820	2,855	2,890	2,925	2,955	2,990	3,020	3,055	3,090	3,125
54	2,495	2,530	2,560	2,590	2,625	2,655	2,690	2,720	2,755	2,790	2,820	2,855	2,885	2,920	2,950	2,985	3,020	3,050	3,085
56	2,460	2,495	2,525	2,560	2,590	2,625	2,655	2,690	2,720	2,755	2,790	2,820	2,855	2,885	2,920	2,950	2,980	3,015	3,045
58	2,430	2,460	2,495	2,525	2,560	2,590	2,625	2,655	2,690	2,720	2,750	2,785	2,815	2,850	2,880	2,920	2,945	2,975	3,010
60	2,400	2,430	2,460	2,495	2,525	2,560	2,590	2,625	2,655	2,685	2,720	2,750	2,780	2,810	2,845	2,875	2,915	2,940	2,970
62	2,370	2,405	2,435	2,465	2,495	2,525	2,560	2,590	2,620	2,655	2,685	2,715	2,745	2,775	2,810	2,840	2,870	2,900	2,935
64	2,340	2,370	2,400	2,430	2,465	2,495	2,525	2,555	2,585	2,620	2,650	2,680	2,710	2,740	2,770	2,805	2,835	2,865	2,895
66	2,310	2,340	2,370	2,400	2,430	2,460	2,495	2,525	2,555	2,585	2,615	2,645	2,675	2,705	2,735	2,765	2,800	2,825	2,860
68	2,280	2,310	2,340	2,370	2,400	2,430	2,460	2,490	2,520	2,550	2,580	2,610	2,640	2,670	2,700	2,730	2,760	2,795	2,820
70	2,250	2,280	2,310	2,340	2,370	2,400	2,425	2,455	2,485	2,515	2,545	2,575	2,605	2,635	2,665	2,695	2,725	2,755	2,780
72	2,220	2,250	2,280	2,310	2,335	2,365	2,395	2,425	2,455	2,480	2,510	2,540	2,570	2,600	2,630	2,660	2,685	2,715	2,745
74	2,190	2,220	2,245	2,275	2,305	2,335	2,360	2,390	2,420	2,450	2,475	2,505	2,535	2,565	2,590	2,620	2,650	2,680	2,710

From: Archives of Environmental Health
February 1966, Vol. 12, pp .146–189
E. A. Gaensler, MD and G. W. Wright, MD

Table IV Predicted vital capacities (in milliliters) for males.

HEIGHT IN CENTIMETERS AND INCHES

AGE	CM 152 / IN 59.8	154 60.6	156 61.4	158 62.2	160 63.0	162 63.7	164 64.6	166 65.4	168 66.1	170 66.9	172 67.7	174 68.5	176 69.3	178 70.1	180 70.9	182 71.7	184 72.4	186 73.2	188 74.0
16	3,920	3,975	4,025	4,075	4,130	4,180	4,230	4,285	4,335	4,385	4,440	4,490	4,540	4,590	4,645	4,695	4,745	4,800	4,850
18	3,890	3,940	3,995	4,045	4,095	4,145	4,200	4,250	4,300	4,350	4,405	4,455	4,505	4,555	4,610	4,660	4,710	4,760	4,815
20	3,860	3,910	3,960	4,015	4,065	4,115	4,165	4,215	4,265	4,320	4,370	4,420	4,470	4,520	4,570	4,625	4,675	4,725	4,775
22	3,830	3,880	3,930	3,980	4,030	4,080	4,135	4,185	4,235	4,285	4,335	4,385	4,435	4,485	4,535	4,585	4,635	4,685	4,735
24	3,785	3,835	3,885	3,935	3,985	4,035	4,085	4,135	4,185	4,235	4,285	4,330	4,380	4,430	4,480	4,530	4,580	4,630	4,680
26	3,755	3,805	3,855	3,905	3,955	4,000	4,050	4,100	4,150	4,200	4,250	4,300	4,350	4,395	4,445	4,495	4,545	4,595	4,645
28	3,725	3,775	3,820	3,870	3,920	3,970	4,020	4,070	4,115	4,165	4,215	4,265	4,310	4,360	4,410	4,460	4,510	4,555	4,605
30	3,695	3,740	3,790	3,840	3,890	3,935	3,985	4,035	4,080	4,130	4,180	4,230	4,275	4,325	4,375	4,425	4,470	4,520	4,570
32	3,665	3,710	3,760	3,810	3,855	3,905	3,950	4,000	4,050	4,095	4,145	4,195	4,240	4,290	4,340	4,385	4,435	4,485	4,530
34	3,620	3,665	3,715	3,760	3,810	3,855	3,905	3,950	4,000	4,045	4,095	4,140	4,190	4,225	4,285	4,330	4,380	4,425	4,475
36	3,585	3,635	3,680	3,730	3,775	3,825	3,870	3,920	3,965	4,010	4,060	4,105	4,155	4,200	4,250	4,295	4,340	4,390	4,435
38	3,555	3,605	3,650	3,695	3,745	3,790	3,840	3,885	3,930	3,980	4,025	4,070	4,120	4,165	4,210	4,260	4,305	4,350	4,400
40	3,525	3,575	3,620	3,665	3,710	3,760	3,805	3,850	3,900	3,945	3,990	4,035	4,085	4,130	4,175	4,220	4,270	4,315	4,360
42	3,495	3,540	3,590	3,635	3,680	3,725	3,770	3,820	3,865	3,910	3,955	4,000	4,050	4,095	4,140	4,185	4,230	4,280	4,325
44	3,450	3,495	3,540	3,585	3,630	3,675	3,725	3,770	3,815	3,860	3,905	3,950	3,995	4,040	4,085	4,130	4,175	4,220	4,270
46	3,420	3,465	3,510	3,555	3,600	3,645	3,690	3,735	3,780	3,825	3,870	3,915	3,960	4,005	4,050	4,095	4,140	4,185	4,230
48	3,390	3,435	3,480	3,525	3,570	3,615	3,655	3,700	3,745	3,790	3,835	3,880	3,925	3,970	4,015	4,060	4,105	4,150	4,190
50	3,345	3,390	3,430	3,475	3,520	3,565	3,610	3,650	3,695	3,740	3,785	3,830	3,870	3,915	3,960	4,005	4,050	4,090	4,135
52	3,315	3,353	3,400	3,445	3,490	3,530	3,575	3,620	3,660	3,705	3,750	3,795	3,835	3,880	3,925	3,970	4,010	4,055	4,100
54	3,285	3,325	3,370	3,415	3,455	3,500	3,540	3,585	3,630	3,670	3,715	3,760	3,800	3,845	3,890	3,930	3,975	4,020	4,060
56	3,255	3,295	3,340	3,380	3,425	3,465	3,510	3,550	3,595	3,640	3,680	3,725	3,765	3,810	3,850	3,895	3,940	3,980	4,025
58	3,210	3,250	3,290	3,335	3,375	3,420	3,460	3,500	3,545	3,585	3,630	3,670	3,715	3,755	3,800	3,840	3,880	3,925	3,965
60	3,175	3,220	3,260	3,300	3,345	3,385	3,430	3,470	3,500	3,555	3,595	3,635	3,680	3,720	3,760	3,805	3,845	3,885	3,930
62	3,150	3,190	3,230	3,270	3,310	3,350	3,390	3,440	3,480	3,520	3,560	3,600	3,640	3,680	3,730	3,770	3,810	3,850	3,890
64	3,120	3,160	3,200	3,240	3,280	3,320	3,360	3,400	3,440	3,490	3,530	3,570	3,610	3,650	3,690	3,730	3,770	3,810	3,850
66	3,070	3,110	3,150	3,190	3,230	3,270	3,310	3,350	3,390	3,430	3,470	3,510	3,550	3,600	3,640	3,680	3,720	3,760	3,800
68	3,040	3,080	3,120	3,160	3,200	3,240	3,280	3,320	3,360	3,400	3,440	3,480	3,520	3,560	3,600	3,640	3,680	3,720	3,760
70	3,010	3,050	3,090	3,130	3,170	3,210	3,250	3,290	3,330	3,370	3,410	3,450	3,480	3,520	3,560	3,600	3,640	3,680	3,720
72	2,980	3,020	3,060	3,100	3,140	3,180	3,210	3,250	3,290	3,330	3,370	3,410	3,450	3,490	3,530	3,570	3,610	3,650	3,680
74	2,930	2,970	3,010	3,050	3,090	3,130	3,170	3,200	3,240	3,280	3,320	3,360	3,400	3,440	3,470	3,510	3,550	3,590	3,630

From: Archives of Environmental Health
February 1966, Vol. 12, pp. 146–189
E. A. Gaensler, MD and G. W. Wright, MD

B Appendix
Tests and Methods

Carbohydrate Differentiation

Several experiments in this book require that sugars and starches be differentiated. An understanding of the application of these tests can be derived from the separation outline at the bottom of this page.

Barfoed's Test   To 5 ml of reagent in test tube, add 1 ml of unknown. Place in boiling water bath for 5 minutes. Interpretation is based on separation outline.

Benedict's Test   This is a semiquantitative test for amounts of reducing sugar present. To 5 ml of reagent in test tube, add 8 drops of unknown. Place in boiling water bath for 5 minutes. A green, yellow, or orange-red precipitate determines the amount of reducing sugar present. See Ex. 82 and 86.

Electronic Equipment Adjustments

Lengthy instructions pertaining to recorder and transducer manipulations that are used in several experiments are outlined here for reference.

Balancing Transducer on Unigraph
(Ex. 29 and 82)

The Unigraph is designed to function in a sensitivity range of 0.1 MV/CM to 2 MV/CM. When an experiment is begun, the sensitivity control is set at 2 MV/CM, its least sensitive position. As the experiment progresses it may become necessary to shift to a more sensitive setting, such as 1 MV/CM or 0.5 MV/CM. If the Wheatstone bridge in the transducer is not balanced, the shifting from one sensitivity position to another will cause the stylus to change position. This produces an unsatisfactory record. Thus, it is essential that one go through the following steps to balance the transducer.

1. Connect the transducer jack to the transducer socket in the end of the Unigraph. The small upper socket is the one to use.
2. Before plugging in the Unigraph, make sure that the power switch is off, the chart control switch is on STBY, the gain selector knob is on 2 MV/CM, the sensitivity knob is turned completely counterclockwise, the mode selector control is on TRANS, and the Hi Filter and Mean switches are positioned toward Normal. Now plug in the Unigraph.

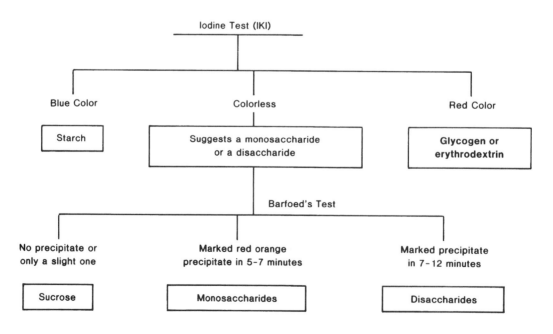

3. Turn on the Unigraph power switch and set the heat control knob at the two o'clock position.
4. Place the speed control lever at the slow position and the c.c. lever at Chart On.
5. As the chart moves along, bring the stylus to the center of the paper by turning the centering knob.
6. Turn the sensitivity control completely clockwise to maximum sensitivity. If the bridge is unbalanced, the stylus will move away from the center. To return the stylus to the center, unlock the TRANS-BAL control by pushing the small lever on the control and turning the knob in either direction to return the stylus to the center of the chart.
7. Set the gain control to 1. If the bridge is unbalanced, the stylus will not be centered. Center with the TRANS-BAL control.
8. Set the gain control to 0.5 and re-center the stylus with the TRANS-BAL control. Repeat this same procedure for 0.2 and 0.1 settings of the gain control. The bridge is now completely balanced at its highest gain and sensitivity. Lock the TRANS-BAL control with the lock lever.

   Note: Although we have gone through this lengthy process to achieve a balanced transducer, we may find that we will still get baseline shifts when going from one setting to another. However, they will be minor compared with one that has not been through this procedure. Any minor deviations should be corrected with the centering control.
9. Return the gain control to 2 and the chart control switch to STBY.

Calibration of Unigraph with Strain Gauge
(Ex. 29 and 82)

When the transducer has been balanced, it is necessary to calibrate the tracing on the Unigraph to a known force on the transducer. Calibration establishes a linear relationship between the magnitude of deflection and the degree of strain (extent of bending) of the transducer leaf before the muscle is attached to the transducer. Calibrations must be made for each gain at maximum sensitivity that one expects to use in the experiment (your muscle preparations will probably require gain settings of 2, 1, or 0.5). After calibration, it is possible to directly determine the force of each contraction by simply reading the maximum height of the tracing.

Materials:
   small paper clip weighing around 0.5 gram

1. Weigh a convenient object, such as a small paper clip, to the nearest 0.1 mg.
2. With the chart control switch on STBY, suspend the paper clip on the two smallest leaves of the transducer.
3. Put the c.c. switch to Chart On and note the extent of deflection on the chart. Record about 1 centimeter on the chart and place the c.c. switch on STBY. Label this deflection in mm/mg.
4. Set the gain control at 1, c.c. to Chart On, and record for another centimeter. Place the c.c. switch on STBY. Label the deflection.
5. Repeat the above procedure for the other two gain settings (0.5, 0.2, and 0.1). Note how the stylus vibrates as you increase the sensitivity of the Unigraph. This is normal.
6. Return the c.c. switch to STBY, gain to 2, and remove the paper clip. The Unigraph is now calibrated and the muscle can now be attached to it.

Calibration for EMG on Unigraph (Ex. 32)

Calibration of the Unigraph for EMG measurements is necessary to establish that one centimeter deflection of the stylus equals 0.1 millivolt. When calibration is accomplished, the recording is made with the EEG mode since there is no EMG mode. An EMG can be made very well in this mode at lower sensitivities than would be used when making an EEG. Proceed as follows to calibrate:

1. Turn the mode selector control to CC/Cal (Capacitor Coupled Calibrate).
2. Turn the small main switch to ON.
3. Turn the stylus heat control to the two o'clock position.
4. Check the speed selector lever to see that it is in the slow position. The free end of the lever should be pointed toward the styluses.
5. Move the chart-stylus switch to Chart On and observe the width of the tracing that appears on the paper. Adjust the stylus heat control to produce the desired width of tracing.
6. Set the gain knob to 0.1 MV/CM.
7. Push down the 0.1 MV button to determine the amount of deflection. Hold the button down for 2 seconds before releasing. If it is released too quickly the tracing will not be perfect. Adjust the deflection so that it deflects exactly 1 centimeter in each direction by turning the sensitivity knob. The unit is now calibrated so that you get 1 cm deflection for 0.1 millivolt.

8. Reset the gain knob to 0.5 MV/CM. This reduces the sensitivity of the instrument.
9. Return the mode selector control to EEG.
10. Return the chart-stylus switch to STBY. Place the speed selector in the fast position. The instrument is now ready for use.

Balancing the Myograph to Physiograph® Transducer Coupler (Ex. 29 and Ex. 72)

Balancing (matching) of the transducer signal with the amplifier is necessary to get the recording pen to stay on a preset baseline when the RECORD button is in either the OFF or the ON position. Proceed as follows:

1. Before starting, make sure that the RECORD switch is OFF and that the myograph jack is inserted into the transducer coupler.
2. Set the sensitivity control (outer knob) to its lowest numbered setting (highest sensitivity).
3. Set the paper speed at 0.5 cm/sec and lower the pen lifter.
4. Adjust the pen position so that the pen is writing *exactly* on the center line.
5. Place the RECORD switch in the ON position. The pen will probably be moving a large distance either up or down.
6. With the BALANCE control, adjust the pen so that it is again writing *exactly* on the center line.
7. Check your match, or balance, by placing the RECORD button in the OFF position. **Your system is balanced if the pen remains on the center line.**
8. If the pen does not remain on the center line, relocate the pen to the center line again with the position control, place the RECORD button in the ON position, and repeat the balancing procedure. You have a balanced system *only* when the pen remains on a preset baseline with the RECORD button in both OFF and ON positions.

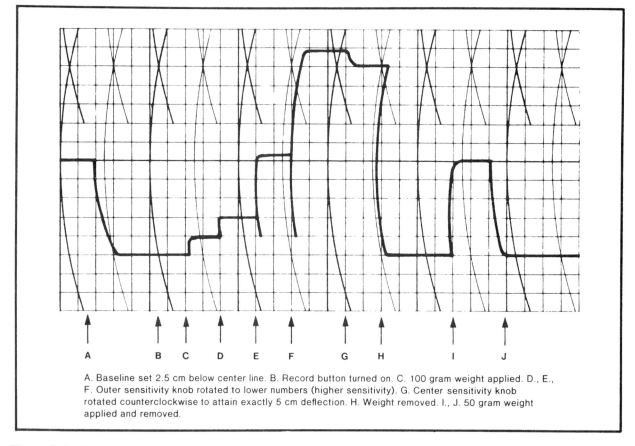

A. Baseline set 2.5 cm below center line. B. Record button turned on. C. 100 gram weight applied. D., E., F. Outer sensitivity knob rotated to lower numbers (higher sensitivity). G. Center sensitivity knob rotated counterclockwise to attain exactly 5 cm deflection. H. Weight removed. I., J. 50 gram weight applied and removed.

Figure B-1 Sample record of calibration at 100 grams per 5 centimeters.

Calibration of Myograph Transducer on Physiograph® (Ex. 29)

Calibration of the transducer is necessary so that we know exactly how much tension, in grams, is being exerted by the muscle during contraction. The channel amplifier must be balanced first. Our calibration here will be to get 5 cm pen deflection with 100 grams. *These instructions apply to couplers that lack built-in calibration buttons.* Figure B-1 on the previous page illustrates the various steps.

Materials:

100 gram weight
50 gram weight

1. Start the paper drive at 0.1 cm/sec and lower the pen lifter so that we are recording on the desired channel.
2. Check to see that the channel amplifier is balanced.
3. With the pen position control, set the baseline exactly 2.5 cm (5 blocks) below the channel center line.
4. Rotate the outer knob of the sensitivity control fully counterclockwise to the lowest sensitivity level (i.e., 1000 setting). Be sure that the inner knob of the sensitivity control is in its fully clockwise "clicked" position.
5. Place the RECORD button in the ON position.
6. Suspend a 100 gram weight to the actuator of the myograph. Note that the addition of the weight has caused the baseline to move upward approximately one block (0.5 cm). See step C, figure B-1.
7. Rotate the *outer* knob of the sensitivity control clockwise until the pen exceeds 5 cm (10 blocks) of deflection from the original baseline. See step F, figure B-1.
8. Rotate the *inner* knob of the sensitivity control counterclockwise to bring the pen back downward until it is *exactly* 5 cm (10 blocks) of deflection from the original baseline.
9. Remove the weight. The pen should return to the original baseline. If it does not return exactly to the baseline, reset the baseline with the position control, reapply the weight, and rotate the inner knob until the pen is writing exactly 5 cm above the original baseline.
10. Attach a 50 gram weight to the myograph to see if you get 5 blocks of pen deflection, as shown in the last deflection in figure B-1. The unit is now properly calibrated.

Calibration of Myograph Transducer with Calibration Button (Ex. 29)

When the coupler has a calibration button, the use of a 100 gram weight is not necessary. To calibrate, all that is necessary is to press the 100 gram button, and follow the steps above to produce the 5 cm pen deflection.

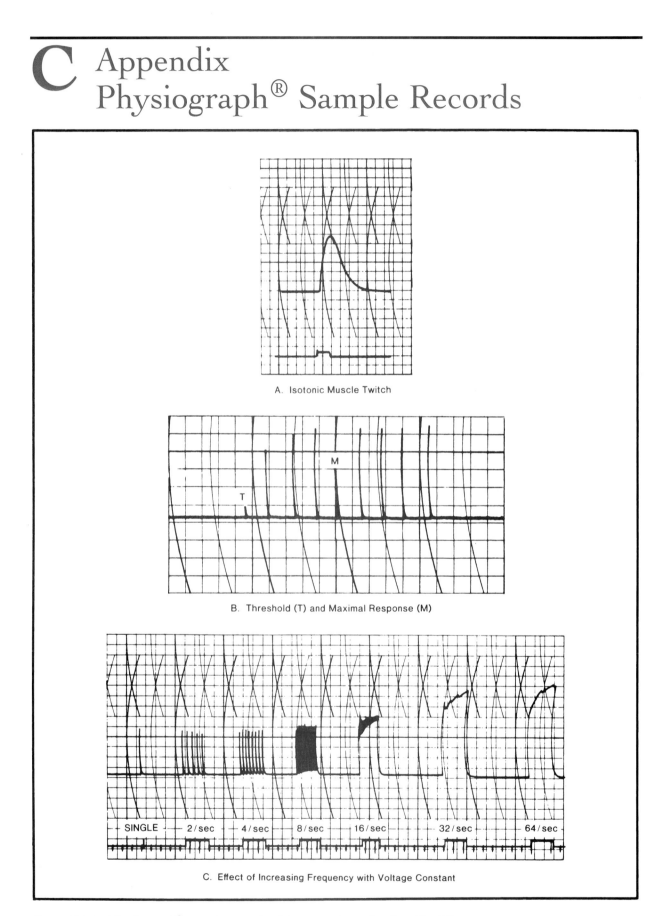

A. Isotonic Muscle Twitch

B. Threshold (T) and Maximal Response (M)

SINGLE | 2/sec | 4/sec | 8/sec | 16/sec | 32/sec | 64/sec

C. Effect of Increasing Frequency with Voltage Constant

Figure SR-1 Physiograph® sample records of frog muscle contraction (Ex. 29).

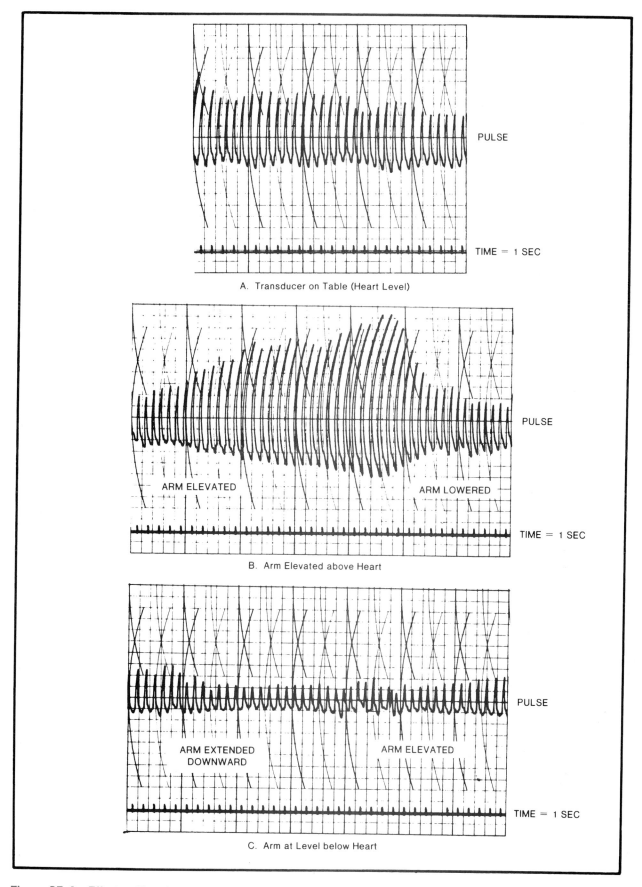

A. Transducer on Table (Heart Level)

B. Arm Elevated above Heart

C. Arm at Level below Heart

Figure SR-2 Effects of hand position on pulse (Ex. 64).

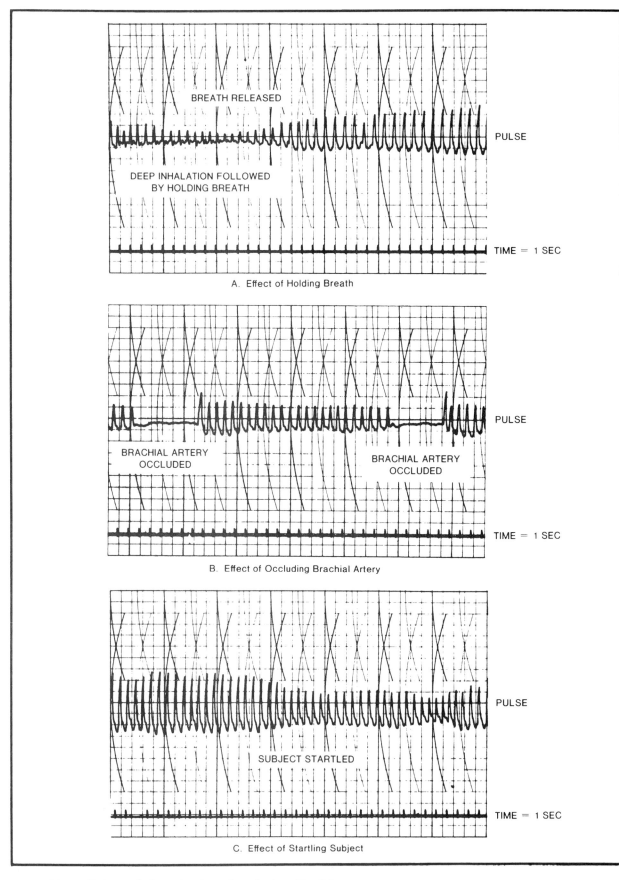

Figure SR-3 Pulse variations due to various factors (Ex. 64).

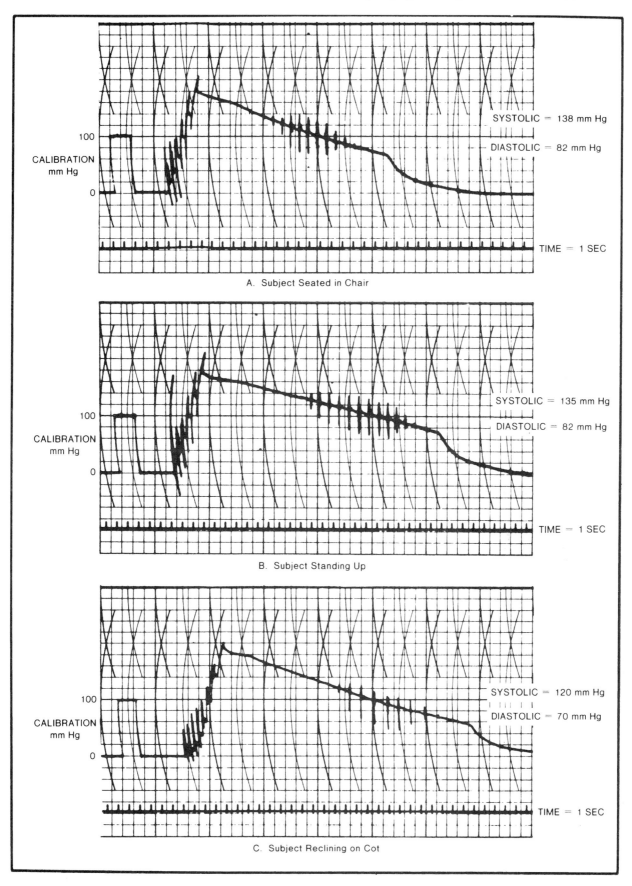

SYSTOLIC = 138 mm Hg

DIASTOLIC = 82 mm Hg

TIME = 1 SEC

A. Subject Seated in Chair

SYSTOLIC = 135 mm Hg

DIASTOLIC = 82 mm Hg

TIME = 1 SEC

B. Subject Standing Up

SYSTOLIC = 120 mm Hg

DIASTOLIC = 70 mm Hg

TIME = 1 SEC

C. Subject Reclining on Cot

Figure SR-4 Effects of posture on blood pressure (Ex. 65).

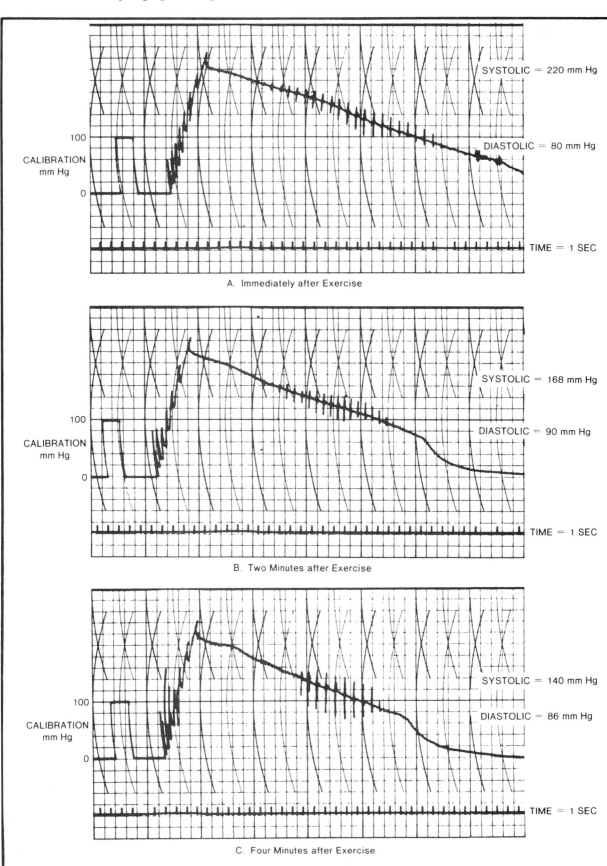

Figure SR-5 Sample records of blood pressure monitoring after exercise (Ex. 65).

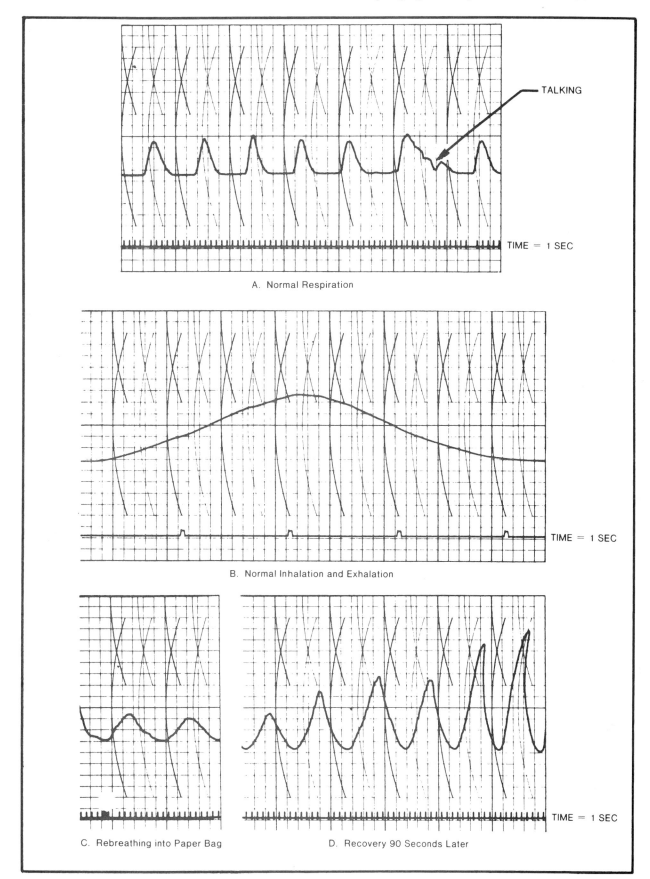

TALKING

TIME = 1 SEC

A. Normal Respiration

TIME = 1 SEC

B. Normal Inhalation and Exhalation

TIME = 1 SEC

C. Rebreathing into Paper Bag

D. Recovery 90 Seconds Later

Figure SR-6 Pneumograph recording records (Ex. 72).

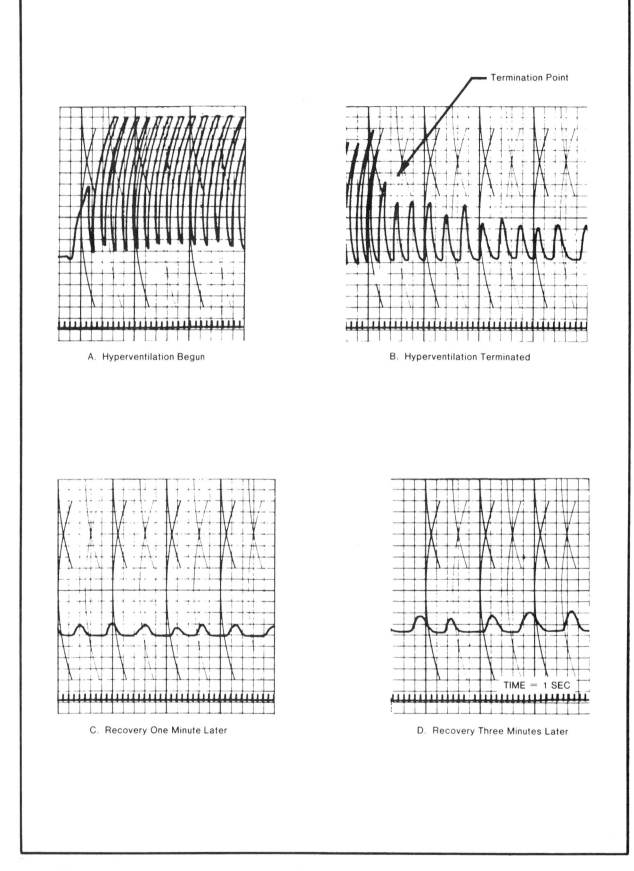

Termination Point

A. Hyperventilation Begun

B. Hyperventilation Terminated

C. Recovery One Minute Later

D. Recovery Three Minutes Later

TIME = 1 SEC

Figure SR-7 Hyperventilation sample record (Ex. 72).

Reading References

Anatomy and Physiology

Hole, John W., Jr., 1994. *Human anatomy and physiology.* 5th ed. Dubuque, IA: WCB/McGraw-Hill.

Jacob, Stanley W., and Ashworth, Clarice A. 1989. *Elements of anatomy and physiology,* 2nd ed. Philadelphia: W. B. Saunders Co.

Tortora, Gerald J., and Garbrowski, Sandra R. 1996. *Principles of anatomy and physiology,* 7th ed. Addison Wesley.

Van De Graaf, Kent M., and Fox, Stuart Ira, 1994. *Concepts of Human Anatomy and Physiology,* 4th ed. Dubuque, IA: WCB/McGraw-Hill.

Anatomy

Clemente, Carmine D., et al, ed, 1987. *Anatomy: A regional atlas of the human body.* 3rd ed. Baltimore: Williams & Wilkins.

Crouch, James E., 1983. *Introduction to human anatomy: A laboratory manual.* 6th ed. Baltimore: Williams & Wilkins.

Crouch, James E. and Carr, Micheline, 1989. *Anatomy and physiology: A laboratory manual.* Baltimore: Williams & Wilkins.

Greenblatt, Gordon M., 1981. *Cat musculature: a photographic atlas.* 2nd ed. Chicago: Univ. of Chicago.

Netter, Frank H., 1983-1990. *The CIBA collection of medical illustrations.* 13 vols. Summit, New Jersey: CIBA Pharmaceutical Products, Inc.

Schlosssberg, Leon, and Zuidema, George D., 1986. *The Johns Hopkins atlas of human functional anatomy.* 3rd ed. Baltimore: Johns Hopkins.

Van De Graaff, Kent M., 1997. *Human anatomy,* 5th ed. Dubuque, IA: WCB/McGraw-Hill.

Physiology

Ganong, William, 1995. *Review of medical physiology,* 17th ed. Los Angeles: Appleton & Lang.

Guyton, Arthur C. and Hall, John E., 1995. *Textbook of medical physiology,* 9th ed. Philadelphia: W. B. Saunders Co.

Histology

Bergman, Ronald A., and Afifi, Adel K., 1974. *Atlas of microscopic anatomy.* Philadelphia: W. B. Saunders Co.

Cormack, David H. *Essential histology,* 1993. Philadelphia: Lippincott-Raven.

Cormack, David H. *Clinically integrated histology,* 1997. Philadelphia: Lippincott-Raven.

Chemistry

Campbell, P. N., and Marshall, R.D., 1985. *Essays in biochemistry,* Vol. 20. Orlando, FL: Academic Press.

Gilman, Alfred G., et al., eds., 1990. *The pharmacological basis of therapeutics,* 8th ed. New York: McGraw-Hill.

Richterich, Roland, and Columbo, J. P., 1981. *Clinical chemistry: theory, practice, and interpretation.* New York: Wiley Interscience.

Tietz, Norbert W., ed, 1997. *Clinical guide to laboratory tests,* 3rd ed. Philadelphia: W. B. Saunders Co.

Instrumentation

Cobbold, Richard S., 1974. *Transducers for biochemical measurements: principles and applications.* New York: Wiley Interscience.

Dewhurst, D.J., 1975. *An introduction to biomedical instrumentation.* 2d ed. Elmsford, NY: Pergamon Press.

DuBorg, Joseph L., 1978. *Introduction to biomedical electronics.* New York: McGraw-Hill.

Geddes, L. A., 1991. *Handbook of blood pressure measurement.* Humana.

Geddes, L. A., and Baker, L. E., 1989. *Principles of applied medical instrumentation,* 3rd ed. New York: Wiley Interscience.

Miller, H., and Harrison, D. C., 1974. *Biomedical electrode technology.* Orlando, FL: Academic Press.

Tischler, W. A., and Geddes, L. A. 1980. *Electrical defibrillation.* Reprint by CRC.

Tischler, Morris, 1981. *Experiments in amplifiers, filters and oscillators.* New York: McGraw-Hill.

_____ 1981. *Experiments in general and biomedical instrumentation.* New York: McGraw-Hill.

_____ 1985, *Optoelectronics: a text-lab manual.* New York: McGraw-Hill.

Muscle Physiology

Basmajian, John V. and Wolf, Steven, 1990. *Therapeutic exercise,* 5th ed. Baltimore: Williams and Wilkins.

Basmajian, John V. and MacConaill, M. A., 1977. *Muscles and movements: a basis for human kinesiology,* rev. ed. Baltimore: Williams and Wilkins.

Jabre, Joe F., and Hackett, Earl R., 1983. *EMG manual.* Springfield, IL: C. C. Thomas.

Murray, J. M., and Weber, A., 1974. *The cooperative action of muscle proteins.* Scientific Offprint #1290. San Francisco: W. H. Freeman Co.

Murray, Jim, and Karpovich, Peter, 1982. *Weight training in athletics.* Englewood Cliffs, NJ: Prentice-Hall.

Shepard, Roy J. l, 1984. *Biochemistry of physical activity.* Springfield, IL: C.C. Thomas.

_____, *Physiology and biochemistry of exercise.* New York: Praeger.

Eye and Ear

Browning, G. G. 1986. *Clinical otology and audiology.* Stoneham, MA: Butterworth.

Corboy, John M., 1995. *The retinoscopy book: an introductory manual for eye care professionals,* 4th ed. Thorofare, NJ: SLACK, Inc.

Masland, Richard H., December, 1986, Scientific American. *The functional architecture of the retina.* New York: Scientific American, Inc.

Schnapf, Julie L., and Baylor, Denis A., April, 1987, Scientific American. *How photoreceptor cells respond to light.* New York: Scientific American, Inc.

Miscellaneous

Adamovich, David R., 1984. *The heart: fundamentals of electrocardiography, exercise physiology, and exercise stress testing.* Hempstead, NY: Sports Medicine Books.

Basmajian, John V., 1989. *Biofeedback: principles and practice for clinicians,* 3rd ed. Oradell, NJ: Medical Economics.

Diggs, L. W., et al. *The morphology of human blood cells,* 4th ed. Chicago: Abbott Laboratories.

Duke, Richard C., Ojcius, D. M., and Young, John, Ding-E, December, 1996, Scientific American. *Cell suicide in health and disease.* New York: Scientific American, Inc.

Hughes, John R., 1994. *EEG in clinical practice.* Stoncham, MA: Butterworth.

Kalashnikov, V., 1986. *Beat the box: the insider's guide to outwitting the lie detector.* Staten Island, NY: Gordon Press.

Keeley, Eloise, 1984. *Lie detector manual.* Marshfield, MA: TelShare Pub. Co.

Matte, James, 1980. *The art and science of the polygraph technique.* Springfield, IL: C. C. Thomas.

Rothman, James E., and Orci, Lelio, March, 1996, Scientific American. *Budding vesicles in living cells.* New York: Scientific American, Inc.

Short, Charles E., 1993. *Alpha-2 agents in animals sedation, analgesia and anesthesia.* Vet Practice.

Wintrobe, Maxwell M., et al., 1997. *Clinical hematology.* 10th ed. Baltimore: Williams and Wilkins.